Jacobi Dynamics

Astrophysics and Space Science Library

For other titles published in this series, go to
http://www.springer.com/series/5664

V.I. Ferronsky • S.A. Denisik • S.V. Ferronsky

Jacobi Dynamics

A Unified Theory with Applications to Geophysics, Celestial Mechanics, Astrophysics and Cosmology

Second edition

Springer

V.I. Ferronsky
Water Problems Institute
of the Russian Academy of Sciences
Gubkin st. 3, 119333 Moscow
Russia
ferron@aqua.laser.ru

S.V. Ferronsky (deceased)

S.A. Denisik
Water Problems Institute
of the Russian Academy of Sciences
Gubkin st. 3, 119333 Moscow
Russia

Revised and updated edition of the book "Jacobi Dynamics: Many Body Problem in Integral Characteristics", by V.I. Ferronsky, S.A. Denisik and S.V. Ferronsky, Reidel D. Publishing Company, 1986.

ISSN 0067-0057
ISBN 978-94-007-3587-3 ISBN 978-94-007-0498-5 (eBook)
DOI 10.1007/978-94-007-0498-5
Springer Dordrecht Heidelberg London New York

Cover illustration: Real picture of motion of a body A in the force field of a body B. Digits identify succession of turns of the body A moving around body B along the open orbit C.

Printed on acid-free paper

Springer is part of Springer Science+Business Media (www.springer.com)

Preface to the Second Edition

The first edition of this book was published in 1987. Since that time new scientific and practical results in application of Jacobi dynamics were obtained. The results, which we include into the second edition, are as follows.

The oldest scientific problem of the origin and evolution of the Solar System bodies on the basis of the Jacobi's oscillating dynamics has found its resolution. We discovered by calculation that the present day orbital velocity of each planet is equal to the fist cosmic velocity of the contracting Protosun having its radius reduced to the semi-major axis of the planet's orbit. And also the orbital velocities of the planet's moons are equal to the first cosmic velocities of the corresponding protoplanets having their radiuses successively reduced to the semi-major axes of the moon's orbits. It looks like the protoplanets and their protosatellites were created from the separating upper shells of their contracting parents. The Sun itself and the small bodies like the comets, asteroids, meteors and so on, have the same history of creation.

The discovered mechanism of the Solar System body creation proves to be the physical basis of the Kepler's laws and the inverse square distance law of the outer force field distribution of a self-gravitating body. It also proves the idea that the 'heavens power of gravity' of body's orbital motion has the electromagnetic nature and is induced by the parental body.

We succeeded in understanding physical meaning of the Jacobi's virial equation obtained from the Newton's equations of motion written for a system of n interacting mass-particles. It was found that in his n-body problem, Jacobi, while transforming the initial equations, has converted the Newton's forces and moments of the interacting particles into their volumetric values, namely, into the energy and oscillating moment of inertia. Doing like this, Jacobi has changed the force as a measure of the particles interaction by energy of their interaction. From physical viewpoint Jacobi's approach opens the way to search the common nature for all the known physical models of the matter interaction. In order to prove this idea, we show in the book that the Jacobi's virial equation, besides the Newton's equations, is also derived from the equations of Euler, Hamilton, Einstein and quantum mechanics. This allowed us to put the subtitle of the book as a 'Unified theory of the interacting matter motion'.

We discuss in the book the main scientific discovery, made by the geodetic artificial satellite study that the Earth and the Moon do not stay in hydrostatic equilibrium. It puzzles us because the hydrostatics, up to now, is accepted as the physical basis in dynamics of stars, planets and other celestial bodies. In addition to this, we found that the Earth has no equilibrium between the kinetic and potential energy ($K/U \approx 1/300$). This is because the existing theories in dynamics do not take into account the kinetic energy of the body's interacting mass particles. It makes incorrect formulation of the problem of the equilibrium state. The conclusion is made that the conditions of the dynamical equilibrium state, as an alternative to the hydrostatics, in celestial mechanics, stellar dynamics and astrophysics should be introduced.

We found that the above fundamental effects are well explained by the theory of Jacobi dynamics which considers just self-gravitating, but not hydrostatically state systems and where relationship between the energy and the polar moment of inertia is the basic physical effect. In this connection this problem and physical meaning of the Jacobi's virial equation in the new Chapter 2 are discussed. In order to demonstrate all new effects in dynamics of stars, planets and satellites, including creation and evolution of the Solar system, Chaps. 6, 7, and 8 were revised and updated by applying the above discoveries. Also, the new Chap. "The Nature of Electromagnetic Field of a Celestial Body and Mechanism of Its Energy Generation" is introduced. The former Chaps. 1, 2, 3, 4, and 5 under new numbers were as a whole preserved. The second edition was prepared by V.I. Ferronsky.

V.I. Ferronsky

Preface to the First Edition

This book sets forth and builds upon the fundamentals of the dynamics of natural systems in formulating the problem presented by Jacobi in his famous lecture series "Vorlesung über Dynamik" (Jacobi 1884).

In the dynamics of systems described by models of discrete and continuous media, the many-body problem is usually solved in some approximation, or the behavior of the medium is studied at each point of the space it occupies. Such an approach requires the system of equations of motion to be written in terms of space co-ordinates and velocities, in which case the requirements of an internal observer for a detailed description of the processes are satisfied.

In the dynamics discussed here we study the time behavior of the fundamental characteristics of the physical system, i.e. the Jacobi function (polar moment of inertia) and energy (potential, kinetic and total), which are functions of mass density distribution, and the structure of a system. This approach satisfies the requirements of an external observer. It is designed to solve the problem of global dynamics and the evolution of natural systems in which the motion of the system's individual elements written in space co-ordinates and velocities is of no interest. It is important to note that an integral approach is made to internal and external interactions of a system which results in radiation and absorption of energy. This effect constitutes the basic content of global dynamics and the evolution of natural systems.

From the standpoint of methodology, the integral approach has an important advantage. In this approach the integral character of the principle of least action – the basic philosophical principle of mechanics and physics – is fully realized. It is achieved by using a canonical pair consisting of the Jacobi function and frequency in writing the basic equation of global dynamics. The practical use of this pair in Jacobi's virial equation made it possible to farther generalize the forms of motion and to show that the non-linear oscillations of a system is such a generalization.

We note that the ten well-known integrals of motion in the many-body problem in its classical formulation should be regarded as historically the earliest equations of the integral type. These integrals, however, reflects not the specific nature of a system under consideration but the general properties of space and time, i.e. homogeneity of space and time and isotropicity of space.

The first non-trivial equation of dynamics in terms of the integral characteristics of a system is Jacobi's virial equation, which describes changes in the moment of inertia (Jacobi function) as a function of time. The next step in this direction was taken by Chandrasekhar (1969). He used and developed for solution of problems in mechanics the method of moments, so called in analogy to the method well known in mathematical physics. However, the problem of non-trivial solution of the non-linearized equations in terms of integral characteristics was not solved in either of these cases.

Our work began in 1974. As a result, a number of articles on the theory of virial oscillations of celestial bodies were published in the journal Celestial Mechanics and other periodicals (Ferronsky et al. 1978, 1979a–c, 1981, 1982, 1984, 1985). The theory was based on solution of Jacobi's virial equation for conservative and dissipative systems.

To solve Jacobi's initial equation, we first used the heuristically found relationship between the potential energy and moment of inertia of a system, which was expressed in terms of the product of the corresponding form factors. It was found that the product depends little on the law of distribution of mass density for a wide range of formal, non-physical systems. It was then demonstrated that in the asymptotic limit of simultaneous collision of the particles constituting a system, the observed constancy of the product of form factors remained valid without any restrictions within the framework of the Newton and Coulomb interaction laws. The invariant found was also demonstrated to be valid for the widely used relativistic and non-relativistic physical models of natural systems. It enabled us to derive from Jacobi's equation a simple form of the equation of virial oscillations with one unknown function and to find its rigorous solution. The equation obtained describes the dynamics of a wide class of physical systems ranging from empty space-time and collapsing stars to the atom. Thus, it was established that the theory of virial oscillations of celestial bodies was valid far beyond the limits of celestial mechanics based on Newton's law of equations.

The work was done on concepts of Professor V.I. Ferronsky and under his supervision. Chaps. 1 and 7 were written by V.I. Ferronsky and S.A. Denisik; Chaps. 2, 3, 4, 5, 6 and 8 were written by S.V. Ferronsky.

This book presents a systematic description of our research work. It is intended for researchers, teachers and students engaged in theoretical and experimental research in the various branches of astronomy (astrophysics, celestial and stellar mechanics and radiophysics), geophysics (physics of the Earth, atmosphere and oceans), planetology and cosmogony, and for students and postgraduates of classical, statistical, quantum and relativistic mechanics and hydrodynamics.

It is our pleasant duty to express sincere gratitude to Professor G.N. Duboshin of the M.V. Lomonosov Moscow State University for his constant support and encouragement. We are indebted to Professor E.P. Aksenov, Director of the Sternberg Institute of Astronomy, Moscow, who organized helpful discussions of our work at a number of seminars. We also wish to express our gratitude to Dr. L. Osipkov from Leningrad University and Drs J. Schmidt, A. Lorenz, M. Mehta and T. Akity from the Division of Research and Laboratories,

International Atomic Energy Agency, who read several chapters of the manuscript and made editorial contributions, and to Miriam Lewis, who edited the final manuscript. We also wish to thank Renate B. Blamhofer for assistance in preparing the book. We are particularly indebted to the International Atomic Energy Agency and its Division of Publications for support and assistance in preparation of the camera-ready manuscript.

The authors

Contents

Chapter 1
Introduction: General Principles and Approaches in Dynamics

In 1842–1843, when Jacobi was a professor at Königsberg University, he delivered a special series of lectures on dynamics. The lectures were devoted to the dynamics of a system of *n* mass points, whose motion depends only on the distances between them and is independent of velocities. In this connection, by deriving the law of conservation of energy from the equations of motion of mass points for a conservative system, where the force function is a homogeneous function of space co-ordinates, Jacobi gave this law an unusual form and a new content. In transforming the equations of motion, he introduces an expression for the system's center of mass. Then, following Lagrange, he separated the motion of the center of mass from the relative motion of the mass points. Making the center of mass coincide with the origin of the co-ordinate system, he obtained the following equation (Jacobi 1884):

$$\frac{d^2}{dt^2}\left(\sum m_i r_i^2\right) = -\,(2k + 4)U + 4E, \tag{1.1}$$

where m_i is the mass point i; $r_i = \sqrt{x_i^2 + y_i^2 + z_i^2}$ the distance between the points and the center of mass; k the degree of homogeneity of the force function; U the system's potential energy; and E its total energy.

When $k = -1$, which corresponds to the interaction of mass points according to Newton's law, and denoting

$$\frac{1}{2}\sum m_i r_i^2 = \Phi,$$

Jacobi obtained

$$\ddot{\Phi} = U + 2T = 2E - U, \tag{1.2}$$

where Φ is the Jacobi function (the polar moment of inertia).

This is the Jacobi's generalized (nonaveraged) virial equation. In the Russian scientific literature it is known as the Lagrange–Jacobi equation since Jacobi, in

V.I. Ferronsky et al., *Jacobi Dynamics*, Astrophysics and Space Science Library 369, DOI 10.1007/978-94-007-0498-5_1, deriving it, used Lagrange's method of separating the motion of the center mass from the relative motion of mass points.

In the right hand side of the virial equation there is a classical expression of the virial theorem, i.e. relation between the potential and kinetic energy. In the case of constancy of its left hand side, when motion of the system happens with a constant velocity, the equation acquires conditions of hydrostatic equilibrium of a system in the outer force field. The left hand side of the equation, i.e. the second derivative with respect to the Jacobi function expresses oscillation of the polar moment of inertia of the system, which, in fact, is kinetic energy of the inner volumetric torques (oscillations) of the interacted mass points moving in accordance with the Kepler's laws.

Jacobi hasn't paid attention to physics of his equation, which expresses kinetic energy of the interacted volumetric particles in the form of oscillation of the polar moment of inertia. He used the equation for a quantitative analysis of stability of the Solar System and noted that the system's potential and kinetic energies should always oscillate within certain limits. In the contemporary literature of celestial mechanics and analytical dynamics the Jacobi's virial equation is used for the same purposes (Whittaker 1937; Duboshin 1975). Since this equation contains two independent variables, it found no any other practical applications.

Our starting point for the farther development of Jacobi dynamics was the formal and physical relationship that we found between the Jacobi function and the potential energy in Eq. 1.2. This functional relationship enabled us to reduce it to an equation with one unknown function and to obtain a rigorous solution. It then appeared, as Jacobi had assumed, that the potential and kinetic energies and the moment of inertia of a conservative system oscillating periodically according to the law of conic sections. From the standpoint of physics, this means that as a result of interaction (collision) of its component parts or particles the system manifests its potential energy in the form of radiation. In a conservative system, the kinetic energy radiated during the period of pulsation is absorbed by the system itself and again converted into potential energy. In a dissipative system, a part of the energy is radiated during each period of pulsation from its surface and is lost irretrievably, entering into interaction with external systems. In this connection Eq. 1.2 is valid for study of dynamics of systems in classical mechanics and the dynamics of continuous media as well as in molecular physics, quantum mechanics and electrodynamics.

We now turn to some principles of and approaches to dynamics related to the results presented here in order to determine their place in solution of the general problem.

1.1 Principle of Mutual Reversibility

The world by its nature is unique and we want to represent it in terms of unified laws. For this purpose, out of the variety of observed natural objects and phenomena, we try to identify those which are most general and which constitute the basis of the existence of the material world. The main obstacles to this end are, however, the limited possibilities and subjective perceptions of the other world. Therefore, the observed outer multitude of natural objects and phenomena continue to be described by the different branches of physics.

In describing nature, physical objects and phenomena are represented in terms of geometrical entities – scalars, vectors and tensors – and correspondingly the representation of some sets are given in terms of others. Moreover, by generalizing the results of experiments, physical laws are established and expressed in mathematical models after excluding in advance subjective effects which are unavoidable in experiments. The results of generalization of experiments should not, of course, depend on the selection of the frame of reference of the observer, who is located somewhere and is moving somehow. Nor should they depend on the choice of the system of co-ordinates in his frame of reference, which is unlimited. In mathematical language this means that in generalizing experiments all assertions should be unique for any observer – or that laws should be covariant with respect to the corresponding classes of representation of sets of mathematical objects.

Vectorial quantities (co-ordinates, momenta, moments, etc.) and the functions representing them, which remain vectorial in character, suffer from one substantial defect: their components depend on the choice of the frame of reference and the co-ordinate system. Scalar functions, however, which remain invariant during the transformation of co-ordinates, are free from this defect. Therefore, we want to find the representation of these scalars. It is then possible to present vectorial representations in the form of gradients of the corresponding scalar functions. However, the transformation inverse to a given transformation remains a transformation which should have its own scalar function. Since a direct representation has no advantages over an inverse representation, its function should by its nature be symmetrical with the initial one. Then the equation of mutual reversibility of representations in the symmetric form can be written

$$L(r) + H(p) = rp, \tag{1.3}$$

where r and p are vectors from the first and second set of vectors, respectively; L and H are scalar functions defined in the set of vectors r and p, respectively.

Here the direct representation will be determined by the scalar function H as $r \Rightarrow H_p(p)$, where $H_p(p)$ is the gradient of function H(p) with respect to p. The inverse representation is determined by the scalar function L as $p \Rightarrow L_r(r)$, where $L_r(r)$ is the gradient of the function L(r) with respect to r. The product of the vector r and p will be a scalar quantity equal to

$$rp = \sum_{i=1}^{n} r_i p_i.$$

Functions of L and H will in this case have non-zero Hessians or matrices of the second derivatives of functions with respect to the components of their vector arguments. This condition is the condition of mutual reversibility of representations which, in other words, can be expressed in the following manner. The Hessian of scalar function is a Jacobian of vector function of representations of r on p or p on r. The condition of uniqueness of representations is a non-zero Jacobian of representations in the whole region of definition of the function.

We have so far mentioned two different sets of vectors of two representations. To represent the unity of the world, we have to make a generalization. For this purpose both the sets under consideration are combined into one, and the representation becomes an automorphism representing the vectors of a given set by vectors of the same set. Hence, if before combination of the sets the representation has transformed vector X into vector Y, it should, by transforming vector Y, transform the latter into vector X. This is called an involute transformation. When applied twice, it becomes an identical transformation of representations. The quantities which are mutually transformed into each other during involution are called associated quantities.

Anticipating what we are going to say later, we note that this property of representations is reflected in the general theory of relativity in the Einstein equations. There the fundamental tensors of Ricci (contraction of the tensor of Riemann curvature of its spur) and Einstein are mutually associated.

The principle of mutual reversibility described above is the most general for representing nature, and does not depend on nature's internal structure. In its most general term it reflects that the structure of nature is independent of who observes it and how.

The corresponding transformations of the generalized co-ordinates of observed objects will be canonical transformations. Following Wintner (1941), we should point out that we are not writing about any specific types of equations of motion: the basic general properties of transformation are established before these equations are chosen.

During the implementation of this program, which was of course carried out in a more complicated manner in practice, two mutually inverse matrices could be constructed, expressed ultimately in terms of the Jacobian transformation matrix. Historically, these matrices were found by Lagrange and Poisson and are called Lagrange brackets and Poisson brackets. We emphasize that the main idea of the Lagrange and Poisson brackets consists in obtaining the relations which are covariant during the transformation of co-ordinates, i.e. relations which do not change their form during the substitution of variables.

Derivation of the brackets can be started by writing the integral invariant of the canonical transformation, the expression for which, according to the Poincaré theorem (Goldstein 1980), will be

$$J = \iint_{s} \sum_{i} dq_i dp_i, \tag{1.4}$$

where q_i and p_i are canonical variables and s is some surface.

Here the integral invariant J is a scalar whose dimension cannot be obtained from general considerations. We shall show that according to experimental data this scalar has the dimension of action.

The Poincaré theorem for two pairs of canonical variables p, q and P, Q is written in the form

$$\iint_s \sum_i dq_i dp_i = \iint_s \sum_i dQ_i dP_i. \tag{1.5}$$

If both pairs of variables p, q and P, Q are expressed in terms of a third pair u, v on a surface s, we can write

$$dq_i dp_i = \frac{\partial(q_i, p_i)}{\partial(u, v)} du dv,$$

$$dQ_i dP_i = \frac{\partial(Q_i, P_i)}{\partial(u, v)} du dv.$$

Here all quantities are defined on the arbitrary surface *s*. The quantities dqdp, dQdP, dudv are by definition areas on this surface. The quantities

$$\frac{\partial(q_i, p_i)}{\partial(u, v)} \text{ and } \frac{\partial(Q_i, P_i)}{\partial(u, v)}$$

are the Jacobians of transformation from one co-ordinate system into another and serve for conversion of scales.

Now the Poincaré equation can be rewritten in the form

$$\iint_s \sum_i \frac{\partial(q_i, p_i)}{\partial(u, v)} du dv = \iint_s \sum_i \frac{\partial(Q_i, P_i)}{\partial(u, v)} du dv. \tag{1.6}$$

The expressions on the left- and right-hand sides of the equation are reduced to common differentials. But since the region of integration is arbitrary, the equation of the sums of the Jacobians follows, i.e.

$$\sum_i \frac{\partial(q_i, p_i)}{\partial(u, v)} = \sum_i \frac{\partial(Q_i, P_i)}{\partial(u, v)},$$

or, in the expanded form

$$\sum_i \left(\frac{\partial q_i}{\partial u} \frac{\partial p_i}{\partial v} - \frac{\partial p_i}{\partial u} \frac{\partial q_i}{\partial v} \right) = \sum_i \left(\frac{\partial Q_i}{\partial u} \frac{\partial P}{\partial v} - \frac{\partial P_i}{\partial u} \frac{\partial Q}{\partial v} \right). \tag{1.7}$$

Equation 1.7 is the Lagrange brackets, which are denoted by $\{qp\}_{uv}$. This equation shows that the Lagrange brackets are the invariants of canonical transformations.

If as canonical variables (u,v) we successively take the pairs (q_i, q_j), (p_i, p_j), (p_i, q_j), we obtain the system of fundamental Lagrange brackets:

$$\{q_i, q_i\} = 0, \quad \{pi, pi\} = 0, \quad \{pi, qi\} = \delta ijC, \tag{1.8}$$

which are not dependent on the choice of canonical pair.

The mutually inverse matrix for the Lagrange brackets is called Poisson brackets, which are written in the form

$$[uv]_{p,q} = \sum_i \left(\frac{\partial u}{\partial q_i} \frac{\partial v}{\partial p_i} - \frac{\partial v}{\partial q_i} \frac{\partial u}{\partial p_i} \right). \tag{1.9}$$

The expression (1.9) is also a canonical invariant conjugate to the Lagrange brackets, and the following relation is valid for it:

$$\sum \{u_i u_j\} [u_i u_k] = \delta_{ik},$$

which holds even for non-canonical transformations.

It is important to emphasize here that if both these brackets were derived from some earlier-existing mechanics, they would still axiomatically form the basis for constructing at list the mechanics from which they were derived. From the mathematical standpoint, these brackets are the so-called skew forms, with the property of anti-symmetry. In our case, they take the form

$$\{uv\} = -\{vu\}, \; [uv] = -[vu].$$

For skew forms the mathematical apparatus of outer calculus or the method of Cartan external forms has been developed, in which the operations of differentiation are reduced to calculations performed along the contour of the region. Hence is the name ‘outer calculus’.

For the Poisson as for the Lagrange brackets, the system of fundamental brackets is written in the form

$$[q_i, q_i] = 0, \quad [p_i, p_j] = 0, \quad [p_i, q_j] = \delta_{ij}. \tag{1.10}$$

In quantum mechanics the Poisson brackets form the basis for directly writing the equations of motion. There the commutator of two quantities multiplied by i/ħ (where ħ is the Planck’s constant) corresponds to the brackets.

In classical mechanics we can write the equations of motion in the same form. If, for example, the Hamiltonian H is taken as one of the quantities in brackets, we obtain

$$[q_i H] = \frac{\partial H}{\partial p_i} = \dot{q}_i,$$

$$[p_i H] = \frac{\partial H}{\partial q_i} = \dot{p}_i.$$

Within the brackets, the Jacobi identity

$$[u[vw]]+[v[wu]]+[w[uw]] = 0$$

well known also in the method of external forms since Grossman's time, will likewise be valid.

Thus, the principle of mutual reversibility forms the methodological basis for the objective representation of nature. The practical method of covariant transformation of co-ordinates with the help of mutually invertable matrices or the Lagrange and Poisson brackets is based on this principle. The entire mathematical apparatus of classical mechanics, statistical physics and quantum mechanics is, in its turn, built on the brackets as axioms.

1.2 Action and Integral Canonical Pairs

It was Kepler who, in his laws of motion along conic sections with a constant velocity of sweep over surfaces, established the law of conservation of angular momentum and thereby established the specific role of a quantity with the dimension of action. The variational principles of constructing mechanics based on the special role of this quantity were also developed quite a long time ago by Fermat and Maupertuis. It is again to Kepler, and to his conic sections as orbits of motion of celestial bodies, that we owe the second order of the corresponding differential equation describing this motion. Although, as Bondi noted, the special role of acceleration for the Earth's motion and its relation to the position of the Sun could be established simply by observation, it was sufficient to look in the direction of the vector of acceleration of the Earth in order to see that the Sun is always in this direction. Thus, it was established even than that the co-ordinates of position of a body in space and its momenta were the necessary and complete set of parameters for describing the motion of a system. Here a given pair of vectors is specified for each element constituting the system so that there are many such pairs for the whole system.

The next important development in the formation of dynamics was associated with the works of Clairaut and Legendre. The Clairaut equation is an example of an involute system which, with the help of the general Legendre transform, establishes the symmetry between the co-ordinates and momenta. The same Legendre transform converts the co-ordinate function into the momentum function and vice verse. The function performing this transformation is the Hamiltonian function of the system whose integral over a time interval has the dimension of action. The relationship between these functions resemble those between evolvent and evolute in mathematics. Here even their geometrical structure is preserved, which is seen in the shape of the family of enveloping lines reflecting the integral properties of the system from the standpoint of the external observer. It is true that the analytical apparatus of variational calculus then leads to the solution of a system of more equations with partial derivatives, and the integral character of the relations is lost. The reason here is that one wants to obtain a detailed description of the system, i.e. a description of the motion of all its elements, but not a change of structure of a system as a whole.

Clairaut obtained his equation by studying the motion of the Moon, so that we can consider his results to be based on observations of a natural system and therefore verified. It is interesting to note that the Clairaut equation is of second order since the first derivative is the second power. It resembles an equation with two branches of an equation combining two independent equations, and is similar to our first integral of the Jacobi virial equation. One branch of this equation is the envelope of the right-hand family of semi-tangents and the other that of the left-hand family. Thus, the idea of canonical conjugation of equations appeared in the early works of astronomers long before it was fully developed in mechanics.

Further developed of mechanics took place in the direction of its axiomatization and minimization of the number of initial postulates, where advances were made. The main achievement was the reduction of the fundamentals of mechanics to one general integral principle: the principle of least action. The fundamental nature of this principle can be traced in both mechanics and physics. For example, in quantum mechanics the Planck constant with the dimension of action is expressed in terms of the Coulomb electric charge and the velocity of light. Combining these quantities gives the expression known as the fine structure constant:

$$\frac{\hbar c}{e^2} = \frac{1}{\alpha} = 137,$$

where α is a dimensionless constant.

In this fundamental relation, which is apparently simple but absolute in character, gravitation, electrodynamics and quantum physics are connected through Planck's constant or the quantum of action.

Just as in the idea of variational calculus, the notion that there is some minimum value characterizing the true course of events in nature has existed since ancient times. However, it was only in 1662 that Fermat clearly formulated the principle of the shortest path in geometrical optics, from which follow the laws of refraction of light. This principle consists in requiring the value of the integral of reciprocal velocity to be minimal with respect to the trajectory of light between two points. A little later, in 1669, Leibniz in his treatise on the problems of dynamics introduced the concept of action, which is expressed in terms of the product of mass, velocity and path length. In other words, since path is the product of velocity and time, the Leibniz action function is written:

$$S = \int_{t_1}^{t_2} 2T dt$$

where T is the kinetic energy.

Descartes also suggested the quantity $m\dot{q}dq$ (where q is a co-ordinate) as an elementary action. However, since $m\dot{q}dq = m\dot{q}\dot{q}dq = 2Tdt$, his action turned out to be equivalent to the Leibniz action.

The principle of least action was developed further by Maupertius, Bernoulli and Euler. Generalizing the results of Euler's studies, Lagrange then extended the principle to an arbitrary system of n mass points m interacting arbitrarily and being situated in a field of central forces which are proportional to arbitrary powers of distances. In this case, the motion of the system was determined by the requirement of the lowest or the highest value of the sum

$$S = \sum_{i=1}^{n} m_i \int v_i dr.$$

These studies were most closely connected with the development of variational calculus, where the following two principles are of primary importance in dynamics. The principle of virtual displacement states that a mechanical system is in equilibrium only when the total infinitesimal work done by active forces during any possible displacement of the system from a given position equals to zero:

$$\sum_i F_i \delta q_i = 0,$$

where F_i is the active force and δq_i the possible or virtual displacement which is consistent at a given instant with the constraints imposed on the system.

The d'Alembert–Lagrange principle states that for the real motion of a system, the total elementary work done by active forces and forces of inertia during any possible displacements at any instant equals zero:

$$\sum_i (F_i - m_i \ddot{q}_i)\delta q_i = 0.$$

On the basis of these principles, Hamilton derived his principle of least action, which he further developed and generalized.

It is obvious that

$$\ddot{q}_i \delta q_i = \frac{d}{dt}(\dot{q}_i \delta q_i) - \dot{q}_i \frac{d}{dt}(\delta q_i) = \frac{d}{dt}(\dot{q}_i \delta q_i) - \dot{q}_i\ \delta(\dot{q}_i) = \frac{d}{dt}(\dot{q}_i \delta q_i) - \frac{1}{2}\ \delta(\dot{q}_i)^2.$$

Then

$$\sum_i (F_i - m_i \ddot{q}_i)\ \delta q_i = 0 => \sum_i F_i\ \delta q_i = \sum_i m_i \ddot{q}_i\ \delta q_i$$
$$= \frac{d}{dt}\left(\sum_i m_i \dot{q}_i\ \delta q_i\right) = \delta T + \delta A$$

or

$$\frac{d}{dt}\left(\sum_i m_i\dot{q}_i\ \delta q_i\right) = \delta T + \ \delta A, \tag{1.11}$$

where T and A are the virtual work done by the reactions of constraints and external forces, respectively.

If we integrate both the parts of Eq. 1.11 over t from t_1 to $t_{2,}$, in which $\delta q_i = 0$, we obtain

$$\int_{t_1}^{t_2} \delta(T + A)dt = 0.$$

For a conservative system $\delta A = -\ \delta U$ (where U is potential) and

$$\int_{t_1}^{t_2} \delta A dt = -\int_{t_1}^{t_2} \delta U dt = -\ \delta \int_{t_1}^{t_2} U dt,$$

where $\delta t = \delta dt$ since time does not vary.

Finally, the expression for Hamilton's principle of least action will take the form

$$\delta \int_{t_i}^{t_2} L dt = \delta \int_{t_1}^{t_2} (T - U)dt = 0, \tag{1.12}$$

where L is the Lagrangian (free energy in thermodynamics).

Jacobi totally excluded time from the principle of least action. Since

$$T = E - U = \frac{1}{2}\frac{\sum_i m_i(dq_i)^2}{(dt)^2}$$

and

$$dt = \sqrt{\frac{\sum_i m_i(dq_i)^2}{2(E - U)}}$$

the principle itself can be written in the form

$$\delta \int_{t_1}^{t_2} \sqrt{2(E-U)}\sqrt{\sum_i m_i(dq_i)^2} = 0. \tag{1.13}$$

It must be emphasized that the principle of least action differs from other variational principles by its integral character. It does not consider the differential properties of motion (velocity at a given point and so on) characterizing it at each point but the properties which characterize motion in a finite segment measured by the integral over path. Hence it is possible in principle not to include the co-ordinates of a point in formulating a problem of dynamics. That is why the principle of least action characterizes motion independently of the choice of a particular coordinate system.

All the basic equations of motion are derived from the principle of least action, which is postulated as the basic principle. Thus, Lagrange equations of motion are derived from the Euler-Lagrange variational principle. Hamilton's equations are obtained directly from the principle of least action. Here, a frame of reference and a co-ordinate system associated with it are always chosen. Incorrect choice of co-ordinate system and frame of reference may result in equations which we are unable to solve. The problem of choosing the best frame of reference and co-ordinate system and the interest in methods of transforming co-ordinates follows from here. This has its own history, as described below.

The concept of relativity of motion was introduced in mechanics as long ago as Galilei's time, if we disregard ancient scientists. It reflected the need in astronomy to compare the results of observations made at different places and times. This is obvious in astronomy since the observers themselves are moving in relation to the objects observed and in relation to each other. For the results of observation to have any practical value, we should be able to convert them from one observation system to another.

In mathematical models of the phenomena under investigation, transformations of the appropriate quantities correspond to such a conversion. In mathematics, however, unlike physics, conversions can be introduced most arbitrarily and at one's discretion. Therefore, from all conceivable transformations it is necessary to select those, the methods of invariants have been developed; these are quantities which do not change their values for some classes of transformation. At a more abstract level, transformations are considered which retain the mathematical form of the equations describing the laws of motion of mechanical systems, or covariant transformations. These transformations, as applied to mechanics and to the principle of least action, are canonical transformations. As we have already pointed out, canonical transformations do not depend on the equations of dynamics and can be studied without them. Therein lies their fundamental strength.

We have already referred to the selection of groups of transformations of co-ordinates or covariant transformations, which do not change the form of the equations. In this case, the quantities enter into these equations as functions of their variables can and generally do change during transformation. Here the question may selects a class of covariant transformations. All kind of transformations could be tried out until the best result is obtained. In physics all defined quantities have a specific tensor rank – scalars, vectors, higher-rank tensor, spinors, etc. This is an experimental fact. The laws of physics are written in the form of equations and it is natural that all terms entering equations should have identical tensor dimensions.

We cannot, for example, add a scalar and a vector and so on. Hence the requirement of covariance. Returning to mechanics, we recall that the basic variables here are the co-ordinates of momentum space. The latter space is abstract; we do not perceive it by our sense organs. Therefore, the greatest general transformation which does not go beyond the specific transformation is only an arbitrary function of the co-ordinates of configuration space and time. These, known as point transformations for, are covariant transformations for the Lagrangian equations. However, transformations which are usually the most general as functions of all conceivable co-ordinates are covariant transformations for the Hamilton equations and are called canonical.

The so-called generating function of transformation is introduced into the theory of transformation of co-ordinates. It is interesting to note that for canonical transformations the action is such a function.

We seek the generating function to solve the problem of finding a transformation leading to a reference and co-ordinate system which is such that in it the equation of motion will have the simplest form. For this, it is required that in the transformed equations the Hamiltonian should generally be a constant and then, together with the Hamiltonian, the particle co-ordinate system should automatically also be constant. The system in such a co-ordinate system obtained after transformation should be in equilibrium. There can obviously be no simpler form. Then we obtain the equation for determining the generation function, known as the Hamilton-Jacobi equation, in partial derivatives. It is also well known because it is the classical limit of the Schrödinger quantum equation. Since this generating function is action, we again note that all the main branches of physics converge on the concept of action.

We now briefly recall the main postulates of the theory of canonical transformations and the derivation of the Hamilton-Jacobi equation.

By φ and ψ we denote the transformation functions of the co-ordinates of configuration space and momenta q and p so that

$$q_0 = \varphi(q, p, t), \quad p_0 = \psi(q, p, t), \tag{1.14}$$

where q_0, p_0 and q, p are the co-ordinates and momenta before and after their transformation, respectively.

In the theory of canonical transformations, the theorem about the existence of the conditions of canonicity of transformations is proven. A necessary and sufficient condition for the canonicity of the transformations under consideration is the condition of existence of function F(q, p, t), which is such that the following equality will be satisfied:

$$\sum_i \psi_i \delta\varphi_{ji} - C\sum_i p_i \delta q_i = -\,\delta F(q,p,t), \tag{1.15}$$

where the variation is performed for fixed time; C is some constant; and F(q, p, t) the generating function of transformation.

We shall not give the proof of this theorem. We only note that, for practical purposes, a very simple and convenient method of verifying whether a randomly chosen transformation is canonical is derived from this theorem. For this purpose, it is sufficient to verify the validity of the following three equations:

$$\left\{q_i q_j\right\} = 0, \quad \left\{p_i p_j\right\} = 0, \quad \left\{p_i q_j\right\} = \delta_{ij} C,$$

where {....} are the Lagrange brackets.

If the verification gives a positive result, the constant C or the valence of transformation is calculated at once. Then the new Hamiltonian will be expressed in terms of the old one in the following manner:

$$H = CH_0 + \frac{\partial F}{\partial t} + \sum_i \psi \frac{\partial \varphi_i}{\partial t}, \tag{1.16}$$

where the variables p_0 and q_0 are expressed in terms of the new p and q with respect to transformations inverse to φ and ψ.

Now if we do what we earlier wanted to do and require that the new Hamiltonian be identically constant, p and q being constant, we shall obtain an equation for generating a function S (according to Jacobi, it is action):

$$\frac{\partial S}{\partial t} + CH_0 = 0.$$

The quantities p and q entering into the formula for the old Hamiltonian can be expressed in terms of the function S by the formula

$$\frac{\partial S}{\partial q_i} = Cp_i.$$

If we introduce the relation $S_n = S/C$, the equation will take the form

$$\frac{\partial S_n}{\partial t} + H\left(\frac{\partial S_n}{\partial q}, t\right) = 0. \tag{1.17}$$

It is the first-order Hamilton-Jacobi equation in partial derivatives. After its solution we obtain the equations

$$\frac{\partial S}{\partial q_i} = p_i,$$

$$\frac{\partial S}{\partial \alpha_i} = -\beta_i,$$

where α, β are the constants we chose at first which determine in the old coordinates the motions of the system in implicit form. These relations are integrals of the initial Hamiltonian system of equations.

It is as the theory of canonical equations and their canonical transformations that modern dynamics is developing. It is true – and we once again emphasize this – that it is developing mainly on the basis of detailed description of the motion of a system from an internal observer's standpoint. Even when the method of outer forms is used, these forms relate to individual elements constituting the system. So for the whole system many forms are considered. However, the integral invariants proposed by Poincaré (1965) and developed by Cartan are used rarely. In approaching the dynamics of a system from an external observer's standpoint, integral invariants should be the main tool of the researcher.

We now come back to the general equation of the principle of mutual reversibility of representations (1.3):

$$L(r) + H(p) = rp.$$

We note once more that in it the canonical pair r and p have the dimension of the scalar function H and L determining the transformation by its gradient. As shown above, the scalar function itself has the dimension of action. Hence the main canonical pairs in mechanics will be action-angle, energy-time and Jacobi function-frequency. The powers with which time enters (t^o, t^{+1}, t^{-1}) in the case of these pairs will be 0, +1, −1, respectively. Higher powers will not be fundamental because of the Jacobi identity. Proof can be work of Misner and co-workers (1973).

The pair used most extensively in mechanics is action-angle. It gives the simplest description of a system and forms the basis for using the methods of perturbation theory in mechanics.

The energy-time pair enables us to take into account processes with discontinuous functions of state such as, for example, collisions and phase transitions.

However, neither of these pairs which use the most general properties of space-time symmetry takes into account the specific effects of interaction. The Jacobi function-frequency pair for systems with a high degree of symmetry retains the merits of both those pairs and, in addition, takes into account the form of the law of interaction of particles constituting a system. From the principle of mutual reversibility one can already see the merits of this canonical pair, which is the basis of the Jacobi equation (1.2).

We can refer to many literature sources where historical interest in this equation has been noted. For example, Singh, in 1963, in his "Classical Dynamics", called Eq. 1.3 a striking result. Even Jacobi himself, in his "Vorlesung über Dynamik" (1884), considered his equation to be exceptionally interesting.

We give another, most recent, example from the general theory of relativity, which points to the importance of the Jacobi function-frequency canonical pair. In this theory, as also in geodesy, the equation of derivation of geodesics is given in the Riemannian form:

$$\nabla_{\vec{u}}\nabla_{\vec{u}}\vec{n} + R(\vec{n},\vec{u})\vec{u} = 0,$$

where $\nabla_{\vec{u}}$ is the covariant derivative along the tangent; $R(\vec{n}\vec{u})$ is the Riemannian curvature operator and $\vec{n}$ and $\vec{u}$ are the unit vectors of the normal and the tangent to the geodesic, respectively.

In the first term of this equation, differentiation operators are applied to the vector $\vec{n}$ and the curvature operator is applied to $\vec{u}$. However, when the Jacobi curvature tensor

$$J^{\mu}_{\nu\alpha\beta} = \frac{1}{2}\left(R^{\mu}_{\alpha\nu\beta} + R^{\mu}_{\beta\nu\alpha}\right),$$

where $R^{\mu}_{\alpha\nu\beta}$ is the Riemannian curvature tensor, was used, the equation of derivation was written in the symmetric form

$$\nabla_{\vec{u}}\nabla_{\vec{u}}\vec{n} + J(\vec{n},\vec{u})\vec{u} = 0,$$

where $J(\vec{n},\vec{u})$ is the Jacobi curvature operator.

The last equation can also be written as one operator:

$$[\nabla_{\vec{u}}\nabla_{\vec{u}}\vec{n} + J(\vec{n},\vec{u})]\vec{u} = 0,$$

in which case we obtain a zero eigensolution problem.

It was these factors that aroused our special interest in the Jacobi function-frequency canonical pair, which had not previously been utilized.

1.3 Integral Characteristics in the Study of Dynamics of Natural Systems

Traditionally two approaches are used to solve the problems of dynamics. In the first approach the object under study is regarded as a system of mass points interacting in accordance with specific laws. In the second it is represented as a continuum model in which the interactions are expressed in terms of volume fields of forces also acting in accordance with physical laws.

In both cases the mathematical description of the dynamics of the object is based on the latter's fundamental integral characteristics – mass and energy – which have a definite physical interpretation. Let us examine how these characteristics are used and consider, first of all, the concept of energy of an object in the different approaches to its dynamics.

In the mechanics of a system of mass points, energy, E, as we know, can be expressed mathematically as follows:

$$E = \sum_{i=1}^{n} \frac{p_i^2}{2m_i} + U(q_1,q_2,\ldots,q_n), \tag{1.18}$$

where q_i and p_i are, respectively, the co-ordinates and moments of the mass points m_i, and U is potential energy.

The expression written in this form is, strictly speaking, a function of 2n arguments p_i and q_i which has a wide range of values. Energy means only those values of this function which, upon substitution into Eq. 1.18 of the definite arguments p_i and q_i as function of time, are the solutions of the corresponding equations of motion. However, for brevity, the same term 'energy' is commonly used to express also its corresponding mathematical form and its magnitude by a narrower set of values of arguments. The fact that the concept 'energy' is used in many senses does not, however, cause ambiguities in describing the dynamics of a system in co-ordinates and velocities.

The variational approach is generally used in constructing the equations of mechanics of mass points. For this purpose, the Lagrangian L with arguments q_i, $\dot{q}_i$, t are introduced. Here, for brevity, $\dot{q}_i$ is called the velocity of the i-th particle. Strictly speaking, this should be so understood that, if $q_i(t)$ as a function of time is the solution of the corresponding equation of motion, $\dot{q}_i$ is the time derivative of this function. Consequently, the differentiation of the function should be performed only after the solution of the equation of motion has been found.

Thereafter, the procedure of variational calculus is used to find the class of functions $q_i(t)$ satisfying the principle of least action. For this purpose, the functional S is written and its extremum is found:

$$S = \int_{t_1}^{t_2} L(q_i, \dot{q}_i, t)dt. \tag{1.19}$$

We write the variation of action S:

$$\delta S = \int_{t_1}^{t_2} L(q_i, \dot{q}_i, t)dt = \int_{t_1}^{t_2} \sum_{i=1}^{n} \left(\frac{\partial L}{\partial q_i} \delta q_i + \frac{\partial L}{\partial q_i} \delta \dot{q}_i \right) dt.$$

Integrating in parts the second terms of the sum, we obtain

$$\int_{t_1}^{t_2} \frac{\partial L}{\dot{q}_i \partial} \delta \left(\frac{\partial q_i}{\partial t} \right) dt = \frac{\partial L}{\partial \dot{q}_i} \delta \dot{q}_i \Big|_{t_1}^{t_2} - \int_{t_1}^{t_2} \left[\frac{d}{dt} \left(\frac{\partial L}{\partial \dot{q}_i} \right) \right] \delta q_i dt.$$

The first term on the right-hand side of the equality is zero since, according to the condition, we have:

$$\partial q_i|_{t_1} = \partial q_i|_{t_2} = 0.$$

Then

$$\delta S = \int_{t_1}^{t_2} \sum_{i=1}^{n} \left(\frac{\delta L}{\delta q_i} - \frac{\delta}{\delta t} \frac{\delta L}{\delta q_i} \delta q_i dt \right).$$

Since the variations δq_i are independent, we obtain the Lagrangian system of equations

$$\frac{\delta L}{\delta q_i} - \frac{d}{dt}\left(\frac{\delta L}{\delta \dot{q}_i} \right) = 0. \tag{1.20}$$

Thus we have obtained the system of equations describing the motion of each point of the system.

The function S in Eq. 1.19 has the dimension of action between fixed points on the time axis. The function L is thus the time density of action, and the density of the Lagrangian L (per unit volume) introduced into the modern field theories, including quantum theory, finally symmetrize the functional S as the density of action in four-dimensional space-time.

However, the variational principle enables the equations for the search functions $q_i(t)$ to be found in the general case for an arbitrarily defined form of the Lagrangian. Therefore, the form of the Lagrangian is not determined from the variational principle. Moreover, even the set of its arguments is not determined. In this connection, the functional is varied with the Lagrangian. $L(q, \dot{q}, \ddot{q}, \ldots, t)$.

In this case, the principle of least action is applied and equations for determining the co-ordinates q_i are obtained. There are, of course, certain restrictions on the form of the Lagrangian, but none whatever on the number of arguments from the standpoint of mathematics. Here differential equations of an arbitrarily high order may be obtained. The type of Lagrangian, and especially the number of its arguments, are chosen from historical tradition from Kepler to Newton, who made their choice from a careful analysis of empirical observations and generalization of their results.

In such an approach we come back to where we started, i.e. to Newton's equations, while the general nature of the principle of least action seems to be unused. As for Newton's system of equations to describe the motion of mass points, the system of equations to describe the motion of mass points, the system of differential equations for three or more masses is unsolvable. Thus, there arose the many-body problem which has had no satisfactory solution even until now, while the mechanics of mass points, lacking a solution of the fundamental problem, reached a dead end.

The main difficulty encountered by Newtonian mechanics in going over to the integral approach was evidently the fact that the non-localizability of the energy of interaction of mass points m_1 and m_2, proportional to $m_1 m_2/r^2$ (where r is distance between masses) cannot in any reasonable way be related to any region or point in space, whereas, according to Einstein, energy is proportional to some mass.

However, it is known where this mass is. According to the equivalence principle, inertia is inherent in every mass, including mass defect occurring during their interaction.

There was only one thing to do – to study the behavior of natural objects in terms of integral (volumetric) characteristic and to write the virial equations with their help. The main reason to do this is that the gravitational interaction of any quantity of mass produces volumetric forces because any size of mass has volume. Jacobi very easily obtained his equation (1.2) from Newton's equations, noting its exceptionally interesting properties bud did not solve this equation in terms of integral (volumetric) characteristics. A century later, in his fundamental and complete work, Wintner (1941) frankly said that it was impossible in principle to solve this equation in terms of integral characteristics. Here a negative role was apparently played by the concept of non-localizability of potential energy in the mechanics of mass points. Nobody seemed to want to approach the problem from the standpoint of the external observer, where energy can be measured experimentally, since the problem had already been formulated from the internal observer's point of view. It was considered that it should in any case be solved in the style in which it had been formulated, without substituting one problem for another.Jacobi, however, who well understood the weak points of deriving the equations of motion of mass points, suggested his variant of the principle of least action, which was written in the form

$$\delta \int_{t_1}^{t_2} \sqrt{(E - U)} ds = 0, \tag{1.21}$$

where E and U are the total and potential energies and ds is a Riemannian linear element. The geometric interpretation of this equation is as follows. The trajectory is the geodesic of the configuration space in which the Riemannian linear element ds is specified, the latter being equal to $ds^2 = m_{ik} dq_i dq_k$, where m_{ik} is a coefficient written in the form

$$L(q, \dot{q}, t) = \frac{1}{2} m_{ik} \dot{q}_i \dot{q}_k - U.$$

The Hertz-Gauss principle or the principle of least curvature follows at once from the Jacobi principle since the geodesic is a curve of least curvature. This principle reduces the many-body problem to the motion of one body along the geodesic in a complex multidimensional configuration space. Moreover, it determines there the trajectory of motion but not the law of motion along this trajectory. This obviously explains why the given principle did not fully solve the problem. However, it is important to note that attempts were made to go over to the one-body problem.

We now consider how mechanics developed within the framework of the continuum model. In continuum mechanics there are fields of forces, and the energy squared over these fields and the mass associated with it are localized over the whole space. Electromagnetic interactions in relation to energy are localized by introducing the concept of pressure. However, in mechanics, pressure is introduced axiomatically and its internal structure is not determined. It is interesting to note that pressure has the same dimension as the volume density of energy, and that, since energy is the time density of action, pressure is the density of action which is four-dimensional in space-time as is the Lagrangian density.

After introducing the distributed quantities, we can carry out integration over the volume of the system and obtain the system's integral parameters, which have a clear physical sense since the corresponding physical quantities can be measured. It is possible for this reason that the virial equations in cosmology are (and were) easily and clearly obtained from the Einstein equations. However, there were more integral parameters than equations here, and the so-called equations of state were therefore used in order to close the system of equations of dynamics. These equations directly connect the integral (or integrable, where is sufficient) quantities: pressure and density (omitting co-ordinates and velocities). The equations of state themselves are taken either from experiment or from auxiliary theories (simplified, as in the molecular kinetic theory, or partly rigorous, like the Pauly principle in quantum mechanics). In any case, these relations exist and demonstrate the validity of the assertion that the virial equations can be solved in terms of integral characteristics.

It was not always the case that after the virial equations were written they were solved in terms of integral characteristics. Since there were more characteristics then equations and since additional relations (of the type of equations of state in continuum mechanics) cannot be found immediately, some authors did not stick to the initially chosen integral direction and reverted to the ordinary differential approach. An example is the work of Chandrasekhar (1969) referred to earlier, where he developed the method of moments. After obtaining the system of virial equations, he calculated for their solution the variations of the integral characteristics for small shifts (or deformation) in order to linearize the virial equations. Consequently, while calculating the variations, he abandoned the integral approach and expanded the domain of definition of functions which expresses the corresponding integral parameter. This work deserves serious attention, and we shall return to it when we discuss the problem of moments.

It is interesting to dwell on the integral approach used in thermodynamics. We have referred earlier to equations of state and noted that they were obtained both experimentally and with the use of special theories. Let us consider one such example. In the theory of non-ideal gases the following formula for the equation of state for a gas is introduced by means of molecular kinetic theory methods:

$$P = \frac{NT_0}{V}\left(1 + \frac{NC_1(T_0)}{V} + \frac{N^2C_2(T_0)}{V^2} + \ldots\right).$$

Here P is the pressure of the gas; T_0 its temperature; V/N the specific volume per molecule and $C_i(T_0)$ coefficients independent of pressure.

Each term here corresponds to a particular type of interaction of molecules: the first term describes an ideal gas; the second defines pair interactions; the third the three-body interactions of molecules, and so on. The coefficients $C_i(T_0)$ in this theory are called virial coefficients. In statistical physics, the following parametrically defined equations of state are derived for their determination:

$$P = T_0 \sum_{n=1}^{\infty} \left(\frac{J_n}{n!} \xi^n \right),$$

$$\frac{N}{V} = \sum_{n=1}^{n} \left(\frac{J_n}{(n-1)!} \xi^n \right),$$

where ξ is a parameter which should be excluded; and J_n is a function dependent only on temperature and calculated for n-fold interaction of molecules from their given interaction function.

It should be noted that, at zero temperature, pressure in classical physics should become zero. The velocity of sound should also become zero. The energy of electromagnetic interaction of charges takes into account only the random motion of these charges. The Madelung energy (the energy of formation of an atomic lattice) determines the purely Coulomb interactions of charges and does not depend on temperature.

We have so far discussed the problem of using the integral characteristics of a system, which determine its behavior in time. It would be interesting to take a look at the properties of the integral invariants derived at different times. As we have already pointed out, the canonical representations retain some integral invariants and this is the basic property. Among the forms which they retain, the most important are Poincaré relative invariant and the forms associated with it.

The Poincaré integral invariant is

$$\oint pdq - Hdt$$

and the relative integral invariant is $\oint pdq$(where p and q are vectors).

It should be borne in mind that the invariant has the dimension of action and its form

$$pdq - Hdt = p_1dq_1 + p_2dq_2 + p_3dq_3 - Hdt,$$

taking into account that H/C is p_4 and that $cdt = dq_4$, shows that it is simply a scalar pseudo-Euclidian product of two four-vectors $p(p_1, p_2, p_3, p_4)$ and $q(q_1, q_2, q_3, q_4)$ – four-momentum and four-radius of the particle vector. Such, for example, is the eikonal determining the phase of a plane traveling wave:

$$kr - \omega t = k_1X + k_2Y + k_3Z - k_4ct,$$

which represents the scalar product (pseudo-Euclidian) of two four-vectors: $k(k_1, k_2, k_3, k_4 = \omega/c)$ – wave four-vector and four-radius of vector q. In the given case, this contraction is dimensionless. The Jacobi tensor, whose three-spur is the Jacobi function, can be contracted once or twice with the wave four-vector. During the first contraction we obtain a vector with the dimension of action so that the Jacobi function-frequency pair is a canonical pair, while the second contraction dives the Einstein function $E = mc^2$ (scalar). The Liouville theorem forming the basis of statistical physics is associated with the Poincaré invariant from which it follows.

The integral invariants considered above are regarded in mathematics as external forms of different orders and, as has been pointed out, they have been studied for a long time by means of the Cartan external forms, so that a well-developed mathematical apparatus for their study already exists.

From this brief description it can be seen that attempts have been made for a long time to approach the description of the dynamics of natural systems in terms of integral (volumetric) characteristics by means of the construction of virial equations. It is only required to introduce some concepts associated with the conditions of the external observer. In particular, the integral characteristics from the external observer's standpoint are functions of time only and consequently they cannot be subjected to partial differential operations. From the physical point of view this means that these characteristics should be measured directly (even if only in principle) and not calculated by formulae. For example, the potential energy in Eq. 1.18 can be emitted during the time of the system's formation. Mass as an integral characteristic of the system should also be determined from measurements performed externally. For example, the mass of the Sun is determined from the parameters of motion of planets around it. In atomic physics the eigenmoment of the atom can be measured, and this is easier to do than adding its eigen- and orbital moments, which is a mathematical problem.

In the general theory of relativity, co-ordinates near singularities are introduced similarly from the external observer's standpoint. The length of the circle surrounding the singularity, divided by 2π, is introduced instead of the Euclidian radius-vector. In this case, the chosen approach is a question of principle and not of convenience.

We note that for application of the integral approach in dynamics we already have the well-developed mathematical apparatus of external differentiation, discussed above, which is also used to study the contours surrounding the region externally. The distinctive feature of this calculus is that it has only the first differentials of function, the higher ones being absent. The parallels between the methods in physics and mathematics from the external observer's standpoint are obvious.

Another new concept is that natural systems from the external observer's standpoint are not conservative since they emit energy. The emitted energy is an integral effect of all possible types of interaction. However, a specific type of interaction in the system appears in the corresponding part of the emitted energy spectrum.

Finally, we introduce the concept of hierarchy of systems and sub-systems. We note that it is not simply and not only a concept for study of motion in nature since the very fact of the existence of such a hierarchy is empirical. For example, the chain: universe – galaxies – stars – planets – satellites – shells – plates – polycrystals – molecules – atoms – nucleons – electrons, etc., really exists so that this approach to the study of motion mirrors the structure of nature.

In studying motion inside a particular system, we should confine ourselves to the study of the influence exerted on the bodies in our system by the nearest neighbors only. We thus declare our system to be closed. Indeed, all laws of conservation are written for closed systems. It is nevertheless possible to describe with some accuracy the motion, say, of celestial bodies even prognostically (ephemeris) on the basis of mechanics with second-order equations of motion (here we are not going to take into account the co-ordinates-momenta parity). What may be the reason for this? Even the stability of such systems (admittedly according to Lyapunov and not according to Duboshin, i.e. with respect to variation of only the initial conditions and not interactions) is also studied to second-order time derivatives. From electrostatics we know both theoretically and experimentally (Coulomb's law and the Faraday cage) that a closed shell consisting of charges shifting freely in it will, if given sufficient time, distribute these charges over itself in such a way that the influence of charges external to the shell on the space inside the shell will be fully balanced by its own charge distribution. It is important to note that the potential inside the shell is not important since only its spatial gradient has an influence. It is known that this potential satisfies a second-order differential equation, the Poisson equation, so that Coulomb's law from the external observer's standpoint can be formulated as the principle of possibility of shielding the inside space of the shell from external actions. It was found that this approach automatically gives Coulomb's law: the law of inverse squares. Any deviation from two in the exponent of this law deprives the shell of such a property.

Coulomb's law is a static law. If it is to operate in the way indicated, the charge in the external actions must be sufficiently slow in comparison with the time needed for charge redistribution. Newton's law of universal gravitation has the same form as Coulomb's law. If we consider that the time of propagation of gravitational waves is the same as that of electromagnetic waves – this is just what the Einstein law states – and if we take into account that the structure of the universe is hierarchical, we have an analogy with electrostatics. The slowness of charges in the external actions of the shell (the latter being the system nearest to us and enveloping our system, e.g., for the Solar System it is our Galaxy) is ensured by the great distance of these outer bodies. Thus, there appears to be justification for the possible rejection of the influence of almost should be taken into account. The principle of shielding will itself give the criterion for the choice of almost-closed system.

1.3.1 Method of Moments: Specific Features in Integral Approach and First Moments

As we know, it is possible in mathematics under certain conditions to reconstruct a continuous function by means of its moments with the help of a generating function. Here each of the moments is an integral characteristic of the function and consequently, as applied to a physical system, is the external parameter of the function. The parameter can be measured in the appropriate manner.

However, although the function can be reconstructed by means of moments, we require for this purpose all moments, which are infinitely large in number. Moreover, in the detailed description of a system from the internal observer's standpoint, it is not possible to comprehend the results owing to the large number of the system's degrees of freedom even if number is finite. For this reason alone, one resorts to describing a system by the moments of the function which is being sought. Because of the artificial break in the infinite chain of moments, such a description is not only incomplete but also inaccurate. The results obtained in this manner require verification by experiment in order to demonstrate that the errors are not too large. Errors do occur since the aim is to obtain a full description of the motion of the whole system, i.e. the motion of each of its elements. Generally speaking, each moment is in itself a characteristic of the system, indeed of the system as a whole, and is a fully accurate quantity which is not determined with an error. For example, the first moment of mass density determines the position of the center of mass of a system and the zero moment determines its total mass. Both these characteristics are its accurate integral characteristics and will not be charged by whether or not we find still higher moments of the system. Since in the integral approach we are searching for canonical equations, the corresponding description of the system's behavior will be accurate.

In statistical physics the break in the chain of linked equations occurs after the introduction of some approximate relationship between the partial distribution functions. Its sense lies in the distortion of the law of interaction of the elements of the whole, since without introducing any law of interaction it is not possible to obtain a full description of the system from the internal observer's standpoint. This distortion is, of course, towards simplification of the model. The first two moments are the most important. They characterize the average value over the total distribution and its dispersion. The true distribution here is replaced by a Gaussian distribution with the same average value and dispersion. Then the virial theorem is used in order to choose the chain of linked equations, and the method of closure is order to choose is to average the corresponding equation over time.

We found a way to close this chain for high-symmetry systems, not by an approximate method but by a rigorous one. This will be discussed later. Here it is sufficient to note that in the most fundamental general theory of relativity the metric is determined quite rigorously either by one (Schwarzschild) or by two (Kerr) moments, and this is natural for their symmetry because the metric considered is external to the sources.

We now recall the basic postulates of the theory of moments in mathematics and make some comments on them from the standpoint of the integral approach considered here to the solution of the problems of dynamics of systems.

In mathematics there is the concept of characteristic function for a given function F(x) which is determined by the expression

$$\varphi(t) = \int_{\Omega} e^{itx} dF(x),$$

where Ω is the domain of definition of the function and F and t belong to this domain.

If the function F(x) has a derivative f(x) with a finite number of discontinuities of the second kind then

$$\varphi(t) = \int_{\Omega} e^{itx} f(x)dx,$$

where $\varphi(x)$ is simply the Fourier transform of the function f(x).

The given relation is single-valued and unique; it is a theorem of mathematics. So is the inversion formula which enables us, with the help of the characteristic function, to reconstruct the function itself

$$F(b) - F(a) = \lim_{c\to\infty} \frac{1}{2n} \int_{-c}^{+c} \frac{e^{-ita} - e^{-1tb}}{it} \varphi(t)dt.$$

it is simply the inversion of the Fourier transform.

In the theorem of moments, a function is reconstructed with the help of its moments. Here it seems that this can be done, although in the general case the solution is non-unique. However, under some sufficient conditions of the form

$$\lim_{n\to\infty} 1 \frac{\sqrt[n]{\mu_n}}{n} < \infty,$$

where μ_n is the n-th moment of function f, the solution will be unique. In this case, the problem is solved by the following procedure. At first the function

$$\varphi(e) = \sum_{n=0}^{\infty} \left[\frac{(it)^n}{n!} \mu_n \right].$$

is calculated with the help of specified moments. This function will be the characteristic function f(x), or, in the more general case, of function F(x). Here, two important points are usually to be noted: (a) if the function to be reconstructed is given in a finite interval, the solution of the problem is generally unique; (b) the condition on sufficiency for uniqueness in the case of an infinite domain of definition can be satisfied only for particular moments. We further note that this condition itself can easily be fulfilled in physics so that we can consider that the problem of inversion of moments in physics is always solved uniquely.

However, we are speaking about all moments of the function, and they are infinitely large in number. In practice, the method of moments is used to find an approximate and not an accurate solution since, as a rule, one rarely succeeds in finding a general form of solution of the moment equation.

The moment method is used in the following manner. The initial equation for the function being sought in integrated successively over the whole domain of definition after prior multiplication by successive power of the argument. In the general case, we obtain an infinite chain on linked algebraic equations. If this infinite system of equations can be solved in the general form, we obtain all the moments and using the above formulae we find what we were looking for. As a rule, however, this cannot be done because of mathematical difficulties. Then we proceed as follows. We artificially break the chain of equations, so that the number of equations will be smaller than the number of unknowns. To solve such a system, we artificially supplement it by an equation constructed on the basis of some considerations, and solve the shortened system obtained. It is considered that this solution will be chosen closer to the truth.

We note the following, however. As will be seen from the above scheme of reconstruction of a function by means of moments, when truncating the series of moments we shell obtain a polynomial as the characteristics function. The polynomial has a finite number of poles, and consequently even an approximate solution will not be just inaccurate; it will not be specified in the whole domain of the function sought.

Let us give an obvious example. We approximate a sine curve by a third-order polynomial. The region where any approximation of the sine curve to a cubic parabola is at all possible is that between the extreme roots of the third-order equation, as shown in Fig. 1.1.

This is not the main point. The main point is that from the standpoint of mathematics, in calculating moments by this method, integration is performed over the whole domain of definition of the function, i.e. over all its arguments. If we speak of physics, this means integration over space and time co-ordinates. In this particular case, the moments will be constant or, more accurately, functions of the parameters of the problem. This is the case for which all the rigorous theorems of the theory of moments have been established in mathematics. In the physical applications, however, the problem is approached differently, so there is a new formulation of the problem itself. In the physical literature, integration over time is not performed and the moments (now these are spatial moments) are functions of time. The moment equations, which are obtained from the basic equation by multiplying by the successive power of space co-ordinates followed by integration over the whole configuration space, are ordinary differential equations. We further note that, in the case of multi-dimensional configuration space (and this is precisely the general case), multiplication by the corresponding n-th power of co-ordinates denotes multiplication by the product of the powers of all spatial arguments, the sum of all exponents of which is equal to n with all

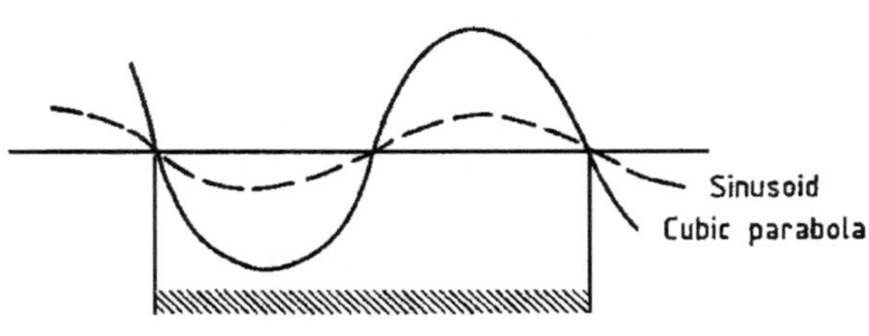

Fig. 1.1 Domain of approximation of the sinusoid to the cubic parabola

such possible combinations, so that ultimately we shell obtain not one moment of power $n(\mu_n)$ but a whole matrix of such values for each n, i.e. a tensor of the n-th rank. Since this is only a matter of mathematical technique, later in this chapter we shell understand by 'moment' the spur of the corresponding matrix, denoting it by μ_n to avoid submerging the physical essence in less substantial details.

The method of moments was developed generally in order to solve problems where insurmountable or almost insurmountable mathematical difficulties were encountered and was not intended for problems which are in principle unsolvable e.g. those in which the number of unknown functions exceeds that of equations for their determination. In the latter case, the method of moments alone cannot help.

In this connection, let us see how Chandrasekhar (1969) evaluates the method. He writes that the virial method is essentially the method of moments and that virial equations of different orders are in fact nothing more than the moments of the corresponding hydrodynamic equations. By moments of equations he means spatial moments, as he makes clear when he says that all these quantities (including volume, V) will, generally speaking, also be functions of time, t_o. So virial equations are differential equations, and it is fortunate that they are only ordinary differential equations and not partial derivatives.

The quantities (moments) entering into these virial equations are in fact integral quantities if we regard them as unknown functions of time (which they are) and do not try to linearize them. Unfortunately, there are more such functions then equations and, possibly, therefore even Chandrasekhar does not see in the virila equations anything new other then that the method of obtaining these equations is identical with the method of moments, and this is obvious from his statements quoted above.

To elucidate our idea, we return to Newton's equations for a system of gravitational mass points of hydrodynamic equations since, is this case, we can more clearly see the role of the integral parameters. Newton's system of equations in the above case is a system of equations of the many-body problem. It is defined accurately and should in principle be solved in the sense that the number of equations and the number of functions sought coincide and that the equations are ordinary differential equations. However, we are unable to solve them for lack of mathematical methods.

But when these equations are replaced by those of spatial moments, it is found (Jacobi) that these equations contain only the integral characteristics of a many-body system: Jacobi function, total energy (system mass) and potential energy (mass defect of the system). They all appear as unknown functions of time. Taking into account the law of conservation of energy, there is only one equation lacking for solving the system (for finding these functions). We are not concerned here with where the equation is to be taken from – this will be dealt with in due course – but we now return with new ideas to the equations of hydrodynamics.

A number of questions arise at ones. The first is what are we seeking as unknown functions in the equations of hydrodynamics before they are converted into the moment form? It is the density fluid for a compressible fluid and the velocity vector at each of its points. But this is indeed an infinite amount of information. Even if it were possible to obtain this information by some unknown means, what more is to be done with it? It appears, just as in the theory of probability, for example, or in quantum mechanics, that this information is used in an integral form. We are seeking the probability density or the probability amplitude in order to find the average values of the finite number of parameters of the system, the so-called measurable or observable quantities. The functions we are seeking cannot be used in their general form even if they existed (we mean precisely the general, i.e. infinite-valued, solution). Then, in reply to the question of how to solve those equations, we pose another question: what do the authors of those equations expect to obtain? They do not in fact consider them unsolvable. It is clear, as they say in mathematical physics, that solving an equation means obtaining it in the form of a finite algebraic combination of either so-called elementary functions or a convergent (although asymptotic) series of such functions, so that in practice, where the criterion is accuracy, one can confine oneself to a finite number of such functions. On the other hand, it may mean that one wants to reduce the solution to quadratures where the methods of finite approximation can also be used.

The main conclusion is that the solution is expected to be obtained in the form of a finite and not an infinite number of 'images'. For example, if as the solution to this kind of problem for a gaseous star we obtain the equation of its rotation (oscillation), one parameter – the angular velocity – is one 'image'. If a baroclinic rotation is obtained, one law – the change in angular velocity (frequency of oscillation) along the radius of the star, given by some elementary function or by their combination or at least by a polynomial approximation – also represents an image, while this time is an optical 'image'. We may also mention, for example, honeycomb convection, where there are more parameters then one to describe motion. But when, in practice, we encounter the phenomenon of turbulence, where pulsations are infinitely large in number, it is clear that they cannot be obtained from the equations of hydrodynamics. Thus, even the authors of the hydrodynamic equations do not expect to obtain from their equations arbitrarily general solutions, and this indicates from the very beginning that these equations contain too much information even in their formulation. It is also apparent that the method of moments can be stronger than the initial equations themselves. This will be the case if the initial equations do not simply possess excess information but also contain the entire integral information corresponding to the given physical theory. This information will then appear, whereas the equations will be accurate equations of the physical problem. Chandrasekhar evidently feels this. On the page where the statements referred to earlier appear, he points out that the advantage of these moment equations lies in the fact that the lowest-order equations often allow a simple physical interpretation. Unfortunately, he immediately makes a reservation to the effect that the purpose of using the moment equations. Later on, for specific problems of equilibrium and stability of configurations of fluids, he himself demonstrates how the method of moments (of finite number) gives a complete solution of these problems. This is done, in principle, by bypassing solution of the problem of obtaining full information on the distribution of density and motion in a system. Moreover, he even states clearly elsewhere in the book that, for solution of the Dirichlet problem, the second-order virial equations contain the necessary and sufficient conditions, and the virial equations therefore give all the necessary information for the study of equilibrium and stability of the permissible ellipsoidal figures.

The actual problems of science and technology set specific tasks for specific purposes rather than general speculative problems. For example, in the examples quoted above, it is necessary to elucidate the conditions of equilibrium and stability of real objects and not to speculate about finding some distribution function. In this connection, it is clear that the virial method will always work.

It is interesting to consider where the classical physicists use the specific formulation of a problem. The Dirichlet problem was first solved fully by Riemann, who used it in selecting the law of dependence on the velocity field at a particular time on the initial Lagrangian co-ordinates of the corresponding points. He selected this law by intuition (which he could afford to do) and expressed it as a linear matrix function with a matrix of coefficients of transformation dependent only on time. Later Likhnerovich and Chandrasekhar specially studied this law in the Dirichlet problem and established its compatibility with free boundary conditions, demonstrating the uniqueness of the selection. Thus, in this case, the specific formulation of the problem (the problem of equilibrium and stability of a given form of the equilibrium figure) manifested itself in requiring certain boundary conditions. These conditions, in their turn, determined the selection of the form of representation to describe internal motion in the system under study, and this form was the linear matrix combination of Lagrangian co-ordinates as the permissible velocity field of the fluid elements of the body

In this specific case it is possible to observe even at the classical level the mechanism of the action of limitation which follows from the specific formulation of the problem. In the general case, this is of course not required, nor is it always possible, and we go over directly to virial equations and their solutions, which are the solutions of the specific problem.

In summary, we emphasize once again that the sense of the virial equations in physics is not the same as that of the method of moments in mathematics. In the latter case, the purpose is to reconstruct the initial function but it is not important what sort of function it is.

In physics, on the other hand, the function which generates moments is the distribution function and it therefore has a precise role. If it is known, that is fine, since with its help we can find other functions of measurable or observable quantities. If it is not known, but if there is way of doing it, that is fine too. The virial equations of provide the means of doing without the intermediate step – the distribution function – but for this purpose, they must of course be correct. That means the moment should be taken for the most general and universal equation such as the Einstein equation in the general theory of relativity.

In concluding this chapter we note that scientific models describing nature are being developed continually in the direction of greater generality and simplicity. Galilei did not take into consideration the need for external causes for free motion; he pointed out that free motion was the natural state of matter. Einstein went further. He introduced motion along the geodesic, thereby excluding gravitation as a force. His motion along the geodesic is now the new form of free motion.

Acceleration of a body occurs only when it is prevented from moving along the geodesic. So, nature has electro-weak and strong interactions.

As to the oscillatory motion of matter, it is also universal. In the absence of forces of friction it also does not need an external force in order to exist since it is a natural form of motion. If we take into account the results obtained recently for empty space, where the oscillatory regime of metric changes also exists (Belinsky et al. 1970), it will be clear that the oscillatory regime of motion is the most general and natural property of nature. Energy is important for free motion, which is described by equations based on the energy-time canonical pair. The action-angle canonical pair is used to describe rotational motion. The Jacobi function-frequency canonical pair is appropriate for describing oscillatory motion. Hence the interest in the Jacobi equation (1.2), which makes practical use of this canonical pair.

Chapter 2
Recent Observations and Understanding Physical Meaning of Jacobi's Virial Equation

Fundamentals of all the planetary and Solar System sciences are tested first of all by the laws of the Earth movement, where the confidence limit to the laws can be checked by observation. More over, all the sense of human being is connected with this planet. As far as the techniques and instruments for observation were developed, then geodesists, astronomers and geophysicists have noticed that in the planet's inertial rotation some irregularities and deviations relative to the accepted standard parameters and hydrostatic state conditions have appeared. Those irregularities that are often called as inaccuracies, number of which is counted by more than dozen, finally were incorporated into two problems, namely, variation of the angular velocity in the daily, monthly, annually and secular time scale, and variation in the poles motion in the same time scales. Just after the problems became obvious and have not find resolution in the frame work of the accepted physical and theoretical conceptions of celestial mechanics the latter lost interest in the problems of the Earth dynamics. In this connection the well known German theoreticians in dynamics, Klein and Sommerfeld, stated that the Earth mechanics appear to be more complicated than the celestial mechanics and represents "*some confused labyrinths of geophysics*" (Klein and Sommerfeld 1903). In order to study irregular velocity of the Earth rotation and the pole motion numerous projects of observation and regular monitoring were organized by the planetary network. As it was always in such cases, the cause of the observed effects was searched in the effects of perturbations coming from the Moon and the Sun, and also in the influence of dynamical effects of the own shells like the atmosphere, the oceans and the liquid core, existence of which is considered by many researchers.

2.1 Dynamical Effects Discovered by Space Study

Artificial satellites, which made a start of space study in the second part of the twentieth century, opened a new page in space sciences. It was determined that the ultimate goal of this scientific program should be an answer to the question of

V.I. Ferronsky et al., *Jacobi Dynamics*, Astrophysics and Space Science Library 369,
DOI 10.1007/978-94-007-0498-5_2, © Springer Science+Business Media B.V. 2011

the Solar system origin. Investigation of the near Earth cosmic space for solution of geodetic and geophysical problems was in the beginning initiated.

The first geodetic satellites for studying dynamic parameters of the planet were launched almost 50 years ago. They gathered vast amounts of data that significantly improved our knowledge of the inner structure and dynamics of the Earth. They made it a real possibility to evaluate experimentally the correctness of basic physical ideas and hypotheses in astronomy, astrophysics, geophysics, geodesy and geology, and to compare theoretical calculations with observations. Success in this direction was achieved in a short period of time.

On the basis of satellite orbit measurements, the zonal, sectorial and tesseral harmonics of gravitational moments in expansion of the gravitational potential by a spherical function, up to tens, twenties and higher degrees were calculated. The calculations have resulted in an important discovery having far-reaching effects. The obtained results proved the long-held assumption of geophysicists that the Earth does not stay in hydrostatic equilibrium, which, in fact, is the basic principle of the theories of dynamics, figure and inner structure of the planet. The same conclusion was made about the Moon.

This conclusion means that the physical conception of hydrostatic equilibrium state of the Earth which was applied for construction of model of the outer and central force field does not satisfy the observed dynamic effects of gravitational interaction of mass particles and should be revised. But the state of scientific knowledge of this phenomenon has been found to be not ready to cope with such a situation. The story of the condition of hydrostatic equilibrium of the planet begins with Newton's consideration, in his famous work *Philosophiae Naturalis Principia Mathematica*, of the Earth's oblateness problem. The investigation based on hydrostatics was further developed by French astronomer and mathematician Clairaut. Later on the hypothesis of hydrostatic equilibrium was extended to all celestial bodies including stars. The authority of Newton was always so high that any other theories for solution of the problem in dynamics and celestial body structure were never proposed. But in current times the problem has arisen of the cause of the discrepancy between theory and observation and a moment has come to take over this crisis in the study of fundamentals of the Earth sciences. A situation like this happened at the beginning of the twentieth century when radioactive and roentgen radiation was discovered and the corpuscular-wave nature of light was proved. This was the starting point for development of quantum mechanics. We seem now to have a similar situation with respect to the planets motion.

We found a still more serious discrepancy related to the Earth hydrostatic equilibrium, which is as follows (Ferronsky and Ferronsky 2007). It is known that the planet's potential energy is almost 300 times more than the kinetic one represented by the body's rotation. This ratio between the potential and kinetic energy contradicts the requirement of the virial theorem according to which the potential energy of a body in the outer uniform force field should be twice as much of the kinetic one.

From point of view of the observed potential energy the Earth's angular velocity should be about 17 times as much as it is. However, the planet has remained for a long time in an equilibrium state. In fact, the Earth appears to have been deprived of its kinetic energy. Some of the other planets, such like Mars, Jupiter, Saturn, Uranus and Neptune, exhibit the same behavior. But for the Mercury, Venus, our Moon and the Sun, the equilibrium states of which are also accepted as hydrostatic, the potential energy exceeds their kinetic energy by 10^4 times. A logical explanation comes to mind that there is some hidden form of motion of the body's interacting mass particles, together with their respective kinetic energy, which has not been taken previously into account. It is known that the hydrostatic equilibrium condition of a body, being stay in the outer force field, satisfies the requirement of the Clausius virial theorem. The same requirement follows also from the Eulerian equations for a liquid-filled uniform sphere. The virial theorem gives an averaged relationship between the potential and kinetic energies of a body. A periodic component of the energy change there during the corresponding time interval is accepted as a constant value and eliminated from consideration. From this evidence it was not difficult to guess that the hidden form of motion and the source of needed kinetic energy of the Earth and the planets including the Moon and the Sun might be found in that eliminated periodic component. In the problem considered by Newton, that component was absent because of his concept of the central gravitational force field, the total sum of which is equal to zero.

Taking into account the relationship between the Earth's gravitational moments and the gravitational potential observed by the satellites, we came back to derivation of the virial theorem in classical mechanics (see below) and obtained its generalized form of the relationship between the energy and the polar moment of inertia of a body. Doing so, we obtained the equation of dynamical equilibrium of a body in its own force field where the hydrostatic equilibrium is a particular case of a uniform body in its outer force field. The equation establishes a relationship between the potential and kinetic energies of a body by means of energy of oscillation of the polar moment of inertia in the form of the energy conservation law. An analytical expression of the derived new form of the virial theorem, based on the Newton's laws of motion, appeared to be the Jacobi's virial equation. In this case the earlier lost of kinetic energy is found by taking into account the oscillating motion of the interacting mass particles, the integral effect of which is expressed through oscillation of the polar moment of inertia. That effect fits the relationship between the potential and kinetic energies in the classical virial theorem. At the same time a novel physical conception about gravitation and electromagnetic interaction is appeared and mechanism of the energy generation becomes clear. The nature of the gravity forces as a derivative of the body's inner energy appears to be discovered.

The obtained, on the basis of Jacobi dynamics, results related to the problem of the Earth are as follows. We found that the new effect, which creates dynamics of the Earth, is its own inner force field. Earlier, the sum of the inner forces and their moments being effected by the outer central force field were considered as equal to zero. We find that the mass forces of interaction being volumetric ones created the inner force field which appears to be the field of power (energy) pressure. That field, according to its definition, can not be equal to zero. The resultant of the field pressure appears to be a space envelope. The envelope has a spherical shape for a sphere and an elliptic shape for an ellipsoid. It was found that dynamic effects of the body's force field occur in oscillation and rotation of the shells according to Kepler's laws. A body that has a uniform mass density distribution realizes all its kinetic energy of the motion in the form of so-called virial oscillations. It was assumed, earlier, that wave properties of this nature, like oscillations for mass particles in mechanics of bodies, are unessential. We found that virial oscillations of a body initiated by the force field of its own interacting mass particles represent the main part of its kinetic energy. Theories based on hydrostatics ignore that energy. But, as it was noted above, in this case the potential energy of the Earth and other celestial bodies by two or more orders exceeds their kinetic energy represented only in the form of axial rotation of the mass. Such an unusual effect has a simple physical explanation. Still in the beginning of the last century French physicist Louis de Broglie expressed an assumption, proved later on, that any micro-particle including electron, proton, atom and molecule, acquires particle-wave properties. The relationship, discovered by the artificial satellites between changes of the Earth's gravitational potential and the moment of inertia, shows that interaction of the planet's masses takes place on their elementary particle levels. It means that the main form of motion of the interacting mass particles is their oscillation. Continuous 'trembling' of the planet's gravitational field, detected by satellites as the gravitational moments change, is another fact proving the de Broglie idea and extending it to the gravitational interaction of celestial body masses.

The dynamical approach to solve the problem under consideration allowed the authors to expand the body's potential energy on its normal, tangential and dissipative components. The differential equations that determine the main body's dynamical parameters, namely its oscillation and rotation, were written. A rigorous solution of the equations was considered on the basis for bodies with spherical and axial symmetry. The solutions of problems relating to rotation, oscillation, obliquity and oblateness of a body's orbit and itself was considered on the basis of the general solution of dynamics of a self-gravitating body in its own force field. It was found that precession and wobbling of the Earth and irregularity of its rotation depends on effects of the polar and equatorial oblateness and the separate rotation of the planet's, the Sun's and the Moon's shells. The induced outer force field of a body follows rotation of the resultant envelope of the shells, but with some delay because of the finite velocity of the energy propagation in the induced outer force field. Also the problems of inner structure of the Earth, the nature of the planet's electromagnetic field and mechanism of the energy generation were considered. The presented theory is applicable not only to the planets and satellites, but also to the stars, where hydrostatic equilibrium is considered as an equation of state. Finally, the theory opens a way to understand the physics of gravitation as the internal power (energy) pressure which occurs at mass interaction on the level of the molecules, atoms and nuclei and elementary mass particles.

The obtained new results which have common relation to all celestial bodies are presented in the corresponding chapters and sections of the book.

2.2 Interpretation of Satellite Orbits and Failure of Hydrostatic Equilibrium of the Earth and the Moon

We recall briefly the conditions of the Earth hydrostatic equilibrium. By definition the hydrostatics is a branch of the hydromechanics, which studies the equilibrium of a liquid and gas and the effects of a stationary liquid on immersed bodies relative to the chosen reference system. For a liquid equilibrated relative to a rigid body, when its velocity of motion is equal to zero and the field of densities is steady the equation of state follows from the Eulerian and Navier–Stokes equations in the form (Sedov 1970)

$$\operatorname{grad} p = \rho F, \tag{2.1}$$

where p is the pressure; ρ is the density; F is the mass force.

In the Cartesian system of reference Eq. 2.1 is written as

$$\frac{\partial p}{\partial x} = \rho F_x, \qquad \frac{\partial p}{\partial y} = \rho F_y, \qquad \frac{\partial p}{\partial z} = \rho F_z \tag{2.2}$$

If the outer mass forces are absent, i.e. $F_x = F_y = F_z = 0$, then

$$\operatorname{grad} p = 0.$$

In this case, in accordance with the Pascal's law the pressure in all liquid points will be the same.

For the uniform incompressible liquid, when $\rho = \text{const}$, its equilibrium can be only in the potential field of the outer forces. For general case of incompressible liquid and potential field of the outer forces from (2.1) one has

$$dp = \rho dU, \tag{2.3}$$

where U is the forces potential.

It follows from Eq. 2.3, that for an equilibrated liquid in the potential force field its density and pressure appear to be a function only of the potential U.

For a gravity force field, when in the steady-state liquid only these forces act, one has

$$F_x = F_y = 0, \quad F_z = -g, \quad U = -gz + \text{const} \quad \text{and} \quad p = p(z), \rho = \rho(z).$$

Here the surfaces of constant pressure and density appear as horizontal planes. Then Eq. 2.3 is written in the form

$$\frac{dp}{dz} = -\rho g < 0. \tag{2.4}$$

It means that with elevation the pressure falls and with depth it grows. From here it follows that

$$p - p_0 = -\int_{z_0}^{z} \rho g dz = -\rho g(z - z_0) \tag{2.5}$$

where g is the acceleration of the gravity force.

If a spherical vessel is filled in by incompressible liquid and rotates around its vertical axis with constant angular velocity ω, then for determination of the equilibrated free surface of the liquid in Eq. 2.2 the centrifugal inertial forces should be introduced in the form

$$\begin{aligned} \frac{\partial p}{\partial x} &= \rho \omega^2 x, \\ \frac{\partial p}{\partial x} &= \rho \omega^2 y, \\ \frac{\partial p}{\partial x} &= -\rho g \end{aligned} \tag{2.6}$$

From here, for the rotating body with radius $r^2 = x^2 + y^2$, one finds

$$p = -\rho g z + \frac{\rho \omega^2 r^2}{2} + C. \tag{2.7}$$

For the points on the free surface $r = 0$, $z = z_o$ one has $p = p_o$. Then

$$C = p_0 + \rho g z_0, \tag{2.8}$$

$$p = p_0 + \rho g(z_0 - z) + \frac{\rho \omega r^2}{2}. \tag{2.9}$$

The equation of the liquid free surface, where $p = p_o$, has a paraboloidal shape

$$z - z_0 = \frac{\omega^2 r^2}{2g}. \tag{2.10}$$

Above relations determine the principal physical conditions and equations of hydrostatic equilibrium of a liquid. They remain a basis for the modern dynamics and theory of the Earth figure. The attempt to harmonize these conditions with the planet's motion conditions has failed, which was proved by observation. It will be shown below that the main obstacle for such harmonization is rejection of the planet's inner force field without which the hydrostatics is unable to provide the equilibrium between the body's interacted forces as Newton's third law requires. The Earth is a self-gravitating body. Its matter moves in the own force field which is generated by elementary mass particle interaction. The mass density distribution, rotation and oscillation of the body's shells result from the inner force field. And the orbital motion of the planet is controlled by interaction of the outer force fields of the planet and the Sun.

Let us look for more specific effects determining the absence of the Earth hydrostatic equilibrium and more realistic conditions of its equilibrium based on the results of the Earth's satellite orbit motion.

The initial factual material for the problem study is presented by the observed orbit elements of the geodetic satellites which move on perturbed Kepler's orbits. The satellite motion is fixed by means of observational stations located within zones of a visual height range of 1,000–2,500 km, which is optimal for the planet's gravity field study. It was found that the satellite's perturbed motion at such a close distance from the Earth's surface is connected with the non-uniform distribution of mass density, the consequences of which are the non-spherical shape in the figure and the corresponding non-uniform distribution of the outer gravity field around the planet. These non-uniformities cause corresponding changes in trajectories of the satellite's motion, which are fixed by tracking stations. Thus, distribution of the Earth's mass density determines an adequate equipotential trajectory in the planet's gravity field, which follows the satellite. The main goal of the geodetic satellites launched under different angles relative to the equatorial plane is in measurement of all deviations in the trajectory from the unperturbed Kepler's orbit.

The satellite orbits data for solving the Earth's oblateness problem are interpreted on the basis of the known (in celestial mechanics) theory of expansion of the gravity potential of a body, the structure and the shape of which do not much differ from the uniform sphere. The expression of the expansion, by spherical functions, recommended by the International Union of Astronomy, is the following equation (Grushinsky 1976):

$$U(r,\varphi,\lambda) = \frac{GM}{r}\left[1 - \sum_{n=2}^{\infty} J_n \left(\frac{R_e}{r}\right)^n P_n(\sin\varphi) + \sum_{n=2}^{\infty}\sum_{m=1}^{n}\left(\frac{R_e}{r}\right)^n P_{nm}(\sin\varphi)(C_{nm}\cos m\lambda + S_{nm}\sin m\lambda)\right], \quad (2.11)$$

where r, φ and λ are the heliocentric polar co-ordinates of an observation point; G is the gravity constant; M and R_e are the mass and the mean equatorial radius of the Earth; P_n is the Legendre polynomial of n order; $P_{nm}(\sin\varphi)$ is the associated spherical functions; J_n, C_{nm}, S_{nm} are the dimensionless constants characterizing the Earth's shape and gravity field.

The first terms of Eq. 2.11 determine the zero approximation of Newton's potential for a uniform sphere. The constants J_n, C_{nm}, S_{nm} represent the dimensionless gravitational moments, which are determined through analyzing the satellite orbits. The values J_n express the zonal moments, and C_{nm} and S_{nm} are the tesseral moments. In the case of hydrostatic equilibrium of the Earth as a body of rotation, in the expression of the gravitational potential (2.11) only the even *n*-zonal moments J_n are rapidly decreased with growth, and the odd zonal and all tesseral moments turn into zero, i.e.

$$U = \frac{GM}{r}\left[1 - J_2\left(\frac{R_e}{r}\right)^2 P_2(\cos\ \theta) - \sum_{n=3}^{\infty} J_n\left(\frac{R_e}{r}\right)^{n+1} P_n(\cos\ \theta)\right], \tag{2.12}$$

where θ is the angle of the polar distance from the Earth's pole.

Here the constant J_2 represents the zonal gravitational moment, which characterizes the axial planet's oblateness and makes the main contribution to correction of the unperturbed potential. That constant determines the dimensionless coefficient of the moment of inertia relative to the polar axis and equal to

$$J_2 = \frac{C - A}{MR_e^2}, \tag{2.13}$$

where C and A are the Earth's moments of inertia with respect to the polar and equatorial axes accordingly, and R_e is the equatorial radius.

For expansion by spherical functions of the Earth's gravity forces potential, the rotation of which is taken to be under action of the outer inertial forces, but not of its own force field, the centrifugal force potential is introduced into Eq. 2.12. Then for the hydrostatic condition with the even zonal moments J_n one has

$$\begin{aligned} W = &\frac{GM}{r}\left[1 - J_2\left(\frac{R_e}{r}\right)^2 P_2(\cos\theta) - \sum_{n=3}^{\infty} J_n\left(\frac{R_e}{r}\right)^{n+1} P_n(\cos\theta)\right] \\ &+ \frac{\omega^2 r^2}{3}[1 - P_2(\cos\theta)], \end{aligned} \tag{2.14}$$

where W is the potential of the body of rotation; $\omega^2 r$ is the centrifugal force. The first two terms and the term of the centrifugal force in Eq. 2.14 express the normal potential of the gravity force

$$W = \frac{GM}{r}\left[1 - J_2\left(\frac{R_e}{r}\right)^2 P_2(\cos\theta)\right] + \frac{\omega^2 r^2}{3}[1 - P_2(\cos\theta)]. \tag{2.15}$$

The potential (2.15) corresponds to the spheroid's surface which within oblateness coincides with the ellipsoid of rotation. Rewriting term $P_2(\cos\theta)$ in this equation through the sinus of the heliocentric latitude and the angular velocity – through the geodynamic parameter q, one can find the relationship of the Earth's oblateness ε with the dynamic constant J_2. Then the equation of the dynamic oblateness ε is obtained in the form (Grushinsky 1976; Melchior 1972)

$$\varepsilon = \frac{3}{2}J_2 + \frac{q}{2}, \tag{2.16}$$

where the geodynamic parameter q is the ratio of the centrifugal force to the gravity force at the equator

$$q = \frac{\omega^2 R}{GM/R^2}. \tag{2.17}$$

Dynamic parameter J_2, found by satellite observation in addition to the oblateness calculation, is used for determination of a mean value of the Earth's moment of inertia. For this purpose the constant of the planet's free precession is also used, which represents one more observed parameter expressing the ratio of the moments of inertia in the form:

$$H = \frac{C - A}{C}. \tag{2.18}$$

This is the theoretical base for interpretation of the satellite observations. But its practical application gave very contradictory results (Grushinsky 1976; Melchior 1972; Zharkov 1978). In particular, the zonal gravitation moment calculated by means of observation was found to be $J_2 = 0.0010827$, from where the polar oblateness $\varepsilon = 1/298.25$ appeared to be short of the expected value and equal to 1/297.3. The all zonal moments J_n, starting from J_3, which relate to the secular perturbation of the orbit, were close to constant value and equal, by an order of magnitude, to the square of the oblateness i.e., $\sim(1/300)^2$ and slowly decreasing with an increase of n. The tesseral moments C_{nm} and S_{nm} appeared to be not equal to zero, expressing the short-term nutational perturbations of the orbit. In the case of hydrostatic equilibrium of the Earth at the found value of J_2, the polar oblateness ε should be equal to 1/299.25. On this basis the conclusion was made that the Earth does not stay in hydrostatic equilibrium. The planet's deviation from the hydrostatic equilibrium evidenced that there is a bulge in the planet's equatorial region with amplitude of about 70 m. It means that the Earth body is forced by normal and tangential forces which develop corresponding stresses and deformations. Finally, by the measured tesseral and sectorial harmonics, it was directly confirmed that the Earth has an asymmetric shape with reference to the axis of rotation and to the equatorial plane.

Because the Earth does not stay in hydrostatic equilibrium, then the above described initial physical fundamentals for interpretation of the satellite observations should be recognized as incorrect and the related physical concepts cannot explain the real picture of the planet's dynamics.

The question is raised of how to interpret the obtained actual data and where the truth should be sought. First of all we should verify correctness of the oblateness interpretation and the conclusion about the Earth's equatorial bulge. It is known from observation that the Earth is a triaxial body (Grushinsky 1976). Theoretical application of the triaxial Earth model was not considered because it contradicts the hydrostatic equilibrium hypothesis. But after it was found that the hydrostatic equilibrium is absent, the alternative with the triaxial Earth should be considered first.

Let us analyze Eq. 2.16. It is known from the observation data, that the constant of the centrifugal oblateness q is equal to

$$q = \frac{\omega_3^2}{GM/R^3} = \left(\frac{1}{17.01}\right)^2 = \frac{1}{289.37}. \tag{2.19}$$

Determine a difference between the centrifugal oblateness constant q and the polar oblateness ε′ found by the satellite orbits, assuming that the desired value has a relationship with the perturbation caused by the equatorial ellipsoid

$$\begin{aligned} \varepsilon' &= \frac{a-c}{a} - \frac{b-c}{a} = \frac{a-b}{a} = \frac{1}{289.37} - \frac{1}{298.25} = \frac{1}{9720} \\ &= 1.713\left(\frac{1}{289.37}\right)^2, \end{aligned} \tag{2.20}$$

where a, b and c are the semi-axes of the triaxial Earth.

The differences between the major and minor equatorial semi-axes can be found from Eq. 2.20. If the major semi-axis is taken in accordance with recommendation of the International Union of Geodesy and Geophysics as a = 6,378,160 m, then the minor equatorial semi-axis *b* can be equal to:

$$a - b = 6378160/9720 = 656\ \text{m}; \quad b = 6377504\,\text{m}.$$

There is a reason now to assume, that the value of equatorial oblateness $\varepsilon' = 1/9{,}720$ is a component in all the zonal gravitation moments J_n, related to the secular perturbations of the satellite orbits including J_2. They are perturbed both by the polar and the equatorial oblateness of the Earth. Experimental result of dependence of the satellite precession of equinoxes on the orbit angle to the equatorial plane proves the above statement. This effect ought to be expected because it was known long ago from observation that the Earth is a triaxial body. If our conclusion is true, then there is no ground for discussion about the equatorial bulge. And also the problem of the hydrostatic equilibrium is closed automatically because in this case the Earth is not a figure of rotation; and the nature of the observing fact of rotation of the Earth should be looked for rather in the action of its own inner force field but not in the effects of the inertial forces. As to the nature of the Earth's oblateness, then for its explanation later on the effects of perturbation arising during separation of the Earth's shells by mass density differentiation and separation of the Earth itself from the Protosun will be considered. In particular, the effect of heredity in creation of the body's oblateness is evidenced by the ratio of kinetic energy of the Sun and the Moon expressed through the ratio of square frequencies of oscillation ε'' of their polar moments of inertia, which is close to the planet's equatorial oblateness:

$$\varepsilon'' = \frac{\omega_c^2}{\omega_\pi^2} = \frac{(10^{-4})^2}{(0.96576 \cdot 10^{-2})^2} = \left(\frac{1}{96.576}\right)^2 = 1.73\left(\frac{1}{289.3}\right)^2,$$

where $\omega_c = 10^{-4}\ s^{-1}$ and $\omega_л = 0.96576 \cdot 10^{-2}\ s^{-1}$ are the frequencies of oscillation of the Sun's and the Moon's polar moment of inertia correspondingly.

By observation the Moon is also a triaxial body. In addition, the retrograde motion of the nodes of the Earth, the Moon and the artificial satellites is registered and is explained by rotation of the bodies' orbits. Later on it will be shown, that the above remarkable phenomenon is explained by rotation of the body's inner masses together with their gravity fields, the periods of which are equal to the periods of the precession of their oblique axes. The observed body rotation is valid only for the upper shells, which were separated during mass density differentiation in their own force fields and stay in that field in a suspended state of equilibrium.

The most prominent effect, which was discovered by investigation of the geodetic satellite orbits, is the fact of a physical relationship between the Earth's mean (polar) moment of inertia and the induced outer gravity field. That fact without exaggeration can be called a fundamental contribution to understanding the nature of the planet's self-gravity. The planet's moment of inertia is an integral characteristic of the mass density distribution. Calculation of the gravitational moments based on measurement of elements of the satellite orbits is the main content of satellite geodesy and geophysics. Short-periodic perturbations of the gravity field fixed at revolution of a satellite around the Earth, the period of which is small compared to the planet's period, provides evidence about oscillation of the moment of inertia or, to be more correct, about oscillating motion of the interacting mass particles. It will be shown, that oscillating motion of the interacting particles forms the main part of a body's kinetic energy and the moment of inertia itself is the periodically changing value.

Oscillation of the Earth's moment of inertia and also the gravitational field is fixed not only during the study by artificial satellites. Both parameters have also been registered by surface seismic investigations. Consider briefly the main points of these observations.

The study of the Earth's eigenoscillation started with Poisson's work on oscillation of an elastic sphere, which was considered in the framework of the theory of elasticity. In the beginning of the twentieth century Poisson's solution was generalized by Love in the framework of the problem solution of a gravitating uniform sphere of the Earth's mass and size. The calculated values of periods of oscillation were found to be within the limit of some minutes to 1 h.

In the middle of the twentieth century during the powerful earthquakes in 1952 and 1960 in Chile and Kamchatka an American team of geophysicists headed by Beneoff, using advanced seismographs and gravimeters, reliably succeeded in recording the an entire series of oscillations with periods from 8.4 min up to 57 min. Those oscillations in the form of seismograms have represented the dynamical effects of the interior of the planet as an elastic body, and the gravimetric records have shown the "tremor" of the inner gravitational field (Zharkov 1978). In fact, the effect of the simultaneous action of the potential and kinetic energy in the Earth's interior was fixed by the above experiments.

About 1,000 harmonics of different frequencies were derived by expansion of the line spectrum of the Earth's oscillation. These harmonics appear to be integral characteristics of the density, elastic properties and effects of the gravity field, i.e. of the potential and kinetic energy of separate volumetric parts of the non-uniform planet. As a result two general modes of the Earth's oscillations were found by the above spectral analysis, namely, spherical with a vector of radial direction and torsion with a vector perpendicular to the radius.

From the point of view of the existing conception about the planet's hydrostatic equilibrium, the nature of the observed oscillations was considered to be a property of the gravitating non-uniform (regarding density) body in which the pulsed load of the earthquake excites elementary integral effects in the form of elastic gravity quanta (Zharkov 1978). Considering the observed dynamical effects of earthquakes, geophysicists came close to a conclusion about the nature of the oscillating processes in the Earth's interior. But the conclusion itself still has not been expressed because it continues to relate to the position of the planet's hydrostatic equilibrium.

2.3 Imbalance Between the Earth's Potential and Kinetic Energy

We discovered the most likely serious cause, for which even formulation of the problem of the Earth's dynamics based on the hydrostatic equilibrium is incorrect. The point is that the ratio of kinetic to potential energy of the planet is equal to ~1/300, i.e. the same as its oblateness. Such a ratio does not satisfy the fundamental condition of the virial theorem, the equation of which expresses the hydrostatic equilibrium condition. According to that condition the considered energies' ratio should be equal to 1/2. Taking into account that kinetic energy of the Earth is presented by the planet's inertial rotation, then assuming it to be a rigid body rotating with the observed angular velocity $\omega_r = 7.29 \cdot 10^{-5}\ s^{-1}$, the mass $M = 6 \cdot 10^{24}$ kg, and the radius $R = 6.37 \cdot 10^6$ m, the energy is equal to:

$$\begin{aligned} T_e &= 0.6MR^2\,\omega_r^2 = 0.6 \cdot 6 \cdot 10^{24} \cdot \left(6.37 \cdot 10^6\right)^2 \left(7.29 \cdot 10^{-5}\right)^2 \\ &= 7.76 \cdot 10^{29}\mathrm{J} = 7.76 \cdot 10^{36}\mathrm{erg}. \end{aligned}$$

The potential energy of the Earth at the same parameters is

$$\begin{aligned} U_e &= 0.6 \cdot GM^2/R \; = 0.6 \cdot 6.67 \cdot 10^{-11} \cdot \left(6 \cdot 10^{24}\right)^2/6.37 \cdot 10^6 \\ &= 2.26 \cdot 10^{32}\mathrm{J} \; = 2.26 \cdot 10^{39}\mathrm{erg}. \end{aligned}$$

The ratio of the kinetic and potential energy comprises

$$\frac{T_e}{U_e} = \frac{7.76 \cdot 10^{29}}{2.26 \cdot 10^{32}} = \frac{1}{291}.$$

One can see that the ratio is close to the planet's oblateness. It does not satisfy the virial theorem and does not correspond to any condition of equilibrium of a really existing natural system because, in accordance with the third Newton's law, equality between the acting and the reacting forces should be satisfied. The other planets, the Sun and the Moon, the hydrostatic equilibrium for which is also accepted as a fundamental condition, stay in an analogous situation. Since the Earth in reality exists in equilibrium and its orbital motion strictly satisfies the ratio of the energies, then the question arises where the kinetic energy of the planet's own motion has disappeared. Otherwise the virial theorem for the Earth is not valid. Moreover, if one takes into account that the energy of inertial rotation does not belong to the body, then the Earth and other celestial bodies equilibrium problem appears to be out of discussion.

Thus, we came to the problem of the Earth equilibrium from two positions. From one side, the planet by observation does not stay in hydrostatic equilibrium, and from the other side, it does not stay in general mechanical equilibrium because there is no reaction forces to counteract to the acting potential forces. The answer to both questions is given below while deriving an equation of the dynamical equilibrium of the planet by means of generalization of the classical virial theorem.

2.4 Generalization of Classical Virial Theorem

The main methodological question arises: in what kind state of equilibrium the Earth exists? The answer to the question results from the generalized virial theorem for a self-gravitating body, i.e. the body which itself generates the energy for its own motion by interaction of the constituent particles having innate moments. The guiding effect which we use here is the observed by artificial satellite functional relationship between changes in the outer gravity field of the Earth and its mean (polar) moment of inertia. The deep physical meaning of this relationship is as follows. The observed planet's polar moment of inertia is an integral (volumetric) parameter, which represents not fixed interacted mass particles, but expresses changes in their motion under the inner body's energy. The Clausius' virial theorem represents relationship between the potential and kinetic energy in averaged form for a non-interacted (ideal) gaseous cloud of particles or a uniform body which stay in the outer force field. In order to generalize the theorem for a uniform and non-uniform body staying in the own force field we introduce there the volumetric moments of interacted particles, taking into account their volumetric nature. Moreover, the interacted mass particles of a continuous medium generate volumetric forces (pressure or capacity of energy) and volumetric moments, which, in fact, produce the motion in the form of oscillation and rotation of matter. The oscillating form of motion of the Earth and other celestial bodies is the dominating part of their kinetic energy which up to now has not been taken into account. We wish to fill in this gap in dynamics of celestial bodies.

The classic virial theorem is the analytical expression of the hydrostatic equilibrium condition and follows from Newton's and the Euler's equations of motion. Let us recall its derivation in accordance with classical mechanics (Goldstein 1980).

Consider a system of mass points, the location of which is determined by the radius vector $\mathbf{r}_i$ and the force $\mathbf{F}_i$ including the constraints. Then equations of motion of the mass points through their moments $\mathbf{p}_i$ can be written in the form

$$\dot{\mathrm{p}}_{\mathrm{i}} = \mathrm{F}_{\mathrm{i}}, \tag{2.21}$$

The value of the moment of momentum is

$$\mathrm{Q} = \sum_{\mathrm{i}} \mathrm{p}_{\mathrm{i}} \cdot \mathrm{r}_{\mathrm{i}},$$

where the summation is done for all masses of the system. The derivative with respect to time from that value is

$$\frac{\mathrm{dQ}}{\mathrm{dt}} = \sum_{\mathrm{i}} \dot{\mathrm{r}}_{\mathrm{i}} \cdot \mathrm{p}_{\mathrm{i}} + \sum_{\mathrm{i}} \dot{\mathrm{p}}_{\mathrm{i}} \cdot \mathrm{r}_{\mathrm{i}}. \tag{2.22}$$

The first term in the right hand side of (2.22) is reduced to the form

$$\sum_{\mathrm{i}} \dot{\mathrm{r}}_{\mathrm{i}} \cdot \mathrm{p}_{\mathrm{i}} = \sum_{\mathrm{i}} \mathrm{m}_{\mathrm{i}} \cdot \dot{\mathrm{r}}_{\mathrm{i}} \cdot \dot{\mathrm{r}}_{\mathrm{i}} = \sum_{\mathrm{i}} \mathrm{m}_{\mathrm{i}} \mathrm{v}_{\mathrm{i}}^2 = 2\mathrm{T},$$

where T is the kinetic energy of particle motion under action of the forces $\mathbf{F}_i$. The second term in the Eq. 2.22 is

$$\sum_{\mathrm{i}} \dot{\mathrm{p}}_{\mathrm{i}} \cdot \mathrm{r}_{\mathrm{i}} = \sum_{\mathrm{i}} \mathrm{F}_{\mathrm{i}} \cdot \mathrm{r}_{\mathrm{i}}.$$

Now Eq. 2.22 can be written as

$$\frac{\mathrm{d}}{\mathrm{dt}} \sum_{\mathrm{i}} \mathrm{p}_{\mathrm{i}} \cdot \mathrm{r}_{\mathrm{i}} = 2\mathrm{T} + \sum_{\mathrm{i}} \mathrm{F}_{\mathrm{i}} \cdot \mathrm{r}_{\mathrm{i}}. \tag{2.23}$$

The mean values in (2.23) within the time interval τ are found by their integration from 0 to t and division by τ:

$$\frac{1}{r}\int_0^t \frac{dQ}{dt}dt = \frac{\overline{dQ}}{dt} = \overline{2T} + \overline{\sum_1 F_1 \cdot r_1}$$

or

$$\overline{2T} + \overline{\sum_i F_i \cdot r_i} = \frac{1}{\tau}[Q(\tau) - Q(0)]. \tag{2.24}$$

For the system, in which the co-ordinates of mass point motion are repeated through the period τ, the right hand side of Eq. 2.24 after its averaging is equal to zero. If the period is too large, then the right hand side becomes a very small quantity. Then, the expression (2.24) in the averaged form gives the following relation

$$-\overline{\sum_i F_i \cdot r_i} = 2T, \tag{2.25}$$

or in mechanics it is written in the form

$$2T = -U$$

Equation 2.25 is known as the virial theorem, and its left hand side is called the virial of Clausius (German *virial* is from the Latin *vires* which means forces). The virial theorem is a fundamental relation between the potential and kinetic energy and is valid for a vide range of natural systems, the motion of which is provided by action of different physical interactions of their constituent particles. Clausius proved the theorem in 1870 when he solved the problem of work of the Carnot thermal machine, where the final effect of the water vapor pressure (the potential energy) was connected with the kinetic energy of the piston motion. The water vapor was considered as a perfect gas. And the mechanism of the potential energy (the pressure) generation at the coal burning in the firebox was not considered and was not taken into account.

The starting point in the above-presented derivation of virial theorem in mechanics is the moment of the mass point system, the nature of which is not considered both in mechanics and by Clausius. By Newton's definition the moment "*is the measure of that determined proportionally to the velocity and the mass*". The nature of the moment by his definition is "*the innate force of the matter*". By his understanding that force is an inertial force, i.e. the motion of a mass continues with a constant velocity.

The observed (by satellites) relationship between the potential and the kinetic energy of the gravitation field and the Earth's moment of inertia evidences, that the kinetic energy of the interacted mass particle motion, which is expressed as a volumetric effect of the planet's moment of inertia, is not taken into account. The evidence of that was given in the previous Sect. 2.3 in the quantitative calculation of a ratio between the kinetic and potential energies, equal to ~1/300.

In order to correct the contradiction, the kinetic energy of motion of the interacted particles should be taken into account in the derived virial theorem. Because of any mass has volume the momentum **p** should be written in volumetric form:

$$p_i = \sum_i m_i \dot{r}_i. \tag{2.26}$$

Now the volumetric moment of momentum acquires the wave nature and is presented as

$$Q = \sum_i p_i \cdot r_i = \sum_i m_i \cdot \dot{r}_i \cdot r_i = \frac{d}{dt}\left(\sum_i \frac{m_i r_i^2}{2}\right) = \frac{1}{2}\dot{I}_p \tag{2.27}$$

where I_p is the polar moment of inertia of the system of interacted particles (for the sphere it is equal to 3/2 of the axial moment).

The derivative from that value with respect to time is

$$\frac{dQ}{dt} = \frac{1}{2}\ddot{I}_p = \sum_i \dot{r}_i \cdot p_i + \sum_i \dot{p}_i \cdot r_i. \tag{2.28}$$

The first term in the right hand part of (2.28) remains without change

$$\sum_i \dot{r}_i \cdot p_i = \sum_i m_i \cdot \dot{r}_i \cdot \dot{r}_i = \sum_i m_i v_i^2 = 2T. \tag{2.29}$$

The second term represents the potential energy of the system

$$\sum_i \dot{p}_i \cdot r_i = \sum_i F_i \cdot r_i = U. \tag{2.30}$$

Equation 2.28 is written now in the form

$$\frac{1}{2}\ddot{I}_p = 2T + U. \tag{2.31}$$

Expression (2.31) represents a generalized equation of the virial theorem for a mass particle system interacted by the Newton's law. Here in the left hand side of (2.31) the ignored up to now inner kinetic energy of interaction of the mass particles appears. Solution of Eq. 2.31 gives a variation of the polar moment of inertia within the period τ. For a conservative (uniform with respect to density) system averaged expression (2.28) by integration from 0 to t within time interval τ gives

$$\frac{1}{T}\int_{0}^{t}\frac{dQ}{dt}dt=\overline{\frac{dQ}{dt}}=2\overline{T}+\overline{U}=\dot{I}_{p}. \qquad (2.32)$$

Equation 2.32 at $\ddot{I}_p = 0$ (conservative system) gives $\dot{I}_p = E = \text{const.}$, where E is the total system's energy. It means that the interacted mass particles of the system move. with constant velocity. In the case of dissipative system, Eq. 2.32 is not equal to zero and the interacted mass particles move with acceleration. Now the ratio between the potential and kinetic energy has a value in accordance with the Eq. 2.31. Kinetic energy of the interacted mass particles in the form of oscillation of the polar moment of inertia in that equation is taken into account. And now in the frame of the law of energy conservation the ratio of the potential to kinetic energy of a celestial body has a correct value.

Expression (2.31) appears to be an equation of dynamical equilibrium of a self-gravitating body (star, planet, satellite). The hydrostatic equilibrium is absent here because the interacted particles are continuously moving by use of inner energy. Integral effect of the moving particles is fixed by the satellite orbits in the form of changing zonal, sectorial and tesseral gravitational moments. For derivation of the generalized virial theorem we used the potential energy generated by interacted particles of the initial moment (2.26). The initial moments form the inner, or "*innate*" by Newton's definition, energy of the body which has an inherited origin.

Thus, we obtained a differential equation of the second order (2.31) which describes the body dynamics and its dynamical equilibrium.

The virial Eq. 2.31 was obtained by Jacobi already one and a half century ago from the Newton's equations of motion in the form (Jacobi 1884)

$$\ddot{\phi} = U + 2T \qquad (2.33)$$

wher где Φ is the Jacobi's function (the polar moment of inertia).

Jacobi has not considered physical meaning of his equation. He assumed that because of two independent variables Φ and U in the equation it can not be resolved.

We succeeded to find an empirical relationship between the two variables and obtained at first an approximate and later on rigorous solution of the equation (Ferronsky et al. 1978, 1987; Ferronsky 2005). The relationship is proved by means of the satellite observation.

Let us tray to explain the cause of discrepancy between the geometric (static) and dynamic oblateness of the Earth. The reason is as follows. The planet's moment of inertia (polar or axial) has changing in time value. The polar moment of inertia of a self-gravitating body has a functional relation with the potential energy, the generation of which results by interaction of the mass particles in regime of periodic oscillations. The hydrostatic equilibrium of a body does not express the real dynamic processes because of loss of energy of the interacting particle oscillation. Because of that it was not possible to understand the nature of the energy. The main part of the body's kinetic energy of the body's oscillation was also lost. As to the rotational motion of the body shells, it appears only in the case of the non-uniform distribution of the mass density. The contribution of rotation to the total body's kinetic energy covers its very small part.

The cause of the accepted incorrect ratio between the Earth potential and kinetic energy lies in the body's hydrostatic equilibrium. Clairaut's equation (1.20), derived for the planet's hydrostatic equilibrium state and applied to determine the geometric oblateness, because of the above discussed reason, has no functional relationship between the force function and the moment of inertia. Therefore for the Earth dynamics problem the equation gives only a first approximation. In formulation of the Earth oblateness problem, Clairaut accepted the Newton's model of action of the centripetal forces from the surface of the planet to its geometric center. In such a physical conception the total effect of the inner force field becomes equal to zero. Below in Sect. 2.5 it will be shown, that the force field of the continuous body's interacted masses represents volumetric pressure, but not a field of vector forces. That is the cause, why the accepted postulate related to the planet's inertial rotation is physically incorrect. It was proved in electrodynamics that the force is not acceptable to be a measure of particle interaction.

The question is raised about how was it happened, that geodynamic problems and first of all the problem of stability of the Earth motion up to now were solved without knowing the planet's kinetic energy. The probable explanation of that seems to lie in the history of the development of science. In Kepler's problem and in the Newton's two body problem solution the transition from the averaged parameters of motion to the real conditions is provided through the mean and the eccentric anomalies, which by geometric procedures indirectly take into account the above energy of motion. In the Earth figure problem this procedure of Kepler is not applicable. Therefore, the so called "inaccuracies" in the Earth motion appear to be the regular dynamic effects of a self-gravitating body, and the hydrostatic model in the problem is irrelevant. The hydrostatic model was accepted by Newton for the other problem, where just this model allowed discovery and formulation the general laws of the planets motion around the Sun. The Newton's centripetal forces in principle satisfy the Kepler's condition when the distance between bodies is mach more than their size accepted as mass points. Such model gives a first approximation in the problem solution. Kepler's laws express the real picture of the planets and satellites motion around their parent bodies in averaged parameters. All the deviations of those averaged values related to the outer perturbations are not considered as it was done in the Clausius' virial theorem for the perfect gas.

Newton solved the two body problem, which has been already formulated by Kepler. The solution was based on the heliocentric world system of Copernicus, on the Galilean laws of inertia and free fall in the outer force field and on Kepler's laws of the planet's motion in the central force field considered as a geometric plane task. The goal of Newton's problem was to find the force by which the planet's motion is resulted. His centripetal attraction and the inertial forces in the two body problem satisfy Kepler's laws.

As it was mentioned, Newton understood the physical meaning of his centripetal or attractive forces as a pressure, which is accepted now like a force field. But by his opinion, for mathematical solutions the force is a more convenient instrument. And in the two body problem the force-pressure is acting from the center (of point) to the outer space.

It is worth to discuss briefly the Newton's preference given to the force but not to the pressure. In mechanics the term "*mass point*" is understood as a geometric point of space, which has no dimension but possesses a finite mass. In physics a small amount of mass is called by the term "*particle*", which has a finite value of size and mass. But very often physicists use models of particles, which have neither size nor mass. A body model like mass point has been known since ancient times. It is simple and convenient for mathematical operations. The point is an irreplaceable geometric symbol of a reference point. The physical point, which defines inert mass of a volumetric body, is also suitable for operations. But the interacted and physically active mass point creates a problem. For instance, in the field theory the point value is taken to denote the charge, the meaning of which is not better understood than is the gravity force. But it is considered often there, that the point model for mathematical presentation of charges is not suitable because operations with it lead to zero and infinite values. Then for resolution of the situation the concept of charge density is introduced. The charge is presented as an integral of density for the taken volume and by this way the solving problem is resolved.

The point model in the two-body problem allowed reduction of it to the one-body problem and for a spherical body of uniform density to write the main seven integrals of motion. In the case when a body has a finite size, then not the forces but the pressure becomes an effect of the body particle interaction. The interacted body's mass particles form a volumetric gravitational field of pressure, the strength of which is proportional to the density of each elementary volume of the mass. In the case of a uniform body, the gravitational pressure should be also uniform within the whole volume. The outer gravitational pressure of the uniform body should be also uniform at the given radius. The non-uniform body has a non-uniform gravitational pressure of both inner and outer field, which has been observed in studying the real Earth field. Interaction of mass particles results in their collision, which leads to oscillation of the whole body system. In general if the mass density value is higher then the frequency of body oscillation has also higher vale.

It was known from the theory of elasticity, that in order to calculate the stress and the deformation of a beam from a continuous load, the latter can be replaced by the equivalent lumped force. In that case the found solution will be approximate because the beam's stress and deformation will be different. The question is what degree of approximation of the solution and what kind of the error is expected. Volumetric forces are not summed up by means of the parallelogram rule. Volumetric forces by their nature can not be reduced for application either to a point, or to a resultant vector value. Their actions are directed to the 4π space and they form inner and outer force field. The force field by its action is proportional to action of the energy. This is because the force is the derivative of the energy.

The centrifugal and Coriolis' forces are also proved to be inertial forces as a consequence of inertial rotation of the body. And the Archimedes force has not found its physical explanation, but it became an observational fact of hydrostatic equilibrium of a body mass immersed into a liquid.

Such is the short story of appearance and development of the hydrostatic equilibrium of the Earth in the outer uniform gravity field. The force of gravity of a body mass is an integral value. In this connection Newton's postulate about the gravity center as a geometric point should be considered as a model for presentation of two interacted bodies, when their mutual distance is much more of the body size. It is shown in the next section, that the reduced physical, but not geometrical, gravity center of a volumetric body is represented by an envelope of the figure, which draws averaged value of radial density distribution of the body.

The problem of dynamics of the Earth as a self-gravitating body, including the figure problem in its formulation and solution needs for a higher degree of approximation. Generalized virial theorem (2.31) satisfies the condition of the Earth dynamical equilibrium state and creates a physical and theoretical basis for farther development of theory. It follows from the theorem that hydrostatic equilibrium state there is the particular case of the dynamics. Solution of problem of the Earth dynamics based on the equation of dynamical equilibrium appears to be the next natural and logistic step from the hydrostatic equilibrium model to a more realistic model without loss of the previous preference.

Below we consider the problem of "decentralization" of the own force field for a self-gravitating body.

2.5 Reduction of Inner Gravitational Field of a Body to the Resultant Envelope of Pressure

As an example, consider the Earth as a self-gravitating sphere with uniform and one-dimensional interacting media. The motion of the Earth proceeds both in its own and in the Sun's force fields. It's known from theoretical mechanics that any motion of a body can be represented by a translation motion of its mass center, rotation around that center and motion of the body mass related to its changes in the shape and structure (Duboshin 1975). In the two-body problem the last two effects are neglected due to their smallness.

In order to study the Earth motion in the own force field the translational (orbital) motion relative to the fixed point (the Sun) should be separated from the two other components of motion. After that one can consider the rotation around the geometric center of the Earth masses under action of the own force field and changes in the shape and structure (oscillation). Such separation is required only for the moment of inertia, which depends on what frame of reference is selected. The force function

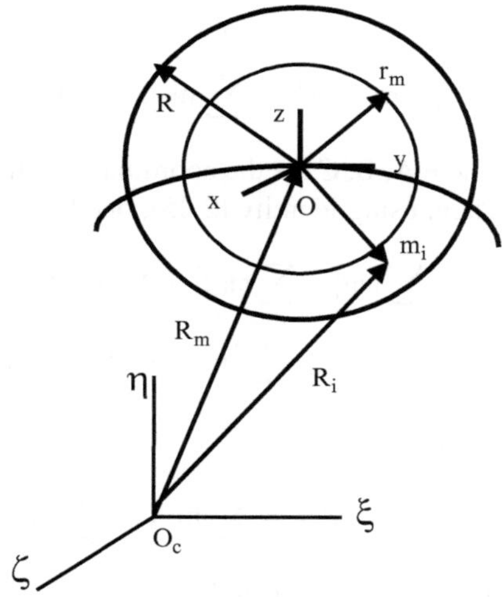

Fig. 2.1 Body motion in own force field

depends on a distance between the interacted masses and does not depend on selection of a frame of reference (Duboshin 1975). The moment of inertia of the Earth relative to the solar reference frame should be split into two parts. The first is the moment of the body mass center relative to the same frame of reference and the second – moment of inertia of the planet's mass relative to the own mass center.

So, set up the absolute Cartesian coordinates $O_c\xi\eta\zeta$ with the origin in the center of the Sun and transfer it to the system Oxyz with the origin in the geometrical center of the Earth's mass (Fig. 2.1).

Then, the moment of inertia of the Earth in the solar frame of reference is

$$I_c = \sum m_i R_i^2, \tag{2.34}$$

where m_i is the Earth mass of particle; R_i is its distance from the origin in the same frame.

The Lagrange's method is applied to separate the moment of inertia (2.34). The method is based on his algebraic identity

$$\left(\sum_{1\leqslant i\leqslant n} a_i^2\right)\left(\sum_{1\leqslant i\leqslant n} b_i^2\right) = \left(\sum_{1\leqslant i\leqslant n} a_i b_i\right)^2 + \frac{1}{2}\sum_{1\leqslant i\leqslant n}\sum_{1\leqslant j\leqslant n}\left(a_i b_j - b_i a_j\right)^2, \tag{2.35}$$

where a_i and b_i are whichever values; n is any positive number.

Jacobi in his "*Vorlesungen über Dynamik*" was the first who performed the mathematical transformation for separation of the moment of inertia of n interacting mass points into two algebraic sums (Jacobi 1884; Duboshin 1975; Ferronsky et al. 1987). It was shown that if we denote (Fig. 2.1)

$$\xi_i = x_i + A; \qquad \eta_i = y + B; \qquad \zeta_i = z + C;$$

$$\sum m_i = M; \quad \sum m_i\ \xi_i = MA; \quad \sum m_i\ \eta_i = MB; \quad m_i\ \zeta_i = MC, \tag{2.36}$$

where A, B, C are the coordinates of the mass center in the solar frame of reference. Then, using identity (2.35), one has

$$\sum m_i r_i^2 = \sum m_i\ \xi_i^2 + \sum m_i\ \eta_i^2 + \sum m_i\ \zeta_i^2 = \sum m_i x_i^2 + 2A\sum m_i x_i A^2 \sum m_i$$
$$+ \sum m_i y_i^2 + 2B\sum m_i y_i + B^2\sum m_i + \sum m_i z_i^2 + 2C\sum m_i z_i + C^2\sum m_i.$$

Since

$$MA = \sum m_i\ \xi_i = \sum m_i x_i + \sum m_i A = \sum m_i x_i + MA,$$

then $\sum m_i x_i = 0$, and also $\sum m_i y_i = 0$, $\sum m_i z_i = 0$. Now, the moment of inertia (2.34) acquires the form

$$\sum m_i R_i^2 = M\left(A^2 + B^2 + C^2\right) + \sum m_i\left(x_i^2 + y_i^2 + z_i^2\right), \tag{2.37}$$

where

$$M\left(A^2 + B^2 + C^2\right) = \ MR^2{}_m, \tag{2.38}$$

$$\sum m_i\left(x_i{}^2 + y_i{}^2 + z_i{}^2\right) = M\ r_m{}^2, \tag{2.39}$$

M is the Earth's mass; R_m and r_m are the radii of inertia of the Earth in the Sun's and the Earth's frame of reference.

Thus, we separated the moment of inertia of the Earth, rotating around the Sun in the inertial frame of reference, into two algebraic terms. The first one (2.38) is the Earth's moment of inertia in the solar reference system $O_c\xi\eta\zeta$. The second term (2.39) presents the moment of inertia of the Earth in the own frame of reference Oxyz. The Earth mass here is distributed over the spherical surface with the reduced radius of inertia r_m. In literature the geometrical center of mass O in the Earth reference system is erroneously identified with the center of inertia and center of gravity of the planet.

For farther consideration of the problem of the Earth's dynamics we accept the polar frame of reference with its origin in center O. Then expression (2.39) for the Earth polar moment of inertia I_p acquires the form

$$I_p = \sum m_i\left(x_i{}^2 + y_i{}^2 + z_i{}^2\right) = \sum m_i r_i{}^2 = \ M\ r_m{}^2 \tag{2.40}$$

Now the reduced radius of inertia r_m, which draws a spherical surface, is

$$r_m{}^2 = \frac{\sum m_i r_i^2}{M}. \tag{2.41}$$

Here $M = \sum m_i$ is the Earth's mass relative to own frame of reference.

Taking into account the spherical symmetry of the uniform and one-dimensional Earth, we consider the sphere as a concentric spherical shell with the mass $dm(r) = 4\pi\, r^2\rho(r)dr$. Then the expression (2.41) in the polar reference system can be rewritten in the form

$$r_m{}^2 = \frac{1}{M}\int_0^R r^2 4\,\pi r^2\rho(r)dr = \frac{4\pi R^2}{MR^2}\int_0^R r^4\rho(r)dr, \qquad (2.42)$$

or

$$\frac{4\pi r_m^2}{4\,\pi R^2} = \frac{4\pi\int_0^R r^4\rho(r)dr}{MR^2} = \frac{\beta^2 MR^2}{MR^2} = \beta^2, \qquad (2.43)$$

from where

$$r_m{}^2 = b^2R^2,$$

where $\rho(r)$ is the law of radial density distribution; R is the radius of the sphere; β^2 is the dimensionless coefficient of the reduced spheroid (ellipsoid) of inertia β^2MR^2.

The value of β^2 depends on the density distribution $\rho(r)$ and is changed within the limits of $1 \geq \beta^2 > 0$. Earlier (Ferronsky et al. 1987) it was defined as a structural form-factor of the polar moment of inertia.

Analogously, the reduced radius of gravity r_g, expressed as a ratio of the potential energy of interaction of the spherical shells with density $\rho(r)$ to the potential energy of interaction of the body mass distributed over the shell with radius R. The potential energy of the sphere is written as

$$U = 4\,\pi G\int_0^R r\rho(r)m(r)dr = \alpha^2\frac{GM^2}{R}, \qquad (2.44)$$

from where

$$\alpha^2 = \frac{4\pi G\int_0^R r\rho(r)m(r)dr}{\frac{GM^2}{R}} = \frac{r_g^2}{R^2}, \qquad (2.45)$$

where in expressions (2.44) and (2.45) $m(r) = 4\,\pi\int_0^r r^2\rho(r)dr$.

The value of β^2 depends on the density distribution $\rho(r)$ and is changed within the limits of $1 \geq \alpha^2 > 0$. Earlier (Ferronsky et al. 1987) it was defined as a structural form-factor of the force function.

Table 2.1 Numerical values of form factors α^2 and β^2 for radial distribution of mass density and for polytropic models

Density distribution law	α^2	$\beta_\perp{}^2$	β^2
Radial distribution of mass density			
$\rho(r) = \rho_0$	0.6	0.4	0.6
$\rho(r) = \rho_0(1-r/R)$	0.74	0.27	0.4
$\rho(r) = \rho_0(1-r^2/R^2)$,	0.71	0.29	0.42
$\rho(r) = \rho_0 \exp(1-kr/R)$	0.16 k	$8/k^2$	$12/k^2$
$\rho(r) = \rho_0 \exp(1-kr^2/R^2)$	$\sqrt{\frac{k}{2\pi}}$	1/k	1.5/k
$\rho(r) = \rho_0\, \delta(1-r/R)$	0.5	0.67	1.0
Politrope models			
0	0.6	0.4	0.6
1	0.75	0.26	0.38
1.5	0.87	0.20	0.30
2	1.0	0.15	0.23
3	1.5	0.08	0.12
3.5	2.0	0.045	0.07

Numerical values of the dimensionless form-factors α^2 and β^2 for a number of density distribution laws $\rho(r)$ are given in Table 2.1. The numerical calculations, done by integration of the numerators in Eqs. 2.43–2.45 for the polar moment of inertia and the force function, can be found in the paper (Ferronsky et al. 1978). Note, that value of the polar I_p and axial I_a moments of inertia of one dimensional sphere are related as $I_p = 3/2I_a$.

It follows from Table 2.1 that for a uniform sphere with $\rho(r) = \text{const}$ its reduced radius of inertia coincides with the radius of gravity. Here both dimensionless structural coefficients α^2 and β^2 are equal to 3/5, and the moments of gravitational and inertial forces are equilibrated and because of that the rotation of the mass is absent (Fig. 2.2a).

Thus

$$\frac{r_m^2}{R^2} = \frac{r_g^2}{R^2} = \frac{3}{5}, \tag{2.46}$$

from where

$$r_m = r_g = \sqrt{3/5R^2} = 0.7745966R. \tag{2.47}$$

For a non-uniform sphere at $\rho(r) \neq \textit{const}$ from Eqs. 2.43–2.45 one has

$$0 < \frac{4\pi r_m^2}{4\,\pi R^2} < \frac{3}{5} < \frac{4\pi r_g^2}{4\,\pi R^2} < 1. \tag{2.48}$$

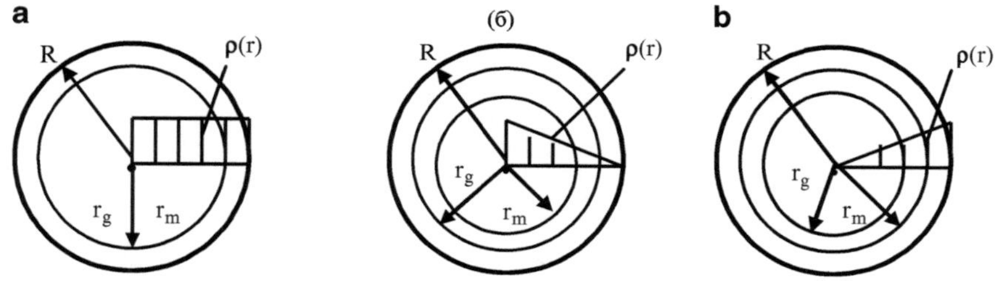

Fig. 2.2 Radius of inertia and radius of gravity for uniform (**a**) and non-uniform sphere with density increased to the center (**b**) and from the center (**c**)

It follows from inequality (2.48) and Table 2.1 that in comparison with the uniform sphere, the reduced radius of inertia of the non-uniform body decreases and the reduced gravity radius increases (Fig. 2.2b). Because of $r_m \neq r_g$ and $r_m < 0.77R < r_g$ the torque appears as a result of an imbalance between gravitational and inertial volumetric forces of the shells. Then from Eq. 2.48 it follows that

$$I_m = I_{mo} - \delta r_{mt} \quad \text{and} \quad r_g = r_{go} + \delta r_{gt} \tag{2.49}$$

where subscripts 0 and t relate to the uniform and non-uniform sphere.

In accordance with (2.48) and (2.49) rotation of shells of a one-dimensional body should be hinged-like and asynchronous. In the case of increasing mass density towards the body surface, then the signs in (2.48) and (2.49) are reversed (Fig. 2.2c). This remark is important because the direction of rotation of a self-gravitating body is function of its mass density distribution.

The main conclusion from the above consideration is that the inner force field of a self-gravitating body is reduced to a closed envelope (spheroid, ellipsoid or more complicated curve) of gravitational pressure, but not to a resulting force passing through the geometric center of the masses. In the case of a uniform body the envelopes have a spherical shape and both gravitational and inertial radii coincide. For a non-uniform body the radius of inertia does not coincide with the radius of gravity, the reduced envelope is closed but has non-spherical (ellipsoidal or any other) shape. Analytical solutions done below justify the above said.

So, we accept the force pressure as an effect of mass particles interaction which is the matter's property to do the work in the form of matter motion.

It follows from this Chapter that physical meaning of the Jacobi's virial equation consists in description of motion of a body (material system) by action of its own force field. This field is formed by the energy of the body's interacting elementary particles and expressed through oscillation of the polar moment of inertia. To the contrary of the failed hydrostatic equilibrium, Jacobi's equation describes the motion in the volumetric forces (energy) and in the volumetric moments (oscillations). The energy here is accepted as the measure of the matter interaction.

We now proceed to derivation of Jacobi's virial equation for the well known physical models of natural systems.

Chapter 3
Derivation of Jacobi's Virial Equation for Description of Dynamics of Natural Systems

Let us begin by deriving Jacobi's virial equation from the equations of Newton, Euler, Hamilton, Einstein and also from equations of quantum mechanics. By doing so we will show that Jacobi's virial equation appears to be a unified instrument for the description of dynamics of natural systems using volumetric (integral) characteristics in the framework of the various physical models of the matter interaction employed. The assumptions under which this equation is derived put only one restriction on the potential energy function to be homogeneous in the co-ordinates. But it will be seen that even this single restriction does not have to be always obligatory. The limitations following from any concrete physical model used for describing dynamics of systems in classical mechanics, hydrodynamics, statistical physics, or the theory of relativity, become unimportant.

We have defined the classical virial theorem for a system moving in the outer uniform force field, which determines the relationship between mean values of the potential and kinetic energy within a certain period of time to be the averaged virial theorem. To the contrary, the virial theorem for a system moving in its own force field and establishing a relationship between the potential and kinetic energy of the oscillating polar moment of inertia, is defined as the generalized (non-averaged) virial theorem or the equation of dynamical equilibrium of a body.

We come to the conclusion that the physical basis of hydrostatic equilibrium does not satisfy the demands of general equilibrium of a body motion. As it was shown in the previous chapter, hydrostatic equilibrium, expressed by the averaged virial theorem, does not take into account kinetic energy of the interacting mass particles of a self-gravitating body and does not provide fundamentals for study of its dynamics. The dynamic equilibrium state based on the generalized virial theorem ensures study of the following physical and dynamical problems:

- Body's shell oscillation and rotation;
- Interpretation of satellite data with respect to the body's precession, nutation and pole wobbling, non-tidal variation in the body's angular velocity, geopotential, sea level changes etc.;

V.I. Ferronsky et al., *Jacobi Dynamics*, Astrophysics and Space Science Library 369,
DOI 10.1007/978-94-007-0498-5_3, © Springer Science+Business Media B.V. 2011

- The Sun, the Earth and the Moon perturbation effects based on analysis of the dynamical equilibrium of the interacting outer force fields of the bodies;
- Relationship between the gravitational and electromagnetic interaction of the mass particles and the nature of the gravity forces;
- Other dynamical effects arising from action of the inner force field, which earlier was not taken into account.

The theory presented in this book can be applied to study the body which, by its structure, presents a system that includes gaseous, liquid and solid shells. For this purpose derivation of the equation of Jacobi's virial equation from the equations of Newton, Euler, Hamilton, Einstein and also from the equations of quantum mechanics is presented. In this part of the work we justify physical applicability of the above fundamental equation for study of the dynamics and structure of stars, planets, satellites and their shells. The main idea of derivation by introduction of volumetric forces and moments into the transformed equations, as was done in Eqs. 2.27–2.28, is to show that the effect of matter interaction in nature is unique, namely, the motion by energy.

3.1 Derivation of Jacobi's Virial Equation from Newtonian Equations of Motion

Throughout this section the term 'system' is defined as an ensemble of material mass points m_i ($i = 1, 2, 3, \ldots, n$) which interact by Newton's law of universal attraction. This physical model of a natural system forms the basis for a number of branches of physics, such as classical mechanics, celestial mechanics, and stellar dynamics.

We shall not present the traditional introduction in which the main postulates are formulated; we shall simply state the problem (see, for example, Landau and Lifshitz 1973a). We start by writing the equations of motion of the system in some absolute Cartesian co-ordinates ξ, η, ζ. In accordance with the conditions imposed, the mass point m_i is not affected by any force from the other n–1 points except that of gravitational attraction. The projections of this force on the axes of the selected co-ordinates ξ, η, ζ can be written (Fig. 3.1):

$$\Xi_i = Gm_i \sum_{1 \leqslant j \leqslant n, i \neq j} \frac{m_j(\xi_j - \xi_i)}{\Delta_{ij}^3},$$

$$H_i = Gm_i \sum_{1 \leqslant j \leqslant n, i \neq j} \frac{m_j(\eta_j - \eta_i)}{\Delta_{ij}^3}, \tag{3.1}$$

$$Z_i = Gm_i \sum_{1 \leqslant j \leqslant n, i \neq j} \frac{m_j(\zeta_j - \zeta_i)}{\Delta_{ij}^3},$$

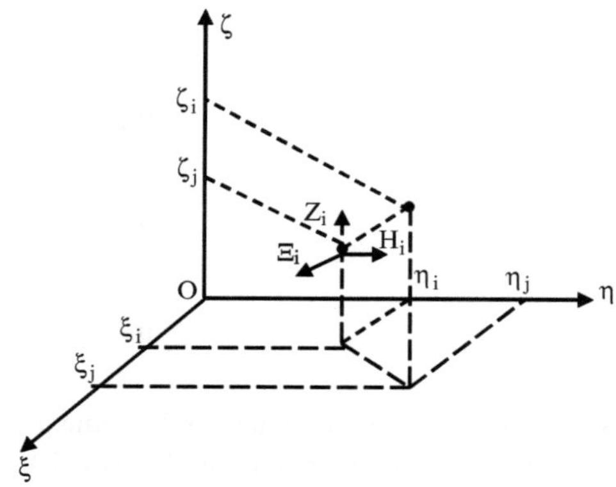

Fig. 3.1 Absolute Cartesian co-ordinate system Oξ η ζ

where G is the gravitational constant and

$$\Delta_{ji} = \sqrt{(\xi_j - \xi_i)^2 \left(\eta_j - \eta_i\right)^2 + (\zeta_j - \zeta_i)^2}$$

is the reciprocal distance between points i and j of the system.

It is easy to check that the forces affect the i-th material point of the system and are determined by the scalar function U, which is called the potential energy function of the system, and is given by

$$U = -G \sum_{1 \leqslant i < j \leqslant n} \frac{m_i m_j}{\Delta_{ij}}. \tag{3.2}$$

Now Eqs. 3.1 can be rewritten in the form

$$\Xi_i = -\frac{\partial U}{\partial \xi_i},$$

$$H_i = -\frac{\partial U}{\partial \eta_i},$$

$$Z_i = -\frac{\partial U}{\partial \zeta_i}.$$

Then Newton's equations of motion for the i-th point of the system take the form

$$m_i \ddot{\xi}_i = \Xi_i,$$

$$m_i \ddot{\eta}_i = H_i, \tag{3.3}$$

$$m_i \ddot{\zeta}_i = Z_i,$$

or

$$m_i\ddot{\xi}_i = -\frac{\partial U}{\partial \xi_i},$$
$$m_i\ddot{\eta}_i = -\frac{\partial U}{\partial \eta_i}, \tag{3.4}$$
$$m_i\ddot{\zeta}_i = -\frac{\partial U}{\partial \zeta_i},$$

where dots over co-ordinate symbols mean derivatives with respect to time.

The motion of a system is described by Eqs. 3.3 and 3.4 and is completely determined by the initial data. In classical mechanics, the values of projections ξ_{i0}, η_{i0}, ζ_{i0} and velocities $\dot{\xi}_{i0}$, $\dot{\eta}_{i0}$, $\dot{\zeta}_{i0}$ at the initial moment of time $t = t_0$ may be known from the initial data .

The study of motion of a system of *n* material points affected by self-forces of attraction forms the essence of the classical many-body problem. In the general case, ten classical integrals of motion are known for such a system, and they are obtained directly from the equations of motion.

Summing all the equations (3.3) for each co-ordinate separately, it is easy to be convinced of the correctness of the expressions:

$$\sum_{1\leqslant i\leqslant n} \Xi_i = 0,$$
$$\sum_{1\leqslant i\leqslant n} H_i = 0,$$
$$\sum_{1\leqslant i\leqslant n} Z_i = 0.$$

From those equations it follows that

$$\sum_{1\leqslant i\leqslant n} m_i\ddot{\xi}_i = 0,$$
$$\sum_{1\leqslant i\leqslant n} m_i\ddot{\eta}_i = 0, \tag{3.5}$$
$$\sum_{1\leqslant i\leqslant n} m_i\ddot{\zeta}_i = 0.$$

Equations 3.5, appearing as a sequence of equations of motion, can be successively integrated twice. As a result, the first six integrals of motion are obtained:

$$\sum_{1\leqslant i\leqslant n} m_i\dot{\xi}_i = a_1,$$

$$\sum_{1\leqslant i\leqslant n} m_i\dot{\eta}_i = a_2,$$

$$\sum_{1\leqslant i\leqslant n} m_i\dot{\zeta}_i = a_3.$$

$$\sum_{1\leqslant i\leqslant n} m_i(\xi_i - \dot{\xi}_i t) = b_1, \tag{3.6}$$

$$\sum_{1\leqslant i\leqslant n} m_i(\eta_i - \dot{\eta}_i t) = b_2,$$

$$\sum_{1\leqslant i\leqslant n} m_i(\zeta_i - \dot{\zeta}_i t) = b_3,$$

where a_1, a_2, a_3, B_1, B_2, B_3 are integration constants.

These integrals are called integrals of motion of the center of mass. The integration constants a_1, a_2, a_3, B_1, B_2, B_3 can be determined from the initial data by substituting their values at the initial moment of time for the values of all the coordinates and velocities.

Let us obtain one more group of first integrals. To do this, the second of Eqs. 3.3 can be multiplied by $-\zeta_i$, and the third третье by η_i. Then all expressions obtained should be added and summed over the index i. In the same way, the first of Eqs. 3.3 should be multiplied by ζ_i, and the third by $-\xi_i$ added and summed over index i. Finally, the second of Eqs. 3.3 should be multiplied by ξ_i, and the first by $-\eta_i$ added and summed over index i. It is easy to show directly that the right-hand sides of the expressions obtained are equal to zero:

$$\sum_{1\leqslant i\leqslant n} (Z_i\eta_i - H_i\zeta_i) = 0,$$

$$\sum_{1\leqslant i\leqslant n} (\Xi_i\zeta_i - Z_i\xi_i) = 0,$$

$$\sum_{1\leqslant i\leqslant n} (H_i\xi_i - \Xi_i\eta_i) = 0.$$

Consequently their left-hand sides are also equal to zero:

$$\sum_{1\leqslant i\leqslant n} m_i(\ddot{\zeta}_i\eta_i - \ddot{\eta}_i\zeta_i) = 0,$$

$$\sum_{1\leqslant i\leqslant n} m_i(\ddot{\xi}_i\zeta_i - \ddot{\zeta}_i\xi_i) = 0, \tag{3.7}$$

$$\sum_{1\leqslant i\leqslant n} m_i(\ddot{\eta}_i\xi_i - \ddot{\xi}_i\eta_i) = 0.$$

Integrating Eqs. 3.7 over time, three more first integrals can be obtained:

$$\sum_{1\leqslant i\leqslant n} m_i(\dot{\zeta}_i\eta_i - \dot{\eta}_i\zeta_i) = c_1,$$
$$\sum_{1\leqslant i\leqslant n} m_i(\dot{\xi}_i\zeta_i - \dot{\zeta}_i\xi_i) = c_2, \quad (3.8)$$
$$\sum_{1\leqslant i\leqslant n} m_i(\dot{\eta}_i\xi_i - \dot{\xi}_i\eta_i) = c_3.$$

The integrals (3.8) are called area integrals or integrals of moments of momentum. Three integration constants c_1, c_2, c_3 are also determined from the initial data by changing over from the values of all the co-ordinates and velocities to their values at the initial moment of time.

The last of the classical integrals can be obtained by multiplying the three Eqs. 3.4 by $\dot{\xi}_i$, $\dot{\eta}_i$ and $\dot{\zeta}_i$ respectively, and adding and summing all the expressions obtained. As a result, the following equation is obtained:

$$\sum_{1\leqslant i\leqslant n} m_i\left(\ddot{\xi}_i\dot{\xi}_i+\ddot{\eta}_i\dot{\eta}_i+\ddot{\zeta}_i\dot{\zeta}_i\right) = -\sum_{1\leqslant i\leqslant n}\left(\frac{\partial U}{\partial \xi_i}\dot{\xi}_i + \frac{\partial U}{\partial \eta_i}\dot{\eta}_i + \frac{\partial U}{\partial \zeta_i}\dot{\zeta}_i\right). \quad (3.9)$$

It is not difficult to see that the right-hand side of Eq. 3.9 is the complete differential over time of the potential energy function U of the system as a whole. The left-hand side of the same equation is also the complete differential of some function T called the kinetic energy function of the system, and equal to

$$T = \frac{1}{2}\sum_{1\leqslant i\leqslant n} m_i\left(\dot{\xi}_i^2 + \dot{\eta}_i^2 + \dot{\zeta}_i^2\right). \quad (3.10)$$

Equation (3.9) can then be written finally in the form

$$\frac{d}{dt}(T) = -\frac{d}{dt}(U),$$

from which, after integration, one finds that

$$E = T + U, \quad (3.11)$$

where E is the integration constant, determined from the initial conditions.

Equation 3.11 is called the integral of motion or the integral of living (kinetic) forces.

To derive the equation of dynamic equilibrium, or Jacobi's virial equation, each of the equations (3.4) should be multiplied by ξ_i, η_i and ζ_i respectively; then, after summing all the expressions, one can obtain

$$\sum_{1\leqslant i\leqslant n} m_i(\xi_i\ddot{\xi}_i + \eta_i\ddot{\eta}_i + \zeta_i\ddot{\zeta}_i) = -\sum_{1\leqslant i\leqslant n}\left(\xi_i\frac{\partial U}{\partial \xi_i} + \eta_i\frac{\partial U}{\partial \eta_i} + \zeta_i\frac{\partial U}{\partial \zeta_i}\right). \quad (3.12)$$

We can take farther advantage of the obvious identities:

$$m_i\xi_i\ddot{\xi}_i = \frac{1}{2}\frac{d^2}{dt^2}\left(m_i\xi_i^2\right) - m_i\dot{\xi}_i^2,$$

$$m_i\eta_i\ddot{\eta}_i = \frac{1}{2}\frac{d^2}{dt^2}\left(m_i\eta_i^2\right) - m_i\dot{\eta}_i^2,$$

$$m_i\zeta_i\ddot{\zeta}_i = \frac{1}{2}\frac{d^2}{dt^2}\left(m_i\zeta_i^2\right) - m_i\dot{\zeta}_i^2$$

from the Eulerian theorem concerning the homogenous functions. For the interaction of the system points, according to Newton's law of universal attraction, the degree of homogeneity of the potential energy function of the system is equal to –1, and hence

$$-\sum_{1\leqslant i\leqslant n}\left(\xi_i\frac{\partial U}{\partial \xi_i} + \eta_i\frac{\partial U}{\partial \eta_i} + \zeta_i\frac{\partial U}{\partial \zeta_i}\right) = U.$$

Substituting the above expressions into the right- and left-hand side of Eq. 3.12, one obtains

$$\frac{d^2}{dt^2}\left[\frac{1}{2}\sum_{1\leqslant i\leqslant n} m_i\left(\xi_i^2 + \eta_i^2 + \zeta_i^2\right)\right] - 2\sum_{1\leqslant i\leqslant n}\frac{1}{2}m_i\left(\dot{\xi}_i^2 + \dot{\eta}_i^2 + \dot{\zeta}_i^2\right) = U.$$

For a system of material points we now introduce the Jacobi function expressed through the moment of inertia of the system and presented in the form

$$\ddot{\Phi} = \frac{1}{2}\sum_{1\leqslant i\leqslant n} m_i\left(\xi_i^2 + \eta_i^2 + \zeta_i^2\right).$$

Then taking into account (3.1), the previous equation can be rewritten in a very simple form as follows:

$$\ddot{\Phi} = 2E - U. \tag{3.13}$$

This is the equation of dynamic equilibrium or Jacobi's virial equation describing both the dynamics of a system and its dynamic equilibrium using integral (volumetric) characteristics Φ and U or T.

Let us derive now another form of Jacobi's virial equation where the translational moment of the center of mass of the system is separated and all the characteristics depend only on the relative distance between the mass points of the system. For this purpose the Lagrangian identity can be used:

$$\left(\sum_{1\leqslant i\leqslant n} a_i^2\right)\left(\sum_{1\leqslant i\leqslant n} b_i^2\right) = \left(\sum_{1\leqslant i\leqslant n} a_i b_i\right)^2 + \frac{1}{2}\sum_{1\leqslant i\leqslant n}\sum_{1\leqslant j\leqslant n}\left(a_i b_j - b_i a_j\right)^2, \quad (3.14)$$

where a_i and b_i may acquire any values and n is any positive number.

Let us now put $a_i = \sqrt{m_i}$, and b_i equal to $\sqrt{m_i}\xi_i$, $\sqrt{m_i}\,\eta_i$ and $\sqrt{m_i}\zeta_i$ respectively. Then three identities can be obtained from (3.7):

$$\left(\sum_{1\leqslant i\leqslant n} m_i\right)\left(\sum_{1\leqslant i\leqslant n} m_i\xi_i^2\right) = \left(\sum_{1\leqslant i\leqslant n} m_i\xi_i\right)^2 + \frac{1}{2}\sum_{1\leqslant i\leqslant n}\sum_{1\leqslant j\leqslant n} m_i m_j\left(\xi_j - \xi_i\right)^2,$$

$$\left(\sum_{1\leqslant i\leqslant n} m_i\right)\left(\sum_{1\leqslant i\leqslant n} m_i\eta_i^2\right) = \left(\sum_{1\leqslant i\leqslant n} m_i\eta_i\right)^2 + \frac{1}{2}\sum_{1\leqslant i\leqslant n}\sum_{1\leqslant j\leqslant n} m_i m_j\left(\eta_j - \eta_i\right)^2,$$

$$\left(\sum_{1\leqslant i\leqslant n} m_i\right)\left(\sum_{1\leqslant i\leqslant n} m_i\zeta_i^2\right) = \left(\sum_{1\leqslant i\leqslant n} m_i\zeta_i\right)^2 + \frac{1}{2}\sum_{1\leqslant i\leqslant n}\sum_{1\leqslant j\leqslant n} m_i m_j\left(\zeta_j - \zeta_i\right)^2.$$

In summing up one finds

$$2m\Phi = \left(\sum_{1\leqslant i\leqslant n} m_i\xi_i\right)^2 + \left(\sum_{1\leqslant i\leqslant n} m_i\eta_i\right)^2 + \left(\sum_{1\leqslant i\leqslant n} m_i\zeta_i\right)^2 + \frac{1}{2}\sum_{1\leqslant i\leqslant n}\sum_{1\leqslant j\leqslant n} m_i m_j \Delta_{ij}^2.$$

Using now Eqs. 3.6, the last equality can be rewritten in the form

$$2m\Phi = \frac{1}{2}\sum_{1\leqslant i\leqslant n}\sum_{1\leqslant j\leqslant n} m_i m_j \Delta_{ij}^2 + (a_1 t + b_1)^2 + (a_2 t + b_2)^2 + (a_3 t + b_3)^2, \quad (3.15)$$

where

$$m = \sum_{1\leqslant i\leqslant n} m_i$$

is the total mass of the system.

Let us put

$$\Phi_0 = \frac{1}{4m}\sum_{1\leqslant i\leqslant n}\sum_{1\leqslant j\leqslant n} m_i m_j \Delta_{ij}^2.$$

The value Φ_0 does not depend on the choice of the co-ordinate system and coincides with the value of the Jacobi function in the barycentric co-ordinate system. Moreover, from Eq. 3.15 it follows that

$$\ddot{\Phi} = \ddot{\Phi}_0 + \frac{a_1^2 + a_2^2 + a_3^2}{m}.$$

Excluding the value Φ from Jacobi's equation (3.13) with the help of the last equality, the same equation can be obtained in the barycentric co-ordinate system:

$$\ddot{\Phi}_0 = 2E_0 - U, \tag{3.16}$$

where $E_0 = T_0 + U_0$ is the total energy of the system in the barycentric co-ordinate system equal to

$$E_0 = E - \frac{a_1^2 + a_2^2 + a_3^2}{2m}.$$

We can now show that the value of E_0 does not depend on the choice of the co-ordinate system. For this purpose we can again use the Lagrangian identity (3.14). In this case $a_i = \sqrt{m_i}$, and $b_i = \sqrt{m_i}\dot{\xi}_i$, $\sqrt{m_i}\dot{\eta}_i$ and $\sqrt{m_i}\dot{\zeta}_i$. Then the following three identities can be justified:

$$\left(\sum_{1\leqslant i\leqslant n} m_i\right)\left(\sum_{1\leqslant i\leqslant n} m_i\dot{\xi}_i^2\right) = \left(\sum_{1\leqslant i\leqslant n} m_i\dot{\xi}_i\right)^2 + \frac{1}{2}\sum_{1\leqslant i\leqslant n}\sum_{1\leqslant j\leqslant n} m_i m_j\left(\dot{\xi}_j - \dot{\xi}_i\right)^2,$$

$$\left(\sum_{1\leqslant i\leqslant n} m_i\right)\left(\sum_{1\leqslant i\leqslant n} m_i\dot{\eta}_i^2\right) = \left(\sum_{1\leqslant i\leqslant n} m_i\dot{\eta}_i\right)^2 + \frac{1}{2}\sum_{1\leqslant i\leqslant n}\sum_{1\leqslant j\leqslant n} m_i m_j\left(\dot{\eta}_j - \dot{\eta}_i\right)^2,$$

$$\left(\sum_{1\leqslant i\leqslant n} m_i\right)\left(\sum_{1\leqslant i\leqslant n} m_i\dot{\zeta}_i^2\right) = \left(\sum_{1\leqslant i\leqslant n} m_i\dot{\zeta}_i\right)^2 + \frac{1}{2}\sum_{1\leqslant i\leqslant n}\sum_{1\leqslant j\leqslant n} m_i m_j\left(\dot{\zeta}_j - \dot{\zeta}_i\right)^2.$$

After summing and using (3.16) one obtains

$$2mT = \left(a_1^2 + a_2^2 + a_3^2\right) + \frac{1}{2}\sum_{1\leqslant i\leqslant n}\sum_{1\leqslant j\leqslant n} m_i m_j \left[\left(\dot{\xi}_i - \dot{\xi}_j\right)^2 + \left(\dot{\eta}_i - \dot{\eta}_j\right)^2 + \left(\dot{\zeta}_i - \dot{\zeta}_j\right)^2\right]$$

or

$$T = \frac{\left(a_1^2 + a_2^2 + a_3^2\right)}{2m} + \frac{1}{2m} \times \left\{\frac{1}{2}\sum_{1\leqslant i\leqslant n}\sum_{1\leqslant j\leqslant n} m_i m_j\left[\left(\dot{\xi}_i - \dot{\xi}_j\right)^2 + \left(\dot{\eta}_i - \dot{\eta}_j\right)^2 + \left(\dot{\zeta}_i - \dot{\zeta}_j\right)^2\right]\right\}. \tag{3.17}$$

Here the second term on the right-hand side of Eq. 3.17 coincides with the expression for the kinetic energy T_o of a system.

Substituting (3.17) into an expression for E_0, one obtains

$$\begin{aligned} E_0 = T_0 + U = \frac{1}{2m} \sum_{1 \leqslant i < j \leqslant n} m_i m_j \left[\left(\dot{\xi}_i - \dot{\xi}_j\right)^2 + \left(\dot{\eta}_i - \dot{\eta}_j\right)^2 + \left(\dot{\zeta}_i - \dot{\zeta}_j\right)^2 \right] \\ - G \sum_{1 \leqslant i < j \leqslant n} \frac{m_i m_j}{\Delta_{ij}}. \end{aligned} \tag{3.18}$$

Thus, the total energy of the system E_0 depends only on the distance between the points of the system and on the velocity changes of these distances. But Jacobi's equation (3.16) appears to be invariant with respect to the choice of the co-ordinate system.

We can show now that the requirement of homogeneity of the potential energy function for deriving Jacobi's virial equation is not always obligatory. For this purpose we consider two examples.

3.2 Derivation of a Generalized Jacobi's Virial Equation for Dissipative Systems

Let us derive Jacobi's virial equation for a non-conservative system. We consider a system of *n* material points, the motion of which is determined by the force of their mutual gravitation interaction and the friction force. It is well known that the friction force always appears in the course of evolution of any natural system. It is also known that there is no universal law describing the friction force (Bogolubov and Mitropolsky 1974). The only general statement is that the friction force acts in the direction opposite to the vector of velocity of a considered mass point.

Consider as an example the simplest law of Newtonian friction when its force is proportional to the velocity of motion of the mass:

$$\begin{aligned} \Xi_f &= -k m_i \dot{\xi}_i, \\ H_f &= -k m_i \dot{\eta}_i, \\ Z_f &= -k m_i \dot{\zeta}_i, \end{aligned} \tag{3.19}$$

where $\dot{\xi}_i$, $\dot{\eta}_i$, $\dot{\zeta}_i$ are the components of the radius-vector of the velocity of the i-th mass point in the barycentric co-ordinate system; k is a constant independent of i; $k > 0$.

Sometimes the friction force is independent of the velocity of the mass point. There are also some other laws describing the friction force.

We derive the equation of dynamical equilibrium for a system of *n* material points using the equations of motion (3.4) and taking into account the friction force expressed by Eqs. 3.19:

$$\begin{aligned} m_i\ddot{\xi}_i &= -\frac{\partial U}{\partial \xi_i} - km_i\dot{\xi}_i, \\ m_i\ddot{\eta}_i &= -\frac{\partial U}{\partial \eta_i} - km_i\dot{\eta}_i, \\ m_i\ddot{\zeta}_i &= -\frac{\partial U}{\partial \zeta_i} - km_i\dot{\zeta}_i, \end{aligned} \tag{3.20}$$

where the value of the system's potential energy is determined by Eq. 3.2.

Multiplying each of Eqs. 3.20 by ξ_i, η_i and ζ_i, respectively, and summing through all i, one obtains

$$\begin{aligned} \sum_{1\leqslant i\leqslant n} m_i\left(\xi_i\ddot{\xi}_i + \eta_i\ddot{\eta}_i + \zeta_i\ddot{\zeta}_i\right) = &- \sum_{1\leqslant i\leqslant n}\left(\frac{\partial U}{\partial \xi_i}\xi_i + \frac{\partial U}{\partial \eta_i}\eta_i + \frac{\partial U}{\partial \zeta_i}\zeta_i\right) \\ &- \sum_{1\leqslant i\leqslant n} km_i\left(\xi_i\dot{\xi}_i + \eta_i\dot{\eta}_i + \zeta_i\dot{\zeta}_i\right). \end{aligned} \tag{3.21}$$

Transforming the right- and left-hand sides of Eq. 3.21 in the same way as in deriving Eq. 3.13, one obtains

$$\ddot{\Phi} - 2T = U - k\dot{\Phi}$$

or

$$\ddot{\Phi} = 2E - U - k\dot{\Phi}. \tag{3.22}$$

Let us show that the total energy E of the system is a monotonically decreasing function of time. For this purpose we multiply each of the equations (3.20) by the vectors $\dot{\xi}_i$, $\dot{\eta}_i$, $\dot{\zeta}_i$, respectively, and sum over all from 1 to n, which results in

$$\begin{aligned} \sum_{1\leqslant i\leqslant n} m_i\left(\ddot{\xi}_i\dot{\xi}_i + \ddot{\eta}_i\dot{\eta}_i + \ddot{\zeta}_i\dot{\zeta}_i\right) = &- \sum_{1\leqslant i\leqslant n}\left(\frac{\partial U}{\partial \xi_i}\dot{\xi}_i + \frac{\partial U}{\partial \eta_i}\dot{\eta}_i + \frac{\partial U}{\partial \zeta_i}\dot{\zeta}_i\right) \\ &- k\sum_{1\leqslant i\leqslant n} m_i\left(\dot{\xi}_i^2 + \dot{\eta}_i^2 + \dot{\zeta}_i^2\right). \end{aligned}$$

The last expression can be rewritten in the form

$$\frac{d}{dt}(T) = -\frac{d}{dt}(U) - 2kT$$

or

$$dE = -2kTdt. \tag{3.23}$$

Since the kinetic energy T of the system is always greater than zero, $dE \leq 0$, i.e. the total energy of a gravitating system is a monotonically decreasing function of time. Thus the expression for the total energy E(t) of the system can be written as

$$E(t) = E_0 - 2k \int_{t_0}^{t} T(t)dt = E_0[1 + q(t)],$$

where q(t) is a monotonically increasing function of time.

Finally, the equation of dynamical equilibrium for a non-conservative system takes the form

$$\ddot{\Phi} = 2E_0[1 + q(t)] - U - k\dot{\Phi}. \tag{3.24}$$

The second example where the requirement of homogeneity of the potential energy function for deriving Jacobi's virial equation is not obligatory is as follows. We derive Jacobi's virial equation for a system whose mass points interact mutually in accordance with Newton's law and move without friction in a spherical homogenous cloud whose density ρ_o is constant in time. Let, also, the geometric center of the cloud coincide with the center of mass of the considered system. The equations of motion for such a system can be written in the form:

$$\begin{aligned} m_i \frac{d^2\xi_i}{dt^2} &= -\frac{4}{3}\pi G\rho_0 m_i\xi_i - \frac{\partial U}{\partial \xi_i}, \\ m_i \frac{d^2\eta_i}{dt^2} &= -\frac{4}{3}\pi G\rho_0 m_i\eta_i - \frac{\partial U}{\partial \eta_i}, \\ m_i \frac{d^2\zeta_i}{dt^2} &= -\frac{4}{3}\pi G\rho_0 m_i\zeta_i - \frac{\partial U}{\partial \zeta_i}, \end{aligned} \tag{3.25}$$

where i =1, 2, . . ., n.

It is obvious that the above system of equations possesses the ten first integrals of motion and that Jacobi's virial equation, written in the form

$$\frac{d^2\Phi}{dt^2} = 2E - U - \frac{8}{3}\pi G\rho_0\Phi. \tag{3.26}$$

is valid for it.

The equation in the form (3.26) was first obtained by Duboshin et al. (1971). Equations 3.24 and 3.26 can be written in a more general form:

$$\ddot{\Phi} = 2E - U + X(t, \Phi, \dot{\Phi}), \tag{3.27}$$

where $X(t, \Phi, \dot{\Phi})$ is a given function of time t, the Jacobi function Φ and first derivative $\dot{\Phi}$. Moreover, we can call Eq. 3.27 a generalized equation of dynamical equilibrium.

The examples considered above justify the statement that for conditions of homogeneity of the potential energy function, required for the derivation of Jacobi's virial equation, is not always necessary. This condition is required for description of dynamics of conservative systems but not for dissipative systems or for systems in which motion is restricted by some other conditions.

3.3 Derivation of Jacobi's Virial Equation from Eulerian Equations

We now derive Jacobi's virial equation by transforming of the hydrodynamic or continuum model of a physical system. As is well known, the hydrodynamic approach to solving problems of dynamics is based on the system of differential equations of motion supplement, in the simplest case, by the equations of state and continuity, and by the appropriate assumptions concerning boundary conditions and perturbations affecting the system.

In this section, we understand by the term 'system' some given mass M of ideal gas localized in space by a finite volume V and restricted by a closed surface S. Let the gas in the system move by the forces of mutual gravitational interaction and of baric gradient. In addition, we accept the pressure within the volume to be isotropic and equal to zero on the surface S bordering the volume V. Then for a system in some Cartesian inertial co-ordinate system ξ, η, ζ, the Eulerian equations can be written in the form

$$\begin{aligned}
&\rho\frac{\partial u}{\partial t} + \rho u\frac{\partial}{\partial \xi}u + \rho v\frac{\partial}{\partial \eta}u + \rho w\frac{\partial}{\partial \zeta}u = -\frac{\partial p}{\partial \xi} + \rho\frac{\partial U_G}{\partial \xi},\\
&\rho\frac{\partial v}{\partial t} + \rho u\frac{\partial}{\partial \xi}v + \rho v\frac{\partial}{\partial \eta}v + \rho w\frac{\partial}{\partial \zeta}v = -\frac{\partial p}{\partial \eta} + \rho\frac{\partial U_G}{\partial \eta},\\
&\rho\frac{\partial w}{\partial t} + \rho u\frac{\partial}{\partial \xi}w + \rho v\frac{\partial}{\partial \eta}w + \rho w\frac{\partial}{\partial \zeta}w = -\frac{\partial p}{\partial \zeta} + \rho\frac{\partial U_G}{\partial \zeta},
\end{aligned} \tag{3.28}$$

where $\rho(\xi, \eta, \zeta, t)$ is the gas density; u, v, w are components of the velocity vector $\bar{v}(\xi, \eta, \zeta, t)$ in a given point of space; $p(\xi, \eta, \zeta, t)$ is the gas pressure; U_G is Newton's potential in a given point of space.

The value U_G is given by

$$U_G = G \int\limits_{(V)} \frac{\rho(x, y, z, t)}{\Delta} dxdydz, \tag{3.29}$$

where G is the gravity constant; $\Delta = \sqrt{(x-\xi)^2 + (y-\eta)^2 + (z-\zeta)^2}$ is the distance between system points.

The potential energy of the gravitational interaction of material points of the system is linked to the Newtonian potential (3.29) by the relation

$$U = -\frac{1}{2} \int\limits_{(V)} U_G \rho(\xi, \eta, \zeta, t) d\xi d\eta d\zeta.$$

To supplement the system of equations of motion we write the equation of continuity:

$$\frac{\partial p}{\partial t} + \frac{\partial}{\partial \xi}(\rho u) + \frac{\partial}{\partial \eta}(\rho v) + \frac{\partial}{\partial \zeta}(\rho w) = 0 \tag{3.30}$$

and the equation of state

$$p = f(\rho) \tag{3.31}$$

assuming at the same time that the processes occurring in the system are barotropic.

Let us obtain the ten classical integrals for the system whose motion is described by Eqs. 3.28.

We derive the integrals of the motion of the center of mass by integrating each of the equations (3.28) with respect to all the volume filled by the system. Integrating the first equation, we obtain

$$\begin{aligned} &\int\limits_{(V)} \rho \frac{du}{dt} d\xi d\eta d\zeta + \int\limits_{(V)} \rho \left(u \frac{du}{d\xi} + v \frac{du}{d\eta} + w \frac{du}{d\zeta} \right) d\xi d\eta d\zeta \\ &\quad = -\int\limits_{(V)} \frac{dp}{d\xi} d\xi d\zeta \eta d + G \int\limits_{(V)} \rho(\xi, \eta, \zeta, t) \left[\int\limits_{(V)} \rho(x,y,z,t) \frac{x - \xi}{\Delta^3} dxdydz \right] d\xi d\eta d\zeta \end{aligned} \tag{3.32}$$

The second term in the right-hand side of Eq. 3.32 disappears because of the symmetry of the integral expression with respect to x and ξ. In accordance with the Gauss-Ostrogradsky theorem the first term in the right-hand side of Eq. 3.32 terns to zero. In fact

$$\int_{(V)} \frac{dp}{d\xi} d\xi d\eta d\zeta = \int_{(S)} p d\eta d\zeta = 0 \tag{3.33}$$

as pressure p on the border of the considered system is equal to zero owing to the absence of outer effects.

Bearing in mind the possibility of passing to a Lagrangian co-ordinate system, and taking into account the law of the conservation of mass $dm = \rho dV = \rho_0 dV_0 = dm_0$, we get

$$\int_{(V)} \rho \frac{du}{dt} d\xi d\eta d\zeta + \int_{(V)} \rho \left(u \frac{du}{d\xi} + v \frac{du}{d\eta} + w \frac{du}{d\zeta} \right) d\xi d\eta d\zeta$$
$$= \int_{(V)} \rho \frac{du}{dt} dV = \int_{(V_0)} \rho_0 \frac{du}{dt} dV_0 = \frac{d}{dt} \int_{(V_0)} u \rho_0 dV_0 = \frac{d}{dt} \int_{(V)} \rho u dV,$$

where V_o and ρ_o are the volume and the density in the initial moment of time t_0.

Finally, Eq. 3.32 can be rewritten as

$$\frac{d}{dt} \int_{(V)} \rho \, u dV = 0. \tag{3.34}$$

Integrating (3.34) with respect to time and writing analogous expressions for two other equations of the system (3.28), we obtain the first three integrals of motion:

$$\int_{(V)} \rho \, u dV = a_1,$$
$$\int_{(V)} \rho \, v dV = a_2, \tag{3.35}$$
$$\int_{(V)} \rho \, w dV = a_3.$$

Equations 3.35 represent the law of conservation of the system moments. Integration constants a_1, a_2, a_3 can be obtained from the initial conditions.

We consider the first equation of the system (3.35) using again the law of conservation of mass. Then it is obvious that

$$\int_{(V)} u\,\rho dV = \int_{(V)} \frac{d\xi}{dt}\rho dV = \int_{(V_0)} \frac{d\xi}{dt}\rho_0 dV_0 = \frac{d}{dt}\int_{(V_0)} \xi\rho_0 dV_0 = \frac{d}{dt}\int_{(V)} \xi\rho dV = a_1. \tag{3.36}$$

Analogous expressions can be written for the two other equations (3.35). Integrating them with respect to time, we obtain integrals of motion of the center of mass of the system in the form

$$\int_{(V)} \xi\rho dV = a_1 t + b_1,$$

$$\int_{(V)} \eta\rho dV = a_2 t + b_2, \tag{3.37}$$

$$\int_{(V)} \zeta\rho dV = a_3 t + b_3.$$

We now derive three integrals of the moment of momentum of motion. For this purpose we multiply the second of Eqs. 3.28 by $-\zeta$, the third by η, and then sum and integrate the resulting expressions with respect to volume V occupied by the system. We obtain

$$\int_{(V)} \rho\left(\eta\frac{dw}{dt} - \zeta\frac{dv}{dt}\right)dV = -\int_{(V)} \left(\eta\frac{\partial p}{\partial \zeta} - \zeta\frac{\partial p}{\partial \eta}\right)dV + \int_{(V)} \rho\left(\eta\frac{\partial U_G}{\partial \zeta} - \zeta\frac{\partial U_G}{\partial \eta}\right)dV. \tag{3.38}$$

Analogously, multiplying the first of Eqs. 3.28 by ζ, the third by $-\xi$, then summing and integrating with respect to volume V, we obtain

$$\int_{(V)} \rho\left(\zeta\frac{du}{dt} - \xi\frac{dw}{dt}\right)dV = -\int_{(V)} \left(\zeta\frac{\partial p}{\partial \xi} - \xi\frac{\partial p}{\partial \zeta}\right)dV + \int_{(V)} \rho\left(\zeta\frac{\partial U_G}{\partial \xi} - \xi\frac{\partial U_G}{\partial \zeta}\right)dV. \tag{3.39}$$

Multiplying the second of Eqs. 3.28 by ξ, the first by −η, and summing and integrating as above, the third equality can be written

$$\int\limits_{(V)} \rho\left(\xi\frac{dv}{dt}-\eta\frac{du}{dt}\right)dV=-\int\limits_{(V)}\left(\xi\frac{\partial p}{\partial\eta}-\eta\frac{\partial p}{\partial\xi}\right)dV+\int\limits_{(V)}\rho\left(\xi\frac{\partial U_G}{\partial\eta}-\eta\frac{\partial U_G}{\partial\xi}\right)dV. \tag{3.40}$$

We write the second integral in the right-hand side of Eq. 3.38 in the form

$$\int\limits_{(V)} \rho\left(\eta\frac{dw}{dt}-\zeta\frac{dv}{dt}\right)dV=G\int\limits_{(V)}\rho(\xi,\eta,\zeta,t)\eta d\xi d\eta d\zeta\int\limits_{(V)}\rho(x,y,z,t)\frac{z-\zeta}{\Delta^3}dx,dy,dz$$
$$-G\int\limits_{(V)}\rho(\xi,\eta,\zeta,t)\zeta d\xi d\eta d\zeta\int\limits_{(V)}\rho(x,y,z,t)\frac{y-\zeta}{\Delta^3}dx,dy,dz.$$

The integral is equal to zero owing to the asymmetry expressed by the integral expressions with respect to z, ζ and y, η. Because the pressure at the border of the domain S is equal to zero, the first term in the right-hand side of Eq. 3.38 is also equal to zero. Actually,

$$\int\limits_{(V)}\left(\eta\frac{\partial p}{\partial\zeta}-\zeta\frac{\partial p}{\partial\eta}\right)dV=\int\limits_{(V)}\left[\frac{d}{d\eta}(\xi p)-\frac{d}{d\xi}(\eta p)\right]dV=\int\limits_{(V)}[\xi pd\xi d\zeta-\eta pd\eta d\zeta]=0.$$

Taking into account the law of mass conservation, the left-hand side of Eq. 3.38 in the Lagrange co-ordinate system can be rewritten as

$$\begin{aligned}\int\limits_{(V)}\rho\left(\eta\frac{dw}{dt}-\zeta\frac{dv}{dt}\right)dV&=\int\limits_{(V)}\rho\frac{d}{dt}(\eta w-\zeta v)dV\\&=\frac{d}{dt}\int\limits_{(V)}\rho(\eta w-\zeta v)dV=0\end{aligned} \tag{3.41}$$

Integrating this equation with respect to time, the first of the three integrals is obtained:

$$\int\limits_{(V)}\rho(\eta w-\zeta v)dV=C_1.$$

The other two integrals can be obtained analogously. Thus the system of integrals of the moment of momentum has the form

$$
\begin{aligned}
&\int_{(V)} p(\eta w - \zeta v)dV = C_1,\\
&\int_{(V)} p(\zeta u - \xi w)dV = C_2, \qquad (3.42)\\
&\int_{(V)} p(\xi v - \eta u)dV = C_3.
\end{aligned}
$$

To derive the tenth integral of motion representing the law of energy conservation, we multiply each of the system of equations (3.28) by u, v, and w accordingly, and then sum and integrate the equality obtained with respect to the system volume

$$
\begin{aligned}
\int_{(V)} \rho\left(\frac{du}{dt}u+\frac{dv}{dt}v+\frac{dw}{dt}w\right)dV = &-\int_{(V)}\left(\frac{\partial p}{\partial \xi}u+\frac{\partial p}{\partial \eta}v+\frac{dp}{d\zeta}w\right)dV\\
&+\int_{(V)} \rho(\xi,\eta,\zeta,t)\left(\frac{\partial U_G}{\partial \xi}u+-\frac{\partial U_G}{\partial \eta}v+\frac{\partial U_G}{\partial \zeta}w\right)dV.
\end{aligned} \qquad (3.43)
$$

Applying the law of mass conservation for an elementary volume, it can easily be seen that the left-hand side of Eq. 3.43 expresses the change of the velocity of kinetic energy of the system:

$$
\int_{(V)} \rho\left(\frac{du}{dt}u+\frac{dv}{dt}v+\frac{dw}{dt}w\right)dV = \frac{d}{dt}\left[\frac{1}{2}\int_{(V)}(u^2+v+w^2)dV\right] = \frac{d}{dt}(T).
$$

The first integral in the right-hand side of Eq. 3.43 can be transferred into

$$
-\int_{(V)}\left(\frac{\partial p}{\partial \xi}u+\frac{\partial p}{\partial \eta}v+\frac{dp}{d\zeta}w\right)dV = 3\frac{d}{dt}\int_{(V)} pdV
$$

and gives the change of velocity of the internal energy of the system.

The second integral in the right-hand side of the same equation expresses the velocity of the potential energy change:

$$
\begin{aligned}
&\int_{(V)} \rho(\xi,\eta,\zeta,t)d\xi d\eta d\zeta\left(\frac{\partial U_G}{\partial \xi}\frac{d\xi}{dt}+-\frac{\partial U_G}{\partial \eta}\frac{d\eta}{dt}+\frac{\partial U_G}{\partial \zeta}\frac{d\zeta}{dt}\right)\\
&= \frac{d}{dt}\left[-\frac{1}{2}\int_{(V)}\rho(\xi,\eta,\zeta,t)d\xi d\eta d\zeta\, U_G\right] = -\frac{d}{dt}(U).
\end{aligned}
$$

Finally, the law of energy conservation can be written in the form

$$T + U + W = E = \text{const.} \tag{3.44}$$

where W is the internal energy of the system.

We now derive Jacobi's virial equation for a system described by Eqs. 3.28–3.31. For this purpose we multiply each of Eqs. 3.28 by ξ, η and ζ respectively, summing and integrating the resulting expressions with respect to the volume of the system:

$$\int\limits_{(V)} \rho\left(\frac{du}{dt}\xi + \frac{dv}{dt}\eta + \frac{dw}{dt}\zeta\right)dV = -\int\limits_{(V)}\left(\frac{\partial p}{\partial \xi}\xi + \frac{\partial p}{\partial \eta}\eta + \frac{dp}{d\zeta}\zeta\right)dV + \int\limits_{(V)} \rho\left(\frac{\partial U_G}{\partial \xi}\xi + -\frac{\partial U_G}{\partial \eta}\eta + \frac{\partial U_G}{\partial \zeta}\zeta\right)dV \tag{3.45}$$

Using the obtained identities considered in the previous section, we have

$$\frac{du}{dt}\xi = \frac{1}{2}\frac{d^2}{dt^2}\left(\xi^2\right) - u^2,$$

$$\frac{dv}{dt}\eta = \frac{1}{2}\frac{d^2}{dt^2}\left(\eta^2\right) - v^2,$$

$$\frac{dw}{dt}\zeta = \frac{1}{2}\frac{d^2}{dt^2}\left(\zeta^2\right) - w^2.$$

Taking into account the law of conservation of mass for elementary volume, we transform the left-hand side of Eq. 3.45 as follows:

$$\int\limits_{(V)} \rho\left(\frac{du}{dt}\xi + \frac{dv}{dt}\eta + \frac{dw}{dt}\zeta\right)dV = \frac{1}{2}\int\limits_{(V)} \rho\frac{d^2}{dt^2}\left(\xi^2 + \eta^2 + \zeta^2\right)dV - \int\limits_{(V)} \rho\left(u^2 + v^2 + w^2\right)dV = \ddot{\Phi} - 2T, \tag{3.46}$$

where

$$\Phi = \frac{1}{2}\int\limits_{(V)} \rho\left(\xi^2 + \eta^2 + \zeta^2\right)dV$$

is the Jacobi function and

$$T = \frac{1}{2} \int_{(V)} \rho(u^2 + v^2 + w^2)dV$$

is the kinetic energy of the system.

We now transform the first integral in the right-hand side of Eq. 3.45. Using the Gauss-Ostrogradsky theorem and the equality with zero pressure at the border of the system, we can write

$$-\int_{(V)} \left(\frac{\partial p}{\partial \xi}\xi + \frac{\partial p}{\partial \eta}\eta + \frac{dp}{d\zeta}\zeta\right)dV = -\int_{(V)} \left[\frac{\partial}{\partial \xi}(p\xi) + \frac{\partial}{\partial \eta}(p\eta) + \frac{d}{d\zeta}(p\zeta)\right]dV + 3\int_{(V)} pdV = 3\int_{(V)} pdV. \tag{3.47}$$

The obtained equation expresses the doubled internal energy of the system.

The second integral in the right-hand side of Eq. 3.45 is equal to the potential energy of the gravitational interaction of mass particles within the system

$$\int_{(V)} \rho\left(\frac{\partial U_G}{\partial \xi}\xi + -\frac{\partial U_G}{\partial \eta}\eta + \frac{\partial U_G}{\partial \zeta}\zeta\right)dV = U. \tag{3.48}$$

Substituting Eqs. 3.46, 3.47 and 3.48 into (3.45), Jacobi's virial equation is obtained in the form

$$\ddot{\Phi} - 2T = 3\int_{(V)} pdV + U. \tag{3.49}$$

Taking into account the law of conservation of energy (3.44), we rewrite Eq. 3.49 in a form which will be used farther:

$$\ddot{\Phi} = 2E - U, \tag{3.50}$$

where $E = T + U + W$ is the total energy of the system.

3.4 Derivation of Jacobi's Virial Equation from Hamiltonian Equations

Let the system of material points be described by Hamiltonian equations of motion. Let also the considered system consist of n material points with masses m_i. Its generalized co-ordinates and moments are q_i and $p_i = m_i(dq_i/dt)$. Hamiltonian equations for such a system can be written as

$$\dot{p}_i = -\frac{\partial H}{dq_i},$$

$$\dot{q}_i = \frac{\partial H}{dp_i}, \tag{3.51}$$

where H *(*p, q*)* is the Hamiltonian; i = 1, 2, . . ., n.

Using values q_i and p_i, we can construct the moment of momentum

$$\sum_{i=1}^{n} p_i q_i = \sum_{i=1}^{n} m_i q_i \dot{q}_i = \frac{d}{dt}\left(\sum_{i=1}^{n} \frac{m_i q_i^2}{2}\right).$$

Now the Jacobi function may be introduced

$$\sum_{i=1}^{n} p_i q_i = \dot{\Phi}. \tag{3.52}$$

Differentiating Eq. 3.52 with respect to time, Jacobi's virial equation is obtained in the form

$$\ddot{\Phi} = \sum_{i=1}^{n} \dot{p}_i q_i + \sum_{i=1}^{n} p_i \dot{q}_i. \tag{3.53}$$

Substituting expressions for $\dot{p}_i$ and $\dot{q}_i$ taken from the Hamiltonian equations (3.51) into the right-hand side of (3.52), we obtain Jacobi's virial equation written in Hamiltonian form:

$$\ddot{\Phi} = \sum_{i=1}^{n} \left(-\frac{\partial H}{\partial q_i} q_i + \frac{\partial H}{\partial p_i} p_i\right). \tag{3.54}$$

The Hamiltonian of the system of material points interacting according to the law of the inverse squares of distance is a homogeneous function in terms of moments p_i with a degree of homogeneity of the function equal to 2, and in terms of co-ordinates q_i with a degree of homogeneity equal to –1. It follows from this

$$H(p, q) = T(p) + U(q)$$

and hence

$$\sum_{i=1}^{n} P_i \frac{\partial H}{\partial q_i} = 2T.$$

$$\sum_{i=1}^{n} q_i \frac{\partial H}{\partial q_i} = -U$$

Taking these relationships into account, Eq. 3.54 acquires the usual form of Jacobi's virial equation (3.50) for the system of mass points interacting according to the law of inverse squares of distance.

Equation 3.54 is more general then Eq. 3.50. The use of generalized co-ordinates and moments as independent variables permits us to obtain the solution of Jacobi's virial equation, taking into account gravitational and electromagnetic perturbations as well as quantum effects, both in the framework of classical physics and in terms of the Hamiltonian written in an operator form. In the general case, Eq. 3.54 can be reduced to (3.50) as the potential energy of interaction of the system's points is a homogenous function of its co-ordinates.

3.5 Derivation of Jacobi's Virial Equation in Quantum Mechanics

It is known that in quantum mechanics some physical value L by definition takes the linear Hermitian operator$\hat{L}$. Any physical state of the system take the normalized wave function ψ . The physical value of L can take the only eigenvalues of the operator $\hat{L}$. The mathematical expectation $\hat{L}$ of the value L at state ψ is determined by the diagonal matrix element

$$\overline{L} = \langle\psi|\hat{L}|\psi\rangle. \tag{3.55}$$

The matrix element of the operators of the Cartesian co-ordinates $\hat{x}_i$ and the Cartesian components of the conjugated moments $\hat{p}_k$ calculated within wave functions f and g of the system satisfy Hamilton's equations of classical mechanics:

$$\frac{d}{dt}\langle f|\hat{p}_i|g\rangle = -\langle f|\frac{\partial\hat{H}}{d\hat{x}_i}|g\rangle, \tag{3.56}$$

$$\frac{d}{dt}\langle f|\hat{x}_i|g\rangle = -\langle f|\frac{\partial\hat{H}}{\partial\hat{p}_i}|g\rangle, \tag{3.57}$$

where $\hat{H}$ is the operator which corresponds to the classical Hamiltonian.

Operators $\hat{p}_i$ and $\hat{x}_k$ satisfy the commutation relations

$$\begin{aligned} [\hat{p}_i,\hat{x}_k] &= i\hbar\,\delta_{ik}, \\ [\hat{p}_i,\hat{p}_k] &= 0, \\ [\hat{x}_i,\hat{x}_k] &= 0, \end{aligned} \tag{3.58}$$

where ħ is Planck's constant; δ_{ik} is the Kronecker's symbol; $\delta_{ik} = 1$ at i = k and $\delta_{ik} = 0$ at i ≠ k.

Operator components of momentum $\hat{p}_i$ for the functions whose arguments are Cartesian co-ordinates $\hat{x}_i$ have the form

$$\hat{p}_i = i\hbar \frac{\partial}{\partial x_i} \tag{3.59}$$

and reverse vector

$$\hat{p} = -i\hbar\nabla.$$

The derivative taken from the operator with respect to time does not depend explicitly on time; it is defined by the relation

$$\hat{L} = -\frac{i}{\hbar}\left[\hat{L}, \hat{H}\right] \tag{3.60}$$

where $\hat{H}$ is the Hamiltonian operator that can be obtained from the Hamiltonian of classical mechanics in accordance with the correspondence principle.

We have already noted that in the classical many-body problem the translational motion of the center of mass can be separated from the relative motion of the mass points if only the inertial forces affect the system. We can show that in quantum mechanics the same separation is possible.

The Hamiltonian operator of a system of n particles which is not affected by external forces in co-ordinates is

$$\hat{H} = -\frac{\hbar^2}{2}\sum_{i=1}^{n}\frac{\nabla_i^2}{m_i} + \frac{1}{2}\sum_{i=1}^{n}\sum_{i=1}^{n} U_{ik}(x_i - x_k, y_i - y_k, z_i - z_k). \tag{3.61}$$

Let us replace in (3.61) the three *n* co-ordinates x_i, y_i, z_i by co-ordinates X, Y, Z of the center of mass and by co-ordinates ξ_λ,η_λ,ζ_λ, which determine the position of a particle λ (λ = 1, 2, ..., n – 1) relative to particle n. We obtain

$$X = \frac{1}{M}\sum_{i=1}^{n} m_i x_i,$$

$$M = \sum_{i=1}^{n} m_i, \tag{3.62}$$

$$\xi_\lambda = x_\lambda - x_n,$$

where λ = 1, 2, ..., n – 1.

Analogously the corresponding relations for Y, Z, η_λ, ζ_λ are obtained.
It is easy to obtain from (3.62) the following operator relations:

$$\frac{d}{dx_p} = \frac{m_p}{M}\frac{\partial}{\partial X} + \frac{\partial}{\partial \xi_p}, \quad p = 1, 2, \ldots, n-1,$$

$$\frac{\partial}{\partial X_n} = \frac{m_n}{M}\frac{\partial}{\partial X} - \sum_{\lambda=1}^{n-1}\frac{\partial}{\partial \xi_\lambda},$$

$$\sum_{\lambda=1}^{n-1}\frac{1}{\partial x_i}\frac{\partial^2}{\partial x_i^2} = \sum_{\lambda=1}^{n-1}\frac{1}{m_\lambda}\left(\frac{m_\lambda^2}{M^2}\frac{\partial^2}{\partial X^2} + 2\frac{m_\lambda}{M}\frac{\partial^2}{\partial X \partial \xi_\lambda} + \frac{\partial^2}{\partial \xi_\lambda^2}\right)$$
$$+\frac{1}{m_n}\left(\frac{m_n^2}{M^2}\frac{\partial^2}{\partial X^2} - 2\frac{m_n}{M}\sum_{\lambda=1}^{n-1}\frac{\partial^2}{\partial X \partial \xi_\lambda} + \sum_{\mu=1}^{n-1}\sum_{\lambda=1}^{n-1}\frac{\partial^2}{\partial \xi_\mu \partial \xi_\lambda}\right)$$
$$= \frac{1}{m_n}\frac{\partial^2}{\partial X^2} + \left(\sum_{\lambda=1}^{n-1}\frac{1}{m_\lambda}\frac{\partial^2}{\partial \xi_\lambda^2} + \frac{1}{m_n}\sum_{\mu=1}^{n-1}\sum_{\lambda=1}^{n-1}\frac{\partial^2}{\partial \xi_\mu \partial \xi_\lambda}\right),$$

where summing on the Greek index is provided from 1 до $n-1$. It is seen that all the combined derivatives $\partial^2/\partial X \partial \xi_\lambda$ were mutually reduced and do not enter into the final expression. This allows the Hamiltonian to be separated into two parts:

$$H = H_o + H_r$$

where, in the right-hand side, the first term

$$H_o = \frac{\hbar^2}{2M}\left(\frac{\partial^2}{\partial X^2} + \frac{\partial^2}{\partial y^2} + \frac{\partial^2}{\partial Z^2}\right)$$

describes the motion of the center of mass, and the second term

$$H_r = -\frac{\hbar^2}{2}\left(\sum_{\lambda=1}^{n-1}\frac{1}{m_\lambda}\nabla_\lambda^2 + \frac{1}{m_n}\sum_{\mu=1}^{n-1}\sum_{\lambda=1}^{n-1}\nabla_\lambda \nabla\ \mu\right) + U \tag{3.63}$$

describes the relative motion of the particles.

The potential energy in (3.63), which is

$$U = \frac{1}{2}\sum_{\mu=1}^{n-1}\sum_{\lambda=1}^{n-1}U_{\lambda\mu}(\xi_\lambda - \xi_\mu, \eta_\lambda - \eta_\mu, \zeta_\lambda - \zeta_\mu) + \sum_{\lambda-1}^{n-1}U_{\lambda\mu}(\xi_\lambda, \eta_\lambda, \zeta_\lambda), \tag{3.64}$$

also certainly does not depend on the co-ordinates of the center of mass.

Now the Schrödinger's equation

$$(\mathrm{H_o} + \mathrm{H_r})\ \psi = \mathrm{E}\ \psi \tag{3.65}$$

permits the separation of variables.

Assuming $\psi = \phi(\mathrm{X}, \mathrm{Y}, \mathrm{Z})$ and $(\xi_\lambda, \eta_\lambda, \zeta_\lambda,)$, we obtain

$$-\frac{\hbar^2}{2\mathrm{V}}\nabla^2\phi = \mathrm{E_0}\phi, \tag{3.66}$$

$$\mathrm{H_r u} = \mathrm{E_r u}, \tag{3.67}$$

$$\mathrm{E_0} + \mathrm{E_r} = \mathrm{E}. \tag{3.68}$$

The solution of Eq. 3.66 has the form of a plane wave:

$$\phi = \mathrm{e}^{\mathrm{ik}\overline{\mathrm{R}}}, \tag{3.69}$$

$$\mathrm{E_0} = \frac{\hbar^2\mathrm{k}^2}{2\mathrm{M}},$$

where R is a vector with co-ordinates X, Y, Z.

The result obtained is in full accordance with the classical law of the conservation of motion of the center of mass. This means that the center of mass of the system moves like a material point with mass m and momentum $\hbar\overline{k}$. The mode of relative motion of the particles is determined by Eq. 3.67, which does not depend on the motion of the center of mass.

The existence in the right-hand side of Eq. 3.63 of the third term restricts further factorization of the function $\mathrm{u}(\xi_\lambda, \eta_\lambda, \zeta_\lambda)$. Only in the two-body problem, where $\mathrm{n} = 2$ and at $\lambda = \mu = 1$, a part of the Hamiltonian connected with the relative motion simplified and takes the form

$$\mathrm{H_r} = -\frac{\hbar^2}{2}\left(\frac{1}{\mathrm{m_1}}\nabla_1^2 + \frac{1}{\mathrm{m_2}}\nabla_2^2\right) + \mathrm{U_{12}}(\xi_1, \eta_1, \zeta_1) \tag{3.70}$$

It seems that choosing the corresponding system of co-ordinates can lead us to an approach for separating the motion of the center of mass to the many-body problem.

Introducing into Eq. 3.70 the reduced mass m*, which is determined as in classical mechanics by the relation

$$\frac{1}{\mathrm{m_1}} + \frac{1}{\mathrm{m_2}} = \frac{1}{\mathrm{m*}}, \tag{3.71}$$

and omitting indices in the notation for relative co-ordinates and potential energy U_{12}, we come to

$$-\frac{\hbar^2}{2m*}\nabla^2 u + U(\xi, \eta, \zeta)u = E_r u. \tag{3.72}$$

This is Schrödinger's equation for the equivalent one-particle problem.

Considering the hydrogen atom in the framework of the one-particle problem, it is assumed that the nucleus is in ground state. In accordance with Eq. 3.72, the normalized mass of the nucleus and electron m* should be introduced. No changes which account for the effect of the nucleus on the relative motion should be introduced. Because of the nucleus, mass *m* is much heavier than electron mass m_e*; instead of Eq. 3.71 we can use its approximation

$$m^* = m\left(1 - \frac{m}{M}\right).$$

Comparing, for example, the frequency of the red line H_α ($n = 3$–$n' = 2$) in the spectrum of a hydrogen atom:

$$\omega(H_\alpha) = \frac{5}{36}\frac{m_H^* e^4}{2\hbar^2 h}$$

with the frequency of the corresponding line in the spectrum of a deuterium atom:

$$\omega(D_\alpha) = \frac{5}{36}\frac{m_D^* e^4}{2\hbar^2 h},$$

and taking into account that $m_D \approx 2m_H$, for the difference of frequencies, we obtain

$$\omega(D_\alpha) - \omega(H_\alpha) = \frac{m_D^* - m_H^*}{m_H^*}\,\omega(H_\alpha) \approx \frac{m}{2M_H}\,\omega(H_\alpha).$$

This difference is not difficult to observe experimentally. At wavelength 6,563 Å it is equal to 4.12 cm^{-1}. Heavy hydrogen was discovered in 1932 by Urey, Brickwedde and Murphy, who observed a weak satellite D_α in the line H_α of the spectrum of natural hydrogen. This proves the practical significance of even the first integrals of motion.

We now show that the virial theorem is valid for any quantum mechanical system of particles retained by Coulomb (outer) forces:

$$2\overline{T} + U = 0.$$

We prove this by means of scale transformation of the co-ordinates keeping unchanged normalization of wave functions of a system.

The wave function of a many-particle system with masses m_i and electron charge e_i satisfies the Schrödinger's equation:

$$-\frac{\hbar^2}{2}\sum_{i=1}^{n-1}\frac{1}{m_i}\nabla_i^2\,\psi+\frac{1}{2}\sum_{i=1}^{n-1}\sum_{k=1}^{n-1}\frac{e_i e_k}{r_{ik}}\psi=E\psi \tag{3.73}$$

and the normalization condition

$$\int d\,\tau_1\ldots\int\,\psi^*\,\psi d\,\tau_n=1. \tag{3.74}$$

The mean values of the kinetic and potential energies of a system at stage ψ are determined by the expressions

$$T=-\frac{\hbar^2}{2}\sum_{i=1}^{n-1}\frac{1}{m_i}\int d\,\tau_1\ldots.\int\,\psi^*\nabla_i^2\,\psi d\,\tau_n, \tag{3.75}$$

$$U=\frac{1}{2}\sum_{i=1}^{n-1}\sum_{i=1}^{n-1}e_i e_k\int d\,\tau_1\ldots\int d\,\tau_n\frac{\psi^*\psi}{r_{ik}}d\,\tau_n. \tag{3.76}$$

The scale transformation

$$\bar{r}_i^{\prime}=\lambda\bar{r}_i, \tag{3.77}$$

keeps in force the condition (3.74) and means that the wave function

$$\psi(\bar{r}_i,\ldots,\bar{r}_n) \tag{3.78}$$

is replaced by the function

$$\psi_\lambda=\lambda^{3n/2}\,\psi(\lambda\bar{r}_i,\ldots,\lambda\bar{r}_n) \tag{3.79}$$

Substituting (3.79) into Eqs. 3.76 and 3.75 and passing to new variables of integration (3.77), and taking into account that

$$\nabla_i^2=\lambda^2\nabla_i^{\prime 2},$$

$$\frac{1}{r_{ik}}=\lambda\frac{1}{r'_{ik}},$$

instead of the true value of the energy, $\overline{E}=\overline{T}+\overline{U}$, we obtain

$$\overline{E}(\lambda)=\lambda^2\overline{T}+\lambda\overline{U}. \tag{3.80}$$

Equation 3.80 should have a minimum value in the case when the function which is the solution of the Schrödinger's equation is taken from the family of functions (3.79), i.e. when $\lambda = 1$. So, at $\lambda = 1$ the expression

$$\frac{\partial \overline{\mathrm{E}}(\lambda)}{\partial \lambda} = 2\,\lambda^2\overline{\mathrm{T}} + \overline{\mathrm{U}}$$

should turn into zero, and thus

$$2\overline{\mathrm{T}} + \overline{\mathrm{U}} = 0,$$

which is what we want to prove.

We now derive Jacobi's virial equation for a particle in the inner force field with the potential U(q) and fulfilling the condition

$$-\,\mathrm{q}\nabla\mathrm{U}(\mathrm{q}) = \mathrm{U} \tag{3.81}$$

using the quantum mechanical principle of correspondence. We shall also show that in quantum mechanics Jacobi's virial equation has the same form and contents as in classical mechanics, the only difference being that its terms are corresponding operators.

In the simplest case the Hamiltonian of a particle is written

$$\hat{\mathrm{H}} = -\frac{\hbar^2}{2\mathrm{m}}\nabla^2 + \hat{\mathrm{U}}, \tag{3.82}$$

and its Jacobi function is

$$\hat{\Phi} = \frac{1}{2}\mathrm{m}\hat{\mathrm{q}}^2. \tag{3.83}$$

It is clear that the following relations are valid:

$$\nabla\hat{\Phi} = \mathrm{m}\hat{\mathrm{q}},$$

$$\nabla^2\hat{\Phi} = \mathrm{m}.$$

Following the definition of the derivative with respect to time from the operator of the Jacobi function of a particle (3.60), we can write $\hat{U}$

$$\dot{\hat{\Phi}} = -\frac{1}{\hbar}\left[\hat{\Phi}, \hat{\mathrm{H}}\right],$$

where, after corresponding simplification, quantum mechanical Poisson brackets can be reduced to the form

$$[\hat{\Phi}, \hat{H}] = \frac{\hbar^2}{2m}\{\nabla^2\hat{\Phi} + 2(\nabla\hat{\Phi})\nabla\} = \frac{\hbar^2}{2m}(m + 2m\hat{q}\nabla). \tag{3.84}$$

The second derivative with respect to time from the operator of the Jacobi function is:

$$\ddot{\hat{\Phi}} = -\frac{1}{\hbar^2}\{[\hat{\Phi}, \hat{H}], \hat{H}\}. \tag{3.85}$$

Substituting the corresponding value of $[\hat{\Phi}, \hat{H}]$ and $\hat{H}$ from (3.84) and (3.82) into the right-hand side of (3.85), we obtain

$$\ddot{\hat{\Phi}} = -\frac{\hbar^2}{2m}\frac{1}{\hbar^2}\left[(m + 2m\hat{q}\nabla), \left(-\frac{\hbar^2}{2m}\nabla^2 + \hat{U}\right)\right]. \tag{3.86}$$

After simple transformation, the right-hand side of (3.86) will be

$$\ddot{\hat{\Phi}} - \frac{1}{2m}\{2\hbar^2\nabla^2 + 2m\hat{q}(\nabla\hat{U})\} = -\frac{2\hbar^2}{2m}\nabla^2 + \hat{U}, \tag{3.87}$$

where, in writing this expression in the right-hand side, we used condition (3.81).

Add and subtract the operator from the right-hand side of Eq. 3.87 and, following the definition of the Hamiltonian of the system (3.82), we obtain the quantum mechanical Jacobi virial equation (equation of dynamical equilibrium of the system), which has the form

$$\ddot{\hat{\Phi}} = 2\hat{H} - \hat{U}. \tag{3.88}$$

From Eq. 3.88, by averaging with respect to time, we obtain the quantum mechanical analogue of the classical virial theorem (equation of hydrostatic equilibrium of the system). In accordance with this theorem the following relation is kept for a particle performing finite motion in space

$$2\hat{H} = \overline{\hat{U}}. \tag{3.89}$$

Analogously, one can derive Jacobi's virial equation and the classical virial theorem for a many-particle system, the interaction potential for which depends on distance between any particle pair and is a homogeneous function of the coordinates. In particular, Jacobi's virial equation for Coulomb interactions will have the form of Eq. 3.88.

3.6 General Covariant Form of Jacobi's Virial Equation

Jacobi's initial equation

$$\ddot{\Phi} = 2E - U,$$

which was derived in the framework of Newtonian mechanics and is correct for the system of material points interacting according to Newton and Coulomb laws, includes two scalar functions Φ and U relates to each other by a differential relation. We draw attention to the fact that neither function, in its structure, depends explicitly on the motion of the particles constituting the body. The Jacobi function Φ is defined by integrating the integrand $\rho(r)r^2$ over the volume (where $\rho(r)$ is the mass density and r is the radius vector of the material point) and is independent in explicit form of the particle velocities. The potential energy U also represents the integral of m(r)dm(r)/r over the volume (where m(r) is the mass of the sphere's shell of radius r; dm(r) is the shell's mass) independent of the motion of the particles for the same reason.

Let us derive Jacobi's equation from Einstein's equation written in the form

$$\Delta G = 3\pi T, \tag{3.90}$$

where ΔG and T are the Einstein tensor and energy-momentum tensor accordingly.

In fact, since the covariant divergence of Einstein's tensor is equal to zero, we consider the covariant divergence of the energy-momentum tensor T only of substance and fields (not gravitational). Moreover, the ordinary divergence of the sum of the tensor T and pseudotensor t of the energy-momentum of the gravitational field can be substituted for the covariant divergence of the tensor T. This ordinary divergence leads to the existence of the considered quantities.

Let us define the sum of the tensor T and pseudotensor t through T_{ij} and derive Jacobi's equation in this notation.

The equation for ordinary divergence of the sum $T_{ij} = (T + t)_{ij}$ can be written

$$T_{0k,k} - T_{00,0} = 0, \tag{3.91}$$

$$T_{jk,k} - T_{j0,0} = 0. \tag{3.92}$$

We multiply Eq. 3.92 by x^j and integrate over the whole space (assuming the existence of a synchronous co-ordinate system). Integrating by parts, neglecting the surface integrals (they vanish at infinity), and transforming to symmetrical form with respect to indices, we obtain

$$\int T_{ij} dV = \frac{1}{2}\left[\int \left(T_{i0}x^j + T_{j0}x^i\right) dV\right] = 0, \tag{3.93}$$

where i, j are spatial indices.

Similarly, multiplying (3.91) by x^i x^j and integrating over the whole space, it follows that

$$\left[\int T_{00}x^ix^jdV\right]_{,0} = -\int \left(T_{i0}x^j + T_{j0}x^i\right)dV. \tag{3.94}$$

From (3.93) and (3.94) we finally get

$$\int T_{ijdV} = \tfrac{1}{2}\left[\int T_{00}x^ix^j\right]_{,0,0}. \tag{3.95}$$

It is worth recalling that T_{00} also includes the gravitational defect of the mass due to the pseudotensor t by definition

The integral $\int T_{00}x^ix^jdV$ represents the generalization of the Jacobi function $\Phi = \frac{1}{2}\int \rho r^2 dV$ introduced earlier, if we take the spur (also commonly known as the trace) of Eq. 3.95. Let us clarify this operation.

In Eq. 3.95 the spur is taken by the spatial co-ordinates. It is therefore necessary either to represent the total zero spur by four indices, as happens in the case of a transverse electromagnetic field, or to represent the relationship between the reduced spur with three indices and the total spur, as happens in the case of the energy-momentum tensor of matter.

Special care should be taken while representing the spur of the pseudotensor of the energy-momentum t. Consider the post-Newtonian approximation. In this approximation, assuming the value of 2u to be --g_{oo} -1, the components of the pseudotensor t are written in the form

$$t^{00} = -\frac{7}{8\,\pi}u_{,j,i},$$

$$t^{ij} = -\frac{1}{4\,\pi}\left(u_{,j,i} - \frac{1}{2}\ \delta_{ij}u_{,k}u_{,k}\right),$$

so that

$$S_pt = t^{00} + S_p\left(t^{ij}\right) = -\frac{1}{\pi}u_{,i}u_{,j} = \frac{1}{7}t^{00},$$

$$S_p\left(t^{ij}\right) = \frac{6}{7}t^{00}.$$

The spur in the left-hand side of Eq. 3.95 can therefore be reduced to the energy of the Coulomb field, the total energy of the transverse electromagnetic field and the gravitational energy (when it can be separated, i.e. post-Newtonian approximation).

Finally, it follows in this case that the scalar form of Jacobi's equation holds:

$$\Phi_{,0,0} = mc^2, \tag{3.96}$$

where m is the mass, accounting for the baryon defect of the mass and the total energy of the electromagnetic radiation. We do not take into account the radiation of the gravitational waves.

The result obtained by Tolman for the spherical mass distribution (Tolman 1969) is of interest:

$$m = 4\,\pi \int \hat{\varepsilon} r^2 dr, \tag{3.97}$$

where r is the radius and $\hat{\varepsilon}$ is the energy density.

The integral (3.97) acquires a form which is also valid in the case of flat space-time. This result can be explained as follows. The curvature of space-time is exactly compensated by the mass defect. This probably explains the fact that Jacobi's virial equation, derived from Newton's equations of motion which are valid in the case of non-relativistic approximation for a weak gravitational field, becomes more universal than the equations from which it was derived.

We shall not study the general tensor of Jacobi's virial equation, since in the framework of the assumed symmetry for the considered problems we are interested only in the scalar form of the equation as applied to electromagnetic interactions. As follows from the above remarks, in this case Jacobi's equation remains unchanged and the energy of the free electromagnetic field is accounted for in the term defining the total energy of the system. Total energy enters into Jacobi's equation without the electromagnetic energy irradiated up to the considered moment of time. Moreover, for the initial moment of time we take the moment of system formation. This irradiated energy appears also to be responsible for the growth of the gravitational mass defect in the system, as was mentioned above.

3.7 Relativistic Analogue of Jacobi's Virial Equation

Let us derive Jacobi's virial equation for asymptotically flat space-time. We write the expression of a 4-moment of momentum of a particle:

$$p^i x_i, \tag{3.98}$$

where $p^i = mcu^i$ is the 4-momentum of the particle; c is the velocity of light; $u^i = dx^i/ds$ is the 4- velocity; x^i is the 4-co-ordinate of the particle; s is the interval of events, and i is the running index with values 0, 1, 2, 3.

In asymptotically flat space-time we write

$$\frac{d}{ds}\left(p^i x_i\right) = mc\frac{d}{ds}\left(u^i x_i\right) = mc\frac{d^2}{ds^2}\left(\frac{x^i x_i}{2}\right). \tag{3.99}$$

Since

$$x^i x_i = c^2 t^2 - r^2 \text{and } \frac{d}{ds} = \frac{\gamma}{c}\frac{d}{dt},$$

where $\gamma = 1/\sqrt{1-(v^2/c^2)}$, and r is the radius of mass particle.

Then we continue transformation of the Eq. 3.99:

$$mc\frac{d^2}{ds^2}\left(\frac{x^i x_i}{2}\right) = mc\frac{\gamma^2}{c^2}\frac{d^2}{dt^2}\left(\frac{c^2t^2 - r^2}{2}\right) = mc\,\gamma^2 - \frac{\gamma^2}{c^2}\frac{d^2}{dt^2}\left(\frac{mr^2}{2}\right),$$

and finally

$$\frac{d}{ds}\left(p^i x_i\right) = mc\,\gamma^2 - \frac{\gamma^2}{c}\ddot{\Phi}, \tag{3.100}$$

where

$$\frac{d^2}{dt^2}\left(\frac{mr^2}{2}\right) = \ddot{\Phi}.$$

is the Jacobi function.

On the other hand, we have

$$\frac{d}{ds}\left(p^i x_i\right) = mc\frac{d}{ds}\left(u^i x_i\right) = mcu^i u_i + mc\frac{du^i}{ds}x_i. \tag{3.101}$$

Using the identity $u_i u^i \equiv 1$ and the geodetic equation

$$\frac{du^i}{ds} = -\Gamma^i_{kl}u^k u^l,$$

where

$$\Gamma^i_{kl} = \frac{1}{2}g^{im}\left(\frac{\partial g_{km}}{\partial x^l} + \frac{\partial g_{lm}}{\partial x^k} + \frac{\partial g_{kl}}{\partial x^m}\right).$$

are the Christoffel's symbols, and the equation (3.101) will be rewritten as

$$\frac{d}{ds}\left(p^i x_i\right) = mc - mcx_i\Gamma^i_{kl}u^k u^l. \tag{3.102}$$

The metric tensor g_{ik} for a weak stationary gravitational field is

$$g_{ik} = \eta_{ik} + \xi_{ik}, \tag{3.103}$$

where in our notation η_{ik} is the Lorentz tensor with signature (+,–,–,–).

For the Schwarzschild metric tensor ξ_{ik} we write

$$\xi_{00} = -\frac{r_g}{r}; \xi_{11} = -\frac{1}{1 - r_g/r} + 1 \approx -\frac{r_g}{r}; \xi_{ik} = 0 \text{ if } i \neq k \text{ and } i \neq 0, 1. \quad (3.104)$$

Here $r_g = 2GV/c^2$ is the Schwarzschild gravitational radius of the mass m'.

Now we can rewrite the second term in the right-hand side of Eq. 3.102, using (3.103) and (3.104)

$$\begin{aligned} mcx_i\Gamma^i_{kl}u^k u^l &= mcx^m u^k u^l \left(\frac{\partial \xi_{km}}{\partial x^m} - \frac{1}{2}\frac{\partial \xi_{kl}}{\partial x^m}\right) \\ &= mc\left(x^0 u^0 u^1 \frac{\partial \xi_{00}}{\partial x^1} + x^1 u^1 u^1 \frac{\partial \xi_{11}}{\partial x^1} - \frac{1}{2} x^1 u^0 u^0 \frac{\partial \xi_{00}}{\partial x^1} - x^1 u^1 u^1 \frac{\partial \xi_{11}}{\partial x^1}\right). \end{aligned} \quad (3.105)$$

But $u^1 \ll u^0 = \gamma$ and $x^1 = r$.

We therefore obtain for Eq. 3.105

$$\begin{aligned} mcx_i\Gamma^i_{kl}u^k u^l &= -\frac{mc}{2} x^1 u^0 u^0 \frac{\partial \xi_{00}}{\partial x^1} \\ &= -\frac{mc}{2} r\, \gamma^2 \frac{r_g}{r^2} = \frac{mc}{2}\, \gamma^2 \frac{2Gm'}{c^2 r} = -\frac{\gamma^2}{c}\frac{Gm'm}{r} \end{aligned} \quad (3.106)$$

Finally, we see that

$$\frac{d}{ds}\left(p^i x_i\right) = mc - \frac{\gamma^2}{c} U, \quad (3.107)$$

where U is the potential energy of the mass in the gravitational field of the mass m'.

Identification of the expression $(d/ds)(p^i x_i)$ obtained from Eqs. 3.100 and 3.107 gives

$$mc\, \gamma^2 - \frac{\gamma^2}{c}\ddot{\Phi} = mc - \frac{\gamma^2}{c} U. \quad (3.108)$$

It is easy to see that

$$mc\left(\gamma^2 - 1\right) = mc\left(\frac{1}{1 - v^2/c^2} - 1\right) = mc\frac{v^2}{c^2}\frac{1}{1 - v^2/c^2} = \frac{\gamma^2}{c} mv^2 = \frac{\gamma^2}{c} 2T.$$

We then obtain

$$\frac{\gamma^2}{c}\ddot{\Phi} = \frac{\gamma^2}{c} U + \frac{\gamma^2}{c} 2T,$$

which gives

$$\ddot{\Phi} = U + 2T,$$

or

$$\ddot{\Phi} = 2E - U, \tag{3.109}$$

where T is the kinetic energy of the particle m and E = U + T is its total energy.

Equations 3.109 are known as classical Jacobi's virial equations, and the expression (3.102) represents its relativistic analogue for asymptotically flat space-time.

3.8 Universality of Jacobi's Virial Equation for Description of Dynamics of Natural Systems

It follows from this derivation of Jacobi's virial equation that it appears to be a universal mathematical expression for consideration of the dynamics of celestial bodies described by equations of motion for a wide range of existing physical models. The derived equation represents not only formal mathematical transformation of the initial equations of motion. Physical quintessence of mathematical transformation of the equations of motion involves change of the vector forces and moment of momentums by the volumetric forces or pressure and the oscillation of the interacted mass particles (inner energy) expressed through the energy of oscillation of the polar moment of inertia of a body. Here the potential (kinetic) energy and the polar moment of inertia of a body have a functional relationship and within the period of oscillation are inversely changed by the same law. Moreover, as it was demonstrated in Sect. 2.2 of Chap. 2. and will be shown in Chap. 6, the virial oscillations of a body represent the main part of the body's kinetic energy, which is lost in the hydrostatic equilibrium model. The change of the vector forces and moment of momentums by the force pressure and the oscillation of the interacting mass particles disclose the physical meaning of the gravitation and mechanism of generation of the gravitational and electromagnetic energy and their common nature, which is considered in Chap. 6. The most important advantage given by Jacobi's virial equation, is its independence from the choice of the co-ordinate system, transformation of which, as a rule, creates many mathematical difficulties.

By averaging for a uniform system the generalized virial equation $\ddot{\Phi} = 2E - U$, when the first derivative over the polar moment of inertia Φ acquires constant value, it becomes the classical virial theorem 2E = U, or –U = 2 T, which expresses the condition of the hydrostatic equilibrium being in the outer force field and without kinetic energy of oscillations of the interacting particles.

The starting point for derivation of the virial theorem is the particle momentum. By Newton's definition this value "*is a certain measure determined proportionally to the velocity and the mass*". This value is defined or it is found experimentally. All the other force parameters are obtained by transformation of the initial momentum and those actions are explained by physical interaction of the mass particles, which are the carrier of the momentum. In fact, we recognize the momentum to be "*innate*", according to Newton's terminology, value, i.e. the hereditary value. Under the "*innate*" value Newton understood "both the resistance and the pressure of the mass" and finally the effect acquires its status of the inertial force. But the essence does not change, because the momentum appears together with the mass. Thus, the circle of the philosophical speculations is locked by the momentum, i.e. by the mass and its oscillation. All other attributes of the motion are formed by mathematical transformations.

One more mainly physical problem that was solved in derivation in this chapter of Jacobi's virial equation by mathematical transformations is an understanding of the nature and dynamical effects of the gravitational interaction of mass particles for a continuous body. Contrary to the interaction of two bodies presented by mass points, when a dynamical effect is developed in the orbital motion by vector force and angular momentum, the dynamical effect of the interacting mass particles of the continuous body is developed in the form of volumetric pressure and volumetric oscillation. The integral effect of the mass interaction is expressed by oscillation of the polar moment of inertia. In the next chapter we consider solution of Jacobi's virial equation.

Chapter 4
Solution of Jacobi's Virial Equation for Conservative Systems

In Chap. 3 we derived Jacobi's virial equation of dynamical equilibrium in the framework of various physical models which are used for describing the dynamics of natural systems. We showed that, instead of the traditional description of a system in co-ordinates and velocities, the problem of dynamics can be studied from the position of an external observer. In this case the system as a whole is described by a compact and elegant equation and is characterized by integral (volumetric) parameters. Such a description of the integral equation does not depend on the choice of the frame of reference. The external observer can estimate by observations only some moments of distribution of mass density, i.e. total mass and energy of a system, which are its integral characteristics. Moreover, in order to solve the problem of a body's motion in the framework of its dynamical equilibrium, we invoked the relationship between its force function and the polar moment of inertia, which is the source of motion. This relationship reveals the nature of the gravitational energy. We also succeeded in reanimating the lost kinetic energy and obtaining both an equation of dynamics and an equation of dynamical equilibrium in the form of the oscillating motion during each period of time and within the whole duration of the system's evolution.

The problem is now to find the general solution of Jacobi's virial equation relative to oscillation and rotation of a body and to apply the solution to study its dynamics. This application is valid for studying the Sun, the Earth, the Moon and other celestial bodies.

In this chapter we show that Jacobi's virial equation provides first of all a solution for the models of natural systems, which have explicit solutions in the framework of the classical many-body problem. We shall give parallel solutions for both the classical and dynamical approaches, and in doing so we shall show that, with the dynamical approach, the solution acquires a new physical meaning. We shall also consider a general case of the solution of Jacobi's virial equation for conservative and dissipative systems.

V.I. Ferronsky et al., *Jacobi Dynamics*, Astrophysics and Space Science Library 369, DOI 10.1007/978-94-007-0498-5_4,

4.1 Solution of Jacobi's Virial Equation in Classical Mechanics

The many-body problem is known to be the key problem in classical mechanics and especially in celestial mechanics. A particular example of this is the unperturbed problem of Keplerian motion, when the system consists of only two material points interacting by Newtonian law. The explicit solution of the problem of unperturbed Keplerian motion permits the many-body problem to be solved with some approximation by varying arbitrary constants. In this case the problem of dynamics, for example that of the Solar System, is transferred into the solution of the problem of dynamics of nine pairs of bodies in each of which one body is always the Sun and the second is each of the nine planets forming the system. Considering each planet-sun sub-system, the influence of the other eight planets of the system is taken into account by introducing the perturbation function. By the virial approach we can obtain for the Sun one characteristic period of circulation with respect to the center of mass of the system which will not coincide with any period of the planets. The dynamical approach evidences that the planet's orbital motion is performed by the central body, i.e. by the Sun, by the energy of its outer force field or by the field of the pressure. Each planet interacts with the solar force field by the energy of its own outer force field. The planet's orbit is the certain curve of its equilibrium motion which results from the two interacting fields of pressure. The planet's own oscillation and rotation perform by action of the inner fields of pressure.

Following these brief physical comments on the dynamical equilibrium motion of a planet, we now present two methods of solving the Keplerian problem: the classical and the integral.

4.1.1 *The Classical Approach*

The traditional way of solving the unperturbed Keplerian problem is excellently described in the university courses for celestial mechanics found in (Duboshin 1978). Here we present only the principle ideas. The method consists in transforming the two-body problem described by the system of equations (3.10) into the one-body problem using six integrals of motion of the center of mass (3.13). The system of equations obtained is sixth order and expresses the change of barycentric co-ordinates of one point with respect to the center of mass of the system as a whole. Let us write it in the form

$$\ddot{x} = -\frac{\mu x}{r^3},$$

$$\ddot{y} = -\frac{\mu y}{r^3}, \tag{4.1}$$

$$\ddot{z} = -\frac{\mu z}{r^3},$$

where μ is the constant depending on the number of the point and for which the second point is equal to

$$\mu = \frac{Gm_1^3}{(m_1+m_2)^2}.$$

We then pass on from that Cartesian system of co-ordinates OXYZ to orbital ξ η ζ, using first integrals of the system of equations (4.1). Those are three integrals of the area,

$$\begin{aligned} y\dot{z} - z\dot{y} &= c_1, \\ z\dot{x} - x\dot{z} &= c_2, \\ x\dot{y} - y\dot{x} &= c_3, \end{aligned} \tag{4.2}$$

the energy integral,

$$\dot{x}^2 + \dot{y}^2 + \dot{z}^2 = \frac{2\mu}{r} + h, \tag{4.3}$$

and the Laplacian integrals,

$$\begin{aligned} -\frac{\mu x}{r} + c_3\dot{y} - c_2\dot{z} &= f_1, \\ -\frac{\mu y}{r} + c_1\dot{z} - c_3\dot{x} &= f_2, \\ -\frac{\mu z}{r} + c_2\dot{x} - c_1\dot{y} &= f_3. \end{aligned} \tag{4.4}$$

As these seven integrals are not independent, we conclude that they cannot form a general solution of the system (4.1). In fact there are two relations for these integrals:

$$c_1f_1 + c_2f_2 + c_3f_3 = 0,$$

$$f_1^2 + f_2^2 + f_3^2 = \mu^2 + h\left(c_1^2 + c_2^2 + c_3^2\right),$$

showing that only five of them are independent. But the last integral needed can be found by simple quadrature. Using these integrals we can pass on to the system of orbital co-ordinates Oξηζ using the transformation relations (see Fig. 4.1):

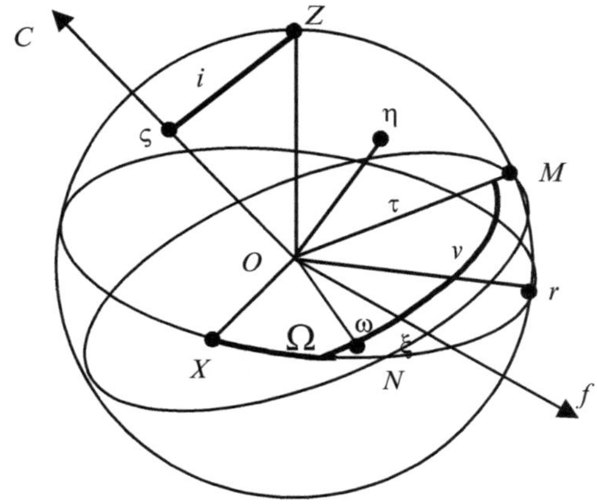

Fig. 4.1 Transition from Cartesian co-ordinate system OXYZ to orbital Oξηζ

$$\xi = \frac{f_1}{f}x + \frac{f_2}{f}y + \frac{f_3}{f}z,$$

$$\eta = \frac{C_2f_3 - C_3f_2}{Cf}x + \frac{C_3f_1 - C_1f_2}{Cf}y + \frac{C_1f_2 - C_3f_1}{Cf}z, \tag{4.5}$$

$$\zeta = \frac{C_1}{C}x + \frac{C_2}{C}y + \frac{C_3}{C}z.$$

The equation of the curve along which the point moves in accordance with (4.1) has the simplest form in the system of initial co-ordinates. The equation is

$$\zeta = 0,$$

$$\mu r = C^2 - f\xi. \tag{4.6}$$

Finally, introducing the polar orbital co-ordinates r and v, which are related to the rectangular orbital co-ordinates ξ and η by the expressions (see Fig. 4.2)

$$\xi = r \cos v$$

and

$$\eta = r \sin v,$$

and using the integral of areas

$$r^2 v = C,$$

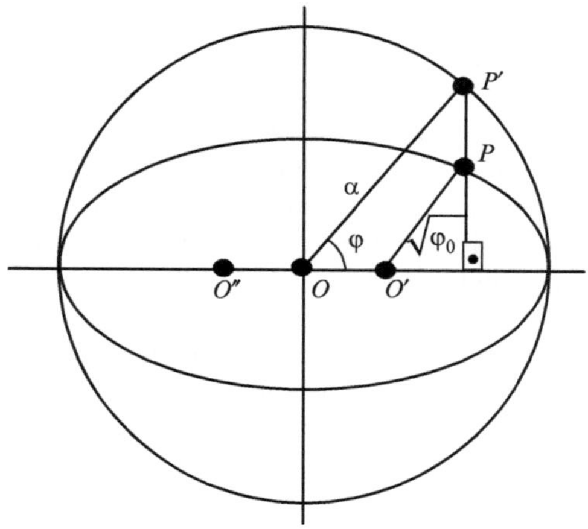

Fig. 4.2 Relationship between the polar and the rectangular co-ordinates

we come to the equation

$$C(t-r) = \left(\frac{C^2}{\mu}\right)^2 \int_0^v \frac{dv}{\left(1+\frac{f}{\mu}\cos v\right)^2}. \tag{4.7}$$

The solution of Eq. 4.7 gives the change of function v with respect to time. Repetition of the transformation in the reverse order leads to solution of the problem. In doing this, we obtain the expression for the change of co-ordinates of the material point with respect to the initial data ξ_{10}, η_{10}, ζ_{10}, ξ_{20}, η_{20}, ζ_{20}, $\dot{\xi}_{10}$, $\dot{\eta}_{10}$, $\dot{\zeta}_{10}$, $\dot{\xi}_{20}$, $\dot{\eta}_{20}$, $\dot{\zeta}_{20}$. It is remarkable that if the total energy (4.3) has negative value, then the solution of Eq. 4.7 leads to the Keplerian equation

$$E' - e\sin E' = n(t-\tau), \tag{4.8}$$

where the function v is related to the variable E' by the expression

$$tg\frac{v}{2} = \sqrt{\frac{1+e}{1-e}}tg\frac{E}{2},$$

and

$$e = \frac{f}{\mu}, \quad n = \frac{\sqrt{\mu}}{a^{3/2}}, \quad p = \frac{C^2}{\mu} = a\left(1-e^2\right).$$

Because energy by definition is the property to do work (motion) and can be only a positive value, then the physical meaning of negative total energy which defines the elliptic orbit of a body moving in the central field of the two-body problem should be revealed. In the presented solution of the two-body problem, the left-hand side of the energy integral (4.3) expresses the kinetic energy and the right-hand side means the potential energy of the mass interaction. The integral of energy (4.3) as a whole, in the co-ordinates and in the velocities, represents the averaged virial theorem, where the potential energy has formally a negative value. Here the physical meaning of the total energy determination consists in comparison of magnitude of the potential and kinetic energy. A negative value of the total energy means that the potential energy exceeds the kinetic one by that value. As it follows from analysis of the inner force field of a self-gravitating body presented in Chap. 2, the potential energy exceeds the kinetic one only in the case of non-uniform distribution of the mass density and cannot be less than it. In the case of equality of both energies the total potential energy is realized into oscillating motion. The excited part of the potential energy is used for rotation of the masses and in the dissipation. The last case is discussed in Chap. 7.

4.1.2 The Dynamic Approach

Let us consider the solution of the problem of unperturbed motion of two material points on the basis of Jacobi's virial equation which in accordance with Eq. 3.16 is written in the form

$$\ddot{\Phi}_0 = 2E_0 - U,$$

where $E_0 = T_0 + U = \text{const}$ is the total energy of the system in a barycentric co-ordinate system;

The Jacobi function Φ_0 is expressed by (3.15):

$$\Phi_0 = \frac{m_1 m_2}{2(m_1 + m_2)}\left[(\xi_1 - \xi_2)^2 + (\eta_1 - \eta_2)^2 + (\zeta_1 - \zeta_2)^2\right],$$

and the potential energy U in accordance with (3.2) is

$$U = \frac{G m_1 m_2}{\sqrt{(\xi_1 - \xi_2)^2 + (\eta_1 - \eta_2)^2 + (\zeta_1 - \zeta_2)^2}}.$$

It is easy to see that between the Jacobi function Φ_0 and the potential energy U exists the relationship

$$|U|\sqrt{\Phi_0} = \frac{G(m_1 m_2)^{3/2}}{\sqrt{2(m_1 + m_2)}} = G^{1/2} m \mu^{3/2} = B = \text{const}, \qquad (4.9)$$

where μ is the generalized mass of the two bodies; m is the total mass of the system; B is a constant value.

The relationship (4.9) is remarkable because it is independent of the initial data, i.e. of its co-ordinates and velocities. Being an integral characteristic of the system and dependent only on the total mass and the generalized mass of the two points, the relationship permits Jacobi's virial equation to be transformed to an equation with one variable as follows:

$$\ddot{\Phi}_0 = 2E_0 + \frac{B}{\sqrt{\Phi_0}}. \tag{4.10}$$

We consider the solution of Eq. 4.10 for the case when total energy E_0 has negative value. Introducing $A = -2E_0 > 0$, Eq. 4.10 can be rewritten:

$$\ddot{\Phi}_0 = -A + \frac{B}{\sqrt{\Phi_0}}. \tag{4.11}$$

We apply the method of change of variable for solution of Eq. 4.11 and show that partial solution of two linear equations (Ferronsky et al. 1984a):

$$\left(\sqrt{\Phi_0}\right)'' + \sqrt{\Phi_0} = \frac{B}{A}, \tag{4.12}$$

$$t'' + t = \frac{4B\lambda}{(2A)^{3/2}}, \tag{4.13}$$

which include only two integration constants, is also the solution of Eq. 4.11.

We now introduce the independent variable λ into Eqs. 4.12 and 4.13, where primes denote differentiation with respect to λ. Note that time here is not an independent variable. This allows us to search for the solution of two linear equations instead of solving one non-linear equation. The solution of Eqs. 4.12 and 4.13 can be written in the form

$$\sqrt{\Phi_0} = \frac{B}{A}[1 - \varepsilon\cos(\lambda - \psi)], \tag{4.14}$$

$$t = \frac{4B}{(2A)^{3/2}}[\lambda - \varepsilon\sin(\lambda - \psi)]. \tag{4.15}$$

Let us prove that the partial solution (4.14) and (4.15) differential Eqs. 4.12 and 4.13 is the solution of Eq. 4.10 that is sought. For this purpose we express the first and second derivatives of the function $\sqrt{\Phi_0}$ with respect to λ through corresponding derivatives with respect to time using Eq. 4.15. From (4.15) it follows that

$$\frac{dt}{d\lambda} = \frac{4B}{(2A)^{3/2}}[1 - \varepsilon\cos(\lambda - \psi)]. \tag{4.16}$$

We can replace the right-hand side of the obtained relationship by $\sqrt{\Phi_0}$ from (4.14)

$$\frac{dt}{d\lambda} = \sqrt{\Phi_0}\sqrt{\frac{2}{A}}. \tag{4.17}$$

Transforming the derivative from $\sqrt{\Phi_0}$ with respect to λ into the form

$$\frac{d\sqrt{\Phi_0}}{d\lambda} = \frac{d\sqrt{\Phi_0}}{dt}\frac{dt}{d\lambda} = \frac{\dot{\Phi}_0}{2\sqrt{\Phi_0}}\frac{dt}{d\lambda}$$

and taking into account (4.17), we can write

$$\left(\sqrt{\Phi_0}\right)' = \frac{\dot{\Phi}_0}{\sqrt{2A}}.$$

The second derivative can be written analogously:

$$\left(\sqrt{\Phi_0}\right)'' = \frac{dt}{d\lambda}\frac{d}{dt}\left(\frac{\dot{\Phi}_0}{\sqrt{2A}}\right) = \frac{\ddot{\Phi}_0}{\sqrt{2A}}\sqrt{\Phi_0}\sqrt{\frac{2}{A}} = \frac{\ddot{\Phi}_0\sqrt{\Phi_0}}{A}. \tag{4.18}$$

Putting Eq. 4.18 into (4.12), we obtain

$$\frac{\ddot{\Phi}_0\sqrt{\Phi_0}}{A} + \sqrt{\Phi_0} = \frac{B}{A}.$$

Dividing the above expression by $\sqrt{\Phi_0}/A$, we can finally write

$$\ddot{\Phi}_0 = -A + \frac{B}{\sqrt{\Phi_0}}.$$

This shows that the partial solution of the two linear differential Eqs. 4.12 and 4.13 appears to be the solution of the non-linear Eq. 4.11.

4.2 Solution of the N-Body Problem in the Framework of Conservative System

After solving Jacobi's virial equation for the unperturbed two-body problem, we come to dynamics of a system of n material particles where $n \to \infty$.

Let us assume that an external observer studying the dynamics of a system of *n* particles in the framework of classical mechanics has the following information. He has the mass of the system, its total and potential energy and can determine the Jacobi function and its first derivative with respect to time in any arbitrary moment. Then he can use Jacobi's virial Eq. 4.9 and, making only the assumption needed for its solution that $|U|\sqrt{\Phi_0} = B = \text{const}$, may predict the dynamics of the system, i.e. the dynamics of its integral characteristics at any moment of time. The assumption $|U|\sqrt{\Phi_0} = \text{const}$will be considered separately in Chap. 6.

If the total energy E_0 of the system has negative value, the external observer can immediately write the solution of the problem of the Jacobi function change with respect to time in the form of (4.14) and (4.15):

$$\sqrt{\Phi_0} = \frac{B}{A}[1 - \ \varepsilon \cos(\lambda - \ \psi)],$$

$$t = \frac{4B}{(2A)^{3/2}}[\ \lambda - \ \varepsilon \sin(\lambda - \ \psi)],$$

where $A = -2E_0$; ε and ψ are constants depending on the initial values of the Jacobi function Φ_0 and its first derivative $\dot{\Phi}_0$ at the moment of time t_0 .

Let us obtain the values of constants ε and ψ, in explicit form expressed through the values Φ_0 and $\dot{\Phi}_0$ at the initial moment of time t_0. For convenience we introduce a new independent variable φ, connected to λ by the relationship соотношением $\lambda - \psi = \varphi$. Then Eqs. 4.14 and 4.15 can be rewritten:

$$\sqrt{\Phi_0} = \frac{B}{A}[1 - \ \varepsilon \cos \varphi], \tag{4.19}$$

$$t - \frac{4B}{(2A)^{3/2}}\psi = \frac{4B}{(2A)^{3/2}}[\varphi - \ \varepsilon \sin \varphi]. \tag{4.20}$$

Using Eq. 4.19 we write the expression for φ:

$$\varphi = \arccos \frac{1 - \frac{A}{B}\sqrt{\Phi_0}}{\varepsilon} \tag{4.21}$$

and taking into account the equality

$$\frac{d\sqrt{\Phi_0}}{d\lambda} = \frac{d\sqrt{\Phi_0}}{d\varphi},$$

substitute Eq. 4.21 into the expression

$$\frac{\dot{\Phi}_0}{\sqrt{2A}} = \frac{B}{A}\varepsilon\sin\varphi.$$

The last equation can be rewritten finally in the form

$$\frac{\dot{\Phi}_0}{\sqrt{2A}} = \frac{B}{A}\,\varepsilon\sqrt{1-\left(\frac{1-\frac{A}{B}\sqrt{\Phi_0}}{\varepsilon}\right)^2}. \tag{4.22}$$

Equation 4.22 allows us to determine the first constant of integration ε as a function of the initial data Φ_0 and $\dot{\Phi}_0$ at $t = t_0$. Solving Eq. 4.22 with respect to ε after simple algebraic transformation, we obtain

$$\varepsilon = \sqrt{1-\frac{A}{2B^2}\left(-\dot{\Phi}_0 + 4B\sqrt{\Phi_0} - 2A\Phi_0\right)}\Big|_{t=t_0} = \text{const.} \tag{4.23}$$

The second constant of integration ψ can be expressed through the initial data after solving Eq. 4.20 with respect to ψ and change of value φ by its expression from Eq. 4.21. Defining

$$t - \frac{4B}{(2A)^{3/2}}\,\psi = \tau,$$

we obtain

$$-\tau = \left\{\frac{4B}{(2A)^{3/2}}\left[\arccos\frac{1-\frac{A}{B}\sqrt{\Phi_0}}{\varepsilon} - \varepsilon\sqrt{1-\left(\frac{1-\frac{A}{B}\sqrt{\Phi_0}}{\varepsilon}\right)^2}\right] - t\right\}\Big|_{t=t_0} = \text{const.} \tag{4.24}$$

The physical meaning of the integration constants ε, τ, and the parameter $T_v = 8\pi B/(2A)^{3/2}$ can be understood after the definitions

$$T_v = \frac{8\pi B}{(2A)^{3/2}},$$

$$n = \frac{2\pi}{T_v} = \frac{(2A)^{3/2}}{4B},$$

$$a = \frac{B}{A}$$

and rewriting Eqs. 4.19 and 4.20 in the form

$$\sqrt{\Phi_0} = a(1 - \varepsilon\cos\varphi), \tag{4.25}$$

$$M = \varphi - \varepsilon\sin\varphi, \tag{4.26}$$

where $M = n(t - \tau)$.

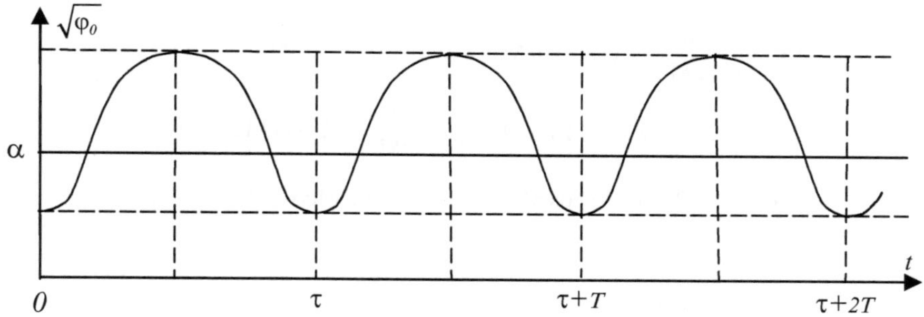

Fig. 4.3 Changes of the Jacobi function over time

The value $\sqrt{\Phi_0}$ draws an ellipse during the period of time $T_v = 8\pi B/(2A)^{3/2}$ (see Fig.4.3). The ellipse is characterized by a semi-major axis *a* equal to B/A and by the eccentricity ε which is defined by expression (4.23). In the case considered, where $E_0 < 0$, the value ε is changed in time from 0 to 1. The value τ characterizes the moment of time when the ellipse passes the pericentre.

Let us obtain explicit expressions with respect to time for the functions $\sqrt{\Phi_0}$, Φ_0 and $\dot{\Phi}_0$. For this purpose we write Eq. 4.24 in the form of a Lagrangian:

$$F(\varphi) = \varphi - \varepsilon \sin\varphi - M = 0. \tag{4.27}$$

It is known (Duboshin 1978) that by application of Lagrangian formulas we can write in the form of a series the expressions for the root of the Lagrange Eq. 4.27 and for the arbitrary function f which is dependent φ:

$$\varphi = \sum_{k=0}^{\infty} \frac{\varepsilon^{k-1}}{k!} \frac{d^{k-1}}{dM^{k-1}} \left[\sin^k M\right] = M + \varepsilon \sin M + \frac{\varepsilon^2}{1\cdot 2} \frac{d}{dM} \left[\sin^2 M\right] + \cdots, \tag{4.28}$$

$$f(\varphi) = \sum_{k=0}^{\infty} \frac{\varepsilon^{k-1}}{k!} \frac{d^{k-1}}{dM^{k-1}} \left[f'(M)\sin^k M\right] = f(M) + \varepsilon f'(M) \sin M + \frac{\varepsilon^2}{1\cdot 2} \frac{d}{dM} \left[f(M)\sin^2 M\right] + \ldots. \tag{4.29}$$

Using Eq. 4.29, we write expressions for cosφ, $\cos^2\varphi$ and sinφ in the form of a Lagrangian series of parameter ε power:

$$\cos\varphi = \sum_{k=0}^{\infty} \frac{\varepsilon^{k-1}}{k!} \frac{d^{k-1}}{dM^{k-1}} \left[(-1)\sin M \sin^k M\right] = \cos M + \varepsilon(-1)\sin M \sin(M) + \frac{\varepsilon^2}{1\cdot 2} \frac{d}{dM} \left[(-1)\sin M \sin^2 M\right] + \cdots = \cos M - \frac{\varepsilon}{2} + \frac{\varepsilon}{2}\cos 2M - \frac{3}{4}\varepsilon^3 \cos M + \frac{3}{8}\varepsilon^2 \cos 3M + \cdots \tag{4.30}$$

$$\cos^2\varphi = \sum_{k=0}^{\infty} \frac{\varepsilon^{k-1}}{k!} \frac{d^{k-1}}{dM^{k-1}} \left[(-2)\sin M \cos M \sin^k M\right] = \cos^2 M$$
$$+\ \varepsilon(-2)\sin M \cos M \sin M + \frac{\varepsilon^2}{1\cdot 2}\frac{d}{dM}\left[(-2)\sin M \cos M \sin^2 M\right] + \cdots$$
$$= \cos^2 M - 2\ \varepsilon \sin^2 M \cos M + \frac{\varepsilon^2}{2}(-2)\left(3\sin^2 M \cos^2 M - \sin^4 M\right) + \cdots \tag{4.31}$$

$$\sin\varphi = \sum_{k=0}^{\infty} \frac{\varepsilon^{k-1}}{k!} \frac{d^{k-1}}{dM^{k-1}} [\cos M \sin M] = \sin M + \ \varepsilon \cos M \sin M$$
$$+ \frac{\varepsilon^2}{1\cdot 2}\frac{d}{dM}\left[\cos M \sin^2 M\right] + \ldots = \sin M + \ \varepsilon \cos M \sin M \tag{4.32}$$
$$+ \frac{\varepsilon^2}{1\cdot 2}\left[2\sin M \cos^2 M - \sin^3 M\right] + \cdots$$

We write the expressions for $\sqrt{\Phi_0}$, Φ_0, $\dot{\Phi}_0$ using Eqs. 4.25 and 4.26 in the form

$$\sqrt{\Phi_0} = a(1 - \ \varepsilon \cos\varphi), \tag{4.33}$$

$$\Phi_0 = a^2\left(1 - 2\ \varepsilon \cos\varphi + \ \varepsilon^2 \cos^2\varphi\right), \tag{4.34}$$

$$\dot{\Phi}_0 = \sqrt{\frac{2}{A}}\ \varepsilon B \sin\varphi. \tag{4.35}$$

Substituting into (4.33)–(4.35) the expressions for $\cos\varphi$, $\cos^2\varphi$ and $\sin\varphi$ in the form of the Lagrangian series (4.30)–(4.32) we obtain

$$\sqrt{\Phi_0} = \frac{B}{A}\left[1 + \frac{\varepsilon^2}{2} + \left(-\ \varepsilon + \frac{3}{8}\ \varepsilon^3\right)\cos M - \frac{\varepsilon^2}{2}\cos 2M - \frac{3}{8}\ \varepsilon^3 \cos 3M + \cdots\right], \tag{4.36}$$

$$\Phi_0 = \frac{B^2}{A^2}\left[1 + \frac{3}{2}\ \varepsilon^2 + \left(-2\ \varepsilon + \frac{\varepsilon^3}{4}\right)\cos M - \frac{\varepsilon^2}{2}\cos 2M - \frac{\varepsilon^3}{4}\cos 3M + \cdots\right], \tag{4.37}$$

$$\dot{\Phi}_0 = \sqrt{\frac{2}{A}}\ \varepsilon B\left[\sin M + \frac{1}{2}\ \varepsilon \sin 2M + \frac{\varepsilon^2}{2}\sin M\left(2\cos^2 M - \sin^2 M\right) + \ldots\right]. \tag{4.38}$$

The series of Eqs. 4.36–4.38 obtained are put in order of increased power of parameter ε and are absolutely convergent at any value of M in the case when the parameter ε satisfies the condition

$$\varepsilon < \bar{\varepsilon} = 0{,}6627\ldots, \tag{4.39}$$

where $\bar{\varepsilon}$ is the Laplace limit.

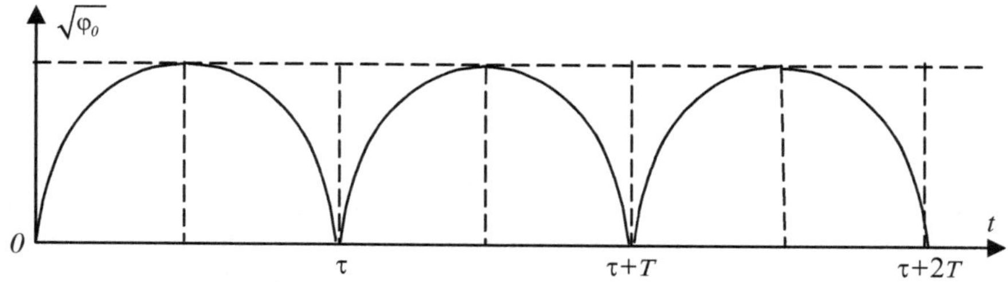

Fig. 4.4 Changes of the value $\sqrt{\Phi_0}$ in time at $\varepsilon = 1$

In some cases it is convenient to expand the values $\sqrt{\Phi_0}$, Φ_0, $\dot{\Phi}_0$ in the form of a Fourier series, using conventional methods (see, for example, Duboshin 1978). Figure 4.4 demonstrates the changes of $\sqrt{\Phi_0}$ in time at $\varepsilon = 1$.

It is also possible to consider the case solution of Jacobi's virial equation for $E_0 = 0$ and $E_0 > 0$. The reader can find here without difficulty a full analogy of these results as well as the solution of the two-body problem.

4.3 Solution of Jacobi's Virial Equation in Hydrodynamics

Let us consider the solution of the problem of the dynamics of a homogeneous isotropic gravitating sphere in the framework of traditional hydrodynamics and the virial approach we have developed.

4.3.1 The Hydrodynamic Approach

The sphere is assumed to have radius R_0 and be filled by an ideal gas with ρ_0. We assume that at the initial time the field of velocities which has the only component is described by equation

$$u = H_0 r, \tag{4.40}$$

where u is the radial component of the velocity of the sphere's matter at the distance r from the center of mass; H is independent of the quantity r and equal to H_0 at time t_0.

We also assume that the motion of the matter of the sphere goes on only under action of the forces of mutual gravitational interaction between the sphere particles. In this case the influence of the pressure gradient is not taken into account, assuming that the matter of the sphere is sufficiently diffused. Then the symmetric spherical shells will move only under forces of gravitational attraction and will not coincide. In this case the mass of the matter of any sphere shell will keep its constant value and the condition (4.40) will be satisfied at any moment of time, and constant H should be dependent on time.

Under those conditions the Eulerian system of equations (3.28) can be written in the form

$$\rho\frac{\partial u}{\partial t} + \rho(u\nabla)u = \rho\frac{\partial U_G}{\partial r}, \tag{4.41}$$

where $\rho(t)$ is the density of the matter of the sphere at the moment of time t; u is the radial component of the velocity of matter at distance r from the sphere's center; U_G is the Newtonian potential for the considered point of the sphere.

The expression for the Newtonian potential U_G (3.29) can be written as follows:

$$U_G = G\frac{4}{3}\pi\rho r^2, \tag{4.42}$$

and the continuity equation will be

$$\frac{\partial\rho}{\partial t} + \rho\frac{\partial U}{\partial r} = 0. \tag{4.43}$$

Within the framework of the traditional approach, the problem is to define the sphere radius R and the value of the constant H at any moment of time, if the radius R_0, density ρ_0 and the value of the constant H_0 at the initial moment of time t_0 are given. If we know the values H(t) and R(t), we can then obtain the field of velocities of the matter within the sphere which is defined by Eq. 4.40, and also the density ρ of matter at any moment of time, using the relationship

$$\frac{4}{3}\pi R_0^3\,\rho_0 = \frac{4}{3}\pi R^3\rho = \text{const} = m.$$

Hence the formulated problem is reduced to identification of the law of motion of a particle which is on the surface of the sphere and within the field of attraction of the entire sphere mass $m = 4/3\ \pi\ \rho_0 R_0{}^3$.

The equation of motion for a particle on the surface of the sphere, which follows from Eq. 4.41 after transforming the Eulerian co-ordinates into a Lagrangian, has the form

$$\frac{d^2R}{dt^2} = -G\frac{m}{R^2}. \tag{4.44}$$

It is necessary to determine the law of change of R(t), resolving Eq. 4.44 at the initial data:

$$R(t_0) = R_0, \tag{4.45}$$

$$\left.\frac{dR}{dt}\right|_{t=t_0} = H_0R_0.$$

We reduce the order of Eq. 4.44. To do so we multiply it by dR/dt:

$$\frac{dR}{dt}\frac{d^2R}{dt^2} = -\frac{dR}{dt}\frac{Gm}{R^2}$$

and integrate with respect to time:

$$\int_{t_0}^{t}\frac{1}{2}\frac{d}{dt}\left(\dot{R}\right)^2 = \int_{t_0}^{t}\frac{d}{dt}\left(\frac{Gm}{R}\right)dt.$$

After integration we obtain

$$\frac{1}{2}\dot{R}^2 - \frac{1}{2}\dot{R}_0^2 = \frac{Gm}{R} - \frac{Gm}{R_0}$$

or

$$\frac{1}{2}\dot{R}^2 = \frac{Gm}{R} + k, \tag{4.46}$$

where the constant k is determined as

$$\begin{aligned} k &= \frac{1}{2}\dot{R}_0^2 - \frac{Gm}{R_0} = \frac{1}{2}H_0^2R_0^2 - G\frac{4\pi}{3}\rho_0\frac{R_0^3}{R_0} \\ &= \frac{1}{2}H_0^2R_0^2\left[1 - \frac{8\pi}{3}\frac{G\rho_0}{H_0^2}\right] = \frac{1}{2}H_0^2R_0^2[1-\Omega] = \text{const} \end{aligned} \tag{4.47}$$

Here the quantity $\Omega = \rho_o/\rho_{cr}$, where $\rho_{cr} = 3H_0{}^2/8\pi G$.

Note that Eq. 4.46 obtained after reduction of the order of the initial Eq. 4.44 is in its substance the energy conservation law. Equation 4.46 permits the variables to be divided and can be rewritten in the form

$$\int_{R_0}^{R}\frac{dR}{\sqrt{\frac{2Gm}{R} + 2k}} = \int_{t_0}^{t}dt. \tag{4.48}$$

The plus sign before the root is chosen assuming that the sphere at the initial time is expanding, i.e., $H_0 > 0$.

The differential Eq. 4.46 has three different solutions at $k = 0, k > 0$ and $k < 0$ depending on the sign of the constant k, which is in its turn defined by the value of the parameter Ω at the initial moment of time. First we consider the case when $k = 0$ which relates, by analogy with the Keplerian problem, to the parabolic model at k = 0. Equation 4.46 is easily integrated and for the expression case, i.e.,$\dot{R}>0$, we obtain

$$\dot{R}^2 = \frac{2Gm}{R},$$

$$\dot{R} = \frac{(2Gm)^{1/2}}{R^{1/2}},$$

from which it follows that

$$R^{1/2}dR = (2Gm)^{1/2}dt$$

or

$$\frac{2}{3}R^{3/2} = (2Gm)^{1/2}t + \text{const.} \tag{4.49}$$

We choose as initial counting time t = 0, the moment when R = 0. In this case the integration constant disappears:

$$R = \left(\frac{9}{2}Gm\right)^{1/3} t^{2/3}. \tag{4.50}$$

The density of the matter changes in accordance with the law

$$\rho(t) = \frac{m}{\frac{4}{3}\pi R^3} = \frac{1}{6\pi G t^2}, \tag{4.51}$$

and the quantity H(t), as a consequence of (4.50), has the form

$$H(t) = \frac{\dot{R}}{R} = \frac{2}{3}\frac{1}{t}. \tag{4.52}$$

For the case when $k > 0$, which corresponds to so-called hyperbolic motion, the solution of Eq. 4.46 can be written in parametric form (Zeldovich and Novikov 1967)

$$R = \frac{Gm}{2k}(\text{ch}\eta - 1), \tag{4.53}$$

$$t = \frac{Gm}{(2k)^{3/2}}(\{\text{ch}\eta - \eta),$$

where the constants of integration in (4.53) have been chosen so that $t = 0, \eta = 0$ at $R = 0$.

Finally we consider the case when $k < 0$, which corresponds to elliptic motion. At $k < 0$ the expansion of the sphere cannot continue for unlimited time and the expansion phase should be changed by attraction of the sphere.

The explicit solution of Eq. 4.46 at $k < 0$ can be written in parametric form (Zeldovich and Novikov 1967)

$$R = \frac{Gm}{2|k|}(1 - \mathrm{ch}\eta), \tag{4.54}$$

$$t = \frac{Gm}{(2|k|)^{3/2}}(\eta - \mathrm{sh}\eta).$$

The maximum radius of the sphere is determined from Eq 4.46 on the condition $dR/dt = 0$ and equals

$$R_{max} = \frac{Gm}{|E|}. \tag{4.55}$$

The time needed for expansion of the sphere from $R_0 = 0$ at $t_0 = 0$ to R_{max} is

$$t_{max} = \frac{\pi Gm}{(2|k|)^{3/2}}. \tag{4.56}$$

So the sphere should make periodic pulsations with period T_p equal to

$$T_p = \frac{2\pi Gm}{(2|k|)^{3/2}}. \tag{4.57}$$

The considered solution has important cosmologic applications.

4.3.2 The Virial Approach

We shall limit ourselves by formal consideration of the same problem in the framework of the condition of the dynamical equilibrium of a self-gravitating body based on the solution of Jacobi's virial equation, which we discussed earlier.

As shown in Chap. 3, Jacobi's virial equation (3.50), derived from Eulerian equations (3.28), is valid for the considered gravitating sphere. It was written in the form

$$\ddot{\Phi} = 2E - U, \tag{4.58}$$

where Φ is the Jacobi function for a homogeneous isotropic sphere and is defined by

$$\Phi = \frac{1}{2}\int_0^R 4\,\pi r^2 \rho r^2 dr = \frac{2\pi\rho R^5}{5} = \frac{3}{10} mR^2. \tag{4.59}$$

The potential gravitational energy of the matter of the sphere is expressed as

$$U = -4\pi G\int_0^R r\rho(r)m(r)dr = -\frac{16\pi^2}{15}G\rho^2R^2 = -\frac{3}{5}G\frac{m^2}{R}. \tag{4.60}$$

The total energy of the sphere E will be equal to the sum of the potential U and kinetic T energies.

The kinetic energy T is expressed as

$$T = \frac{1}{2}\int_0^R 4\,\pi u^2\rho r^2 dr = \frac{1}{2}\int_0^R 4\,\pi H^2 r^2\rho r^2 dr = \frac{4\pi\rho H^2R^5}{10} = \frac{3}{10}mH^2R^2. \tag{4.61}$$

For a homogeneous isotropic gravitating sphere, the constancy of the relationship between the Jacobi function (4.59) and the potential energy (4.60) can be written:

$$|U|\sqrt{\Phi} = B = \frac{3}{5}G\frac{m^2}{R}\sqrt{\frac{3}{10}mR^2} = \frac{1}{\sqrt{2}}\left(\frac{3}{5}\right)^{3/2}Gm^{3/2}, \tag{4.62}$$

where B has constant value because of the conservation law of mass m of the considered sphere.

The total energy E of the sphere also has a constant value:

$$E = T + U = \frac{A}{2}. \tag{4.63}$$

Then, if the total energy of the sphere has a negative value, Jacobi's virial equation can be written in the form:

$$\ddot{\Phi} = -A + \frac{B}{\sqrt{\Phi}}. \tag{4.64}$$

Let us consider the conditions under which the total energy of the system will have a negative value. For this purpose we write it explicitly:

$$E = T + U = -\frac{16}{15}\pi^2G\rho^2R^5 + \frac{2\pi\rho H^2R^5}{5} = \frac{2}{5}\pi\rho H^2R^5\left[1 - \frac{8\pi G\rho}{3H^2}\right]. \tag{4.65}$$

It is clear from Eq. 4.65 that the total energy E has a negative value, when $\rho > \rho_c$, where $\rho_c = 3H^2/8\pi G$.

The general solution of Eq. 4.64 has the form of Eqs. 4.14 and 4.15:

$$\sqrt{\Phi_0} = \frac{B}{A}[1 - \varepsilon\cos(\lambda - \psi),] \tag{4.66}$$

$$t = \frac{4B}{(2A)^{3/2}}[\lambda - \varepsilon \sin(\lambda - \psi)], \tag{4.67}$$

where ε and ψ are constants dependent on the initial values of the Jacobi function Φ_0 and its first derivative $\dot{\Phi}_0$ at the moment of time t_o. The constants ε and ψ are determined by Eqs. 4.23 and 4.24 accordingly.

If we express all the constants in Eq. 4.23:

$$\varepsilon = \sqrt{1 - \frac{A}{2B^2}\left(-\dot{\Phi}_0 + 4B\sqrt{\Phi_0} - 2A\Phi_0\right)}\Big|_{t = t_0} = \text{const.} \tag{4.68}$$

through mass m of the system, it is not difficult to see that

$$-\dot{\Phi}_0^2 + 4B\sqrt{\Phi_0} - 2A\Phi_0 = 0.$$

Then the constant ε will be equal to zero. Hence the solutions (4.28) and (4.29) coincide with the solution (4.54), which was obtained in the framework of the traditional hydrodynamic approach. In this case the period of eigenpulsations of the Jacobi function (the polar moment of inertia) of the sphere $T = 8\pi R/(2A)^{3/2}$ will be equal to the period of change of its radius $T_p = 2\pi Gm/(2|k|)^{3/2}$ obtained from Eq. 4.54.

4.4 The Hydrogen Atom as a Quantum Mechanical Analogue of the Two-Body Problem

Let us consider the problem concerning the energy spectrum of the hydrogen atom, which is a unique example of the complete conformity of the analytical solution with experimental results. The problem consists of a study of all the forms of motion using the postulates of quantum mechanics and based on the solution of Jacobi's virial equation.

The classical Hamiltonian in the two-body problem is written as

$$H = \frac{\bar{p}_1^2}{2m_1} + \frac{\bar{p}_2^2}{2m_2} + U(|\bar{r}_1 - \bar{r}_2|), \tag{4.69}$$

where

$$\bar{p}_1 = \frac{\partial H}{\partial \dot{\vec{r}}_1} = m_1\dot{\bar{r}}_1,$$

$$\bar{p}_2 = \frac{\partial H}{\partial \dot{\vec{r}}_2} = m_2\dot{\bar{r}}_2,$$

which after separation of the center of mass can be transformed into the form

$$H = \frac{\bar{P}^2}{2M} + \frac{\bar{p}^2}{2m} + U(r), \tag{4.70}$$

where $r = |\bar{r}_1 - \bar{r}_2|$ is the distance between two particles and

$$P = M\dot{\bar{R}}; \quad p = m\dot{\bar{r}}; \quad M = m_1 + m_2;$$

$$R = \frac{m_1\bar{r}_1 + m_2\bar{r}_2}{m_1 + m_2}; \quad m = \frac{m_1 m_2}{m_1 + m_2}.$$

We obtain the Hamiltonian operator for the quantum mechanical two-body problem through changing the pulses and radii by the corresponding operators with the communication relations

$$[\hat{p}_i, \hat{p}_0] = -i\hbar\ \delta_{ik},$$

$$[\hat{p}_i, \hat{r}_k] = -i\hbar\ \delta_{ik}.$$

Then

$$\hat{H} = -\frac{\hbar^2}{2M}\Delta_R - \frac{\hbar^2}{2m}\Delta_r + \hat{U}(r).$$

The wave functionu$(\bar{r}_1,\bar{r}_2) = \varphi(\overline{R})\psi(r)$, which satisfies the Schrödinger equation

$$\hat{H}u = \varepsilon u,$$

describes the motion of the inertia center (the free motion of the particle of mass m_c is described by the function φ(R) and the motion of the particle of mass m in the U(r) is described by the wave function $\Psi(r)$). Subsequently we consider only the wave function of the motion of particle m.
The Schrödinger equation

$$\Delta\Psi + \frac{2m}{\hbar^2}[E - U(r)]\Psi = 0$$

written here for the stationary state in a central symmetrical field in spherical coordinates, has the form

$$\frac{1}{r^2}\frac{\partial}{\partial r}\left(r^2\frac{\partial\Psi}{\partial r}\right) + \frac{1}{r^2}\left[\frac{1}{\sin\Theta}\frac{\partial}{\partial\Theta}\left(\sin\Theta\frac{\partial\Psi}{\partial\Theta}\right) + \frac{1}{\sin^2\Theta}\frac{\partial^2\Psi}{\partial\varphi^2}\right] + \frac{2m}{\hbar^2}[E - U(r)]\ \psi = 0. \tag{4.71}$$

Using the Laplacian operator $\hat{\ell}^2$:

$$\hat{\ell}^2 = \left[\frac{1}{\sin\Theta}\frac{\partial}{\partial\Theta}\left(\sin\Theta\frac{\partial}{\partial\Theta}\right) + \frac{1}{\sin^2\Theta}\frac{\partial^2}{\partial\varphi^2}\right],$$

we obtain

$$\frac{\hbar^2}{2m}\left[-\frac{1}{r^2}\frac{\partial}{\partial r}\left(r^2\frac{\partial\Psi}{\partial r}\right) + \frac{\hat{\ell}^2}{r^2}\Psi\right] + U(r)\Psi = E\Psi.$$

The operators $\hat{\ell}^2$ and $\hat{\ell}_z$ ($\hat{\ell}_z = -i\partial/\partial\varphi$) commutate with the Hamiltonian $\hat{H}(r)$ and therefore there are common eigenfunctions of the operators $\hat{H}$, $\hat{\ell}^2$ и $\hat{\ell}_z$. We consider only such solutions of Schrödinger equations. This condition determines the dependence of the function Ψ on the angles

$$\Psi(r, \Theta, \varphi) = R(r)Y_{\ell k}(\Theta, \varphi),$$

where the quantity $Y_{\ell k}(\Theta, \varphi)$ is determined by the expression

$$Y_{\ell k}(\Theta, \varphi) = \frac{1}{\sqrt{2}\,\pi}e^{ik\varphi}(-1)^k i^\ell \sqrt{\frac{(2\ell+1)(\ell-k)!}{2(\ell+k)!}}P_\ell^k(\cos\Theta),$$

and $P_\ell^k(\cos\Theta)$ is the associated Legendre polynomial, which is

$$P_\ell^k(\cos\Theta) = \frac{1}{2^\ell \ell!}\sin^k\Theta\frac{d^{r+\ell}}{d\cos\Theta^{r+\ell}}\left(\cos^2\Theta - 1\right)^\ell.$$

Since

$$\hat{\ell}^2 Y_{\ell k} = \ell(\ell+1)Y_{\ell k},$$

we obtain for the radial part of the wave function R(r)

$$\frac{1}{r^2}\frac{\partial}{\partial r}\left(r^2\frac{dR}{dr}\right) - \frac{\ell(\ell+1)}{r^2}R + \frac{2m}{\hbar^2}[E - U(r)]R = 0. \qquad (4.72)$$

Equation 4.72 does not contain the value $\ell_z = m$, i.e. at the given ℓ the energy level E corresponds to $2\ell + 1$ states differing by the value ℓ_z.
The operator

$$\frac{1}{r^2}\frac{\partial}{\partial r}\left(r^2\frac{d\Psi}{dr}\right).$$

is equivalent to the expression

$$\frac{1}{r}\frac{d^2}{dr^2}(rR)$$

and thus it is convenient to make the change of variables, assuming that

$$X(r) = rR(r).$$

So that Eq. 4.71 can be rewritten in the form

$$\frac{d^2X}{dr^2} - \frac{\ell(\ell+1)}{r^2}X + \frac{2m}{\hbar^2}[E - U(r)]X = 0. \tag{4.73}$$

We now consider the demand following from the boundary conditions and related to the behavior of the wave function X(r). At $r \to 0$ and the potentials satisfying the condition

$$\lim_{r \to 0} U(r)r^2 = 0, \tag{4.74}$$

only the first two terms play an important role in Eq. 4.73. $X(r) \sim r^v$ and we obtain

$$v(v-1) = \ell(\ell+1).$$

This equation has roots $v_1 = \ell + 1$ and $v_2 = -\ell$.

The requirement of normalization of the wave function is incompatible with the values $v = -\ell$ at $\ell \neq 0$ because the normalization integral

$$\int_0^\infty |X_r^2(r)dr|$$

will be divergent for the discrete spectrum, and the condition

$$\int \Psi(\lambda, \xi)\Psi(\lambda, X)d\lambda = \delta(X - \xi)$$

does not hold for the continuous spectrum.

At $\ell = 0$ the boundary conditions are determined by the demand for the finiteness of the mean value of the kinetic energy which is satisfied only at $v = 1$. So, when the condition (4.74) is satisfied, then the wave function of a particle is everywhere finite and at any ℓ

$$X(0) = 0.$$

Let us consider the energy spectrum and the wave function of the bounded states of a system of two charges. The bounded states exist only in the case of the attracted particles. Such a system defines the properties of the hydrogen atom and hydrogen-like ions.

The equation for the radial wave function is

$$\frac{d^2R}{dt^2}+\frac{2}{r}\frac{dR}{dr}-\frac{\ell(\ell+1)}{r^2}R+\frac{2m}{\hbar^2}\left(E+\frac{\alpha}{r}\right)R=0, \tag{4.75}$$

where $\alpha = Ze^2$ is constant, characterizing the potential; e is the electron charge; Z is the whole number equal to the nucleus charge in the charge units.

The constants e^2, m and $\hbar$ allow us to construct the value with the dimension of length

$$a_0=\frac{\hbar^2}{me^2}=0.529\cdot 10^{-8}\text{cm}$$

known as the Bohr radius, and the time

$$t_0=\frac{\hbar^3}{me^4}=0.242\cdot 10^{-11}\text{s}.$$

These quantities define the typical space and time scale for describing a system, and it is therefore convenient to use these units as the basic system of atomic units. Equation 4.75 in atomic units (at $Z = 1$) takes the form

$$\frac{d^2R}{dt^2}+\frac{2}{r}\frac{dR}{dr}-\frac{\ell(\ell+1)}{r^2}R+2\left(E+\frac{1}{r}\right)R=0. \tag{4.76}$$

At $E < 0$ the motion is finite and the energy spectrum is discrete. We need the solutions (4.76) quadratically integrable with r^2. Let us introduce the specification

$$n=\frac{1}{\sqrt{-2E}} \qquad \rho=\frac{2r}{n}.$$

Equation 4.76 can be written as

$$\frac{d^2R}{dt^2}+\frac{2}{\rho}\frac{dR}{d\rho}+\left[\frac{n}{\rho}-\frac{1}{4}-\frac{\ell(\ell+1)}{\rho^2}\right]R=0. \tag{4.77}$$

We find the asymptotic forms of the radial function R(r). At $\rho\to\infty$ and omitting the terms $\sim\rho^{-1}$ and $\sim\rho^{-2}$ in (4.77), we obtain

$$\frac{d^2R}{d\rho^2}=\frac{R}{4}.$$

Therefore at high values of ρ, $R \propto e^{\pm\rho/2}$. The normalization demand is satisfied only by $R(\rho) \propto e^{-\rho/2}$. The asymptotic forms at $r \to 0$ have already been determined.

Substituting

$$R(\rho) = \rho^{\ell} e^{-\rho/2}\, \omega(\rho),$$

Equation 4.77 is reduced to the form

$$\rho \frac{d^2\omega}{d\rho^2} + (2\ell + 2 - \rho)\frac{d\omega}{d\rho} + (n - \ell - 1)\,\omega = 0. \tag{4.78}$$

To solve this equation in the limit of $\rho = 0$, we substitute $\omega(\rho)$ in the form of a power series

$$\omega(\rho) = 1 + \frac{(0-v)}{(0+\lambda)}\rho + \frac{(0-v)(1-v)}{(0+\lambda)(1+\lambda)}\frac{\rho^2}{2!} + \frac{(0-v)(1-v)(2-v)}{(0+\lambda)(1+\lambda)(2+\lambda)}\frac{\rho^3}{3!} + \dots, \tag{4.79}$$

where $\lambda = 2\ell + 2$ and $-v = -n + \ell + 1$.

At $\rho \to \infty$, the function $\omega(\rho)$ should increase, but not faster than the limiting power ρ. Then $\omega(\rho)$ has to be a polynomial of v power. So, $-n + \ell + 1 = -k$, and $n = \ell + 1 + k$ $(k = 0, 1, 2, \dots.)$ at a given value of ℓ. Hence, using the definition for n, we can find the expression for the energy spectrum

$$E_n = -\frac{1}{2n^2}. \tag{4.80}$$

The number n is called the principal quantum number. In general units it has the form

$$E = -Z^2 \frac{me^4}{2\hbar^2 n^2}. \tag{4.81}$$

This formula was obtained by Bohr in 1913 on the basis of the old quantum theory, by Pauli in 1926 from matrix mechanics, and by Schrödinger in 1926 by solving the differential equations.

Let us solve the problem of the spectrum of the hydrogen atom using the equation of dynamical equilibrium of the system. In Chap. 3 we obtained Jacobi's virial equation for a quantum mechanical system of particles whose interaction is defined by the potential being a homogeneous function of the co-ordinates. This equation in the operator form is

$$\ddot{\hat{\Phi}} = 2\hat{H} - \hat{U}\nu \tag{4.82}$$

where $\hat{\Phi}$ is the operator of the Jacobi function, which, for the hydrogen atom, is written

$$\hat{\Phi} = \frac{1}{2}m\hat{r}^2, \tag{4.83}$$

The Hamiltonian operator is

$$\hat{H} = -\frac{\hbar^2}{2m}\Delta_r + \hat{U} \tag{4.84}$$

and the operator of the function of the potential energy for the hydrogen atom is

$$\hat{U} = -\frac{e^2}{r}. \tag{4.85}$$

We solve the problem with respect to the eigenvalues of Eq. 4.82, using the main idea of quantum mechanics. For this we use the Schrödinger equation

$$\hat{H}\Psi = E\Psi$$

and rewrite Eq. 4.82 in the form

$$\ddot{\hat{\Phi}} = 2E - \hat{U}v \tag{4.86}$$

This equation includes two (unknown in the general case) operator functions $\hat{\Phi}$ and $\hat{U}$. In the case of the interaction, the potential is determined by the relation (4.85), and we can use a combination of the operators $\hat{\Phi}$ and $\hat{U}$ in the form

$$|\hat{U}|\sqrt{\hat{\Phi}} = \frac{e^2 m^{1/2}}{\sqrt{2}} = B. \tag{4.87}$$

We now transform (4.86) into the form which was considered in classical mechanics:

$$\ddot{\hat{\Phi}} = 2E + \frac{B}{\sqrt{\hat{\Phi}}}. \tag{4.88}$$

Equation 4.88 is a consequence of Eq. 4.86 when the Schrödinger equation and the relationship (4.87) are satisfied. Its solution for the bounded state, i.e. when total energy E is determined in parametric form, can be written

$$\sqrt{\Phi} = \frac{B}{2|E|}(1 - \varepsilon\cos\varphi), \tag{4.89}$$

$$\varphi - \varepsilon \sin \varphi = \mathrm{M}, \tag{4.90}$$

where the parameter M is defined by the relation

$$\mathrm{M} = \frac{(4|\mathrm{E}|)^{3/2}}{4\mathrm{B}}(\mathrm{t} - \tau), \tag{4.91}$$

where ε and τ are integration constants and where

$$\varepsilon = \sqrt{1 - \frac{\mathrm{AC}}{2\mathrm{B}^2}},$$

$$C = -\dot{\hat{\Phi}}_0^2 + 4\mathrm{B}\sqrt{\hat{\Phi}_0} - 2\mathrm{A}\hat{\Phi}_0.$$

Moreover, the solution can be written in the form of Fourier and Lagrange series. Thus, the expression (4.37) describes the expansion of the operator $\hat{\Phi}$ into a Lagrange series including the accuracy of ε^3, and has the form

$$\hat{\Phi}_0 = \frac{\mathrm{B}^2}{\mathrm{A}^2}\left[1 + \frac{3}{2}\varepsilon^2 + \left(-2\varepsilon + \frac{\varepsilon^3}{4}\right)\cos \mathrm{M} - \frac{\varepsilon^2}{2}\cos 2\mathrm{M} - \frac{\varepsilon^3}{4}\cos 3\mathrm{M} + \ldots.\right] \tag{4.92}$$

Using the general expression for the mean values of the observed quantities in quantum mechanics

$$<\Psi|\hat{\Phi}|\Psi> = \bar{\Phi}\mathrm{v}$$

and taking into account that the mean value of the Jacobi function of the hydrogen atom should be different from zero, we find that our system has multiple eigen-frequencies $\nu_n = \mathrm{n}\nu_o$ with respect to the basic ν_o which corresponds to the period

$$\mathrm{T_v} = \frac{8\pi\mathrm{B}}{(4|\mathrm{E}|)^{3/2}}. \tag{4.93}$$

In accordance with the expression

$$\mathrm{E}_n = \hbar\, \omega_\mathrm{n} = \frac{\hbar 2\pi\mathrm{n}}{\mathrm{T}_0} \tag{4.94}$$

each of these frequencies corresponds to the energy level $\mathrm{E_n}$ of the hydrogen atom. We substitute the expression (4.93) for $\mathrm{T_v}$ into Eq. 4.94 and resolve it in relation to $\mathrm{E_n}$:

$$|\mathrm{E_n}| = \frac{\hbar 2\pi\mathrm{n}(4|\mathrm{E_n}|)^{3/2}}{8\,\pi\mathrm{B}} = \frac{\hbar\mathrm{n}(4|\mathrm{E_n}|)^{3/2}}{\frac{4\mathrm{e}^2\mathrm{m}^{1/2}}{\sqrt{2}}} = \frac{\hbar\mathrm{n}2\sqrt{2|\mathrm{E_n}|}}{\mathrm{e}^2\mathrm{m}^{1/2}}. \tag{4.95}$$

The expression obtained by Bohr follows from (4.95):

$$E_n = \frac{e^4 m}{2\hbar^2 n^2}. \tag{4.96}$$

This equation solves the problem.

4.5 Solution of a Virial Equation in the Theory of Relativity (Static Approach)

We consider now the solution of Jacobi's virial equation in the framework of the theory of relativity, showing its equivalence to Schwarzschild's solution.

Let us write down the known expression for the radius of curvature of space-time as a function of mass density:

$$\frac{1}{R^2} = \frac{8\pi}{3}\frac{G\rho}{c^2}, \tag{4.97}$$

where R is the curvature radius; ρ is the mass density; G is the gravitational constant and c is the velocity of light.

Equation 4.97 can also be rewritten in the form

$$\rho R^2 = \frac{3}{8\pi}\frac{c^2}{G}. \tag{4.98}$$

If the product ρR^2 in Eq. 4.98 is the Jacobi function ($\Phi = \rho R^2$ is the density of the Jacobi function) then, from (4.98):

$$\Phi = \frac{3}{8\pi}\frac{c^2}{G}. \tag{4.99}$$

and it follows that the Jacobi function is a fundamental constant for the Universe. (In general relativity, the spatial distance does not remain invariant. Therefore, instead of this the Gaussian curvature is used, which has the dimension of the universe distance and is the invariant or, more precisely, the covariant.)

The constancy of the Jacobi function in this case reflects the smoothness of the description of motion in general relativity. The oscillations relative to this smooth motion described by Jacobi's equation are the gravitational waves and horizons, in particular the collapse and all types of singularity up to the process of condensation of matter in galaxies, stars etc.

Now we can show that Schwarzschild's solution in general relativity is equivalent to the solution of Jacobi's equation when $\ddot{\Phi} = 0$. Let us write the expression for the energy-momentum tensor

$$T_i^k = (\rho + p)u_i u^k + p\delta_i^k. \tag{4.100}$$

In the corresponding co-ordinate system, we obtain

$$u^i = \left(0, 0, 0, \frac{1}{\sqrt{-g_{00}}}\right), \tag{4.101}$$

where $\rho = \rho(r)$ and $p = p(r)$.

The independent field equations are written

$$G_1^1 = T_1^1, \quad G_0^0 = T_0^0, \tag{4.102}$$

$$R^{-2} = \frac{1}{3}G\rho c^2.$$

The expression for the metric is written in the form

$$ds^2 = \frac{dr^2}{1 - \frac{r^2}{R^2}} + r^2(d\Omega)^2 - \left\{A - B\sqrt{1 - \frac{r^2}{R^2}}\right\}^2 c^2 r^2, \tag{4.103}$$

where

$$\frac{dr^2}{1 - \frac{r^2}{R^2}} + r^2(d\Omega)^2.$$

is the spatial element.

In this case the expression for the volume occupied by the system is written

$$V = \int_0^r \int_0^\pi \int_0^{2\pi} \frac{r^2 \sin\Theta}{\sqrt{1 - \frac{r^2}{R^2}}} dr d\Theta d\Psi = \frac{4\pi R^3}{3}\left[\arcsin\frac{r}{R} - \frac{r}{R}\sqrt{1 - \frac{r^2}{R^2}}\right] \tag{4.104}$$

It can be easily verified that the right-hand side of Eq. 4.104 coincides with solution (4.14) and (4.15) of the equation of virial oscillations (4.11) at $\ddot{\Phi} = 0$, i.e.,

$$\arcsin x - x\sqrt{1 - x^2} = \arccos\left(\frac{\frac{A}{B}\sqrt{\Phi} - 1}{\sqrt{1 - \frac{AC}{2B^2}}}\right) - \sqrt{1 - \frac{AC}{2B^2}}$$
$$\times \sqrt{1 - \left(\frac{\frac{A}{B}\sqrt{\Phi} - 1}{\sqrt{1 - \frac{AC}{2B^2}}}\right)^2}. \tag{4.105}$$

In fact, Eq. 4.105 is satisfied for

$$x = \frac{\frac{A}{B}\sqrt{\Phi} - 1}{\sqrt{1 - \frac{AC}{2B^2}}} \text{ and } x = \sqrt{1 - \frac{AC}{2B^2}},$$

i.e.

$$\frac{A}{B}\sqrt{\Phi} - 1 = 1 - \frac{AC}{2B^2}, \text{ or } \frac{AC}{2B^2} + \frac{A\sqrt{\Phi}}{B} = 2.$$

At $\ddot{\Phi} = 0$, the parameter of virial oscillations,

$$e = \sqrt{1 - \frac{AC}{2B^2}} \text{ and } \sqrt{\Phi} = \frac{B}{A},$$

so the last condition is satisfied.

Schwarzschild's solution is rigorous and unique for Einstein's equation for a static model of a system with spherical symmetry.

Since this solution coincides with the solution of virial oscillations at the same conditions, the solutions (4.14) and (4.15) of Eq. 4.11, obtained in this chapter, should be considered rigorous. Thus we can conclude that the constancy of the product $U\sqrt{\Phi}$ in the framework of the static system model is proven. In Chap. 6 we will come back to this condition and will obtain another proof of the same very important relationship which is applied for study of the Earth's dynamics.

Chapter 5
Perturbed Virial Oscillations of a System

In the previous chapter we have considered a number of cases of explicitly solved problems in mechanics and physics for the dynamics of an n-body system and shown that all those classical problems have also explicit solution in the framework of the virial approach. But in the latter case, the solutions acquire a new physical meaning because the dynamics of a system is considered with respect to new parameters, i.e. its Jacobi function (polar moment of inertia) and potential (kinetic) energy. In fact, the solution of the problem in terms of co-ordinates and velocities specifies the changes in location of a system or its constituents in space. The solution, with respect to the Jacobi function and the potential energy, identifies the evolutionary processes of the structure or redistribution of the mass density of the system. Moreover, the main difference of the two approaches is that the classical problem considers motion of a body in the outer central force field. The virial approach considers motion of a body both in the outer and in its own force field applying, instead of linear forces and moments, the volumetric forces (pressure) and moments (oscillations).

It appears from the cases considered that the existence of the relationship between the potential energy of a system and its Jacobi function written in the form

$$U\sqrt{\Phi} = B = \text{const.} \tag{5.1}$$

is the necessary condition for the resolution of Jacobi's equation.

This is the only case when the scalar equation

$$\ddot{\Phi} = 2E - U$$

is transferred into a non-linear differential equation with one variable in the form

$$\ddot{\Phi} = 2E + \frac{B}{\sqrt{\Phi}}. \tag{5.2}$$

V.I. Ferronsky et al., *Jacobi Dynamics*, Astrophysics and Space Science Library 369, DOI 10.1007/978-94-007-0498-5_5, © Springer Science+Business Media B.V. 2011

It was shown in Chap. 4 that if the total energy of a system $E_o = -A/2 < 0$, then the general solution for Eq. 5.2 can be written as

$$\sqrt{\Phi_0} = \frac{B}{A}[1 - \varepsilon\cos(\lambda - \psi)] \tag{5.3}$$

$$t = \frac{4B}{(2A)^{3/2}}[\lambda - \varepsilon\sin(\lambda - \psi)],$$

where ε and ψ are integration constants, the values of which are determined from initial data using Eqs 4.23 and 4.24.

Equation 5.2 was called the equation of virial oscillation because its solution discovers a new physical effect – periodical non-linear change of the Jacobi function and hence the potential energy of a system around their mean values determined by the generalized virial theorem. Thus, in addition to the static effects determined by the hydrostatic equilibrium, in the study of dynamics of a system the effects, determined by a condition of dynamical equilibrium expressed by the Jacobi function, are introduced.

The equation of virial oscillations (5.2) reflects physics of motion of the interacting mass particles of a body or masses of bodies themselves by the inverse square law. Its application opens the way to study the nature and the mechanism of generation of the body's energy, which performs its motion, and to search the law of change for the system's configuration, i.e. a mutual change location of particles or the law of redistribution of the mass density for the system's matter during its oscillations. This problem was considered earlier in our work (Ferronsky et al. 1987). We continue its study in the next chapter.

As described in Chap. 4, cases of solution of Eq. 5.2 relate to unperturbed conservative systems. But in reality, in nature all systems are affected by internal and external perturbations which, from a physical point of view, are developed in the form of dissipation or absorption of energy. In this connection, as shown in Chap. 3 in the right–hand side of the equation of virial oscillations (5.2), an additional term appears which is proportional to the Jacobi function Φ (indicating the presence of gravitational background or the existence of interaction between the system particles in accordance with Hook's law) and its first derivative $\dot{\Phi}$ depending on time t (indicating the existence of energy dissipation). All these and other possible cases can be formally described by a generalized equation of virial oscillations (3.27):

$$\ddot{\Phi} = 2E + \frac{B}{\sqrt{\Phi}} + X(t,\Phi,\dot{\Phi}), \tag{5.4}$$

where $X(t, \Phi, \dot{\Phi})$ is the perturbation function, the value of which is small in comparison with the term $B/\sqrt{\Phi} \neq \text{const}$.

In this chapter we consider general as well as some specific approaches to the solution of Eq. 5.4 in the framework of different physical models of a system.

5.1 Analytical Solution of a Generalized Equation of Virial Oscillations

The equation of perturbed virial oscillations is generalized in the form

$$\ddot{\Phi} = -A + \frac{B}{\sqrt{\Phi}} + X(t, \Phi, \dot{\Phi}), \tag{5.5}$$

where $A = -2E$; B is constant; $X(t, \Phi, \dot{\Phi})$ is the perturbation function which we assume is given and dependent in general cases on time t, the Jacobi function Φ and its first derivative $\dot{\Phi}$.

We consider two ways for analytical construction of the solution of Eq. 5.5. In addition, let the function $X(t, \Phi, \dot{\Phi})$ in Eq. 5.5 depend on some small parameter e in relation to which the function can be expanded into absolutely convergent power series of the form

$$X(t, \Phi, \dot{\Phi}) = \sum_{r=1}^{\infty} e^k X^k(t, \Phi, \dot{\Phi}). \tag{5.6}$$

Let the series be convergent in some time interval t absolute for all values of e which are satisfied to condition $|e| < \bar{e}$. Then Eq. 5.5 can be rewritten in the form

$$X(t, \Phi, \dot{\Phi}) = -A + \frac{B}{\sqrt{\Phi}} \sum_{r=1}^{\infty} e^k X^k(t, \Phi, \dot{\Phi}). \tag{5.7}$$

We look for the solution of Eq. 5.7 also in the form of the power series of parameter e. For this purpose we write the function $\Phi(t)$ in the form of a power series, the coefficients of which are unknown:

$$\Phi(t) = \sum_{k=0}^{\infty} e^k \Phi^{(k)}(t). \tag{5.8}$$

Putting (5.8) into (5.7), the task can be reduced to the determination of such functions $\Phi^{(k)}(t)$ which identically satisfy Eq. 5.7. In this case, the coefficient $\Phi^{(o)}(t)$ becomes the solution of the unperturbed oscillation equation (5.2), which can be obtained from (5.7) by putting $e = 0$.

One can consider the series (5.8) as a Taylor series expansion in order to determine all the other coefficients $\Phi^{(k)}(t)$, i.e.,

$$\Phi^{(k)} = \frac{1}{k!} \left(\frac{d^k \Phi}{de^k} \right) \Big|_{e=0},$$

$$\dot{\Phi}^{(k)} = \frac{1}{k!} \left(\frac{d^k \dot{\Phi}}{de^k} \right) \Big|_{e=0}. \tag{5.9}$$

Accepting the series (5.8) for introduction into Eq. 5.7, it becomes identical with respect to the parameter *e*. Thus we have justified the differentiation of the identity with respect to the parameter *e* several times assuming that the identity remains after repeated differentiation.

We next obtain

$$\frac{d^2}{dt^2}\left(\frac{d\Phi}{de}\right) = -\frac{1}{2}\frac{B}{\Phi^{3/2}}\left(\frac{d\Phi}{de}\right) + \sum_{k=1}^{\infty} ke^{k-1}X^{(k)} + \sum_{k=1}^{\infty} e^k\left(\frac{dX^{(k)}}{de}\right), \tag{5.10}$$

where $dX^{(k)}/de$ is the total derivative of the function $X^{(k)}$ with respect to parameter e, expressed by

$$\frac{dX^{(k)}}{de} = \frac{\partial X^{(k)}}{\partial \Phi}\left(\frac{d\Phi}{de}\right) + \frac{\partial X^{(k)}}{\partial \dot{\Phi}}\left(\frac{d\dot{\Phi}}{de}\right).$$

Now let $e = 0$ in (5.10). Then by taking into account (5.8) and (5.9), we obtain

$$\frac{d^2\Phi^{(1)}}{dt^2} + p_1\Phi^{(1)} = X_1, \tag{5.11}$$

where

$$p_1 = \frac{1}{2}\frac{B}{\Phi^{3/2}}|_{e=0} = \frac{1}{2}\frac{B}{\Phi^{(0)3/2}},$$

$$X_1 = X^1\left(t, \Phi^{(0)}, \dot{\Phi}^{(0)}\right)$$

are known functions of time, since the solution of the equation in the zero approximation (unperturbed oscillation equation (5.3)) is known.

Carrying out differentiation of Eq. 5.7 with respect to parameter e for the second, third and so on $(k-1)$ times, and assuming after each differentiation that $e = 0$, we will step by step obtain equations determining second, third and so on approximations. It is possible to show that in each succeeding approximation the equation will have the same form and the same coefficient p_1 as in Eq. 5.11. If so, the equation determining the functions $\Phi^{(k)}$ and $\dot{\Phi}^{(k)}$ has the form

$$\frac{d^2\Phi^{(k)}}{dt^2} + p_1\Phi^{(k)} = X_k\left(t, \Phi^{(0)}, \dot{\Phi}^{(0)}, ..., \Phi^{(k-1)}, \dot{\Phi}^{(k-1)}\right), \tag{5.12}$$

where the function X_k depends on $\Phi^{(0)}$, $\dot{\Phi}^{(0)}$,... $\Phi^{(k-1)}$, $\dot{\Phi}^{(k-1)}$, which were determined earlier and are the functions of t and unknown functions $\Phi^{(0)}$ and $\dot{\Phi}^{(0)}$.

It is known that there is no general way of obtaining a solution for any linear differential equation with variable coefficients, but in our case we can use the following theorem of Poincaré (Duboshin 1975). Let the general solution of the unperturbed virial oscillation equation be determined by the function $\Phi^{(0)} = f(t, C_1, C_2)$, where C_1 and C_2 are, for instance, arbitrary constants ε and Ψ in the solution (5.3) of Eq. 5.2. Then, Poincaré's theorem confirms that the function determined by the equalities

$$\Phi_1 = \frac{\partial f}{\partial C_1},$$

$$\Phi_2 = \frac{\partial f}{\partial C_2}$$

satisfies the linear homogeneous differential equation reduced by omission of the right-hand side of Eq. 5.12.

Thus, the general solution of the linear homogeneous equation

$$\frac{d^2\Phi^{(k)}}{dt^2} + p_1\Phi^{(k)} = 0$$

has the form

$$\Phi_1 C_1^{(k)} + \Phi_2 C_2^{(k)} = \Phi^{(k)} \tag{5.13}$$

and the general solution of Eq. 5.12 can be obtained by the method of variation of arbitrary constants, i.e. assuming that $C_2^{(k)}$ are functions of time. Then, using the key idea of the method of variation of arbitrary constants, we obtain a system of two equations:

$$\dot{C}_1^{(k)}\Phi_1 + \dot{C}_2^{(k)}\Phi_2 = 0,_k. \tag{5.14}$$

$$\dot{C}_1^{(k)}\dot{\Phi}_1 + \dot{C}_2^{(k)}\dot{\Phi}_2 = X_k.$$

Solving this system with respect to $\dot{C}_1^{(k)}$ and $\dot{C}_2^{(k)}$ and integrating the expression obtained, we write the general solution of Eq. 5.12 as follows:

$$\Phi^{(k)}(t) = \Phi_2 \int_{t_0}^{t} \frac{\Phi_1 X_k dt}{\Phi_1\dot{\Phi}_2 - \Phi_2\dot{\Phi}_1} - \Phi_1 \int_{t_0}^{t} \frac{\Phi_2 X_k dt}{\Phi_1\dot{\Phi}_2 - \Phi_2\dot{\Phi}_1},$$

where

$$\Phi_1 = \frac{\partial f(t, C_1, C_2)}{\partial C_1}$$

and

$$\Phi_2 = \frac{\partial f(t, C_1, C_2)}{\partial C_2}.$$

Thus, we can determine any coefficient of the series (5.8), reducing Eq. 5.7 into an identity, and therefore write the general solution of Eq. 5.5 in the form

$$\Phi = \sum_{k=0}^{\infty} e^k \Phi^{(k)} = \sum_{k=0}^{\infty} e^k \left[\Phi_2 \int_{t_0}^{t} \frac{\Phi_1 X_k dt}{\Phi_1 \dot{\Phi}_2 - \Phi_2 \dot{\Phi}_1} - \Phi_1 \int_{t_0}^{t} \frac{\Phi_2 X_k dt}{\Phi_1 \dot{\Phi}_2 - \Phi_2 \dot{\Phi}_1} \right]. \quad (5.15)$$

Let us consider the second way of approximate integration of the perturbed virial equation (5.5), based on Picard's method (Duboshin 1975). It is convenient to apply this method of integrating the equations which was obtained using the Lagrange method of variation of arbitrary constants.

We assume that the first integrals (4.23) and (4.24)

$$\varepsilon = \sqrt{1 - \frac{A}{2B^2}\left(-\dot{\Phi}_0 + 4B\sqrt{\Phi_0} - 2A\Phi_0\right)}, \quad (5.16)$$

$$-\tau = \left\{ \frac{4B}{(2A)^{3/2}} \left[\arccos \frac{1 - \frac{A}{B}\sqrt{\Phi_0}}{\varepsilon} - \varepsilon \sqrt{1 - \left(\frac{1 - \frac{A}{B}\sqrt{\Phi_0}}{\varepsilon} \right)^2} \right] - t \right\} \quad (5.17)$$

of the unperturbed virial oscillation equation (5.2) are also the first integrals of the perturbed oscillation equation (5.5). But constants ε and τ are now unknown functions of time. Let us derive differential equations which are satisfied by these functions, using the first integrals (5.16) and (5.17). For convenience, we replace the integration constant ε by C, using the expression

$$\varepsilon = \sqrt{1 - \frac{AC}{2B^2}}.$$

Now we rewrite Eq. 5.16 in the form

$$C = -\dot{\Phi}_0^2 + 4B\sqrt{\Phi_0} - 2A\Phi_0. \quad (5.18)$$

Then using the main idea of the Lagrange method, after variation of the first integrals (5.17) and (5.18) and replacement of $\ddot{\Phi}$ by

$$\left(-A + \frac{B}{\sqrt{\Phi}} + X\left(t, \Phi, \dot{\Phi}\right)\right)$$

we write

$$\dot{C} = -2\dot{\Phi}X\left(t, \Phi, \dot{\Phi}\right), \quad (5.19)$$

$$\dot{\tau} = \Psi(\Phi, C)\dot{C} = -2\dot{\Phi}X\left(t, \Phi, \dot{\Phi}\right)\Psi(\Phi, C), \quad (5.20)$$

where

$$\Psi(\Phi, C) = -\frac{4B}{(2A)^{3/2}} \frac{d}{dC}\left[\arccos\frac{1-\frac{A}{B}\sqrt{\Phi}}{\sqrt{1-\frac{AC}{2B^2}}} - \sqrt{1-\frac{AC}{2B^2}}\sqrt{1-\left(\frac{1-\frac{A}{B}\sqrt{\Phi}}{\sqrt{1-\frac{AC}{2B^2}}}\right)^2}\right].$$

We now express Φ and $\dot{\Phi}$ in explicit form through C, τ and t using, for example, the Lagrangian series (4.36) and (4.37)

$$\Phi(t) = \frac{B^2}{A^2}\left[1 + \frac{3}{2}\varepsilon^2 + \left(-2\varepsilon + \frac{\varepsilon^3}{4}\right)\cos M - \frac{\varepsilon^2}{2}\cos 2M - \frac{\varepsilon^3}{4}\cos 3M + \ldots\right], \tag{5.21}$$

$$\dot{\Phi}(t) = \sqrt{\frac{2}{A}}\,\varepsilon B\left[\sin M + \frac{1}{2}\varepsilon \sin 2M + \frac{\varepsilon^2}{2}\sin M\left(2\cos^2 M - \sin^2 M\right) + \ldots\right]. \tag{5.22}$$

Thus, taking into account Eqs. 5.21 and 5.22 for the functions Φ and $\dot{\Phi}$, Eqs. 5.19 and 5.20 can be rewritten as

$$\frac{dC}{dt} = F_1(t, C, \tau), \tag{5.23}$$

$$\frac{d\tau}{dt} = F_2(t, C, \tau).$$

To solve the system of differential equations (5.23), we use Picard's successive approximation method, obtained in the k-th approximation expressions for $C^{(k)}$ and $\tau^{(k)}$ in the form

$$C^{(k)} = C^{(0)} + \int_{t_0}^{t} F_1\left(t, C^{(k-1)},\ \tau^{(k-1)}\right) dt, \tag{5.24}$$

$$\tau^{(k)} = \tau^{(0)} + \int_{t_0}^{t} F_2\left(t, C^{(k-1)},\ \tau^{(k-1)}\right) dt, \tag{5.25}$$

where $C^{(0)}$ and $\tau^{(0)}$ are the values of arbitrary constants C and τ at initial time t_0, and k = 1, 2, ...

Then, in the limit of $k \to \infty$, we obtain the solution of the system (5.23):

$$C = \lim_{k\to\infty} C^{(k)},$$

$$\tau = \lim_{k\to\infty} \tau^{(k)}. \tag{5.26}$$

Consider now two possible cases of the perturbation function behavior. First, assume that the perturbation function X does not depend explicitly on time. Then, since it is possible to expand functions Φ and $\dot{\Phi}$ into a Fourier series in terms of sine and cosine of argument M, the right-hand sides of the system (5.23) can also be expanded into a Fourier series in terms of sine and cosine of M.

Finally we obtain

$$\frac{dC}{dt} = \left[A_0 + \sum_{k=1}^{\infty} (A_k \cos kM + B_k \sin kM)\right], \tag{5.27}$$

$$\frac{d\tau}{dt} = \left[a_0 + \sum_{k=1}^{\infty} (a_k \cos kM + b_k \sin kM)\right], \tag{5.28}$$

where A_0, A_k, B_k, a_0, a_k, b_k are the corresponding coefficients of the Fourier series which are

$$A_0 = \frac{2}{\pi}\int_0^{2\pi} F_1(M, C)dM,$$

$$A_k = \frac{2}{\pi}\int_0^{2\pi} F_1(M,C) \cos kMdM,$$

$$B_k = \frac{2}{\pi}\int_0^{2\pi} F_1(M,C) \sin kMdM,$$

$$a_0 = \frac{2}{\pi}\int_0^{2\pi} F_2(M,C)dM,$$

$$a_k = \frac{2}{\pi}\int_0^{2\pi} F_2(M,C) \cos kMdM,$$

$$b_k = \frac{2}{\pi}\int_0^{2\pi} F_2(M,C) \sin kMdM.$$

Following Picard's method, in order to solve Eqs. 5.27 and 5.28 in the first approximation, we introduce into the right-hand side of the equations the values of arbitrary constants C and τ corresponding to the initial time t_0 . Then we obtain

$$C^{(1)}(t) = C^{(0)} + A_0^{(0)}(t - t_0) + \sum_{k=1}^{\infty} \frac{1}{kn} \left\{ A_k^{(0)}[\sin kM - \sin kM_0] + B_k^{(0)}[\cos kM - \cos kM_0] \right\} \quad (5.29)$$

$$\tau^{(1)}(t) = \tau^{(0)} + a_0^{(0)}(t - t_0) + \sum_{k=1}^{\infty} \frac{1}{kn} \left\{ a_k^{(0)}[\sin kM - \sin kM_0] + b_k^{(0)}[\cos kM - \cos kM_0] \right\} \quad (5.30)$$

Thus, when the function X does not depend explicitly on time t, solutions (5.29) and (5.30) of Eq. 5.5 have three analytically different parts. The first is a constant term, depending on the initial values of the arbitrary constants. It is usually called the constant term of perturbation of the first order. The second part is a function monotonically increasing in time. It is called the secular term of a perturbation of the first order. The third part consists of an infinite set of trigonometric terms. All of them are periodic functions of M and consequently of time t. This is called periodic perturbation

Similarly, we can obtain solutions in the second, third etc., orders. Here we limit our consideration only within the first order of perturbation theory. In practice, few terms of the periodic perturbation can be taken into account and the solution obtained becomes effective only for a short period of time.

When the perturbation function X is a periodic function of some argument M',

$$M' = n'(t - \tau'),$$

the right-hand side of the system of Eqs. 5.23 are periodic functions of the two independent arguments M and M'. Therefore, they can be expanded into a double Fourier series in terms of sine and cosine of the linear combination of arguments M and M'. Then in the first approximation of perturbation theory we obtain the following system of equations:

$$\frac{dC^{(1)}}{dt} = A_{00}^{(0)} + \sum_{k',k=-\infty}^{\infty} \left[A_{k,k'}^{(0)} \cos(kM + k'M') + B_{k,k'}^{(0)} \sin(kM + k'M') \right], \quad (5.31)$$

$$\frac{d\tau^{(1)}}{dt} = a_{00}^{(0)} + \sum_{k',k=-\infty}^{\infty} \left[a_{k,k'}^{(0)} \cos(kM + k'M') + b_{k,k'}^{(0)} \sin(kM + k'M') \right]. \quad (5.32)$$

Integrating equations (5.31) and (5.32) with respect to time, we obtain a solution of the system:

$$\begin{aligned} C^{(1)}(t) = C^{(0)} + A^{(0)}_{00}(t - t_0) + \sum_{k',k=-\infty}^{\infty} \frac{1}{kn + k'n'} \Big\{ A^{(0)}_{k,k'}[\cos(kM + k'M') \\ - \cos(kM_0 + k'M'_0)] + B^{(0)}_{k,k'}[\sin(kM + k'M') \\ - \sin(kM_0 + k'M'_0)] \Big\} \end{aligned} \tag{5.33}$$

$$\begin{aligned} \tau^{(1)}(t) = \tau^{(0)} + a^{(0)}_{00}(t - t_0) + \sum_{k',k=-\infty}^{\infty} \frac{1}{kn + k'n'} \Big\{ a^{(0)}_{k,k'}[\cos(kM + k'M') - \\ - \cos(kM_0 + k'M'_0)] + b^{(0)}_{k,k'}[\sin(kM + k'M') \\ - \sin(kM_0 + k'M'_0)] \Big\} \end{aligned} \tag{5.34}$$

Equations 5.33 and 5.34 have the same analytical structure as (5.29) and (5.30). At the same time, in this case, the periodic part of the perturbation can be divided into two groups, depending on the value of the divisor $kn + k'n'$. If the values of k and k' are such that the divisor is sufficiently large, then period $T_{k,k'} = 2\pi/(kn + k'n')$ of the corresponding inequality will be rather small. Such inequalities are called short-periodic. Their amplitudes are also rather small, and they can play a role only within short periods of time.

If the values of k and k' are such that the divisor $kn + k'n'$ is sufficiently small but unequal to zero, then the period of the corresponding inequality will become large. The amplitude of such terms could also be large and play a role within large periods of time. Such terms form series of long-periodic inequalities. In the case of such k and k', when $kn + k'n' = 0$, the corresponding terms are independent of t and change the value of the secular term in the solutions (5.33) and (5.34).

5.2 Solution of the Virial Equation for a Dissipative System

In Chap. 3 we derived Jacobi's virial equation for a non-conservative system in the form

$$\ddot{\Phi} = 2E_0[1 + q(t)] - U - k\dot{\Phi}. \tag{5.35}$$

At $k \ll 1, t \gg t_o$, $|U|\sqrt{\Phi} = B = \text{const}$, $2E_o = -A_o$, and when the magnitude of the term $k\dot{\Phi}$ is sufficiently small, Eq. 5.35 can be rewritten in a parametric form

$$\ddot{\Phi} = -A_0[1 + q(t)] + \frac{B}{\sqrt{\Phi}}, \tag{5.36}$$

where q(t) is a monotonically increasing function of time due to dissipation of energy during 'smooth' evolution of a system within a time interval $t \in [0, \tau]$.

Using the theorem of continuous solution depending on the parameter, we write the solution of Eq. 5.36 as follows:

$$\begin{aligned} &-\arccos W + \arccos W_0 - \sqrt{1 - \frac{A_0[1 + q(t)]C}{2B^2}}\sqrt{1 - W^2} \\ &+ \sqrt{1 - \frac{A_0 C}{2B^2}}\sqrt{1 - W_0^2} = \sqrt{\frac{(2A_0[1 + q(t)])^{3/2}}{4B}}(t - t_0), \end{aligned} \tag{5.37}$$

$$\begin{aligned} &\arccos W - \arccos W_0 + \sqrt{1 - \frac{A_0[1 + q(t)]C}{2B^2}}\sqrt{1 - W^2} \\ &- \sqrt{1 - \frac{A_0 C}{2B^2}}\sqrt{1 - W_0^2} = \sqrt{\frac{(2A_0[1 + q(t)])^{3/2}}{4B}}(t - t_0), \end{aligned} \tag{5.38}$$

where

$$W = \frac{\frac{A_0[1 + q(t)]}{B}\sqrt{\Phi} - 1}{\sqrt{1 - \frac{A_0[1 + q(t)]C}{2B^2}}}; \quad W_0 = \frac{\frac{A_0}{B}\sqrt{\Phi} - 1}{\sqrt{1 - \frac{A_0 C}{2B^2}}};$$

$$A_0[1 + q(t)] > 0; \quad C < \frac{2B^2}{A_0[1 + q(t)]};$$

$$|-A_0[1 + q(t)]\sqrt{\Phi} + B| < B\sqrt{1 - \frac{A_0[1 + q(t)]C}{2B^2}};$$

$$C = -2A_0\Phi_0 + 4B\sqrt{\Phi_0} - \dot{\Phi}_0^2.$$

Equations of discriminant curves which bound oscillations of the Jacobi function Φ by analogy with the case of the conservative system can be written as

$$\sqrt{\Phi_1} = \frac{B}{A_0[1 + q(t)]}\left[1 + \sqrt{1 - \frac{A_0[1 + q(t)]C}{2B^2}}\right], \quad t \in [0, \tau], \tag{5.39}$$

$$\sqrt{\Phi_2} = \frac{B}{A_0[1 + q(t)]}\left[1 - \sqrt{1 - \frac{A_0[1 + q(t)]C}{2B^2}}\right], \quad t \in [0, \tau]. \tag{5.40}$$

It is obvious that the solution of Jacobi's virial equation for a non-conservative system is quasi-periodic with period

$$T_v(q) = \frac{8\pi B}{(2A_0[1 + q(t)])^{3/2}}, \tag{5.41}$$

and an amplitude of Jacobi function oscillations

$$\Delta\sqrt{\Phi} = \frac{B}{A_0[1 + q(t)]}\left(1 - \frac{A_0[1 + q(t)]C}{2B^2}\right)^{1/2}. \tag{5.42}$$

As q(t) is a monotonically and continuously increasing parameter confined in time, the period and the amplitude of the oscillations will gradually decrease and tend to zero in the time limit.

In Fig. 5.1a the integral curves (5.37) and (5.38) and the discriminant curves (5.39) and (5.40) are shown in a general case when $0 < C < 2B^2/A_0$. At the point O_b, the integral and discriminant curves tend to coincide and the value of the amplitude of the Jacobi function (polar moment of inertia) oscillations of the system goes to zero.

When $C = 0$ (Fig. 5.1b) the discriminant line (5.39) coincides with the axis of abscissa, $\Phi_2 = 0$. In the accepted case of constancy of the system mass, the point O_b, where the integral and discriminant curves coincide, will be reached in the time limit $t \to \infty$.

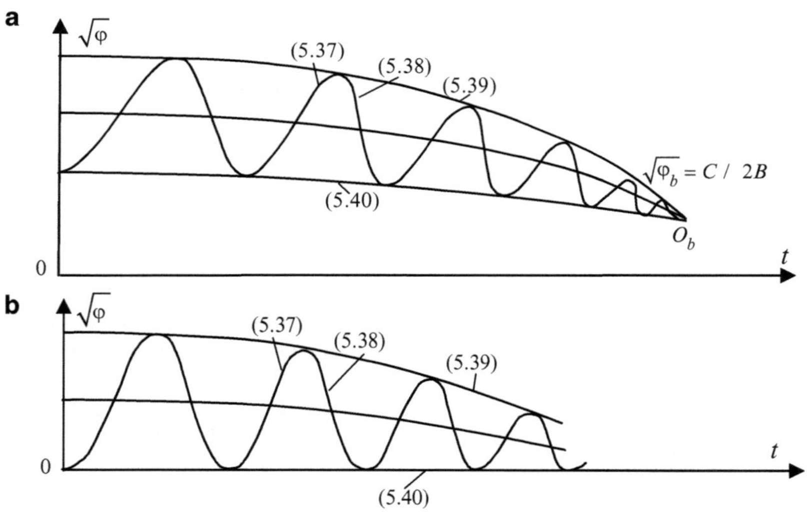

Fig. 5.1 Virial oscillations of the Jacobi function in time for a non-conservative system (**a**) and for the general (Wintner's) case (**b**)

Where $2B^2/A_o \to C$ and $C < 0$, the solutions (5.37)–(5.39), (4.40) could be complex so the processes considered are not physical.

We note that, by analogy with the case for a conservative system, considered in Chap. 4, we can show here that the asymptotic relations (4.30)–(4.32) for the solutions (5.37) and (5.38) of Jacobi's Eq. 5.36 in the points of contact of the discriminant line $\Phi_2 = 0$, are justified. In the points of contact for the integral curves (5.37) and (5.38) and the discriminant curves (5.39) and (5.40) for which Φ_1 and Φ_2 are not equal to zero, the following asymptotic relations are also justified:

$$\left(\sqrt{\Phi_1} - \sqrt{\Phi}\right) \propto (t' - t)^2, \tag{5.43}$$

$$\left(\sqrt{\Phi} - \sqrt{\Phi_2}\right) \propto (t - t')^2, \tag{5.44}$$

where t' is time of a tangency point for the corresponding integral curve of the discriminant lines $\Phi_{1,2}$ when $\Phi_{1,2} \neq 0$.

5.3 Solution of the Virial Equation for a System with Friction

Let us consider the solution of Jacobi's virial equation for conservative systems, but let the relationship between its potential energy and the Jacobi function be as follows:

$$U\sqrt{\Phi} = B + k\dot{\Phi}. \tag{5.45}$$

In this case, the equation of virial oscillations (5.2) can be written

$$\ddot{\Phi} = -A + \frac{B}{\sqrt{\Phi}} - k\frac{\dot{\Phi}}{\sqrt{\Phi}}. \tag{5.46}$$

The term $-k\dot{\Phi}/\sqrt{\Phi}$ in (5.46) plays the role of perturbation function, reflecting the effect of internal friction of the matter while the system is oscillating.

In principle, Eq. 5.46 can be solved using the above perturbation theory methods. However, we can show that a particular solution exists for the system of two differential equations of the second order, which satisfies Eq. 5.46. These differential equations are as follows:

$$\left(\sqrt{\Phi}\right)'' + \sqrt{\frac{2}{A}}k\left(\sqrt{\Phi}\right)' + \sqrt{\Phi} = \frac{B}{A}, \tag{5.47}$$

$$t'' + \sqrt{\frac{2}{A}}kt' + t = \frac{4B}{(2A)^{3/2}}\lambda. \tag{5.48}$$

In Eqs. 5.47 and 5.48 we introduced a new variable λ, so the primes at Φ and t mean differentiation with respect to λ. Note also that time t here is not an independent variable. This allows us to transfer the non-linear equation into two linear equations. The partial solution of Eqs. 5.47 and 5.48 containing two integration constants is

$$\sqrt{\Phi} = \frac{B}{A}\left[1 - \varepsilon e^{-r/2\sqrt{2/A}\lambda}\cos\left(\sqrt{\frac{4A - 2k^2}{4A}}\lambda + \psi + \tau\right)\right], \tag{5.49}$$

$$t = \frac{4B}{(2A)}\left[\lambda - \varepsilon e^{-r/2\sqrt{2/A}\lambda}\sin\left(\sqrt{\frac{4A - 2k^2}{4A}}\lambda + \psi\right)\right] - \frac{4B}{(2A^{3/2})}\sqrt{\frac{2}{A}}k, \tag{5.50}$$

where ε and ψ are arbitrary constants and

$$\tau = \operatorname{arctg}\sqrt{\frac{2}{A}}k\left(\frac{4A - 2k^2}{4A}\right)^{-1/2}.$$

To show that Eqs. 5.49 and 5.50 of the two linear differential equations (5.47) and 5.48 are also general solutions of (5.46), let us do as follows.

Differentiating (5.50) with respect to λ, we obtain

$$t' = \sqrt{\frac{2}{A}}\sqrt{\Phi}. \tag{5.51}$$

We write the derivative from function $\sqrt{\Phi}$ with respect to λ using Eq. 5.51 in the form

$$\left(\sqrt{\Phi}\right)' = \frac{\dot{\Phi}}{\sqrt{2A}}. \tag{5.52}$$

Then the second derivative from $\sqrt{\Phi}$ with respect to λ can be obtained analogously

$$\left(\sqrt{\Phi}\right)'' = \frac{\ddot{\Phi}}{\sqrt{2A}}t' = \frac{\ddot{\Phi}\sqrt{\Phi}}{A}. \tag{5.53}$$

Substituting Eqs. 5.52 and 5.53 for $\left(\sqrt{\Phi}\right)'$ and $\left(\sqrt{\Phi}\right)''$ into Eq. 5.47, we obtain

$$\frac{\ddot{\Phi}\sqrt{\Phi}}{A} + \sqrt{\frac{2}{A}}k\frac{\dot{\Phi}}{\sqrt{\Phi}} + \sqrt{\Phi} = \frac{B}{A}. \tag{5.54}$$

Dividing Eq. 5.54 by $\sqrt{\Phi}/A$ we have

$$\ddot{\Phi} + k\frac{\dot{\Phi}}{\sqrt{\Phi}} + A = \frac{B}{\sqrt{\Phi}},$$

which is in fact our Eq. 5.46. This means that Eqs. 5.49 and 5.50 are the general solution of Eq. 5.46.

Note that Eq. 5.50 differs in general from Kepler's equation both by the exponential factor before the sine function and by the constant term in the right-hand side of Eq. 5.50. In addition, it follows from Eq. 5.49 that the period of virial oscillations of the Jacobi function depends on the parameter k. Therefore, when λ changes its value by $2\pi/\left[\sqrt{(4A - 2k^2)/4A}\right]$ the value of $\sqrt{\Phi}$ remains unchanged (we neglect the changes of the amplitude of virial oscillations due to existence of the exponential factor) assuming that

$$\frac{k}{2}\sqrt{\frac{2}{A}}2\pi \Bigg/ \sqrt{\frac{4A - 2k^2}{4A}} << 1.$$

It follows from Eq. 5.50 that time t changes by the relationship of $T = 8\pi B/(2A)^{3/2}\sqrt{(4A - 2k^2)/4A}$ defining the period of the damping virial oscillations. Therefore, from solutions (5.49) and (5.50) of Eq. 5.46 it follows that if during the evolution of the system the value $U\sqrt{\Phi}$ varies only slightly around the constant, this leads to damping of the virial oscillations of the integral characteristics of the system around their averaged virial theorem value..

In conclusion we have to note that derivation of the equation of dynamical equilibrium and its solution for conservative and dissipative systems shows that dynamics of celestial bodies in their own force field puts forward a wide class of geophysical, astrophysical and geodetic problems which can be solved by the methods of celestial mechanics and with the new physical concepts we have introduced.

Chapter 6
Relationship Between Jacobi Function and Potential Energy

In the previous chapters we have considered the general approach to the formulation and solution of global dynamics problems for a self-gravitating system in terms of volumetric (integral) characteristics. For this purpose we have transformed Jacobi's virial equations for conservative and non-conservative systems:

$$\ddot{\Phi} = 2E - U, \tag{6.1}$$

$$\ddot{\Phi} = 2E - U + X(t, \Phi, \dot{\Phi}) \tag{6.2}$$

into equations of virial oscillations in the form

$$\ddot{\Phi} = -A + \frac{B}{\sqrt{\Phi}}, \tag{6.3}$$

$$\ddot{\Phi} = -A + \frac{B}{\sqrt{\Phi}} + X(t, \Phi, \dot{\Phi}). \tag{6.4}$$

The transfer from Eqs. 6.1 and 6.2 to Eqs. 6.3 and 6.4 has been made by using the following relationship between the Jacobi function and potential energy:

$$U\sqrt{\Phi} = B = \text{const}. \tag{6.5}$$

As shown in Chap. 4, the validity of the relationship (6.5) for explicitly solved cases of the many-body problem in mechanics and physics is an obvious fact. Consequently, for example, in the case of two-body problem which represents the conservative system, the solutions of Eq. 6.3 will be analogous to Keplerian equations of conic sections according to which the Jacobi function (or potential energy) changes with time. In the same manner the solution of the generalized equation of virial oscillations (6.4) in celestial mechanics will correspond to the solution for the periodic motion in the two-body problem obtained by perturbation theory methods.

V.I. Ferronsky et al., *Jacobi Dynamics*, Astrophysics and Space Science Library 369, DOI 10.1007/978-94-007-0498-5_6, © Springer Science+Business Media B.V. 2011

The validity of Eq. 6.5 for a many-body system in a general case is not obvious despite the fact that both volumetric integral characteristics considered are functions of the distribution of mass density of a system.

In this chapter we consider in detail the main physical aspect of the relationship between the Jacobi function and the potential energy of a system

6.1 Asymptotic Limit of Simultaneous Collision of Mass Points for a Conservative System

We take advantage of the results presented by Wintner (1941) in order to study the many-body problem. From such a study it follows that a conservative system of n mass points of arbitrary configuration interacting according to Newton's law, the following statement is valid.

If the motion of the material points of a system of arbitrary configuration has the consequence that all of them tend to simultaneous collision then the relationship $U\sqrt{\Phi}$ approaches a constant value. This result obtained by Wintner supplements the general properties of conservative systems of material points interacting according to Newton's law when their number remains constant all the time. The condition of constancy of the number of mass points of a system is equivalent to that of the distance $\Delta_{ij} = |r_i - r_j|$ between any pair of points at any moment of time and should be $\Delta_{ij}>0$, where r_i and r_j denote the 3-vectors of the co-ordinates of mass points in the barycentric co-ordinate system.

For such a system, from the analysis of Jacobi's virial equation (6.1) and the expression for the Jacobi function Φ,

$$\Phi = \frac{1}{2m} \sum_{1\leq i<j\leq n} m_i m_j \Delta_{ij}^2 \tag{6.6}$$

for kinetic energy T

$$T = \frac{1}{2m} \sum_{1\leq i<j\leq n} m_i m_j (\dot{r}_i - \dot{r}_j)^2 \tag{6.7}$$

and for potential energy U

$$U = -G \sum_{1\leq i<j\leq n} \frac{m_i m_j}{\Delta_{ij}} \tag{6.8}$$

three inequalities were obtained which produce restrictions on the Jacobi function (or potential energy) and its derivatives. These inequalities can be written in the form

$$|\dddot{\Phi}| \leq \eta(|\ddot{\Phi}| + 2|E|)^{5/2} \tag{6.9}$$

$$(\ddot{\Phi} - 2E)\Phi^{1/2} \geq \mu > 0 \tag{6.10}$$

$$\ddot{\Phi} - E - \frac{1}{4}\frac{\dot{\Phi}^2}{\Phi} \geq \frac{M^2}{4\Phi} \tag{6.11}$$

where constants

$$\eta = \frac{\sqrt{2m}}{G} \sum_{1\leq i<j\leq n} (m_i m_j)^{-3/2} > 0,$$

$$\mu = \frac{G}{\sqrt{2m}} \sum_{1\leq i<j\leq n} (m_i m_j)^{3/2} > 0,$$

$$M^2 = C_1^2 + C_2^2 + C_3^2,$$

$$m = \sum_{i=1} m_i$$

and m_i is the mass of the i-th point; $E = T + U$ is the total energy; C_1, C_2, C_3, are projections of the angular momentum M on the axes.

The third inequality (6.11) is more complicated than the others as is contains, besides the constant E, which is the total energy of the system, the value M of the constant angular momentum.

It has been shown by Wintner (1941) that if the motion of material points of an arbitrary configuration system provides their simultaneous collision, then the system possesses zero angular momentum and a simultaneous collision will occur in the finite interval of time. In addition, the behavior of the Jacobi function in the vicinity of the time moment t_o of simultaneous collision is defined by the following asymptotics:

$$\Phi \propto (t - t_0)^{4/3}, \tag{6.12}$$

$$\Phi \propto (t - t_0)^{1/3}, \tag{6.13}$$

$$\Phi \propto (t - t_0)^{-2/3}. \tag{6.14}$$

Following Wintner (1941), we introduce the definition of a central configuration which is needed for further consideration of the problem. If the positions of the material points in the system are such that the following relation is satisfied:

$$U_\eta = \sigma m_i r_i \tag{6.15}$$

then the configuration of the system is called central.

Here, in Eq. 6.15

$$\sigma = -\frac{U}{2\Phi}.$$

The definition (6.15) of the central configuration can be rewritten in equivalent form:

$$(U^2\Phi)_\eta = 0. \tag{6.16}$$

As proved by Wintner (1941), the important relation follows from asymptotics (6.12)–(6.14) at $t \to t_o$:

$$(U^2\Phi)_\eta \to 0 \tag{6.17}$$

which, together with the definition for the central configuration, leads to the following theorem:

Any arbitrary configuration of material points in the asymptotic time limit of simultaneous collisions of all the mass points tends to the central configuration.

It follows from this that

$$\lim_{t \to t_0} |U|\sqrt{\Phi} = \text{const.} \tag{6.18}$$

This theorem justifies the transformation of Jacobi's virial equation (6.1) and (6.2) into equation of virial oscillations (6.3) and (6.4) within the framework of Newton's law of interaction of material points of a conservative system.

6.2 Asymptotic Limit of Simultaneous Collision of Mass Points for Non-conservative Systems

The model of a conservative system permits a limited number of problems to be solved. In reality all natural systems are non-conservative. Study of the dynamics of such systems is the main object of the problem of evolution.

It is well known from the observations described in the general course of physics by Kittel et al. (1965) that the gravitating systems in nature are contracting while losing part of their total energy through friction and electromagnetic radiation. From the kinematics point of view this gravitational contraction is equivalent to the simultaneous collision of all n mass points of the system. We consider below the validity of the theorem expressed by Eq. 6.18 for non-conservative systems.

Let the motion of a system of n mass points occur by means of the gravitational interaction and Newtonian friction of the mass points. Then Jacobi's virial equation, in agreement with (3.22), can be written as

$$\ddot{\Phi} = 2E(t) - U(t) - k\dot{\Phi}, \tag{6.19}$$

where E(t) is the value of the total energy of the system at the moment of time t. From analyses of the equations of motion resulting in (3.23) it follows that

$$E(t) = E_0 - 2k\int_{t_0}^{t} T(t)dt = E_0[1 + q(t)],$$

where E_0 is the value of the total energy of the system at the initial moment of time t_o; q(t) is a monotonically increasing function of time.

We also accept the condition of the constancy of the number of mass particles in the system, from which it follows that the distance between any pairs of points $\Delta_{ij}>0$ and the following relation is correct:

$$\left|\frac{d}{dt}\Delta_{ij}\right| \le |\dot{r}_i - \dot{r}_j|$$

In the framework of this essentially important condition which forbids paired, threefold and higher-fold collisions, we obtain three inequalities analogous to (6.9)–(6.11). The inequalities are valid at any stage of the system's evolution and place restrictions on the Jacobi function and its derivatives.

From expression (6.8) for the potential energy of the system, the following inequalities can be written:

$$|U| = \left|G \sum_{1\le i<j\le n} \frac{m_i m_j}{\Delta_{ij}^2}\dot{\Delta}_{ij}\right| \le G \sum_{1\le i<j\le n} \frac{m_i m_j}{\Delta_{ij}^2}|\dot{r}_i - \dot{r}_j| \tag{6.20}$$

and

$$\frac{G m_i m_j}{\Delta_{ij}} < -U,$$

where r_i and r_j are 3-vectors of co-ordinates of mass points in the barycentric co-ordinate system.

Substituting the last inequality into (6.20) we obtain

$$|\dot{U}| \le \frac{U^2}{G} \sum_{1\le i<j\le n} \frac{|\dot{r}_i - \dot{r}_j|}{m_i m_j}.$$

Since $m_i m_j(\dot{r}_i - r_j) \leq 2mT$, and assuming

$$\eta = \frac{1}{G} \sum_{1 \leq i < j \leq n} \frac{m^{1/2}}{(m_i m_j)^{3/2}},$$

we obtain

$$|\dot{U}| \leq U^2 \eta (2T)^{1/2}. \tag{6.21}$$

Then using Eq. 6.19 in the form

$$U = 2E_0[1 + q(t)] - \ddot{\Phi} - k\dot{\Phi} \tag{6.22}$$

and the law of conservation of energy for a dissipative system

$$U + T = E_0[1 + q(t)] \tag{6.23}$$

We rewrite the inequality (6.21) in the form

$$\begin{aligned} |\dot{U}| &\leq \{2|E_0|[1 + q(t)] + |\ddot{\Phi}\} + k|\dot{\Phi}|\}^2 \eta \sqrt{2} \{2|E_0|[1 + q(t)] + |\ddot{\Phi}\} + k|\dot{\Phi}|\}^{1/2} \\ &= \sqrt{2}\, \eta \{2|E_0|[1 + q(t)] + |\ddot{\Phi}\} + k|\dot{\Phi}\}^{5/2} \end{aligned} \tag{6.24}$$

Differentiating (6.22) with respect to time and substituting this into (6.24), we finally obtain the first inequality:

$$|\dddot{\Phi} + k\ddot{\Phi} - 2E_0\dot{q}(t)| \leq \sqrt{2}\, \eta \{2|E_0|[1 + q(t)] + |\ddot{\Phi}| + k|\dot{\Phi}|\}^{5/2}. \tag{6.25}$$

In the same way, it follows from (6.6) that

$$\Phi^{1/2} \geq \frac{1}{(2m)^{1/2}} (m_i m_j)^{1/2} \Delta_{ij}.$$

Then

$$\frac{\Phi^{1/2} m_i m_j}{\Delta_{ij}} \geq (2m)^{1/2} (m_i m_j)^{1/2}.$$

By virtue of (6.4) and (6.8),

$$\ddot{\Phi} + k\dot{\Phi} - 2E[1 - q(t)] = G \sum_{1 \leq i < j \leq n} \frac{m_i m_j}{\Delta E}.$$

The second inequality has the form

$$\ddot{\Phi} + k\dot{\Phi} - 2E[1 + q(t)]\Phi^{1/2} \geq \mu > 0 \tag{6.26}$$

where

$$\mu = \frac{G}{(2m)^{1/2}} \sum_{1 \leq i < j \leq n} (m_i m_j)^{3/2}.$$

Now let us derive the third inequality following from the Cauchy-Bunjakowski inequality, which is

$$\left(\sum_{i=1}^{n} a_i b_i\right)^2 \leq \left(\sum_{i=1}^{n} a_i^2\right)\left(\sum_{i=1}^{n} b_i^2\right).$$

Since

$$r_i^2 = |r_i|^2 \quad \text{and} \quad |(r_i \cdot \dot{r}_i)| = (|r_i| \cdot |\dot{r}_i|)$$

and from the definition of the Jacobi function, one obtains

$$\Phi = \sum_{i=1}^{n} m_i(|r_i| \cdot |\dot{r}_i|).$$

Applying the Cauchy-Bunjakowski inequality to this expression at

$$a_i = m_i^{1/2}|r| \quad \text{and} \quad b_i = m_i^{1/2}|r|$$

we can write

$$\dot{\Phi}^2 \leq 2\Phi \sum_{i=1}^{n} m_i|\dot{r}_i{}^2| = 2\Phi \sum_{i=1}^{n} \frac{m_i(r_i \cdot \dot{r}_i)^2}{r_i^2}.$$

Assuming

$$a_i = m_i^{1/2}|r_i|, \quad A_i = \frac{m_i^{1/2}[r_i X \dot{r}_i]}{|r_i|},$$

the vector of the angular momentum M is

$$M = \sum_{i=1}^{n} a_i A_i.$$

Then in a similar may we write

$$M^2 \leq \left(\sum_{i=1}^{n} a_i^2\right)\left(\sum_{i=1}^{n} A_i^2\right) \equiv 2\Phi \sum_{i=1}^{n} \frac{m_i[r_i X \dot{r}_i]}{r_i^2}.$$

The addition of the last two inequalities yields

$$\dot{\Phi}^2 + M^2 \leq 2\Phi \sum_{i=1}^{n} \frac{m_i\left\{(r_i \cdot \dot{r}_i)^2 + [r_i X \dot{r}_i]^2\right\}}{r_i^2}.$$

But since

$$\left\{(r_i \cdot \dot{r}_i)^2 + [r_i X \dot{r}_i]^2\right\} = r_i^2 \cdot \dot{r}_i^2,$$

we have

$$\dot{\Phi}^2 + M^2 \leq 2\Phi \sum_{i=1}^{n} m_i \dot{r}_i.$$

As Jacobi's equation can be written in the form

$$\ddot{\Phi} + k\dot{\Phi} - E_0[1 + q(t)] = \frac{1}{2} \sum_{i=1}^{n} m_i \dot{r}_i^2,$$

then after substitution of this into the right-hand side of the last inequality, we obtain

$$\dot{\Phi}^2 + M^2 \leq 4\Phi\left\{\ddot{\Phi} + k\dot{\Phi} - E_0[1 + q(t)]\right\}.$$

Hence the third inequality can be written

$$\ddot{\Phi} + k\dot{\Phi} - E_0[1 + q(t)] - \frac{\dot{\Phi}^2}{4\Phi} \geq \frac{M^2}{4\Phi}. \tag{6.27}$$

Let us now analyze the behavior of the Jacobi function Φ and its derivatives. For this purpose we introduce the auxiliary function $Q = Q(t)$, equal to

$$Q = k\dot{\Phi}\,\Phi^{1/2} - E_0[1 + q(t)]\Phi^{1/2} + \frac{1/4\dot{\Phi}^2 + 1/4M}{\Phi^{1/2}}, \tag{6.28}$$

where $\Phi^{1/2} > 0$.

Then differentiating (6.28) and using

$$\frac{d}{dt}\left(\Phi^{1/2}\right) = \frac{\dot{\Phi}}{2\Phi^{1/2}},$$

we obtain

$$\dot{Q} = \frac{1}{2}\frac{\dot{\Phi}}{\Phi^{1/2}}\left\{\ddot{\Phi} + k\dot{\Phi} - E_0[1 + q(t)] - \frac{1}{4}\frac{M^2}{\Phi} - \frac{1}{4}\frac{\dot{\Phi}^2}{4\Phi}\right\} + \Phi^{1/2}\left[k\ddot{\Phi} - E_0\dot{q}(t)\right]$$

where $\dot{q}(t)>0$ and, in agreement with (6.27),

$$\left\{\ddot{\Phi} + k\dot{\Phi} - E_0[1 + q(t)] - \frac{1}{4}\frac{M^2}{\Phi} - \frac{1}{4}\frac{\dot{\Phi}^2}{\Phi}\right\} \geq 0.$$

Let t_0 be the time of simultaneous collision of all the particles of the system. Then for $t \to t_o$ $(t \to t_o)$, $\Phi \to 0$. Let us show that the necessary condition for existence of such t_0 for which $\Phi \to 0$ (if $t \to t_o$) is that the constant angular momentum M must be zero.

Note that if, for $t \to t_o$, $\Phi \to 0$, then all mutual $\Delta_{ij} = \left|r_i - r_j\right|$ also tend to zero, and the potential energy $U \to -\infty$.

Since

$$\ddot{\Phi} = 2E_0[1 + q(t)] - U - k\dot{\Phi},$$

where $E_0 = \text{const}$, $|\dot{\Phi}| \to \infty$, $|q(t)|$, $|\dot{q}(t)| < \infty$, then, for $t \to t_o$, $\ddot{\Phi} \to \infty$. Thus for t sufficiently close to t_0 we have $\ddot{\Phi} > 0$ and therefore the derivative $\dot{\Phi}$ increases and does not change its sign. Since $\Phi > 0$ and $\Phi \to 0$, then Φ is monotonically decreasing function. It therefore follows from the expression for $\dot{Q}$ that the function Q in (6.28) for t sufficiently close to t_0 must decrease and its time limit for $t \to t_o$ might be $-\infty$, but cannot be $+\infty$. Moreover, it follows from the above statement that for $t \to t_0$ the limit of function (6.28) is

$$\lim_{t\to t_0} Q = \lim_{t\to t_0} \frac{1}{4}\frac{\dot{\Phi}^2 + M^2}{\Phi^{1/2}}, \tag{6.29}$$

but since $\Phi^{1/2} > 0$, the time limit (6.29) must be finite and non-negative. Hence for $t \to t_o$ and $\Phi \to 0$ the value $M^2/\Phi^{1/2}$ must remain limited. Therefore, since $M^2 = \text{const}$, then $M \equiv 0$ and proof is completed.

The above analysis shows that, at $t \to t_o$, $\ddot{\Phi} \to \infty$, and it therefore follows from (6.25) that

$$|\dddot{\Phi} = 2E_0\dot{q}(t) + k\ddot{\Phi}| \leq \text{const}\left(|\ddot{\Phi}| + k|\dot{\Phi}|\right)^{5/2}. \tag{6.30}$$

Using the second inequality (6.26), it can be shown that if t_o is the time moment of simultaneous collision of all the particles of the system, then as $\Phi^{1/2} > 0$ at $t \to t_o$, the ratio $\dot{\Phi}/\Phi^{1/2}$ tends to a finite and positive limit.

In fact, as has been shown above, the limit (6.29) of the function (6.28) for $t \to t_o$ has a finite value. Since $M = 0$

$$\lim_{t - t_0} \frac{\dot{\Phi}^2}{\Phi^{1/2}}$$

will also be finite and non-negative. Let us show that this limit cannot be equal to zero.

Since for $t \to t_o$, $M = 0$, $\Phi^{1/2} \to 0$, then the function (6.28) and its limit (6.29) may be written in the form

$$Q = k\dot{\Phi}\Phi^{1/2} - E_0[1 - q(t)]\Phi^{1/2} + \frac{1}{4}\frac{\dot{\Phi}^2}{\Phi^{1/2}}, \tag{6.31}$$

$$\mu_0 = \frac{1}{4} \lim_{t - t_0} \frac{\dot{\Phi}^2}{\Phi^{1/2}}, \tag{6.32}$$

where

$$\mu_0 = \lim_{t - t_0} Q.$$

From (6.31) we find that

$$2Q\Phi^{1/2} = k\dot{\Phi}\Phi - 2E_0[1 - q(t)]\Phi + \frac{1}{2}\dot{\Phi}^2.$$

Hence

$$\frac{d}{dt}\left(2Q\Phi^{1/2}\right) = \ddot{\Phi}\dot{\Phi} + k\ddot{\Phi}\Phi + 2k\dot{\Phi}^2 - 2E_0[1 - q(t)]\dot{\Phi} - 2E_0\Phi\dot{q}.$$

Let us catty out the integration between the limit t_o and $\bar{t}$ of the last relation where t_o has a fixed value and $\bar{t} \to t_o$. We take into account that

$$\lim_{t - t_0} \Phi^{1/2} = 0,$$

$$\mu_0 = \frac{1}{4} \lim_{t - t_0} \frac{\dot{\Phi}^2}{\Phi^{1/2}} < \infty.$$

Then we write

$$2Q\Phi^{1/2}=\int_{t_0}^{\bar{t}}\{[\ddot{\Phi}-2E_0(1-q(t))+2k\dot{\Phi}]\dot{\Phi}+[2k\ddot{\Phi}-2E_0\dot{q}]\Phi\}dt.$$

As shown above, the derivative Φ retains its sign in the sufficiently small neighborhood of point t_o. Since $\Phi \geq 0$ and $q > 0$, the positive constant μ in the inequality (6.26) will be such that in the sufficiently small neighborhood of t_o we have

$$2|Q|\Phi^{1/2} \geq \int_{t_0}^{\bar{t}}\left\{\frac{\mu}{\Phi^{1/2}}\dot{\Phi}+[2k\ddot{\Phi}-2E_0\dot{q}]\Phi\right\}dt.$$

The first integral to the right of this inequality being equal to $2\mu\Phi^{1/2}$, and $\Phi^{1/2} \to 0$ with $t \to t_o$, then, in the sufficient small neighborhood of t_o, we have

$$2|Q|\Phi^{1/2} \geq 2\mu\Phi^{1/2} \quad \text{or} \quad |Q| \geq \mu.$$

Since $\mu > 0$, and taking into account the existence of the time limit (6.32), we have finished the proof of correctness of the inequality

$$\lim_{t\to t_0}\left(\frac{\dot{\Phi}}{\Phi^{1/2}}\right) > 0.$$

The above analysis allows us to obtain the following asymptotic relations for the Jacobi function when $t \to t_o$.

Since the limit

$$\mu_0 = \frac{1}{4}\lim_{t\to t_0}\frac{\dot{\Phi}^2}{\Phi^{1/2}}$$

has a non-zero value, the function $\Phi = \Phi(t) > 0$ tends to zero as $t \to t_o$ in such a way that, in neighborhood of t_o, it is proportional to $(t - t_o.)^{4/3}$ with a coefficient of proportionality of $((9/4)\mu_o)^{2/3}$, and one can differentiate this asymptotic relation with respect to t. Hence the following asymptotic relations are satisfied:

$$\Phi \propto \left(3/2\mu_o^{1/2}\right)^{4/3}(t-t_0)^{4/3}, \tag{6.33}$$

$$\Phi \propto \left(12\mu_o^2\right)^{1/3}(t-t_0)^{1/3}. \tag{6.34}$$

In fact (6.34) follows from (6.33) not only from groundless differentiation, but actually from (6.33), if (6.32) is taken into account. The asymptotic relation (6.33) itself follows from (6.32), if we write the last relation in the form

$$\pm \frac{dt}{d\Phi} \propto \frac{1}{2}\,\mu_0^{-1/2}\Phi^{-1/4}$$

and then integrate it between the limits $\Phi = 0$ and $\Phi > 0$, but sufficiently close to $\Phi = 0$. Integration (but not differentiation) of such an asymptotic relation is always as allowed procedure and hence the asymptotic relations (6.33) and (6.34) are satisfied.

Let us show that besides (6.32), (6.33) and (6.34), the following asymptotic relations are also available:

$$\mu_0 = \pm \lim_{t \to t_0} \Phi^{1/2}\ddot{\Phi}, \tag{6.35}$$

$$\Phi \propto \left(2/3\mu_o^{1/2}\right)^{2/3}(t - t_0)^{-2/3}. \tag{6.36}$$

To prove relation (6.35), we multiply (6.27) by $\Phi^{1/2}$. Assuming for $t \to t_o$ and $\Phi^{1/2} \to 0$, $|E_0|(1 + q(t)) < \infty$, $M \equiv 0$ and using (6.32), we find that the lower limit $\underline{\lim}\, \Phi^{1/2}\ddot{\Phi} \geq \mu_0$. Since (6.35) is equivalent to (6.36), this asymptotic relation will be proved, if the upper limit $\overline{\lim}\, \Phi^{1/2}\ddot{\Phi} \leq \mu_0$.

For the proof we assume $F = (\dot{\Phi})^3$, so that

$$\ddot{F} = 6\dot{\Phi}\ddot{\Phi}^2 + 3\dot{\Phi}^2\dddot{\Phi}.$$

Then, with the aid of (6.30)

$$\left|\dddot{\Phi} - 2E_0\dot{q} + k\ddot{\Phi}\right| \leq \text{const}\left(|\ddot{\Phi}| + k|\dot{\Phi}|\right)^{5/2}$$

and expressing $\dot{\Phi}$ and $\ddot{\Phi}$ through the function $\dot{F} = \Phi^3$ and $\dot{F} = 3\dot{\Phi}^2\ddot{\Phi}$, we find

$$\left|\ddot{F} + 6\dot{q}(t)F^{2/3}\right| < \text{const}\frac{\dot{F}^2 + \left(|\dot{F}|\right)^{5/2}}{|F|}.$$

On the right-hand side of this inequality, we find from (6.34) where $\dot{\Phi} = F^{1/3}$ that for $t \to t_o$

$$\left|\ddot{F} + 6\dot{q}(t)F^{2/3}\right| < \text{const}\frac{\dot{F}^2 + \left(|\dot{F}|\right)^{5/2}}{t - t_0}. \tag{6.37}$$

Finally, if ν_o is a positive constant equal to $(12\mu_0)^2$, then for $t \to t_o$

$$F \propto \nu_0(t \to t_0), \tag{6.38}$$

$$\underline{\lim}\dot{F} \geq \nu_0. \tag{6.39}$$

In fact, $F = \dot{\Phi}^3$ then (6.38) is equivalent to (6.34). At the same time, by virtue of the relation $\nu_o = (12\mu_0)^2$, $F = \dot{\Phi}^3$, $\dot{F} = 3\dot{\Phi}^2\ddot{\Phi}$ and (6.32), the inequality (6.39) is another form of the inequality $\underline{\lim}\ \Phi^{1/2}\ddot{\Phi} \geq \mu_0$ which we have already proved. Therefore, we are bound to prove the inequality which can be written in the form $\overline{\lim}\,\dot{\Phi} \leq \mu_0$ by analogy with (6.39). Hence we must prove that the asymptotic relations (6.38) and (6.39) with the aid of the 'Trauberian condition' (6.37), yields the inequality $\lim \dot{F} \leq \nu_o$ which denotes that $F \to \nu_o$. From this inequality and from (6.39) the existence of the succession of time intervals follows:

$$t_1^I < t < t_1^{II}, \ldots, t_k^I < t < {}_k^{II}$$

which tends to t_0 as $k \to \infty$ in such a way that whenever $t_k^I < t < {}_k^{II}$

$$0 < \upsilon_0 < p = \dot{F}\left(t_k^I\right) < \dot{F}(t) < \dot{F}\left(t_k^{II}\right) < q \tag{6.40}$$

where p and q are some fixed numbers which are chosen between the limits $\underline{\lim}\dot{F}$, $\overline{\lim}\,\dot{F}$ $(\leq \infty)$ of the conditions function $\lim \dot{F}(t)$. It is obvious that we can assume that $t_o = 0$. If we accept const $=$ const $(p^2 + p^{5/2})$, then for any t in any of the time intervals $t_k^I < t < {}_k^{II}$, by virtue of (6.37) and (6.40), we find that the following inequality holds:

$$\left|\ddot{F}(t) + 6\dot{q}(t)F^{2/3}(t)\right| < \frac{\text{const}}{|t|}.$$

Since t tends to $t_o = 0$, increasing or decreasing, all t_k^I and t_k^{II} lie on the same side of $t_o = 0$. Integration of the inequality (6.40) between the limits t_k^I and t_k^{II} yields

$$\left|\dot{F}\left(t_k^{II}\right) - \dot{F}\left(t_k^I\right) + \int_{t_k^I}^{t_k^{II}} 6\dot{q}(t)F^{2/3}(t)dt\right| < \text{const}\ \log\left|\frac{t_k^{II}}{t_k^I}\right|.$$

By virtue of (6.40) the difference $\dot{F}\left(t_k^{II}\right) - \dot{F}\left(t_k^I\right)$ is equal to a positive constant $p - q$ and

$$\int_{t_k^I}^{t_k^{II}} 6\dot{q}(t)F^{2/3}(t)dt > 0.$$

Hence the limit $\log |t_k^{II}/t_k^{I}|$, as $k \to \infty$, is greater than a certain positive number. For this reason, when $k \to \infty$, there exists a certain positive number λ which satisfies the relation

$$\frac{t_k^{II}}{t_k^{I}} > \lambda > 0. \tag{6.41}$$

Then with the aid of (6.38) it follows that

$$\frac{|F(t_k^{I})|}{|t_k^{I}|} \to v_0 \quad \text{and} \quad \frac{|F(t_k^{II})|}{|t_k^{II}|} \to v_0$$

since

$$t_k^{I} \to t_0, \quad t_k^{II} \to t_0, \quad t_0 = 0, \quad v = 0.$$

On the other hand, if k is sufficiently large, the following inequality is valid:

$$\frac{|F(t_k^{II})|}{|t_k^{II}|}\left|\frac{|t_k^{II}|}{t_k^{I}}\right| - \frac{|F(t_k^{I})|}{|t_k^{II}|}\left|\frac{|t_k^{II}|}{F|t_k^{II}|}\right| > p\left|\frac{|t_k^{II}|}{|t_k^{I}|} - 1\right|. \tag{6.42}$$

In fact, all t_k^{I} and t_k^{II} lie on the same side of t_o, and then

$$\left||t_k^{II}| - |t_k^{I}|\right| = t_k^{II} - t_k^{I}.$$

Since $t_k^{I} \to t_0$ and $t_k^{II} \to t_0$, then for sufficiently large k all $F(t_k^{I})$, $F(t_k^{II})$ have the same sign. Hence, (6.42) can be written in the form

$$\left||F(t_k^{II})| - |F(t_k^{I})|\right| > p\left||t_k^{II}| - |t_k^{I}|\right|$$

and is equivalent to the inequality

$$\left|F(t_k^{II}) - F(t_k^{I})\right| > p\left|t_k^{II} - t_k^{I}\right|.$$

The validity of the last inequality is obvious, since by virtue of (6.40) for $t_k^{I} < t < {}_k^{II}$ we have $\dot{F}(t) > p > o$. Therefore, inequality (6.42) also holds.

From (6.42) in the limit $k \to \infty$ and with the aid of (6.41) where $v_o > 0$, we obtain the following inequality:

$$v_0|\lambda - v_0 v_0^{-1}| = p|\lambda - 1|.$$

Finally, by virtue of (6.41)

$$|\lambda - v_0 v_0^{-1}| = |\lambda - 1| > 0,$$

and hence $\nu_o \geq p$. On the other hand, by virtue of (6.40) $p \geq \nu_o$. The observed contradiction that the supposition we made at the beginning ($\overline{\lim}\,\dot{F} >> \nu_0$) is false. Thus we have proved the validity of the increase inequality $\overline{\lim}\,\dot{F} \leq \nu_0$ and this completes the proof of the relations (6.35) and (6.36).

Let us now show that if the motion of n points with masses m_i in the time limit $t \to t_o$ produces their simultaneous collision, then the configuration of these n particles tends to central configuration (6.15) as $t \to t_o$. In the proof, we shell use the asymptotic relations (6.33), (6.34) and (6.36) and the Tauberian lemma, which states that if the function g(u) has continuous derivatives $\dot{g}(u)$ and $\ddot{g}(u)$ for $u \to \infty$ and tends, as $u \to \infty$, to a finite limit and $\ddot{g}(u)$ <const, then $\dot{g}(u) \to 0$.

There is no loss of generality in assuming that $t \to t_o \to 0$, so that $t \to t_o > 0$. Then the asymptotic relations (6.33), (6.34) and (6.36) are simply equivalent to

$$t^{-4/3}\Phi \to \mu_1 > 0, \tag{6.43}$$

$$t\left(t^{-4/\dot{3}}\Phi\right) \to 0, \tag{6.44}$$

$$t\left(t^{-4/\ddot{3}}\Phi\right) \to 0 \tag{6.45}$$

where

$$\mu_1 = \left(\frac{3}{2}\,\mu_0^{1/2}\right)^{4/3} \quad \text{and} \quad t \to 0.$$

Since

$$\Phi = \frac{1}{2}\sum_{i=1}^{n} m_i r_i^2,$$

it follows from (6.43) that the time limit $t \to 0$ all n mass particles collide at the origin of the barycentric co-ordinate system OXYZ in such a way that, for sufficiently small t, the linear dimensions of the configuration will be proportional to $t^{2/3}$. For this reason we eliminate this factor $t^{2/3}$ simply by multiplying the unit of length by the factor $t^{-2/3}$. Then we consider, instead of the values:

$$r_i, \Delta_{ij} = \left|r_i - r_j\right|,$$

$$\Phi = \frac{1}{2}\sum_{i=1}^{n} m_i r_i^2, \tag{6.46}$$

$$U = -G \sum_{1 \leq i < j \leq n} \frac{m_i m_j}{\Delta_{ij}}, \tag{6.47}$$

the corresponding values:

$$\bar{r} = t^{-2/32} r_i, \quad \bar{\Delta}_{ij} = \left|\bar{r}_i - \bar{r}_j\right| = t^{-2/3} \Delta_{ij},$$

$$\bar{\Phi} = t^{-4/3} \Phi = \frac{1}{2} \sum_{i=1}^{n} m_i \bar{r}_i^2,$$

$$U = t^{2/3} U = -G \sum_{1 \le i < j \le n} \frac{m_i m_j}{\Delta_{ij}}.$$

The procedure is permissible since the definition of the central configuration is an invariant relative scale transformation of all the co-ordinates $r_i \to \delta r_i$ where δ is an arbitrary non-zero factor. Then, the relation (6.15) is invalid for the fixed $t \neq 0$, but

$$\left(\bar{\Phi}\bar{U}^2\right)_{\bar{r}_i} = 0 \tag{6.48}$$

where $I = 1, 2, \ldots, n$ in the same limit $t \to 0$.

The proof of this theorem, the mathematically precise formulation of which is expressed by (6.48), has several stages.

First, we show that in the time limit $t \to 0$

$$\frac{4}{9}\bar{\Phi} + U \to 0 \tag{6.49}$$

and

$$\bar{\Delta}_{ij} > \mathrm{const} > 0. \tag{6.50}$$

Let us introduce a time transformation, changing t to $\bar{t} = -\ln t$ in such a way to have $\bar{t} \to \infty$ for $t \to 0$. Let this transformation be

$$t = e^{-\bar{t}}. \tag{6.51}$$

Then, if the arbitrary function f depends on time t, we have

$$t\frac{df}{dt} = -\frac{df}{dt}, \tag{6.52}$$

$$t^2\frac{d^2f}{dt^2} = \frac{d^2f}{dt^2} + \frac{df}{dt}. \tag{6.53}$$

With the aid of (6.51)–(6.53) we rewrite the equation of motion

$$m_i \ddot{r}_i = -U_\eta$$

in the form

$$m_i\left(\ddot{\bar{r}}_i - \frac{1}{3}\dot{\bar{r}}_i - \frac{2}{9}\bar{r}_i\right) = -\overline{U}_{r_i} - k\dot{\bar{r}}_i \tag{6.54}$$

where derivatives are written with respect to $\bar{t}$ and $\overline{U}_{\bar{r}_i} = t^{4/3}U_{\bar{r}_i}$.

Similarly, let us rewrite the energy conservation law and Jacobi's virial equation in the form

$$\frac{1}{2}\sum_{i=1}^{n} m_i\left(\dot{\bar{r}}_i - \frac{2}{3}\bar{r}_i\right)^2 + \overline{U} = E_0[1 + q(t)]e^{-2/3\bar{t}}, \tag{6.55}$$

$$\ddot{\overline{\Phi}} - \frac{5}{3}\dot{\overline{\Phi}} - \frac{4}{9}\overline{\Phi} = -U + 2E_0[1 + q(t)]e^{-2/3\bar{t}}. \tag{6.56}$$

Assuming $f = \Phi$ in (6.52) and (6.53), we obtain relations which are valid in the time limit $\bar{t} \to \infty$ and similar to (6.43)–(6.45) as $t \to 0$:

$$\overline{\Phi} \to \mu_1 > 0, \tag{6.57}$$

$$\dot{\overline{\Phi}} \to 0, \tag{6.58}$$

$$\ddot{\overline{\Phi}} \to 0. \tag{6.59}$$

In the limit $\bar{t} \to \infty$ from (6.56), where $E_0(1 + q(t))$ is finite, with the aid of (6.57)–(6.59), it follows that (6.49) is valid. Moreover, it is obvious from (6.49) to (6.57) that the potential energy U tends to a finite value and hence (6.50) follows from (6.47).

Secondly, let us show that the time limit $\bar{t} \to +\infty$ $(t \to 0)$:

$$\dot{\bar{r}} \to 0, \tag{6.60}$$

$$\ddot{\bar{r}} < \text{const}, \tag{6.61}$$

$$\dddot{\bar{r}} < \text{const}. \tag{6.62}$$

Note that (6.46) yields

$$\dot{\overline{\Phi}} = \sum_{i=1}^{n} m_i \dot{\bar{r}}_i \bar{r}_i. \tag{6.63}$$

Then in the time limit $\bar{t} \to \infty$ and with the aid of (6.49) and (6.63), we obtain

$$\sum_{i=1}^{n} m_i \dot{\bar{r}}_i^2 \to 0$$

which gives (6.60). Furthermore,

$$\bar{r} < \text{const}, \tag{6.64}$$

$$\left|\overline{U}_{r_i}\right| < \text{const}. \tag{6.65}$$

In fact Eq. 6.64 follows from (6.57) by virtue of Eq. 6.46. At the same time, Eq. 6.65 follows from (6.47) to (6.50). Equation 6.56 follows from (6.54), (6.60), (6.64) and (6.65). Finally, by differentiating (6.56) with respect to $\bar{t}$ and then using (6.60) and (6.61), it is easy to see that for the proof of (6.62) it is sufficient to show the boundedness of the second derivatives of the functions $\overline{U}(\bar{r}_1, r_2, \ldots, r_n)$ in the time limit $t \to \infty$. But the boundedness of these derivatives follows obviously from (6.47), (6.50) and (6.64).

Finally, in accordance with (6.60) and (6.62), the Tauberian lemma is valid if we consider the function $g(u) = \dot{\bar{r}}_i$, where $u = \bar{t}$. Hence, not only $\dot{\bar{r}}_i \to 0$, but $\ddot{\bar{r}}_i \to 0$.

It follows therefore from (6.54) that

$$\frac{2}{9} m_i \bar{r}_i - U_{\bar{r}_i} \to 0.$$

Then by virtue of (6.46)

$$\frac{2}{9} \overline{\Phi}_{\bar{\eta}} - U_{\bar{r}_i} \to 0.$$

From the last expression, with the aid of (6.49) and (6.57), it follows that

$$\left(\overline{\Phi}\,\overline{U}^2\right)_{\bar{r}_i} = \overline{\Phi}_{\bar{r}_i} \overline{U}^2_{\bar{r}_i} + 2\bar{\Phi}\overline{U}\overline{U}_{\bar{r}_i} \to 0$$

and therefore

$$\left(\Phi U^2\right)_\eta \to 0$$

as $t \to 0$.

The last expression completes the proof of the theorem that an arbitrary non-conservative system tends to central configuration in the asymptotic limit of simultaneous collision of all its particles.

6.3 Asymptotic Limit of Simultaneous Collision of Charged Particles of a System

The following analysis is given for a system consisting of a large number of charged material particles. The particles considered are positively charged nuclei of atoms and electrons.

The objective is to prove the statement that the arbitrary configuration of a system of charged particles interacting according to an inverse law (i.e. gravitational or Coulomb) in the asymptotic time limit of simultaneous collision of all the particles (for $t \to t_o$) tends to a central configuration.

Using the definition of central configuration (6.15), (Wintner 1941), and assuming its uniqueness, the statement to be proved can be written in the form

$$\lim_{t \to t_0} \left(|U_\Sigma| \sqrt{\Phi} \right) = \text{const} \tag{6.66}$$

where $U_\Sigma = U + U_c$ is the potential energy of the system, which is equal to the sum of the gravitational potential energy of Coulomb interactions.

Using Wintner's method (Wintner 1941), we have previously studied the asymptotic time limit of (6.66) for conservative and non-conservative systems whose particles are interacting according to the law of gravitation. Since the relationship (6.66) is linear as a function of potential energy, we have to prove it only for Coulomb interactions of system particles. The proof given below for a non-conservative system is also based on Wintner's method, modified for the case of charged particles.

So, for a non-conservative system of n particles interacting according to the Coulomb law, let us write down in an inertial barycentric co-ordinate system the Jacobi function, functions of the potential and kinetic energies as well as the energy conservation law and Jabobi's virial equation as follows:

$$\Phi = \frac{1}{2m} \sum_{1 \le i < j \le n} m_i m_j \Delta_{ij}^2, \tag{6.67}$$

$$T = \frac{1}{2m} \sum_{1 \le i < j \le n} m_i m_j (\dot{r}_i - \dot{r}_j)^2, \tag{6.68}$$

$$U = -G \sum_{1 \le i < j \le n} \frac{q_i q_j}{\Delta_{ij}}, \tag{6.69}$$

$$E = E(t) = E_0 - E_\gamma = T + U_c, \tag{6.70}$$

$$\ddot{\Phi} = 2E(t) - E_c \tag{6.71}$$

where $q_i = eZ_i$ is the charge of i-th particle with mass m_i; $Z_i = -1, 1, +2, \ldots, N \le n$; m is the total mass of the system; $E_\gamma < \infty$; $\dot{E}\gamma < \infty$, i.e. the total energy and the luminosity of the system at any time t are functions, monotonically bounded from above.

The proof of the relationship (6.66) can easily be obtained from the asymptotic expressions for Jacobi function and its first and second derivatives as

$$\Phi \propto (t - t_0)^{4/3}, \tag{6.72}$$

$$\dot{\Phi} \propto (t - t_0)^{1/3}, \tag{6.73}$$

$$\ddot{\Phi} \propto (t - t_0)^{-2/3}. \tag{6.74}$$

where $t \to t_o$, and t_o is the moment of simultaneous collision of the charged particles of the system.

From the expressions (6.72)–(6.74), the limit (6.66), which we are proving, follows from exact repetition of Wintner's arguments (Wintner 1941). However, Eqs. 6.72–6.74 follows from the existence of the limits

$$\lim_{t - t_0} \frac{\dot{\Phi}^2}{\Phi^{1/2}} = \mu_0 = \text{const} > 0, \tag{6.75}$$

$$\lim_{t \to t_0} \ddot{\Phi}\Phi^{1/2} = \eta_0 = \text{const} > 0. \tag{6.76}$$

The limits (6.75) and (6.76) may be obtained in future from analysis of the Jacobi function in the neighborhood of t_o, using the auxiliary function

$$Q = -\left(E - E_\gamma\right)\Phi^{1/2} + \frac{1}{4}\frac{\dot{\Phi}^2 + M^2}{\Phi^{1/2}}$$

and the three inequalities, correct in the most general case, i.e. not especially in the close neighborhood of the point of simultaneous collision of particles. These inequalities are

$$\left|\dddot{\Phi} + 2E_\gamma\right| \leq \left(\left|\ddot{\Phi}\right| + 2\left|E - E_\gamma\right|\right)^{5/2}\eta_0, \tag{6.77}$$

$$\left[\ddot{\Phi} - 2\left(E - E_\gamma\right)\Phi^{1/2}\right] \geq \eta_0 > 0, \tag{6.78}$$

$$\ddot{\Phi} - E + E\gamma - \frac{\dot{\Phi}^2}{4\Phi} \geq \frac{M^2}{4\Phi} \tag{6.79}$$

where M is the angular moment of the system.

Let us prove inequalities (6.77)–(6.79) for a system of particles interacting according to Coulomb law.

To prove the inequality (6.77), it is essential that the absolute value of the total potential energy of the system of particles is less than the absolute value of the energy of mutual interactions of any pair of charged particles, i.e.

$$\frac{q_i q_j}{\Delta_{ij}} \leq |U_c|. \tag{6.80}$$

Since

$$|\dot{U}_c| = \left| \sum_{1\le i<j\le n} \frac{q_i q_j}{\Delta_{ij}^2} \frac{d}{dt}\Delta_{ij} \right| \le \sum_{1\le i<j\le n} \frac{|q_i q_j|}{\Delta_{ij}^2} |\dot{r}_i - \dot{r}_j|,$$

and

$$\frac{1}{\Delta_{ij}} \le \frac{|U_c|}{|q_i q_j|^2},$$

then

$$|\dot{U}_c| \le |U_c|^2 \sum_{1\le i<j\le n} \frac{|\dot{r}_i - \dot{r}_j|}{|q_i q_j|}.$$

Analogously, since

$$m_i m_j |\dot{r}_i - \dot{r}_j|^2 \le 2mT$$

and

$$m_i \ge \frac{|q_i|}{e} \mu_e,$$

then

$$2mT \ge \frac{|q_i q_j|}{e^2} \mu_e^2 |\dot{r}_i - \dot{r}_j|^2$$

and therefore

$$|\dot{U}_c| \le |\dot{U}_c|^2 T^{1/2} \frac{(2m)^{1/2}}{\mu_e} \sum_{1\le i<j\le n} \frac{1}{|q_i q_j|^{3/2}}$$

where μ_e is the electron mass.

From Jacobi's equation and the law of conservation of energy it follows that

$$|\dot{U}_c| = |\dddot{\Phi} + 2\dot{E}_\gamma|,$$

$$|U_c| \le (|\ddot{\Phi}| + 2|E - E_\gamma|),$$

$$|T| \le (|\ddot{\Phi}| + 2|E - E_\gamma|),$$

and finally we obtain the first inequality:

$$\left|\dddot{\Phi} + 2\dot{E}_\gamma\right| \leq \left(\left|\ddot{\Phi}\right| + 2\left|E - E_\gamma\right|\right)^{5/2} \eta_0,$$

$$\eta_0 = \frac{(2m)^{1/2}}{\mu_c} e \sum_{1 \leq i < j \leq n} \frac{1}{\left(q_i q_j\right)^{3/2}} > 0.$$

The second inequality (6.78) may be derived from Jacobi's equation:

$$\ddot{\Phi} - 2\left(E - E_\gamma\right) = -U_c = - \sum_{1 \leq i < j \leq n} \frac{q_i q_j}{\Delta_{ij}} = |U_c| \geq \frac{\left|q_i q_j\right|}{\Delta_{ij}}$$

and the inequality following from the definition of the Jacobi function:

$$2m\Phi \geq m_i m_j \Delta_{ij},$$

$$\frac{1}{\Delta_{ij}} \geq \frac{\left(m_i m_j\right)^{1/2}}{(2m)^{1/2} \Phi^{1/2}}.$$

Thus, finally we have

$$\left[\ddot{\Phi} - 2\left(E - E_\gamma\right)\right]\Phi^{1/2} \geq \frac{\left|q_i q_j\right|\left(m_i m_j\right)^{1/2}}{(2m)^{1/2}} = \mu_0 > 0.$$

The derivation of the third inequality (6.79) is based on the Cauchy-Bunjakowski inequality:

$$\left(\sum_{1 \leq i \leq n}^{n} a_i b_i\right)^2 \leq \left(\sum_{1 \leq i \leq n}^{n} a_i^2\right)\left(\sum_{1 \leq i \leq n}^{n} b_i^2\right).$$

Substituting into it

$$a_i = m_i^{1/2} |r_i|, \quad b_i = m_i^{1/2} \frac{d}{dt} |r_i|$$

we have

$$\dot{\Phi} = \sum_{1 \leq i \leq n} m_i |r_i| \frac{d}{dt} |r_i|,$$

$$(\dot{\Phi})^2 \le 2\Phi \sum_{1\le i\le n} m_i \frac{\left(r_i \frac{d}{dt} r_i\right)^2}{|r_i|^2}.$$

Substituting as before

$$a_i = m_i^{1/2}|r_i|, \quad b_i = m_i^{1/2}\frac{[r_i\dot{r}_i]}{|r_i|},$$

we obtain

$$M^2 \le 2\Phi \sum_{1\le i\le n} \frac{m_i|r_i\dot{r}_i|^2}{|r_i|^2}$$

where M is the angular momentum of the system equal to

$$M = \sum_{1\le i\le n} m_i[r_i\dot{r}_i].$$

Summing up the two inequalities just obtained, we have

$$\begin{aligned}(\dot{\Phi})^2 + M^2 &\le 2\Phi \sum_{1\le i\le n} \frac{m_i}{|r_i|^2}\left\{(r_i\dot{r}_i)^2 + [r_i\dot{r}_i]^2\right\} = 2\Phi \sum_{1\le i\le n} m_i(\dot{r}_i)^2 = 4T\Phi \\ &= 4\Phi\left[\ddot{\Phi} - (E - E_\gamma)\right].\end{aligned}$$

We finally obtain an expression for the third inequality (6.79):

$$\ddot{\Phi} - E + E_\gamma - \frac{\dot{\Phi}^2}{4\Phi} \ge \frac{M^2}{4\Phi}.$$

This ends the proof of the expression (6.66) for the Coulomb interactions of charged particles of the system in the asymptotic time limit of their simultaneous collision.

6.4 Relationship Between Jacobi Function and Potential Energy for a System with High Symmetry

If the value

$$|U|\sqrt{\Phi} = B \tag{6.81}$$

does not change for different mass density distribution laws and configurations of the system, the problem of its dynamics would be solved in the framework of an integral (volumetric) approach and a dynamic equilibrium model. In this case, using Jacobi's virial equation,

$$\ddot{\Phi} = 2E - U \tag{6.82}$$

we transfer it into the equation of virial oscillations:

$$\ddot{\Phi} = -A + \frac{B}{\sqrt{\Phi}} \tag{6.83}$$

and obtain three different forms of solutions: elliptic, parabolic or hyperbolic. The form of the solution depends on the total energy which can respectively be negative, equal to zero or positive. The problem is solved with respect to the changes in the time of the simplest moment of distribution of mass density $\rho(\bar{x})$ of the system.

However, even for a system with spherical symmetry and fixed mass, the value of (6.81) changes for different laws of distribution of the mass density $\rho(r)$ (where r is the radius of the shell with density $\rho(r)$; $r \in [0,R]$). In this connection, transformation of Eq. 6.82 into 6.83 is possible only after special study, which is described below.

We pay special attention to the systems with high symmetry, namely spherical and elliptical. This is because most of the natural systems from galaxies to atoms possess such a symmetry. We consider below the conditions which allow us to transform Eq. 6.82 into 6.83 for systems with spherical and elliptical symmetry.

6.4.1 Systems with Spherical Symmetry

Let us begin by consideration the value of Eq. 6.81 for a spherical system. It is convenient to start such a study after rewriting the expressions for the Jacobi function and the potential energy in the form

$$\Phi = \frac{1}{2}\beta^2 mR^2, \tag{6.84}$$

$$U = -\alpha^2 \frac{Gm^2}{R}, \tag{6.85}$$

where α^2 and β^2 are dimensionless form-factors independent of radius R and mass m of the spherical system (See Sect. 2.6).

We now rewrite (6.81), using (6.84) and (6.85), as

$$B = \alpha^2 \beta G m^{5/2}. \tag{6.86}$$

Use of form factors α2 and β2 allows us to show that the parameter B in (6.81) does not depend on radius of the spherical system. The product of α2 and β depends on mass density distribution law ρ(r) and does not depend on the total mass of the system. Hence the problem of the study of the changes of parameter B in (6.81) for an arbitrary spherical system is reduced to consideration of the dependence of the product of the α2 and β form factors on the mass density distribution law for the sphere with radius unity and mass unity. Let us consider such a sphere and calculate the value

$$a = \alpha^2 \beta \tag{6.87}$$

For the arbitrarily given law of density distribution ρ(k), k ∈ [0, 1], satisfying the condition

$$\int_{(V)} \rho(k) dV(k) = 1.$$

The volume of the sphere with radius unity is

$$V = \iiint_{(V)} dx, dy, dz = \int_0^1 k^2 dk \int_0^\pi \sin\theta d\theta \int_0^{2\pi} d\varphi = \frac{4}{3}\pi.$$

The volume of the sphere with radius k is

$$V(k) = \int_0^k k'^2 dk' \int_o^\pi \sin\theta d\theta \int_0^{2\pi} d\varphi = \frac{4}{3}\pi k^3. \tag{6.88}$$

The volume of the spherical shell with radius k and thickness dk is

$$dV(k) = k^2 dk \int_o^\pi \sin\theta d\theta \int_0^{2\pi} d\varphi = 4\pi k^2 dk. \tag{6.89}$$

The mass of the spherical shell with radius k and thickness dk is

$$dm(k) = 4\pi\rho(k)k^2 dk.$$

The mass of the sphere with radius k is

$$m(k) = 4\pi \int_0^k \rho\left(k'\right)\left(k'\right)^2 dk'. \tag{6.90}$$

The mass of the sphere as a whole is

$$m = 4\pi \int_0^1 \rho(k) k^2 dk = 1. \tag{6.91}$$

The polar moment of inertia of the shell with radius k and thickness dk is

$$dI(k) = k^2 dm(k) = 4\pi\rho(k) k^4 dk.$$

The Jacobi function of the sphere is

$$\Phi = \frac{4\pi}{2} \int_0^1 \rho(k) k^4 dk. \tag{6.92}$$

We can write the expression for the form factor β from (6.84) using (6.91) and (6.92):

$$\beta = \sqrt{\frac{\Phi}{\frac{1}{2}m}} = \frac{\sqrt{\int_0^1 \rho(k) k^4 dk}}{\int_o^1 \rho(k) k^2 dk}. \tag{6.93}$$

The potential energy of the shell with radius k and thickness dk in the gravitational field of the sphere of radius k is

$$dU(k) = -G\frac{m(k)dm(k)}{k} = -G\frac{16\pi^2 \rho(k) k^2 dk \int_o^k \rho(k')(k')^2 dk'}{k}.$$

The potential energy of the sphere as a whole is

$$U = -16\,\pi^2 G \int_0^1 \rho(k) k dk \int_0^k \rho(k')(k')^2 dk'. \tag{6.94}$$

We can write the expression for the form factor α using (6.85), (6.91) and (6.94) as

$$\alpha^2 = -\frac{U}{Gm^2} \frac{\int_0^1 \rho(k) k dk \int_o^k \rho(k')(k')^2 dk'}{\left(\int_o^1 \rho(k) k^2 dk\right)^2}. \tag{6.95}$$

Finally, the product of form factors α^2 and β represents the functional of the function of mass density distribution $\rho(k)$:

$$a = \alpha^2\beta = \frac{\int_0^1 k\rho(k)dk \int_o^k \rho(k')(k')^2 dk' \sqrt{\int_0^1 \rho(k)k^4 dk}}{\left(\int_o^1 \rho(k)k^2 dk\right)^{5/2}}. \tag{6.96}$$

The values of the form factors α^2 and β^2 and of their product $\alpha^2\beta$ for different formal laws of mass density distribution are given in Table 6.1. The numerical calculations of this table can be found in our paper (Ferronsky et al. 1978).

It can be seen from Table 6.1 that the form factor β changes from 0 to 1: $\beta \in [0, 1]$. It reaches the value of unity in the case when the entire mass of the sphere is distributed within its outer shell (at $k = 1$). The minimal value of the form factor β must be when the entire mass concentrates in the centre of the sphere (at $k = 0$). But if we do not place any strong restrictions on the function $\rho(k)$, i.e. in the general case, nothing can be said about the changing interval of the value $a = \alpha^2\beta$ (6.85). It is only possible to note that $a = \alpha^2\beta$ always has a positive value. From Table 6.1 it can also be assumed that the value of a is more then $(3/5)^{3/2} \approx 0.46$, which corresponds to the homogeneous distribution of the mass density within the sphere. It is known also from Chap. 4 that the homogeneous sphere, while contracting under gravitational forces, conserves its homogeneity up to the moment of simultaneous collision of all its particles.

The sphere expands and then (the time is reversible in classical physics) becomes homogeneous again. So, in accordance with the definitions given in the previous section, the homogeneous sphere appears to be the central configuration. Applying the main idea of the central configuration theorem discussed above in the general case, we assume the following qualitative picture of the evolution of a heterogeneous spherical system. During the contraction of the system the $\alpha^2\beta$ decreases and tends to the quantity $(3/5)^{3/2}$, reaching this value at the moment

Table 6.1 Numerical values of form factors α and β and their product $\alpha^2\beta$ for various formal laws of radial mass density distribution of the spherical system

Law of mass density distribution $\rho(k)$, $k \in [0, 1]$	α^2	$\beta^2\perp$	β^2	$\alpha^2\beta$
$\rho(r) = \rho_0$	0.6	0.4	0.6	0.46
$\rho(r) = \rho_0(1-k)$	0.728	0.27	0.4	0.47
$\rho(r) = \rho_0(1-k^2)$	0.7142	0.29	0.42	0.46
$\rho(r) = \rho_0(1-k)^n$	$\frac{(5T+8)(T+3)^2}{8(2n+3)(2n+5)}$	$\frac{8}{(n+4)(n+5)}$	$\frac{12}{(n+4)(n+5)}$	At $n \to \infty$, 0.54
$\rho(r) = \rho_0 k^n$	$\frac{n+3}{2n+5}$	$\frac{2n+9}{6n+15}$	$\frac{n+3}{2n+5}{}_0$	At $n \to \infty$,0,5
$\rho(r) = \rho_0\, \delta(1-k)$	0.5	0.67	1.0	0.5

of simultaneous collision of all the particles. If the expansion starts before the moment of simultaneous collision of the matter (at the neighborhood of singularity), then the value of $\alpha^2\beta$ again increases. Thus, there is a case of perturbed virial oscillations of the system. This case is known in the literature as 'stormy relaxation' of a gaseous sphere and is described quantitatively by the following equation of change of value of $|U|\sqrt{\Phi}$ (Ferronsky 1984):

$$U\sqrt{\Phi} = B - k\Phi$$

where $B = \text{const}$, and k is also constant.

This law of change of value of $|U|\sqrt{\Phi}$ will be considered in detail in Chap. 8, which is devoted to astrophysics applications. Here we only note that mechanism that drives the matter of a system towards simultaneous collision is the loss of energy through radiation. So, for conservative systems, the equation of virial oscillations has the form:

$$\ddot{\Phi} = -A + \frac{B}{\sqrt{\Phi}} - \frac{k\dot{\Phi}}{\sqrt{\Phi}}.$$

The term $k\dot{\Phi}/\sqrt{\Phi}$ is part of the perturbation function. It does not lead to the loss of total energy of the system, and we can call it internal friction.

6.4.2 Polytropic Gas Sphere Model

The laws of mass density distribution in the previous section were considered formally, neglecting the requirement of hydrodynamic stability of the system. However, it is well known that for the many really existing celestial gas bodies, a polytropic model in the central domain, is a good one.

Let us study the value of the form factors α and β and their product $\alpha^2\beta$ for the polytropic gas sphere model at various quantities of polytropic index. The equation of state for a gas sphere is

$$\frac{dp(k)}{dk} = -G\frac{m(k)\rho(k)}{k^2}, \tag{6.97}$$

where p(k) is the gas pressure; $\rho(k)$ is the mass density of the gas, and G is the gravitational constant.

Using Eq. 6.97 we can rewrite it for the sphere with radius k and mass m in the form

$$\frac{1}{k}\frac{d}{dk}\left|\frac{k^2}{\rho(k)}\frac{dp(k)}{dk}\right| = -4\pi G\rho(k). \tag{6.98}$$

This is one of the basic equations in the theory of the internal structure of the stars used up to now.

It is assumed that for polytropic models, the two independent characteristics in Eq. 6.98, namely pressure p(k) and mass density ρ(k), are linked by the relationship

$$p(k) = C\rho^b(k), \tag{6.99}$$

where C and b are constants.

From (6.99) it follows that

$$\frac{1}{\rho(k)}\frac{dp(k)}{dk} = C\frac{b}{b-1}\frac{d\rho^{b-1}(k)}{dk}. \tag{6.100}$$

Substituting (6.100) into (6.98) and introducing specification

$$\rho^{b-1}(k) = u(k), \quad n = \frac{1}{b-1} \tag{6.101}$$

we obtain

$$C(1+n)\frac{1}{k^2}\frac{d}{dk}\left|k^2\frac{du(k)}{dk}\right| = 4\pi G u^n(k). \tag{6.102}$$

Equation 6.102 can be simplified if dimensionless variables $\Theta(x) = u(x)/u_o$ and $x = \lambda k$ are introduced. Here u_o is the value u(k) in the center of the sphere, i.e. at $k = 0$. The coefficient λ is selected with the condition that, after substitution of the function Θ(x) into (6.102), all the constants should be cancelled. Then the following relationship for λ can be obtained:

$$C(1+n)\lambda^2 = 4\pi G u_0^{n-1} \tag{6.103}$$

and Eq. 6.102, known as the Emden equation, takes the form:

$$\frac{1}{x^2}\frac{d}{dx}\left|x^2\frac{d\Theta(x)}{dx}\right| = -\Theta^n(x). \tag{6.104}$$

It is obvious that for $x = 0$ the function Θ(x), known as the Emden function, should satisfy two conditions:

$$\Theta(x)|_{x=0} = 1, \quad \frac{d\Theta(x)}{dx}|_{x=0} = 0. \tag{6.105}$$

We now obtain the expression for the form factor α^2 for a sphere with polytropic index n. For this purpose we write the expression of potential energy in the form

$$U = -G\int \frac{m(k)dm(k)}{k}.$$

Using Eq. 6.97 for the gas sphere and the expression for dm(k), we rewrite (6.105) as follows:

$$U = \int \frac{k}{\rho(k)}\frac{dp(k)}{dk}dm(k) = 4\pi\int k^3 dp(k). \tag{6.106}$$

After integration by parts of the right-hand side of (6.105) we obtain

$$U = -12\ \pi\int_0^1 k^2 p(k)dk. \tag{6.107}$$

On the other hand, (6.105) can be rewritten in the form

$$U = -\frac{G}{2}\int \frac{dm^2(k)}{k}.$$

Integrating the right-hand side of the last relationship by parts, we obtain

$$U = -\frac{G}{2}\frac{m^2(k)}{k}\Big|_{k=0}^{k=1} - \frac{G}{2}\int \frac{m^2(k)dk}{k^2}. \tag{6.108}$$

The integral in the right-hand side of (6.108) is transformed with the help of (6.97) as follows

$$-\frac{G}{2}\int \frac{m^2(k)dk}{k^2} = \frac{1}{2}\int \frac{m(k)}{\rho(k)}\frac{dp(k)}{dk}dk.$$

Thus, using (6.100), we obtain

$$-\frac{G}{2}\int \frac{m^2(k)dk}{k^2} = \frac{1}{2}\int m(k)C\frac{b}{b-1}d\rho^{b-1}(k)$$

and, integrating by parts, we have

$$-\frac{G}{2}\int\frac{m^2(k)dk}{k^2}=\frac{1}{2}C\frac{b}{b-1}\rho^{b-1}(k)m(k)\Big|_{k=0}^{k=1}-\frac{1}{2}\int C\frac{b}{b-1}\rho^{b-1}(k)4\pi k^2\rho(k)dk$$

$$=-\frac{1}{2}\int(n+1)4\pi k^2 p(k)dk. \tag{6.109}$$

Substituting (6.109) into (6.108), we obtain the second expression for the potential energy:

$$U=-\frac{G}{2}-\frac{4\pi(n+1)}{2}\int_0^1 k^2\rho(k)dk, \tag{6.110}$$

where the condition $m(1)=1$ has been taken into account.

Solving the system of Eqs. 6.110 and 6.107 with respect to U, we find that

$$U=-G\frac{3}{5-n}$$

and hence

$$\alpha^2=\frac{3}{5-n}. \tag{6.111}$$

Now we derive the expression for the form factor β. For this purpose we write the Jacobi function expression for a polytropic sphere:

$$\Phi=\frac{4\pi}{2}\int_0^1 k^4\rho(k)dk=\frac{4\pi}{2}\int_0^{x_1}\frac{\Theta^n(x)x^4dx}{\lambda^5}, \tag{6.112}$$

where x_1 is the first root of the equation $\Theta(x)=0$.

Let us specify

$$\nu=\int_0^{x_1}\Theta^n(x)x^4dx$$

and, taking into account (6.103), we write

$$C(1+n)\lambda^2=4\pi G u_0^{n-1}$$

Then

$$\Phi = \frac{4\pi\nu}{2}\frac{u_0^n}{\lambda^5} = \frac{4\pi\nu}{2}\frac{[C(1+n)]n/n-1}{(4\pi G)n/n-1}\lambda^{(5-3n)/n-1}. \tag{6.113}$$

Now we obtain the second expression for the Jacobi function using the condition of Eq. 6.99 at the border surface of the sphere, i.e. at $k = 1$. Then

$$\frac{1}{\rho(k)}\frac{dp(k)}{dk}|_{k=1} = -\frac{Gm(k)}{k^2}|_{k=1} \tag{6.114}$$

and

$$m(k)k^2\Big|_{k=1} = -\frac{k^4}{G}\frac{1}{\rho(k)}\frac{dp(k)}{dk}|_{k=1}.$$

The left-hand side of Eq. 6.114, taking into account (6.100) and (6.101), is

$$\frac{1}{\rho(k)}\frac{dp(k)}{dk}|_{k=1} = C\frac{b}{b-1}\frac{d\rho^{b-1}(k)}{dk} = C(n-1)\frac{du(k)}{dk}. \tag{6.115}$$

Finally we obtain

$$\begin{aligned}\Phi &= \frac{1}{2}\beta^2 m(k)k^2|_{k=1} = -\frac{1}{2}\beta^2\frac{C(n+1)}{G}k^4\frac{du(k)}{dk}|_{k=1}\\ &= -\frac{1}{2}\beta^2\frac{C(n+1)}{G}u_0\frac{x^4}{\lambda^3}\frac{d\Theta(x)}{dk}|_{x=x_1}.\end{aligned}$$

Or when using (6.103),

$$\Phi = \frac{1}{2}\pi\beta^2\frac{C(1+n)^{n/n-1}}{(4\,\pi G)^{n/n-1}}\lambda^{(5-3n)/n-1}\left|x^4\frac{d\Theta(x)}{dk}\right||_{x=x_1}. \tag{6.116}$$

Dividing (6.116) by (6.113), we obtain

$$\beta = \sqrt{\frac{\nu}{\left[-x^4\frac{d\Theta(x)}{dx}\right]|_{x=x_1}}}. \tag{6.117}$$

We calculated the values of α^2 and β and their product $\alpha^2\beta$ using the data for ν, x_1, and

$$\frac{-x^2 d\Theta(x)}{dx}|_{x=x_1}$$

Table 6.2 Numerical values of form factors α and β and their product αβ for different values of polytropic index n

Index n	α^2	x_1	$-x^2\frac{B\Theta(x)}{dx}$	ν	β	$\alpha^2\beta$
0	0.6	2.42	4.9	17.63	0.77	0.46
1	0.75	3.14	3.14	12.15	0.62	0.465
1.5	0.87	3.63	2.71	11.12	0.55	0.475
2	1.0	4.35	2.41	10.61	0.48	0.482
3	1.5	6.89	2.01	10.85	0.34	0.502
3.5	2.0	9.53	1.89	11.74	0.26	0.52

at different polytropic index values, taken from Chandrasekhar (1939). The calculated data are shown in Table 6.2. It is interesting to note that in the framework of the really existing physical laws of mass density distribution ρ(k), the quantity $\alpha^2\beta$ changes within the narrow limits despite the fact that each of the form factors α and β varies almost three times more the variation of the polytropic index from 0 to 3,5.

6.4.3 System with Elliptical Symmetry

We have shown in the previous section that the property of the central configurations consisting in the constancy of the product $\alpha^2\beta$ holds for system with spherical symmetry.

Now we prove that this property holds for elliptical symmetry with an ellipsoidal mass distribution. Moreover, we show that among all the configurations only ellipsoidal mass distribution possess this property of central configurations.

Let us write the equation of the general ellipsoid with semi-axes a, b, c:

$$\frac{x^2}{a^2}+\frac{y^2}{b^2}+\frac{z^2}{c_2}=1, \tag{6.118}$$

where x, y, z are the Cartesian co-ordinates of the surface of this ellipsoid.

The equation of a set of similar ellipsoidal shells of this ellipsoid with the ellipsoidal mass distribution ρ(x) is

$$\frac{x^2}{a^2}+\frac{y^2}{b^2}+\frac{z^2}{c_2}=k^2, \tag{6.119}$$

where k ∈ [0, 1] is a parameter of the homogeneous ellipsoidal shell.

The gravitational potential inside this ellipsoidal shell is equal to a constant at an arbitrary point (x, y, z)

$$F(x,y,z)=-\frac{Gm_s}{2}\int_0^\infty \frac{du}{\sqrt{(a^2+u)(b^2+u)(c^2+u)}}, \tag{6.120}$$

where m_s is the mass of the shell; u is a parameter of integration.

We write down the form factor α_e of the potential energy U of this ellipsoid as

$$\alpha_e^2 = -\frac{au}{Gm^2}, \tag{6.121}$$

where a is semi-major axis in the equatorial plane; m is total mass

The volume of an ellipsoid bounded by the surface (6.119) with the parameter k is

$$V(k) = \frac{4}{3}\pi abck^3. \tag{6.122}$$

The volume of the thin shell bounded by ellipsoidal surfaces with the parameters k and k + dk is

$$dV(k) = 4\pi abck^2 dk. \tag{6.123}$$

The mass of this shell is expressed as

$$dm_s(k) = 4\pi abck^2 \rho(k) dk. \tag{6.124}$$

Then the total mass of the ellipsoid is

$$m = 4\pi abc \int_0^1 k^2 \rho(k) dk. \tag{6.125}$$

The mass of an ellipsoid bounded by the surface with the parameter k is

$$m(k) = 4\pi abc \int_o^k \left(k'\right)^2 \rho(k') dk'. \tag{6.126}$$

Using the reciprocation theorem (Duboshin 1975), we write the potential energy of the ellipsoid in the form

$$U = -\int_0^1 m(k) dF(k). \tag{6.127}$$

The gravitational potential inside the thin shell bounded by elliptical surface with parameters k and k + dk (6.120) is

$$dF(k) = 2\pi Gabck\rho(k)dk \int_0^\infty \frac{du}{\sqrt{(a^2+u)(b^2+u)(c^2+u)}}. \tag{6.128}$$

Now we write the expression for the form factor α_e using the corresponding values of U and m as

$$a_e = -\frac{aU}{Gm^2} = \frac{a}{2}\frac{\int_0^1 k\rho(k)dk\int_0^k (k')^2\rho(k')dk'}{\left[\int_0^1 k^2\rho(k)dk\right]^2}\int_0^\infty \frac{du}{\sqrt{(a^2+u)(b^2+u)(c^2+u)}}$$

$$= \alpha\frac{a}{2}\int_0^\infty \frac{du}{\sqrt{(a^2+u)(b^2+u)(c^2+u)}}, \tag{6.129}$$

where α is the potential energy form factor corresponding to the radial mass distribution law $\rho(k)$.

It is easy to see from Eq. 6.129 that when $a = b$ we obtain the value of the form factor α_e for the ellipsoid of rotation

$$\alpha_e^2 = \alpha^2\frac{\arcsin e}{e}. \tag{6.130}$$

Since

$$e = \sqrt{\frac{a^2 - c^2}{a^2}} \in |0, 1|$$

then

$$\alpha_e^2 \in \left[\alpha^2, \frac{\pi}{2}\alpha^2\right].$$

When $a > b > c$, Eq. 6.129 be (Janke et al. 1960)

$$\alpha_e^2 = \alpha^2\int_0^\infty \frac{du}{\sqrt{(a^2+u)(b^2+u)(c^2+u)}}$$

$$= \alpha^2\frac{a}{\sqrt{a^2-c^2}}F\left(\arcsin\sqrt{\frac{a^2-c^2}{a^2}}, \sqrt{\frac{a^2-b^2}{a^2-c^2}}\right).$$

Denoting

$$\arcsin\sqrt{\frac{a^2-c^2}{a^2}} = \arcsin e_1 = \varphi \quad \text{and} \quad \sqrt{\frac{a^2-b^2}{a^2-c^2}} = \frac{e_2}{e_1} = \sin\ \alpha^2 = f,$$

we obtain

$$\alpha_e^2 = \alpha^2 \frac{F(\varphi, f)}{\sin\varphi}, \tag{6.131}$$

where F(φ, f) is an incomplete elliptical integral of the first degree in the normal Legendre form. If $e_1 < 0.999$ and $0 < e_2 < e_1$, the function $F(\varphi, f) \cdot \sin^{-1}\varphi \in [1.000; 3.999]$ (Janke et al. 1960). When the arguments φ and f increase, the function $F(\varphi, f) \cdot \sin^{-1}\varphi$ also increases continuously.

Let us now consider the form factor β, which may be written

$$\beta = \left[\frac{\Phi}{ma^2}\right]^{1/2}. \tag{6.132}$$

Obviously β can be obtained by corresponding integration over the parameter $k \in [0, 1]$, if one writes the Jacobi function for the homogeneous thin shell bounded by the surfaces within the parameters k and k + dk and with mass distribution ρ(k) in the integrand.

Since the Jacobi function for a homogeneous ellipsoid with mass density ρ_o is

$$\Phi = \frac{2}{15}\pi abc\rho_0\left(a^2+b^2+c^2\right) \tag{6.133}$$

the Jacobi function for a thin ellipsoid shell may be written

$$d\Phi(k) = \frac{2}{3}\pi abc\rho(k)k^4dk\left(a^2+b^2+c^2\right). \tag{6.134}$$

Consequently, the Jacobi function Φ of the ellipsoid is equal to

$$\Phi = \frac{2}{3}\pi abc\left(a^2+b^2+c^2\right)\int_0^1 \rho(k)k^4dk. \tag{6.135}$$

Finally, using (6.135) and (6.125), Eq. 6.132 for the form factor β will be

$$\beta_e = \left[\frac{a^2+b^2+c^2}{3a^2}\,\frac{\int_0^1 \rho(k)k^4dk}{\int_0^1 \rho(k)k^2dk}\right]^{1/2} = \beta\left[\frac{a^2+b^2+c^2}{3a^2}\right]^{1/2}, \tag{6.136}$$

where β is a form factor of the Jacobi function of the system with radial mass distribution ρ(k) and the expression

$$\left[\frac{a^2+b^2+c^2}{3a^2}\right]^{1/2} \in \left[\frac{1}{\sqrt{3}}, 1\right].$$

So the value a_e is equal to

$$a_e = \alpha_e^2 \beta_e = a \frac{F(\varphi, f)}{\sin \varphi} \left[\frac{a^2+b^2+c^2}{3a^2}\right]^{1/2}. \tag{6.137}$$

Now it can be shown that the property (6.137) of the product $\alpha^2\beta$ constancy is possessed only by systems with elliptical symmetry and ellipsoidal mass density distribution. This means that for such systems the form factors α and β may be expressed as a product of corresponding for factors of the sphere and terms depending on the form of the boundary surface.

For this proof we consider an arbitrary system with a similar law of mass distribution ρ(k), k ∈ |0, 1| and the boundary surface S. Then, since we consider only one-dimensional ρ(k), mass density will be constant on any surface with a fixed parameter k and similar to S. The area of this surface is

$$S'(k) = Sk^2. \tag{6.138}$$

If the volume of the body is equal to V, then the volume of the part of the body bounded by the surface S′(k) is

$$V'(k) = Vk^3 \tag{6.139}$$

And its mass is

$$m(k) = V \int_0^1 k^2 \rho(k) dk. \tag{6.140}$$

Let us introduce the Cartesian co-ordinate system OXYZ with an origin coinciding with the center of similarity. Let us denote by h in the equatorial plane OXY the longest distance from the center of similarity to the boundary and assume that the form factor α_e^2 of the body can be expressed as a product of the form factors of the potential energy α for the radial mass density distribution law and some term Δ(S) depending on the form of the boundary surface

$$\alpha_e = -\frac{Uh}{Gm^2} = \alpha\Delta(S) = \frac{\int_0^1 k\rho(k)dk \int_0^k (k')^2 \rho(k')dk'}{\left(\int_0^1 k^2\ \rho(k)dk\right)^2} \Delta(S). \tag{6.141}$$

From Eq. 6.141 we can obtain the potential energy in the for

$$U = -\frac{Gm^2}{h}\alpha\Delta(S) = -\frac{GV}{h}\int_0^1 k\rho(k)dk\Delta(S). \tag{6.142}$$

Since the term G, V, H, Δ(S) do not depend on the parameter k, let us put them into the integrand and denote

$$\frac{GV}{h}k\rho(k)\Delta(S)dk = F(k).$$

Then Eq. 6.142 may be written

$$U = -\int_0^1 m(k)dF(k). \tag{6.143}$$

Comparing Eqs. 6.143 and 6.127, one can see that Eq. 6.143 an equation for the reciprocation theorem, whose validity is based on the constancy of the gravitational potential dF(k) inside the thin shell bounded by the similar and similarly situated surfaces with parameters k and k + dk. But as shown in the work of Dive (1931), where one can find rigorous proof of the reverse Newton theorem, only ellipsoidal shells possess such a property. Therefore the body with the one-dimensional mass distribution law ρ(k) for which the form factor α_e is equal to the product of the form factor of the sphere and some term depending on the form of the boundary surface Δ(S) must satisfy the equation of the ellipsoid (6.118).

6.4.4 System with Charged Particles

We shell now show, with the help of a model solution, that for the Coulomb interactions of charged particles, constituting a system, Eq. 6.5 holds with the same conditions as for the previous models discussed above.

The derivation of the expression for the potential energy of the Coulomb interactions of a celestial body is based on the concept of an atom following, for example, from the Tomas-Fermi model (Flügge 1971). In our problem this approach does not result in limited conclusions since the expression for the potential energy, which we shell write, will be correct within a constant factor.

Let us consider a one-component, ionized quasineutral and gravitating gaseous cloud with spherically symmetrical mass distribution and radius of sphere R. We shell not consider the problem of its stability, assuming that the potential energy of interaction of charged particles is represented by the Coulomb energy. Therefore, in order to prove the relationship (6.5), it is necessary to obtain the energy of the Coulomb interactions of positively charged ions with their electron clouds.

Assume that each ion of the gaseous cloud has mass number A_i and order number Z, and that the function ρ(r) expresses the law of mass distribution inside the gaseous cloud. The mass of the ion will be $A_i\, m_p$ (where $m_p = 1.66 \times 10^{-24}$ g is the mass of the proton) and its total charge will be +Ze (where $e = 4.8 \times 10^{-10}$ GCSE is elementary charge). Then let the total charge of the electron cloud, which is equal to −Ze, be distributed around the ion in the spherically symmetrical volume of radius r_i with charge density $q_e(r_e)$, $r_e \in [0, 1]$. Radius r_i of the effective volume of the ion may be expressed through the mass density distribution ρ(x) by

$$\frac{4}{3}\pi r_i^3 = \frac{A_i m_p}{\rho(r)}. \tag{6.144}$$

Then

$$r_i = \sqrt[3]{\frac{3A_i m_p}{4\pi\rho(r)}}. \tag{6.145}$$

Let us calculate the Coulomb energy V_c per ion, using Eq. 6.145. Assuming that the charge distribution law in the effective volume of radius r_i is given, we may write U_c in the form

$$U_c' = U_c^{(+)} + U_c^{(-)}, \tag{6.146}$$

where $U_c^{(-)}$ is the potential energy of the Coulomb repulsion of electrons inside the effective volume of radius r_i;$U_c^{(+)}$ is the potential energy of attraction of the electron cloud to positive ion.

Let us assume that the charge distribution law inside the electron cloud is $q_e(r_e)$. Then normalization of the electron charge of the cloud surrounding the ion may be written

$$-Ze = \int_0^{r_1} 4\pi q_e(r_e) r_e^2 dr_e. \tag{6.147}$$

From the Eq. 6.147 we may obtain the normalization constant q_0, which will depend on the given law of charge distribution, as

$$q_0 = -\frac{Ze}{4\pi \int_0^{r_1} r_e f(r_e) dr_e.} \tag{6.148}$$

Now it is easy to obtain expressions for $U_c^{(-)}$ and $U_c^{(+)}$ in the form

$$U_c^{(-)} = (4\pi)^2 q_0^2 \int_0^{r_1} r_e f(r_e) dr_e \int_0^{r_e} (r'_e)^2 f(r'_e) dr'_e, \tag{6.149}$$

$$U_c^{(+)} = 4\pi Zeq_0 \int_0^{r_1} r_e f(r_e) dr_e. \tag{6.150}$$

Finally, Eq. 6.146 for the potential energy U_c corresponding to one ion may be rewritten using Eqs. 6.148–6.150 in the form

$$U'_c = -e^2 Z \left\{ \frac{\int_0^{r_1} r_e f(r_e) dr_e}{\int_0^{r_1} r_e^2 f(r_e) dr_e} - \frac{\int_0^{r_1} r_e f(r_e) dr_e \int_0^{r_e} (r'_e)^2 f(r'_e) dr'_e}{\left(\int_0^{r_1} r_e^2 f(r_e) dr_e \right)^2} \right\}. \tag{6.151}$$

It is easy to see in the right-hand side of Eq. 6.151 that the expression enclosed in brackets determines the inverse value of some effective diameter of the electron cloud, which may be expressed through the form factor α_i of the ion and the radius r_i, i.e.

$$\frac{\int_0^{r_1} r_e f(r_e) dr_e}{\int_0^{r_1} r_e^2 f(r_e) dr_e} - \frac{\int_0^{r_1} r_e f(r_e) dr_e \int_0^{r_e} (r'_e)^2 f(r'_e) dr'_e}{\left(\int_0^{r_1} r_e^2 f(r_e) dr_e \right)^2} = -\frac{d_i}{r_i}. \tag{6.152}$$

Thus, Eq. 6.151, using (6.152), yields

$$-U'_c = \alpha_i \frac{e^2 Z}{r_i}. \tag{6.153}$$

The numerical values of the form factor α_i, depending on the charge distribution $q_e(r_e)$ inside the electron cloud, are given in Table 6.3; they were calculated in our work (Ferronsky et al. 1981).

Using Eq. 6.153, the total energy of the Coulomb interaction of particles may be written

$$-U_c = 4\pi \int_0^R \frac{\rho(r)}{A_i m_p} U'_c r^2 dr = \frac{3\alpha_i^2 e^2 Z^2}{R} \int_0^R R r^2 \left(\frac{4\pi \rho(r)}{3 A_i m_p} \right)^{4/3} dr. \tag{6.154}$$

Table 6.3 Numerical values of form factors α for different radial charge distribution of the electron cloud around the ion

The law of charge distribution[a]	α_i^2
$q_e(r_e) = q_o = $ const	$\frac{9}{10}$
$q_e(r_e) = q_o\left(1 - \frac{r_e}{r_i}\right)$	$\frac{44}{35}$
$q_e(r_e) = q_o\left(1 - \frac{r_e}{r_i}\right)^n$	$\frac{(n+3)(11n^2+41n+36)}{8(2n+3)(2n+5)}$
$q_e(r_e) = q_o\left(\frac{r_e}{r_i}\right)$	$\frac{16}{21}$
$q_e(r_e) = q_0\left(\frac{r_e}{r_i}\right)^n$	$\frac{(n+3)^2}{(n+2)(2n+5)}$
The same for $n \to \infty$	$\alpha_{ii}^2 \to \frac{1}{2}$

[a] Here q_o is the charge value in the center of the sphere; t_e is the parameter of radius, $r_e \in [0, 1]$; n is an arbitrary number, n = 0, 1, 2, ...

Introducing in Eq. 6.154 the form factor of the Coulomb energy α_c^2, depending on the mass distribution in the gaseous cloud and on the charge distribution inside the effective volume of the ion, we obtain

$$-U_c = \alpha_c^2 \frac{e^2 Z}{R}\left(\frac{m}{A_i m_p}\right)^{4/3}, \tag{6.155}$$

where

$$\alpha_c^2 = \frac{3\alpha_c^2 \int_0^R \left[\left(\frac{4\pi}{3}\rho(r)\right)\right]^{4/3} Rr^2 dr}{m^{4/3}},$$

$$m = \sum_{i=1}^{n} m_i = 4\pi \int_0^R r^2 \rho(r) dr,$$

Since the total number of ions N in the gaseous cloud is equal to

$$N = \frac{m}{A_i m_p}$$

and the relation between the radius of the cloud and the radius of the ion may be obtained from the relationship of the corresponding volumes

$$\frac{4}{3}\pi R^3 = N\frac{4}{3}\pi r_i^3,$$

then Eq. 6.155 may be rewritten in the form

$$-U'_c = \alpha_c^2 \frac{N^{4/3} e^2 Z^2}{R} = \alpha_c^2 N \frac{e^2 Z^2}{r_i}. \qquad (6.156)$$

Hence, the form factor entering the expression of the potential energy of the Coulomb interaction acquires a simple meaning. It may be represented as a ratio of the average radius of all spherical volumes per ion to the average effective distance between electrons, disposed on some spherical shell of radius r_i, i.e.

$$\alpha_c^2 = \frac{r_i}{2r_{ei}}. \qquad (6.157)$$

Let us now examine by means of numerical data the relationship (6.5), assuming different law of mass distribution. The expression for Jacobi function of the system, which we have previously derived (Ferronsky et al. 1978), is

$$\Phi = \frac{4\pi}{3} \int_0^R r^4 \rho(r) dr = \beta^2 m R^2. \qquad (6.158)$$

Thus, Eq. 6.5, using (6.156) and (6.158), may be written

$$-U_c \sqrt{\Phi} = \alpha_c^2 N^{4/3} \frac{Z^2 e^2}{R} \sqrt{\frac{\beta^2 m R^2}{2}} = \frac{1}{\sqrt{2}} \alpha_c^2 \, \beta N^{4/3} m^{1/2} e^2 Z^2. \qquad (6.159)$$

Since we have assumed that the mass of the system and its ion composition are constants, examination of Eq. 6.5 will be equivalent to analysis of the product of the form factors α_c and β. Equation 6.5 holds if

$$\alpha_c^2 \, \beta = \frac{r_i}{2r_{ei}} \approx \text{const}. \qquad (6.160)$$

The results of the numerical calculations of the form factors α_c and β for different mass distribution in the cloud are shown in Table 6.4, and calculations were carried out in our work (Ferronsky et al. 1981). The values of the form factor α_i of the ion, the numerical value of which depends on the choice of charge distribution $q_e(r_e)$, are shown in Table 6.3.

In the Table 6.4 the numerical values of the form facto α_c^2 and the product of the form factors $\alpha_c^2 \, \beta$ are given for the case of homogeneous distribution of the electron charge around ion, i.e. when $q_e(r_e) = \text{const}$. From Table 6.4 it follows that for different laws of mass distribution, when the mass increases to the center, the product of form factors α_c^2 and β remains constant, and therefore Eq. 6.5 holds, with the same comments as were made previously.

Table 6.4 Numerical values of the form factors α_c^2 and β product for different laws of radial mass distribution

The law of mass distribution[a]	$\frac{\alpha_c^2}{\alpha_i^2}$	α_c^2 for $\alpha_i^2 = 0.9$[b]	β	$\alpha_c^2\beta$ for $\alpha_i^2 = 0.9$[b]
$\rho(r) = \text{const}$	1	0.9	0.6324	0.5692
$\rho(r) = \rho_0(1 - r/R)$	1.1303	1.0173	0.5163	0.5253
$\rho(r) = \rho_0(1 - r/R)^2$	1.3331	1.1998	0.4364	0.5236
$\rho(r) = \rho_0(1 - r/R)^3$	1.5510	1.3959	0.3779	0.5276
$\rho(r) = \rho_0(1 - r/R)^n$	$\frac{27}{\sqrt[3]{6}} \frac{[(n+1)(n+2)(n+3)]^{4/3}}{(4n+3)(4n+6)(4n+9)}$	$\frac{243}{10\sqrt[3]{6}} \frac{[(n+1)(n+2)(n+3)]^{4/3}}{(4n+3)(4n+6)(4n+9)}$	$\sqrt{\frac{8}{(n+4)(n+5)}}$	for $n \to \infty$, 0.5909
$\rho(r) = \rho_0(r/R)$	1.0159	0.9143	2/3	0.6095
$\rho(r) = \rho_0(r/R)^2$	1.0461	0.9415	0.6900	0.6497
$\rho(r) = \rho_0(r/R)^n$	$\frac{27}{10\sqrt[3]{3}} \frac{(n+3)^{4/3}}{(4n+9)}$	$\frac{81}{100\sqrt[3]{3}} \frac{(n+3)^{4/3}}{(4n+9)}$	$\sqrt{\frac{2n+3}{3n+5}}$	for $n \to \infty$, $\to \infty$
$\rho(r) = \rho_0 e^{-K(r/R)}$	0.2321 k	0.2089K	$2\sqrt{2}/k$	0.5909
$\rho(r) = \rho_0 e^{-K(r/R)^2}$	$0.5907k^{1/2}$	$0.5316K^{1/2}$	$(1/k)^{1/2}$	0.5316

[a] Here, ρ_0 is the normalization constant; r is the parameter of the radius $r \in [0, R]$; n and k are arbitrary numbers, n = 0, 1, 2, . . .

[b] The value α_i corresponds to the homogeneous charge distribution in the electron cloud, surrounding the ion $q_e(r_e)$ = constant (see Table 6.3)

From Eq. 6.157 it follows, however, that the form factor of the Coulomb energy α_c becomes infinite, when the volume occupied by the ions tends to zero. Correspondingly, the Coulomb energy in this case will also tend to infinity. In Table 6.4 there are two laws of mass distribution for which the last condition holds. They are $\rho(r) = \rho_o(1 - (r/R)^n)$ for $n \to \infty$. When the particles of the system are gathering at the shell of the finite radius, the energy of the Coulomb interaction tends to infinity whereas the energy of gravitational interaction has a finite value. When the mass distribution is $\rho(r) = \rho_0(1 - (r/R)^n)$, the form factor of gravitational and Coulomb energies are both finite. But the form factors of the Jacobi function of the system in this case tends to zero, a circumstance which provides the constancy of the product of the form factors α_c^2 and β. This difference might play a decisive role in the evolution of the system.

In conclusion, we note that the results of the study on the relationship between the Jacobi function and the potential energy allows us to consider that the transfer from Jacobi's equation (6.1) and (6.2) into the equations of virial oscillations (6.3) and (6.4) is form the point of view of physics justified. This justification has been achieved in the framework of Newton and Coulomb interactions of the particles of the system. At the same time, the observed deviations of the value of parameter B in (6.5) form some constant quantity can be accounted for by small perturbation when studying the evolution of a heterogeneous system.

Chapter 7
Applications in Celestial Mechanics and Geodynamics

In the previous chapters we presented physical and theoretical fundamentals of the unified theory for study of unperturbed and perturbed motion of a self-gravitating celestial body which generates energy by interaction of its constituent particles. The theory is based on the functional relationship between the polar moment of inertia and the energy (potential, kinetic and total) of the natural conservative and dissipative system in the form of Jacobi's virial equation or the generalized virial theorem The remarkable property of Jacobi's virial equation is its ability to be simultaneously both the equation of dynamical equilibrium and equation of motion. As it was shown in Chap. 3, the equation is valid for all the known physical models of the matter interaction describing dynamics of natural systems. The functional relationship between the potential energy and the polar moment of inertia was revealed by analyzing of orbits of the artificial satellites. The potential energy, which is generated by interaction of the elementary mass particles, is the force function of the body, i.e. the active component of its motion. This energy induces the inner and outer force field of the body. The kinetic energy appears to be the reactive component of the force field which is developed in the form of motion of the same body's mass particles and the body as a whole. Thus, Jacobi dynamics of a body is based on its own inner forces, which are developed by interaction of the elementary particle. The induced inner and outer force field of the body is the main recoded dynamical effect of the interacted elementary particles.

Now we would like to apply the obtained results of the general solution of Jacobi's virial equation presented in Chaps. 3, 4 and 5 to study dynamics of the Sun, the Earth and the Moon, which represent celestial bodies of the Solar System, and to obtain some quantitative data concerning the concrete elements of their non-perturbed and perturbed motion. As in celestial mechanics, under non-perturbed motion we understand the motion under action of the body's own force field generated by interaction of its own masses. The perturbation motion of a body is considered to be the effects developed by the outer force fields of the Sun, the Earth and the Moon. Interplanetary perturbations are not considered here because of their indirect effect and this is a specific problem. The study of dynamics of a body in its own force field starts first of all with investigation of the basic modes, namely, the oscillation and rotation of the shells, which are developed by interaction of their own masses. The energy which is emitted from the body's surface forms the outer

V.I. Ferronsky et al., *Jacobi Dynamics*, Astrophysics and Space Science Library 369, DOI 10.1007/978-94-007-0498-5_7, © Springer Science+Business Media B.V. 2011

force field of the body. Dynamical equilibrium between the Sun and the Earth and between the Earth and the Moon is achieved by interaction of their outer force fields and guaranties stability of their orbital motion. The effect of differential rotation of shells and changes in the slope of the body axis of rotation is also examined. Herein, the observed rotation of the planet as a rigid body in reality is not proved. The effect of perturbation of the upper shell and the integral effect of rotation of all other shells is discussed as well.

In order to find a quantitative solution of the task it is necessary to have data about the mass, radius, moment of inertia and radial density distribution of the planet's mass. We have sufficiently reliable data about the mass and radius for the Sun, the Earth and the Moon. The reliable mean value of the moment of inertia was found now by artificial satellites only for the Earth. The radial density distribution data appears to be unreliable even for the Earth. This is because the existing methodology of interpretation of seismic data by the Williamson-Adams equation is based on the planet's hydrostatic equilibrium and needs to be reconsidered. That is why we have to find such law of density distribution which satisfies to the experimentally found moment of inertia and the condition of the observed differentiation of masses on the shells with respect to density.

We start the study dynamics of a body with its own oscillations. After that we move to determine the properties of the interacting masses which have not been earlier considered and which are needed. Among them are the structure of the potential and kinetic energy, the nature of the Archimedes' and Coriolis' forces and the electromagnetic component of the body's potential energy. They are the basis for consideration of body's dynamical effects. Finally, we turn to solution of problem of the body shells rotation, the nature of precession, nutation, and the orbit plane dynamics.

7.1 The Problem of Eigenoscillations of a Celestial Body

In order to demonstrate application of the new theory to study of a body dynamics we consider both traditional hydrostatic and proposed dynamic approaches being applied to the Earth. Applying the volumetric forces and volumetric moments we show in this solution that the eigenoscillations of a body is a natural kinetic integral effect of its interacting particles.

7.1.1 Hydrostatic Approach

We consider first the problem of radial oscillations of a gravitating elastic sphere within the framework of the traditional hydrostatic equilibrium approach which is used to study eigenoscillations of a celestial body

The equation of motion of a deformable body in the presence of the mass forces of the outer uniform force field is written in the form (Landau and Lifshitz 1954)

$$\frac{\partial \sigma_{ik}}{\partial x_k} + \rho F_i = \rho \frac{\partial^2 u}{\partial t^2}, \tag{7.1}$$

where ρF_i is the i-th component of the mass force; u_i is the i-th component of the displacement vector; σ_{ik} is the stress vector; ρ is the mass density of the sphere. We write the vector components of the mass force in the spherical system of co-ordinates:

$$\rho F_0 = 0,$$

$$\rho F_\lambda = 0,$$

$$\rho F_r = -G\,\rho \frac{m(r)}{r^2},$$

where m(r) is the body's mass within the sphere with radius r.

Because of radial deformation of the sphere the only radial component of the displacement vector differs from zero, i.e.

$$u_r = 0,$$

$$u_\theta = 0,$$

$$\frac{\partial^2 u_k}{\partial r^2} = \frac{\partial^2 r}{\partial t^2}.$$

For isotropic media and for small deformations the stress tensor σ_{ik} and the deformation tensor u_{ik} have the linear relationship, according to Hooke's law:

$$\sigma_{ik} = k u_{ik}\,\delta_{ik} + 2\mu\left(u_{ik} - \frac{1}{3}\,\delta_{ik} u_{ik}\right) = \lambda u_{ik} + 2\,\mu u_{ik}, \tag{7.2}$$

where k is the displacement modulus; μ is the shear modulus; $\lambda = k - 2/3\ \mu$ is the Lamé constant; $\sigma_{ik} = 0$ at $i \neq k$ and $\sigma_{ik} = 1$ at $i = k$.

In the case of radial deformations, the components of the deformation tensor are equal to

$$u_{rr} = \frac{du_r}{dr},$$

$$u_{\theta\theta} = u_{\phi\phi}\frac{u_r}{r}, \tag{7.3}$$

$$u_{\theta\phi} = u_{\lambda r} = u_{r\theta}.$$

The components of the stress tensor are:

$$\sigma_{rr} = (\lambda + 2\mu)\frac{du_r}{dr} + 2\lambda\frac{u_r}{r},$$

$$\sigma_{\theta\theta} = \sigma_{\phi\phi} = \lambda\frac{du_r}{dr} + (2\lambda + \mu)\frac{u_r}{r}, \tag{7.4}$$

$$\sigma_{\theta\phi} = \sigma_{\phi r} = \sigma_{r\theta} = 0.$$

The general equation (7.1) of motion now takes the form

$$\frac{d\sigma_{rr}}{dr} + \frac{1}{r}\left(2\sigma_{rr} - \sigma_{\theta\theta} - \sigma_{\varphi\varphi}\right) + \rho F_i = \rho\frac{d^2u_r}{dt^2}. \tag{7.5}$$

Putting Eq. 7.4 into Eq. 7.5, we obtain an equation describing the radial displacement of matter in the sphere:

$$(\lambda + 2\mu)\left(\frac{d^2u_r}{dr^2} + \frac{2}{r}\frac{du_r}{r^2}\right) + \left(\frac{du_r}{r^2} + 2\frac{u_r}{r}\right)\frac{d\lambda}{dr} + 2\frac{d\mu}{dr}\frac{du_r}{dr} + \rho F_r = \rho\frac{d^2u_r}{dt^2}. \tag{7.6}$$

Equation 7.6 is used to study the problem of the radial oscillations of the body in the traditional hydrostatic approach. This equation is solved at boundary conditions of uniform radial displacement of matter or at uniform pressure over a spherical surface enveloping the body.

The normal stress at the sphere's surface which is formed by the outer layer of the body with radius r is

$$T_r = \overline{\Sigma}\cdot\frac{r}{r}. \tag{7.7}$$

where $\overline{\Sigma}$ is the stress tensor, the components of which are

$$\left|t_r, t_\theta, t_\varphi\right| = |100|\begin{vmatrix}\sigma_{rr} & 0 & 0\\ 0 & \sigma_{\theta\theta} & 0\\ 0 & 0 & \sigma_{\varphi\varphi}\end{vmatrix} \tag{7.8}$$

Here only $t_r = \sigma_{rr} \neq 0$, but tensor T_r is the purely normal stress, which is

$$T_r = \left|(\lambda + 2\mu)\frac{du_r}{dr} + 2\lambda\frac{u_r}{r}\right|\frac{\bar{r}}{r}. \tag{7.9}$$

Let us consider a uniform sphere with $\lambda = \text{const}$ and $\mu = \text{const}$. Then

$$\frac{d}{dt}\left(\frac{du_r}{dr} + 2\frac{u_r}{r}\right) = \frac{d\theta}{dr} = 0, \tag{7.10}$$

where $\theta = u_{rr} + u_{\theta\theta} + u_{\varphi\varphi}$ is the dilation of the body.

The general solution of Eq. 7.10 is

$$u_r = Ar + \frac{B}{r^2}. \tag{7.11}$$

Constants A and B can be defined from the following boundary conditions: at the centre of the body (r = 0) the displacement $u_r = 0$ and the value B = 0; on the surface of the sphere radius r = a and $T_r = -p$.Then

$$\left|(\lambda + 2\mu)\frac{du_r}{dr} + 2\lambda\frac{u_r}{r}\right|_{r=A} = -p \tag{7.12}$$

from which it follows that

$$A = -\frac{p}{3\lambda + 2\mu}.$$

Now the general solution (7.11) takes the form

$$u_r = -\frac{pr}{3\lambda + 2\mu} \tag{7.13}$$

$$\theta = -\frac{3}{3\lambda + 2\mu} = -\frac{p}{\lambda + 2/3\,\mu} = -\frac{p}{k}.$$

Substituting the solution (7.13) into (7.4), we obtain the expression for the components of the stress tensor:

$$\sigma_{rr} = \sigma_{\theta\theta} = \sigma_{\phi\phi} = -p. \tag{7.14}$$

It is seen from (7.14) that the value of the stress components is reduced to the constant hydrostatic pressure of the body matter.

To solve the problem of the eigenoscillations of a uniform spherical body, we assume

$$u_r = U(r)e^{i\omega t}, \tag{7.15}$$

where ω is the eigenoscillation frequency of the sphere.

Substituting (7.15) into Eq. 7.6, we obtain

$$(\lambda + 2\mu)\left(\frac{d^2U}{dr^2} + \frac{2}{r}\frac{dU}{dr} - \frac{2U}{r^2}\right) + \rho\,\omega^2 U + \frac{4}{3}\pi r G\,\rho^2 = 0. \tag{7.16}$$

We introduce the new variable x:

$$x = \sqrt{\frac{p}{\lambda + 2\,\mu}}\,\omega r = hr. \tag{7.17}$$

Then Eq. 7.16 can be rewritten as

$$\frac{d^2U}{dr^2} + \frac{2}{r}\frac{dU}{dr} - \frac{2U}{dr} + \frac{x^2}{r^2}U = 0.$$

or

$$\frac{d^2U}{dx^2}\left(\frac{dx}{dr}\right)^2 + \frac{2}{r}\frac{dU}{dx}\left(\frac{dx}{dr}\right) - \frac{2U}{r^2} + \frac{x^2}{r^2}U = 0.$$

Considering that

$$\frac{dx}{dr} = \sqrt{\frac{p}{x + 2\,\mu}}\,\omega = \frac{x}{r},$$

we obtain

$$\frac{d^2U}{dx^2} + \frac{2}{r}\frac{dU}{dx} - \frac{2}{x^2}U = 0. \tag{7.18}$$

Equation 7.18 is known as the Riccati equation. Its solution is

$$U(r) = \frac{d}{dx}\left(\frac{A\sin x + B\cos x}{x}\right) \tag{7.19}$$

The arbitrary integration constants A and B can be found from the boundary conditions. In the centre of the body (r = 0, x = 0) we have U(r) = 0. Hence, B = 0. Moreover, the outer surface of the body should be in equilibrium. This means that the surface pressure should be equal to zero:$(\sigma_{rr})|_{r=A} = 0$and

$$(\lambda + 2\,\mu)\frac{dU}{dt}e^{i\omega t} + 2\,\lambda\frac{U}{r}e^{i\omega t} = 0,$$

or

$$(\lambda + 2\,\mu)\frac{d}{dx}\left|\frac{d}{dx}\left(A\frac{\sin x}{x}\right)\right| + 2\frac{\lambda}{r}\frac{d}{dx}\left(A\frac{\sin x}{x}\right) = 0. \tag{7.20}$$

Differentiating this, we have

$$-(\lambda+2\mu)x^2\sin x-4\mu x\cos x+4\mu\sin x=0. \tag{7.21}$$

After dividing (7.21) by 4μ sin x, we obtain

$$x\,\mathrm{ctg}\,x=1-\frac{\lambda+2\mu}{4\mu}x^2.$$

The value x at the outer surface is equal to ah. Finally, we obtain the equation for determining the eigenoscillation values of the body:

$$ah\,\mathrm{ctg}\,ah=1-\frac{\lambda+2\mu}{4\mu}a^2h^2. \tag{7.22}$$

7.1.2 Dynamic Approach

Now we consider the problem of the eigenoscillations of the Earth as a uniform body on the basis of its dynamical equilibrium, i.e. under an action of its own internal force field or within the framework of the dynamic approach. For this purpose we use Eq. 7.5 assuming that the internal pressure in the body is isotropic, i.e.

$$\sigma_{rr}=\sigma_{\theta\theta}=\sigma_{\varphi\varphi}=-p.$$

Then we have from (7.5)

$$\rho\frac{\partial^2 r}{\partial t^2}=-\frac{\partial p}{\partial r}+\rho F_r \tag{7.23}$$

Now let us derive Jacobi's virial equation from Eq. 7.23 for the spherical symmetric model of the body. For this purpose we multiply the right-hand sides of Eq. 7.23 by $4\pi r^3 dr$ and integrate it with respect to dr from 0 to R:

$$\int_0^R 4\pi r^3\rho(r)\frac{\partial^2 r}{\partial t^2}dr=-\int_0^R\frac{\partial p}{\partial r}4\pi r^3 dr-4\pi G\int_0^R r\rho(r)m(r)dr \tag{7.24}$$

The left-hand side of Eq. 7.24 gives

$$\int_0^R 4\pi r^3\rho(r)\frac{\partial^2 r}{\partial t^2}dr=\frac{\partial^2}{\partial t^2}\left[\frac{1}{2}\int_0^R 4\pi r^4\rho(r)dr-\int_0^R r\pi r^2\left(\frac{\partial r}{\partial t}\right)^2\rho(r)dr\right]$$
$$=\ddot{\Phi}-2T, \tag{7.25}$$

where Φ is the Jacobi function of the body and T is the kinetic energy of the displacements of the matter.

The first term in the right-hand side of (7.24) is equal to

$$\int_0^R \frac{\mathrm{dp}}{\mathrm{dr}}\frac{\mathrm{dr}}{d\rho}\mathrm{d}\rho 4\pi \mathrm{r}^3 = -\int_{\rho_0}^{\rho_s} \frac{\mathrm{dp}}{d\rho} 4\,\pi \mathrm{r}^3 d\rho, \tag{7.26}$$

where ρ_o and ρ_s are the mass densities in the centre and on the surface of the body respectively.

Taking into account that, within the framework of the model of an elastic medium, the system reaches its mechanical equilibrium faster than its thermal equilibrium, we assume that the entropy of the system is equal to the constant value, and therefore we write

$$\frac{\mathrm{dp}}{d\rho} = \left(\frac{\mathrm{dp}}{d\rho}\right)_{\mathrm{s}} = \frac{\mathrm{k}}{\rho} = \mathrm{c}_{\mathrm{s}}^2, \tag{7.27}$$

where $\mathrm{c_s}$ is the velocity of sound in elastic media and $\mathrm{c_s^2} = v_\mathrm{p}^2 - 4/3\; v_\mathrm{g}^2$; $v_\mathrm{p} = \sqrt{(\mathrm{k} + 4/3\mu)/\rho}$ is the velocity of the longitudinal waves in an elastic medium; $v_\mathrm{g} = \sqrt{\mu/\rho}$ is the velocity of the transverse waves in the elastic medium.

Finally, Eq. 7.26 can be rewritten in the form

$$-\int_0^R \frac{\partial \mathrm{p}}{\partial \mathrm{r}} 4\,\pi \mathrm{r}^3 \mathrm{dr} = -\int_0^R \mathrm{c}_{\mathrm{s}}^2 4\,\pi \mathrm{r}^3 \mathrm{dr} \tag{7.28}$$

If the velocity of sound does not depend on the radius of the body, then (7.28) is

$$\begin{aligned} -\int_0^R \frac{\partial \mathrm{p}}{\partial \mathrm{r}} 4\pi \mathrm{r}^3 \mathrm{dr} &= -\int_0^R \mathrm{c}_{\mathrm{s}}^2 4\pi \mathrm{r}^3 \mathrm{d}\rho = -\rho(\mathrm{r})\mathrm{c}_{\mathrm{s}}^2 4\pi^3 \big|_0^R \\ &+ \int_0^R 3\mathrm{x}4\pi \mathrm{r}^3 \rho(\mathrm{r})\mathrm{c}_{\mathrm{s}}^2 \mathrm{dr} = 3\mathrm{c}_{\mathrm{s}}^2 \mathrm{m}, \end{aligned} \tag{7.29}$$

where $\rho(\mathrm{R}) = 0,\;\; \rho(0) \neq \infty$.

In the general case when the velocity of sound depends on the radius, the expression for the energy of elastic deformations can be written by the expression for the velocity of sound as a mean value through the mass body, i.e.,

$$2\mathrm{E_e} = -\int_0^R \frac{\partial \mathrm{p}}{\partial \mathrm{r}} 4\pi \mathrm{r}^3 \mathrm{dr} = 3\mathrm{m}\bar{\mathrm{c}}_{\mathrm{s}}^{\,2}, \tag{7.30}$$

where

$$\bar{c}_s^2 = -\frac{4\pi}{3m}\int_0^R c_c^2(r)r^3 d\rho(r). \qquad (7.31)$$

The phenomenological parameter c_s takes into account the different aggregative states of the substance of a body, i.e. gaseous, liquid, solid, and plasma (Ferronsky et al. 1981a).

The second term in the right-hand side of Eq. 7.24 is the potential energy of the sphere:

$$4\,\pi G\int_0^R r\rho(r)m(r)dr = U. \qquad (7.32)$$

Finally, Eq. 7.24 can be rewritten in the form of the Jacobi virial equation:

$$\ddot{\Phi} = 2T + 3m\bar{c}_s^2 + U = 2E - U, \qquad (7.33)$$

where $E = T + 3/2m\bar{c}_s^2 + U$ is the total energy of the body.

One may see now that Eq. 7.33 represents the generalized virial equation (2.31) obtained from Euler's equation of motion for a deformable body (7.1) by means of transformation of the kinetic energy of the mass interaction (7.25) through the polar moment of inertia (the Jacobi function Φ). The polar moment of inertia in (7.33) has a functional relation with the potential energy (7.32). Here the polar moment of inertia physically represents the body's structure and shows its changes with a change in the potential energy.

Averaging Jacobi's virial equation (7.33) with respect to a sufficiently long period of time gives the classical averaged virial theorem of the body and expresses the condition of its hydrostatic equilibrium in the outer force field without taking into account kinetic energy of the interacted masses:

$$2E = U,$$

or

$$-U = 3m\overline{c_s^2}. \qquad (7.34)$$

Equation 7.34 follows also from the condition of hydrostatic equilibrium of the sphere matter when the left-hand side of the equation of motion (7.23) is equal to zero:

$$\frac{\partial p}{\partial r} = \rho F_r.$$

In accordance with (7.30), the left-hand side of this equation represents a double value of the total body's energy, the matter of which stays in hydrostatic equilibrium in the outer uniform force field. The right-hand side of the equation determines the potential energy of the matter interaction.

It follows from Eq. 7.34 that the velocity of sound in the elastic media determines not only the potential energy of the mass interaction, but also the velocity of propagation of the potential energy flux in the media. This relationship between the potential energy and the sound velocity is used now in seismic studies for determination of mass density. We also will apply it for interpretation of the radial density distribution of the Earth.

To obtain the equation of virial oscillations from (7.33), we accept the following assumptions. We assume that the total energy E has a constant value and the relationship between the Jacobi function Φ and the gravitational potential energy U is held in the form

$$\left|U\right|\sqrt{\Phi} = B = \text{const} \tag{7.35}$$

It follows from Eqs. 2.31 and 2.32 and Table 1 of Chap. 2 that, for a density-uniform sphere, the expression (7.35) is strict. Then, for the uniform body Eq. 7.33 with the help of (7.35) can be written in the form of a non-linear differential equation of the second order with respect to the variable Φ:

$$\ddot{\Phi} = -A + \frac{B}{\sqrt{\Phi}} \tag{7.36}$$

where$A = -2E = |U|$

Expression (7.36) is the equation of virial oscillations of the uniform Earth. As it is shown in Chap. 4, the solution of Eq. 7.36 represents the periodic change of polar moment of inertia, i.e. oscillation of the interacting mass particles, and synchronously with this change of the potential (kinetic) energy. The solution of (7.36) is described by Eqs. 4.14 and 4.15 and we rewrite them in the same form:

$$\sqrt{\Phi_0} = \frac{B}{A}[1 - \ \varepsilon\cos(\lambda - \psi)], \tag{7.37}$$

$$t = \frac{4B}{(2A)^{3/2}}[\ \lambda - \ \varepsilon\sin(\lambda - \psi)] \tag{7.38}$$

Here ε and ψ are the integration constants depending on the initial conditions of the Jacobi function Φ_o and its first derivative at the first moment of time t_o .

Equations 7.37 and 7.38 and the integration constants after corresponding generalization were obtained in Chap. 4 in the explicit form:

$$\sqrt{\Phi_0} = a[1 - \ \varepsilon\ \cos\varphi], \tag{7.39}$$

$$\omega = \frac{2\pi}{T_V} = \frac{(2A)^{3/2}}{4B} = \frac{(2{\cdot}2E)^{3/2}}{4U\sqrt{\Phi}} = \sqrt{\frac{GM}{r^3}} = \sqrt{\frac{4}{3}\pi G\rho}, \tag{7.40}$$

$$a = \frac{B}{A}, \tag{7.41}$$

$$M_c = \varphi - \varepsilon \sin\varphi, \tag{7.42}$$

where $M_c = n(t - \tau)$; T_v is the period of the virial oscillations; ω is the frequency of oscillations; ρ is the mass density of the body.

Physical meaning of Eqs. 7.39–7.42 is expressed by Kepler's motion laws, in particular, Eq. 7.39 describes the first and second laws, and Eq. 7.40 describes its third law. And now, due to the functional relationship between the potential energy and the Jacobi function (polar moment of inertia), the Kepler laws express dynamic but not static equilibrium of the planet. In fact, the Jacobi function $\sqrt{\Phi_0}$, which represents the polar moment of inertia, traces a second-order curve with period of oscillations T_v and frequency ω. The curve of the uniform body is the circle having the semi-major axis a and the eccentricity ε. The expression (7.42) represents the Kepler equation. Equations 7.39–7.42 as a whole describe an oscillating motion of the body in accordance with Kepler's laws. But their significance grows owing to the involved volumetric polar moment of inertia and its relationship with the potential energy of the body determining its dynamical effects. Figure 5.3 demonstrates the graphic picture of this effect and Eqs. 4.36–4.38 show the effect in the explicit form by the Lagrange series:

$$\sqrt{\Phi_0} = \frac{B}{A}\left[1 + \frac{\varepsilon^2}{2} + \left(-\varepsilon + \frac{3}{8}\varepsilon^3\right)\cos M_c - \frac{\varepsilon^2}{2}\cos 2M_c - \frac{3}{8}\varepsilon^3\cos 3M_c + \ldots\right] \tag{7.43}$$

$$\Phi_0 = \frac{B^2}{A^2}\left[1 + \frac{3}{2}\varepsilon^2 + \left(-2\varepsilon + \frac{\varepsilon^3}{4}\right)\cos M_c - \frac{\varepsilon^2}{2}\cos 2M_c - \frac{\varepsilon^3}{4}\cos 3M_c + \ldots\right] \tag{7.44}$$

Note that because the polar moment of inertia of the body has a functional relationship with the potential energy, Eqs. 7.43 and 7.44 and Fig. 5.3 express also the effect of a change of the potential energy in time. This fact is important for understanding of the nature and mechanism of the mass particle interaction, the result of which is generation of the potential energy. Later on we will discuss this problem in more details.

Quantitative values of the parameters of the body's virial oscillations are considered together with solution of the problem of oscillation and rotation of the non-uniform body. In order to do this the potential and kinetic energies of a non-uniform body need to be expanded and some other effects of its non-uniform structure have to be understood. It was noted in Section 2.5 of Chap. 2 that the non-uniformities play an important role in dynamical processes of bodies. Let us start consideration of the effect of the non-uniformities with separation of the potential and kinetic energies.

7.2 Structure of Potential and Kinetic Energies of a Non-uniform Body

In fact, all the celestial bodies of the Solar System, including the Sun, are non-uniform creatures. They have a shell structure and the shells themselves are also non-uniform elements of a body. It was shown in Section 2.2 of Chap. 2 that according to the artificial satellite data all the measured gravitational moments of the Earth, including tesseral ones, have significant values. In geophysics this fact is interpreted as a deviation of the Earth from the hydrostatic equilibrium and attendance of the tangential forces which are continuously developed inside the body. From the point of view of the planet's dynamical equilibrium, the fact of the measured zonal and tesseral gravitational moments is a direct evidence of permanent development of the normal and tangential volumetric forces which are the components of the inner gravitational force field. In order to identify the above effects the inner force field of the body should be accordingly separated.

The expressions (2.46)–(2.49) in Chap. 2 indicate that the force function and the polar moment of a non-uniform self-gravitating sphere can be expanded with respect to their components related to the uniform mean density mass and its non-uniformities. In accordance with the superposition principle these components are responsible for the normal and tangential dynamical effects of a non-uniform body. Such a separation of the potential energy and polar moment of inertia through their dimensionless form-factors α^2 and β^2 was done by Garcia Lambas et al. (Garcia et al. 1985) with our interpretation (Ferronsky et al. 1996). Taking into account that the observed satellite irregularities are caused by the non-uniform distribution of the mass density, an auxiliary function relative to the radial density distribution was introduced for the separation:

$$\Psi(s) = \int_0^s \frac{(\rho_r - \rho_0)}{\rho_0} x^2 dx, \tag{7.45}$$

where $s = r/R$ is the ratio of the running radius to the radius of the sphere R; ρ_o is the mean density of the sphere of radius r ; ρ_r is the radial density; x is the running coordinate; the value $(\rho_r - \rho_o)$ satisfies $\int_0^R (\rho_r - \rho_0) r^2 dr = 0$ and the function $\Psi(1) = 0$.

The function $\Psi(s)$ expresses a radial change in the mass density of the non-uniform sphere relative to its mean value at the distance r/R. Now we can write expressions for the force function and the moment of inertia by using the structural form-factors α^2 and β^2 which were introduced in Section 2.5:

$$U = 4\pi G \int_0^R r\rho(r)m(r)dr = \alpha^2 \frac{Gm^2}{R}, \tag{7.46}$$

$$I = 4\pi \int_0^R r^4 \rho(r) dr = \beta^2 m R^2 \tag{7.47}$$

By (7.45) we can do the corresponding change of variables in (7.46) and (7.47). As a result, the expressions for the potential energy U and polar moment of inertia I are found in the form of their components composed of their uniform and non-uniform constituents (Garcia Lambas et al. 1985; Ferronsky et al. 1996):

$$U = \alpha \frac{Gm^2}{R} = \left[\frac{3}{5} + 3\int_0^1 \psi x dx + \frac{9}{2}\int_0^1 \left(\frac{\psi}{x}\right)^2 dx\right] \frac{Gm^2}{R}, \tag{7.48}$$

$$I = \beta^2 m R^2 = \left[\frac{3}{5} - 6\int_0^1 \psi x dx\right] m R^2 \tag{7.49}$$

It is known that the moment of inertia multiplied by the square of the frequency ω of the oscillation-rotational motion of the mass is the kinetic energy of the body. Then Eq. 6.49 can be rewritten

$$K = I\,\omega^2 = \beta^2 M R^2\,\omega^2 = \left[\frac{3}{5} - 6\int_0^1 \psi x dx\right] m R^2 \omega^2. \tag{7.50}$$

Let us clarify the physical meaning of the terms in expressions (7.48) and (7.50) of the potential and kinetic energy

As it follows from (2.46) and Table 1, the first terms in (7.48) and (7.50), numerically equal to 3/5, represent α_0^2 and β_0^2 being the structural coefficients of the uniform sphere with radius r, the density of which is equal to its mean value. The ratio of the potential and kinetic energies of such a sphere corresponds to the condition of the body's dynamical equilibrium when its kinetic energy is realized in the form of oscillations.

The second terms of the expressions can be rewritten in the form

$$3\int_0^1 \psi x dx \equiv 3\int_0^1 \left(\frac{\psi}{x}\right) x^2 dx, \tag{7.51}$$

$$-6\int_0^1 \psi x dx \equiv -6\int_0^1 \left(\frac{\psi}{x}\right) x^2 dx \tag{7.52}$$

One can see that there are written here the additive parts of the potential and kinetic energies of the interacting masses of the non-uniformities of each sphere shell with the uniform sphere having a radius r of the sphere shell. Note that the structural coefficient β^2 of the kinetic energy is twice as high as the potential energy and has the minus sign. It is known from physics that interaction of mass particles, uniform and non-uniform with respect to density is accompanied by their elastic and inelastic scattering of energy and appearance of a tangential component in their trajectories of motion. In this particular case the second terms in Eqs. 7.48 and 7.50 express the tangential (torque) component of the potential and kinetic energy of the body. Moreover, the rotational component of the kinetic energy is twice as much as the potential one.

The third term of Eq. 7.48 can be rewritten as

$$\frac{9}{2}\int_0^1 \left(\frac{\psi}{x}\right)^2 dx \equiv \frac{9}{2}\int_0^1 \left(\frac{\psi}{x^2}\right)^2 x^2 dx. \tag{7.53}$$

Here, there is another additive part of the potential energy of the interacting non-uniformities. It is the non-equilibrated part of the potential energy which does not have an appropriate part of the reactive kinetic energy and represents a dissipative component. Dissipative energy represents the electromagnetic energy which is emitted by the body and it determines the body's evolutionary effects. This energy forms the electromagnetic field of the body.

Thus, by expansion of the expression of the potential energy and the polar moment of inertia we obtained the components of both forms of energy which are responsible for oscillation and rotation of the non-uniform body. Applying the above results we can write separate conditions of the dynamical equilibrium for each form of the motion and separate virial equations of the dynamical equilibrium of their motion.

7.3 Equations of Dynamical Equilibrium of Oscillation and Rotation of a Body

Equations 7.48 and 7.50 can be written in the form

$$U = (\alpha_0^2 + \alpha_t^2 + \alpha_\gamma^2)\frac{Gm^2}{R}, \tag{7.54}$$

$$K = (\beta_0^2 - 2\beta_t^2)mR^2\omega^2, \tag{7.55}$$

where $\alpha_0^2 = \beta_0^2$; $2\alpha_t^2 = \beta_t^2$; the subscripts o, t, γ define the radial, tangential, and dissipative components of the considered values.

Because the potential and kinetic energies of the uniform body are equal ($\alpha_0^2 = \beta_0^2 = 3/5$) then from (7.48) and (7.50) one has

$$U_o = K_o, \tag{7.56}$$

$$E_o = U_o + K_o = 2U_o \tag{7.57}$$

In order to express dynamical equilibrium between the potential and kinetic energies of the non-uniform interacting masses we can write, from (7.48) and (7.50),

$$2U_t = K_t, \tag{7.58}$$

$$E_t = U_t + K_t = 3U_t, \tag{7.59}$$

where E_o, E_t, U_o, K_o, U_t, K_t are the total, potential and kinetic energies of oscillation and rotation accordingly. Note, that the energy is always a positive value.

Equations 7.56–7.59 present expressions for uniform and non-uniform components of an oscillating system which serves as the conditions of their dynamical equilibrium. Evidently, the potential energy U_γ of interaction between the non-uniformities, being irradiated from the body's outer shell, is irretrievably lost and provides a mechanism of body's evolution.

In accordance with classical mechanics, for the above-considered non-uniform gravitating body, being a dissipative system, the torque N is not equal to zero, the angular momentum L of the sphere is not a conservative parameter, and its energy is continuously spent during the motion, i.e.,

$$N = \frac{dL}{dt} > 0, \quad L \neq \text{const.}, \quad E \neq \text{const.} > 0$$

A system physically cannot be conservative if friction or other dissipation forces are present, because F·ds due to friction is always positive and an integral cannot vanish (Goldstein 1980), i.e.: $\oint F \cdot ds > 0$.

7.4 Equations of Oscillation and Rotation of a Body and Their Solution

After we have found that the resultant of the body's gravitational field is not equal to zero and the system's dynamical equilibrium is maintained by the virial relationship between the potential and kinetic energies, the equations of a self-gravitating body motion can be written.

Earlier (Ferronsky et al. 1987) we used the obtained virial equation (7.33) for describing and studying the motion of both uniform and non-uniform self-gravitating spheres. Jacobi (1884) derived it from Newton's equations of motion of n mass points and reduced the n-body problem to the particular case of the one-body task with two independent variables, namely, the force function U and the polar moment of inertia Φ, in the form

$$\ddot{\Phi} = 2E - U. \tag{7.60}$$

Equation 7.60 represents the energy conservation law and describes the system in scalar U and Φ volumetric characteristics. In Chap. 3 it was shown that Eq. 7.60 is also derived from Euler's equations for a continuous medium, and from the equations of Hamilton, Einstein, and quantum mechanics. Its time-averaged form gives the Clausius virial theorem for a system with outer source of forces. It was earlier mentioned that Clausius was deducing the theorem for application in thermodynamics and, in particular, as applied to assessing and designing of Carnot's machines. As the machines operate in the Earth's outer force field, Clausius introduced the coefficient 1/2 to the term of "living force" or kinetic energy, i.e.,

$$K = \frac{1}{2}\sum_i m_i v_i^2$$

As Jacobi has noted, the meaning of the introduced coefficient was to take into account only the kinetic energy generated by the machine, but not by the Earth's gravitational force. That was demonstrated, for instance, by the work of a steam hammer for driving piles. The machine raises the hammer, but it falls down under the action of the force of the Earth's gravity. That is why the coefficient 1/2 of the kinetic energy of a uniform self-gravitating body in Eqs. 7.48–7.50 has disappeared. In its own force field the body moves due to release of its own energy.

Earlier by means of relation $U\sqrt{\Phi} \approx$ const, an approximate solution of Eq. 7.60 for a non-uniform body was obtained (Ferronsky et al. 1987). Now, after expansion of the force function and polar moment of inertia, at $U_\gamma = 0$ and taking into account the conditions of the dynamical equilibrium (7.57) and (7.59), Eq. 7.60 can be written separately for the radial and tangential components in the form

$$\ddot{\Phi}_0 = \frac{1}{2}E_0 - U_0, \tag{7.61}$$

$$\ddot{\Phi}_t = \frac{1}{3}E_t - U_t. \tag{7.62}$$

Taking into account the functional relationship between the potential energy and the polar moment of inertia

$$|U|\sqrt{\Phi} = B = \text{const},$$

and taking into account that the structural coefficients $\alpha_0^2 = \beta_0^2$ and $2\alpha_t^2 = \beta_t^2$, both Eqs. 7.61 and 7.62 are reduced to an equation with one variable and have a rigorous solution:

$$\ddot{\Phi}_n = -A_n + \frac{B_n}{\sqrt{\Phi_n}}, \tag{7.63}$$

where A_n and B_n are the constant values and subscript n defines the non-uniform body.

The general solution of Eq. 7.62 is (4.14) and (4.15):

$$\sqrt{\Phi_n} = \frac{B_n}{A_n}[1 - \varepsilon\cos(\xi - \varphi)], \tag{7.64}$$

$$\omega = \frac{2\pi}{T_v} = \frac{4B}{(2A)^{3/2}}[\xi - \varepsilon\sin(\xi - \varphi)], \tag{7.65}$$

where ε and φ are, as previously, the integration constants depending on the initial values of Jacobi's function Φ_n and its first derivative Φ_n at the time moment t_0 (the time here is an independent variable); T_v is the period of virial oscillations; ω is the oscillation frequency; ξ is the auxiliary independent variable; $A_n = A_0 - 1/2 E_0 > 0$; $B_n = B_0 = U_0\sqrt{\Phi_0}$ for radial oscillations; $A_n = A_t = -1/3E_t, > 0$; $B_n = B_t = U_t\sqrt{\Phi_t}$ for rotation of the body.

The expressions for the Jacobi function and its first derivative in an explicit form can be obtained after transforming them into the Lagrange series:

$$\sqrt{\Phi_n} = \frac{B}{A}\left[1 + \frac{\varepsilon^2}{2} + \left(-\varepsilon + \frac{3}{8}\varepsilon^3\right)\cos M_c - \frac{\varepsilon^2}{2}\cos 2M_c - \frac{3}{8}\varepsilon^3\cos 3M_c + \ldots\right],$$

$$\Phi_n = \frac{B^2}{A^2}\left[1 + \frac{3}{2}\varepsilon^2 + \left(-2\varepsilon + \frac{\varepsilon^3}{4}\right)\cos M_c - \frac{\varepsilon^2}{2}\cos 2M_c - \frac{\varepsilon^3}{4}\cos 3M_c + \ldots\right], \tag{7.66}$$

$$\Phi_n = \sqrt{\frac{2}{A}}\,\varepsilon B\left[\sin M_c + \frac{1}{2}\varepsilon\sin 2M_c + \frac{\varepsilon^2}{2}\sin M_c(2\cos^2 M_c - \sin^2 M_c) + \ldots\right].$$

Radial frequency of oscillation ω_{or} and angular velocity of rotation ω_{tr} of the shells of radius r can be rewritten from (7.40) as

$$\omega_{0r} = \frac{(2A_0)^{3/2}}{4B_0} = \sqrt{\frac{U_{or}}{J_{or}}} = \sqrt{\frac{\alpha_{or}^2 Gm_r}{\beta_{or}^2 r^3}} = \sqrt{\frac{4}{3}\pi G\rho_{0r}}, \tag{7.67}$$

$$\omega_{tr} = \frac{(2A_t)^{3/2}}{4B_t} = \sqrt{\frac{2U_{tr}}{J_{tr}}} = \sqrt{\frac{2\alpha_{tr}^2 Gm_r}{\beta_{tr}^2 r^3}} = \sqrt{\frac{4}{3}\pi G\rho_{0r}k_{er}}, \tag{7.68}$$

where U_{or} and U_{tr} are the radial and tangential components of the force function (potential energy); J_{or} and $J_{tr} = 2/3J_{or}$ are the polar and axial moment of inertia; $\rho_{0r} = \frac{1}{V_r}\int\limits_{V_r} \rho(r)dV_r$; ρ(r) is the law of radial density distribution; ρ_{0r} is the mean density value of the sphere with a radius r; V_r is the sphere volume with a radius r; $2\alpha_{tr}^2 = \beta_{tr}^2$; k_{er} is the dimensionless coefficient of the energy dissipation or tidal friction of the shells equal to the shell's oblateness.

The relations (7.64)–(7.65) represent Kepler's laws of body rotation in dynamical equilibrium. In the case of uniform mass density distribution the frequency (7.57) of oscillation of the sphere's shells with radius r is $\omega_{or} = \omega_o = \text{const}$. It means that here all the shells are oscillating with the same frequency. Thus, it appears that only non-uniform bodies are rotating systems.

Rotation of each body's shell depends on the effect of the potential energy scattering at the interaction of masses of different density. As a result, a tangential component of energy appears which is defined by the coefficient k_{er}. In geodynamics the coefficient is known as the geodynamical parameter. Its value is equal to the ratio of the radial oscillation frequency and the angular velocity of a shell and can be obtained from Eqs. 7.67 and7.68), i.e.

$$k_e = \frac{\omega_t^2}{\omega_0^2} = \frac{\omega_t{}^2}{\frac{4}{3}\pi G\,\rho_0}. \tag{7.69}$$

It was found that in the general case of a three-axial (a, b, c) ellipsoid with the ellipsoidal law of density distribution, the dimensionless coefficient $k_e \in [0, 1]$ is equal to (Ferronsky et al. 1987)

$$k_r = \frac{F(\varphi, f)}{\sin\varphi} \bigg/ \frac{a^2 + b^2 + c^2}{3a^2},$$

where$\varphi = \arcsin\sqrt{\frac{a^2-c^2}{a^2}}$, $f = \sqrt{\frac{a^2-b^2}{a^2-c^2}}$, and F(φ,f) is an incomplete elliptic integral of the first degree in the normal Legendre form.

Thus, in addition to the earlier obtained solution of radial oscillations (Ferronsky et al. 1987), now we have a solution of its rotation. It is seen from expression (7.67) that the shell oscillation do not depend on the phase state of the body's mass and are determined by its density.

It follows from Eqs. 7.67 and 7.68 that in order to obtain the frequency of oscillation and angular velocity of rotation of a non-uniform body, the law of radial density distribution should be revealed. This problem will be considered later on. But before that the problem of the nature of a body shells separation with respect to their density needs to be solved.

7.5 Application of Roche's Tidal Approach for Separation of the Body Shells

It is well known that celestial bodies have a quasi-spherical shell structure. This phenomenon has been confirmed by recording and interpretation of seismic longitudinal and transversal wave propagation during earthquakes. In order to understand the physics and mechanism of a body mass differentiation with respect to its density, we apply Roche's tidal dynamics.

Newton's theorem of gravitational interaction between a material point and a spherical layer states that the layer does not affect a point located inside the layer. On the contrary, the outside-located material point is affected by the spherical layer. Roche's tidal dynamics is based on the above theorem. His approach is as follows (Ferronsky et al. 1996).

There are two bodies of masses M and m interacting in accordance with Newton's law (Fig. 7.1a).

Let $M \gg m$ and $R \gg r$, where r is the radius of the body m, and R is the distance between the bodies M and m. Assuming that the mass of the body M is uniformly distributed within a sphere of radius R, we can write the accelerations of the points A and B of the body m as

$$q_A = \frac{GM}{(R-r)^2} - \frac{Gm}{r^2}, \quad q_B = \frac{GM}{(R+r)^2} + \frac{Gm}{r^2}.$$

The relative tidal acceleration of the points A and B is

$$\begin{aligned} q_{AB} &= G\left[\frac{M}{(R-r)^2} - \frac{M}{(R+r)^2} - \frac{2m}{r^2}\right] \\ &= \frac{4\pi}{3}G\left[\rho_M R^3 \frac{4Rr}{(R^2-r^2)^2} - 2\,\rho_m r\right] \approx \frac{8\pi}{3}Gr(2\,\rho_M - \,\rho_m). \end{aligned} \tag{7.70}$$

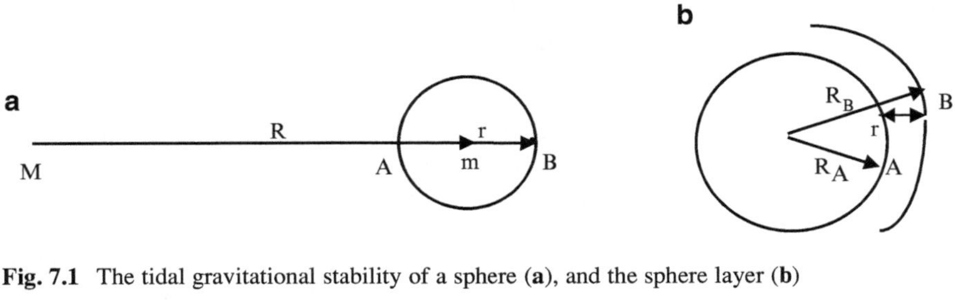

Fig. 7.1 The tidal gravitational stability of a sphere (**a**), and the sphere layer (**b**)

Here $\rho_M = M/\frac{4}{3}\pi R^3$ and $\rho_m = m/\frac{4}{3}\pi r^3$ are the mean density distributions for the spheres of radius R and r. Roche's criterion states that the body with mass m is stable against the tidal force disruption of the body M if the mean density of the body m is at least twice as high as that of the body M in the sphere with radius R. Roche considered the problem of the interaction between two spherical bodies without any interest in their creation history and in how the forces appeared. From the point of view of the origin of celestial bodies and of the interpretation of dynamical effects, we are interested in the tidal stability of separate envelopes of the same body. For this purpose we can apply Roche's tidal dynamics to study the stability of a non-uniform spherical envelope.

Let us assess the tidal stability of a spherical layer of radius R and thickness $r = R_B - R_A$ (Fig. 2.1b). The layer of mass m and mean density $\rho_m = m/4\pi R_A{}^2 r$ is affected at point A by the tidal force of the sphere of radius R_A. The mass of the sphere is M and mean density $\rho_M = M/\frac{4}{3}\pi R_A^3$ The tidal force in point B is generated by the sphere of radius R + r and mass M + m. Then the accelerations of the points A and B are

$$q_A = \frac{GM}{R_A^2}, \quad q_B = \frac{G(M+m)}{(R_A+r)^2}.$$

The relative tidal acceleration of the points A and B is

$$q_{AB} = GM\left[\frac{1}{R_A^2} - \frac{1}{(R_A+r)^2}\right] - \frac{Gm}{(R_A+r)^2}$$

$$= \left(\frac{8}{3}\pi G\rho_M - 4\pi G\rho_m\right)r = 4\pi Gr\left(\frac{2}{3}\rho_M - \rho_m\right), (R>>r). \qquad (7.71)$$

Equations 7.70 and 7.71 give the possibility to understand the nature of a body shell separation including some other dynamical effects.

7.6 Physical Meaning of Archimedes and Coriolis Forces and Separation of the Earth's Shells

The Archimedes principle states: *The apparent loss in weight of a body totally or partially immersed in a liquid is equal to the weight of the liquid displaced.* We saw in the previous section that the principle is described by Eqs. 7.70 and 7.71 and the forces that sink down or push out the body or the shell are of a gravitational nature. In fact, in the case of $\rho_n = \rho_M$ the body immersed in a liquid (or in any other medium) is kept in place due to equilibrium between the forces of the body's weight and the forces of the liquid reaction. In the case of $\rho_n > \rho_M$ or $\rho_n < \rho_M$ the body is sinking or floating up depending on the resultant of the above forces. Thus, the Archimedes forces seem to have a gravity nature and are the radial component of the Earth's inner force field.

It is assumed that the Coriolis forces appeared as an effect of the body motion in the rotational system of co-ordinates relative to the inertial reference system. In this case rotation of the body is accepted as the inertial motion and the Coriolis forces appear to be the inertial ones. It follows from the solution of Eq. 7.62 that the Coriolis' forces appear to be the tangential component of body's inner force field, and the body rotation is caused by the moment of those forces that are relative to the three-dimensional centre of inertia which also does not coincide with the three-dimensional gravity centre.

In accordance with Eq. 7.71 of the tidal acceleration of an outer non-uniform spherical layer at $\rho_M \neq \rho_m$, the mechanism of the gravitational density differentiation of masses is revealed. If $\rho_M < \rho_m$, then the shell immerses (is attracted) up to the level where $\rho_M = \rho_m$. At $\rho_M > \rho_m$ the shell floats up to the level where $\rho_M = \rho_m$ and at $\rho_M > 2/3\rho_m$ the shell becomes a self-gravitating one. Thus, in the case when the density increases towards the sphere's center, which is the Earth's case, then each overlying stratum appears to be in a suspended state due to repulsion by the Archimedes forces which, in fact, are a radial component of the gravitational interaction forces.

The effect of the gravitational differentiation of masses explains the nature of creation of shell-structured celestial bodies and corresponding processes (for instance, the Earth's crust and its oceans, geotectonic, orogenic and seismic processes, including earthquakes). All these phenomena appear to be a consequence of the continuous gravitational differentiation in density of the planet's masses. We assume that creation of the Earth and the Solar system as a whole was resulted by this effect. For instance, the mean value of the Moon's density is less than 2/3 of the Earth's, i.e., $\rho_M < 2/3\rho_m$. If one assumes that this relation was maintained during the Moon's formation, then, in accordance with Eq. 7.71, this body separated at the earliest stage of the Earth's mass differentiation. Creation of the body from the separated shell should occur by means of the cyclonic eddy mechanism, which was proposed in due time by Descartes and which was unjustly rejected. If we take into account existence of the tangential forces in the non-uniform mass, then the above mechanism seems to be realistic.

7.7 Self-similarity Principle and the Radial Component of a Non-uniform Sphere

It follows from Eq. 7.71 that in the case of the uniform density distribution ($\rho_m = \rho_M$), all spherical layers of the gravitating sphere move to the centre with accelerations and velocities which are proportional to the distance from the centre. It means that such a sphere contracts without loss of its uniformity. This property of self-similarity of a dynamical system without any discrete scale is unique for a uniform body (Ferronsky et al. 1996).

A continuous system with a uniform density distribution is also ideal from the point of view of Roche's criterion of stability with respect to the tidal effect. That is why there is a deep physical meaning in separation of the first term of potential energy in expression (7.48). A uniform sphere is always similar in its structure in spite of the fact that it is a continuously contracting system. Here, we do not consider the Coulomb forces effect. For this case we have considered the specific proton and electron branches of the evolution of the body (see Chap. 8).

Note that in Newton's interpretation the potential energy has a non-additive category. It cannot be localized even in the simplest case of the interaction between two mass points. In our case of a gravitating sphere as a continuous body, for the interpretation of the additive component of the potential energy we can apply Hooke's concept. Namely, according to Hooke there is a linear relationship between the force and the caused displacement. Therefore the displacement is in square dependence on the potential energy. Hooke's energy belongs to the additive parameters. In the considered case of a gravitating sphere, the Newton force acting on each spherical layer is proportional to its distance from the centre. Thus, here from the physical point of view, the interpretations of Newton and Hooke are identical.

At the same time in the two approaches there is a principal difference even in the case of uniform distribution of the body density. According to Hooke the cause of displacement, relative to the system, is the action of the outer force. And if the total energy is equal to the potential energy, then equilibrium of the body is achieved. The potential energy plays here the role of elastic energy. The same uniform sphere with Newton's forces will be contracted. All the body's elementary shells will move without change of uniformity in the density distribution. But the first terms of Eqs. 7.48–7.50 show that the tidal effects of a uniform body restrict motion of the interacting shells towards the centre. In accordance with Newton's third law and the d'Alembert principle the attraction forces, under the action of which the shells move, should have equally and oppositely direct forces of Hooke's elastic counter-action. In the framework of the elastic gravitational interaction of shells, the dynamical equilibrium of a uniform sphere is achieved in the form of its elastic oscillations with equality between the potential and kinetic energy. The uniform sphere is dynamically stable relative to the tidal forces in all of its shells during the time of the system contraction. Because the potential and kinetic energies of a sphere are equal, then its total energy in the framework of the averaged virial theorem within one period of oscillation is accepted formally as equal to zero. Equality of the potential and kinetic energy of each shell means the equality of the centripetal (gravitational) and centrifugal (elastic constraint) accelerations. This guarantees the system remaining in dynamical equilibrium. On the contrary, all the spherical shells will be contracted towards the gravity centre which, in the case of the sphere, coincides with the inertia centre but does not coincide with the geometric centre of the masses. Because the gravitational forces are acting continuously, the elastic constraint forces of the body's shells are reacting also continuously.

The physical meaning of the self-gravitation of a continuous body consists in the permanent work which applies the energy of the interacting shell masses on one side and the energy of the elastic reaction of the same masses in the form of oscillating motion on the other side. At dynamical equilibrium the body's equality of potential and kinetic energy means that the shell motion should be restricted by the elastic oscillation amplitude of the system. Such an oscillation is similar to the standing wave which appears without transfer of energy into outer space. In this case the radial forces of the shell's elastic interactions along the outer boundary sphere should have a dynamical equilibrium with the forces of the outer gravitational field. This is the condition of the system to be held in the outer force field of the mother's body. Because of this, while studying the dynamics of a conservative system, its rejected outer force field should be replaced by the corresponding equilibrated forces as they do, for instance, in Hooke's theory of elasticity.

Thus, from the point of view of dynamical equilibrium the first terms in Eqs. 7.48–7.50 represent the energy which provides the field of the radial forces in a non-uniform sphere. Here, the potential energy of the uniform component plays the role of the active force function, and the kinetic energy is the function of the elastic constraint forces.

7.8 Charges-Like Motion of Non-uniformities and Tangential Component of the Force Function

Let us now discuss the tidal motion of non-uniformities due to their interactions with the uniform body. The potential and kinetic energies of these interactions are given by the second terms in Eqs. 7.48 and 7.50. In accordance with (7.71), the non-uniformity motion looks like the motion of electrical charges interacting on the background of a uniform sphere contraction. Spherical layers with densities exceeding those of the uniform body (positive anomalies) come together and move to the centre in elliptic trajectories. The layers with deficit of the density (negative anomalies) come together, but move from the centre on the parabolic path. Similar anomalies come together, but those with the opposite sign are dispersed with forces proportional to the layer radius. In general, the system tends to reach a uniform and equilibrium state by means of redistribution of its density up to the uniform limit. Both motions happen not relative to the empty space, but relative to the oscillating motion of the uniform sphere with a mean density. Separate consideration of motion of the uniform and non-uniform components of a heterogeneous sphere is justified by the superposition principle of the forces action which we keep here in mind. The considered motion of the non-uniformities looks like the motion of the positive and negative charges interacting on the background of the field of the uniformly dense sphere (Ferronsky et al. 1996). One can see here that in the case of gravitational interaction of mass particles of a continuous body, their motion is the consequence not only of mutual attraction, but also mutual repulsion by the same law $1/r^2$. In fact, in the case of a real natural non-uniform body it appears that the Newton and Coulomb laws are identical in details. Later on, while considering a body's by-density differentiated masses, the same picture of motion of the positive and negative anomalies will be seen.

If the sphere shells, in turn, include density non-uniformities, then by means of Roche's dynamics it is possible to show that the picture of the non-uniformity motion does not differ from that considered above.

In physics the process of interaction of particles with different masses without redistribution of their moments is called elastic scattering. The interaction process resulting in redistribution of their moments and change in the inner state or structure is called inelastic scattering. In classical mechanics while solving the problems of motion of the uniform conservative systems (like motion of the material point in the central field or motion of the rigid body), the effects of the energy scattering do not appear. In the problem of dynamics of the self-gravitating body, where interaction of the shells with different masses and densities are considered, the elastic and inelastic scattering of the energy becomes an evident fact following from consideration of the physical meaning of the expansion of the energy expressions in the form of (7.48) and (7.50). In particular, their second terms represent the potential and kinetic energies of gravitational interaction of masses having a non-uniform density with the uniform mass and express the effect of elastic scattering of density-different shells. Both terms differ only in the numeric coefficient and sign. The difference in the numerical coefficient evidences that the potential energy here is equal to half of the kinetic one ($U_t = 1/2K_t$). This part of the active and reactive force function characterizes the degree of the non-coincidence of the volumetric centre of inertia and that of the gravity centre of the system expressed by Eqs. 2.48 and 2.49. This effect is realized in the form of the angular momentum relative to the inertia centre.

Thus, we find that inelastic interaction of the non-uniformities with the uniform component of the system generates the tangential force field which is responsible for the system rotation. In other words, in the scalar force field of the by-density uniform body the vector component appears. In such a case, we can say that, by analogy with an electromagnetic field, in the gravitational scalar potential field of the non-uniform sphere U(R, t) the vector potential A(R, t) appears for which U = rot A and the field U(R, t) will be solenoidal. In this field the conditions for vortex motion of the masses are born, where div A = 0. This vector field, which in electrodynamics is called solenoidal, can be represented by the sum of the potential and vector fields. The fields, in addition to the energy, acquire moments and have a discrete-wave structure. In our case the source of the wave effects appears to be the interaction between the elementary shells of the masses by means of which we can construct a continuous body with a high symmetry of forms and properties. The source of the discrete effects can be represented by the interacting structural components of the shells, namely, atoms, molecules and their aggregates. We shell continue discussion about the nature of the gravitational and electromagnetic energy in Chap. 8.

7.9 Radial Distribution of Mass Density and the Body's Inner Force Field

At present only the Earth has experimental data which allow to interpret them with respect to radial distribution of the body's mass density. Taking into account our consideration of dynamics of celestial bodies as self-gravitating systems we assume that formation of the Earth's mass density distribution is typical at least for all the planets and satellites.

The existent idea about the radial mass density distribution of the Earth is based on interpretation of transmission velocity of the longitudinal and transverse seismic waves. Figure 7.2 presents the classic curve of transmission velocities of the longitudinal and transverse seismic waves in the Earth plotted after generalization of numerous experimental data (Jeffreys 1970; Melchior 1972; Zharkov 1978).

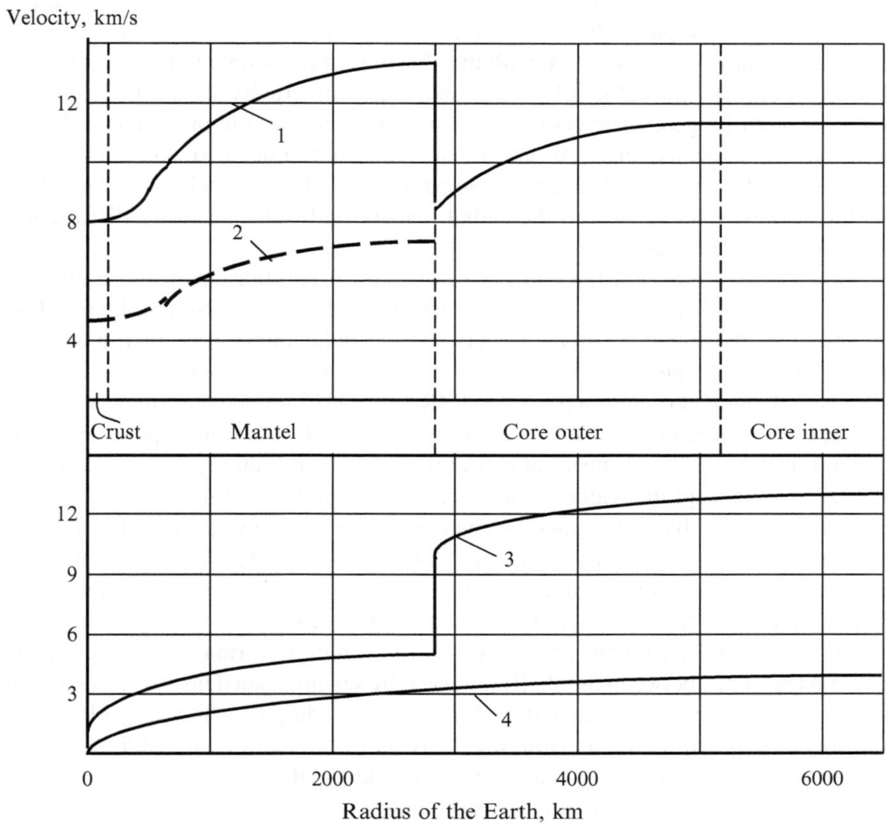

Fig. 7.2 Present-day interpretation of the curves of transmission velocities of longitudinal (1) and transverse (2) seismic waves, density (3), and hydrostatic pressure (4) in the Earth

The curves of the radial density and hydrostatic pressure distribution based on interpretation of the velocities of the longitudinal and transverse seismic waves are also shown.

The picture of the transmission velocities of the seismic waves was obtained by observations and therefore is realistic and correct. But interpretation of the obtained data was based on the idea of hydrostatic equilibrium of the Earth. It leads to incredibly high pressures in the core and high values of the mass density.

In accordance with Bullen's approach for interpretation of the seismic data, the density distribution is characterized by the following values (Bullen 1974; Melchior 1972; Zharkov 1978). The density of the crust rocks is 2.7–2.8 g/cm^3 and increases towards the centre by a certain curve up to ~13.0 g/cm^3 with jumps at the Mohorovičić-Guttenberg discontinuity, between the upper and lower mantle, and on the border of the outer core. Within the core the values of the transverse seismic waves are equal to zero. Despite the jump of the longitudinal seismic wave velocity at the outer core border dropping down, Bullen accepted that the density increases toward the center. It was done after his unsuccessful attempt to approximate the seismic data of the parabolic curve which gives a decrease of density in the core. Such a tendency is not consistent with the idea of iron core content. Bullen certainly had no idea that the radius of inertia and radius of gravity of the body do not coincide with its geometric centre of mass and, therefore, the maximum value of density is not located there. In accordance with our concept of the equilibrium condition of the planet and its dynamical parameters, the approach to interpretation of the seismic data related to the radial density and radial pressure distribution should be done on a new basis.

Now, when we accept the concept of dynamical equilibrium of the Earth and refuse its hydrostatic version, the basic idea to search for a solution of the problem seems to be the found relationship between the polar moment of inertia and the potential (kinetic) energy. The value of the structural form-factor of the Earth's mean axial moment of inertia $\beta^2{}_\perp = J_\perp/MR^2 = 0.3315$ found by artificial satellites (Zharkov 1978) should be taken as a starting point. The mean polar moment of inertia of the assumed spherical non-uniform planet is equal to $\beta^2 = (3/2)\beta^2{}_\perp = 0.49725$. We accept this value for development of the methodology.

Let us take as a basis the found mechanism of the shell separation with respect to the mass density which was presented in Sects. 7.5–7.8. The conditions and mechanism of the shell separation into radial and tangential components of the inner force field (by the Archimedes and Coriolis forces) represent continually acting effects and create physics for the Earth's structure formation. These effects explain the jumps between the shells observed by seismic data density. We take also into account the effect, expressed by Eq. 7.30, according to which the velocity of the sound recorded by the transmission velocity of the longitudinal and transverse seismic waves quantitatively characterize the energy of the elastic deformation of the media and velocity of its transmission there.

Applying the conception of Sect. 7.8, we accept that the non-uniformities of the spherical shells come together and, after their density becomes lower than that of the mean density of the inner sphere, move from the center by the parabolic law because they interact according to the law $1/r^2$. So, we can find a probable law of the radial density distribution in the form

$$\rho(r) = \rho_0(ax^2 + bx + c), \qquad (7.72)$$

where $x = r/R$ is the ratio of the running and the final radius of the planet; ρ_0 is the body's mean density; a, b, c are the numerical coefficients.

The numerical coefficients were selected for different densities for the upper shell and in such a way that the planet's total mass M would be constant, i.e.,

$$M = 4\pi \int_0^R r^2 \rho(r) dr = 4\pi \int_0^R r^2 \rho_0\left(-a\frac{r^2}{R^2} + b\frac{r}{R} + c\right)dr$$
$$= \frac{4}{3}\pi R^3 \rho_0\left(-\frac{3}{5}a + \frac{3}{4}b + c\right)$$

Here the term $(3/5)a + (3/4)b + c = 1$ in the right-hand side of the expression allows us to calculate and plot the distribution density curves in a dimensionless form.

We have selected three most typical parabolas (7.73) which satisfy the condition of equality of their moment of inertia, found by artificial satellite data, namely, the axial moment of inertia $J_\perp = \beta_\perp^2 mR^2 = 0.3315\,mR^2$ or the polar moment of inertia $J = \beta^2 mR^2 = 0.4973\,mR^2$. In addition, the first relation in (7.73) represents the straight line for which the surface mass density and that in the center correspond to the present-day version and to the form-factor $\beta_\perp^2$. The fifth straight line represents the uniform spherical planet. The curve equations with selected numerical coefficients a, b, and c are as follows:

$$\begin{aligned}
&1.\quad \rho(r) = \rho_0\left(-2\frac{r}{R} + 2.495\right), \quad a = 0,\ \rho_s = 2.73\ \Gamma/\text{CM}^3;\\
&2.\quad \rho(r) = \rho_0\left(-1.51\frac{r^2}{R^2} + 0.016\frac{r}{R} + 1.894\right),\ \rho_s = 2.08\ \Gamma/\text{CM}^3;\\
&3.\quad \rho(r) = \rho_0\left(-3.26\frac{r^2}{R^2} + 2.146\frac{r}{R} + 1.3465\right),\ \rho_s = 1.28\ \Gamma/\text{CM}^3; \qquad (7.73)\\
&4.\quad \rho(r) = \rho_0\left(-5.24\frac{r^2}{R^2} + 5.132\frac{r}{R} + 0.295\right),\ \rho_s = 1.03224\ \Gamma/\text{CM}^3;\\
&5.\quad \rho(r) = \rho_0 = \text{const.}
\end{aligned}$$

Figure 7.3 shows all the curves of (7.73). They intersect the straight line 5 of the mean density in the common point which corresponds to the value $r/R = 0.61475$.

Using Eq. 7.73 and the found (by observations) form-factor $\beta_\perp^2 = 0.3315$, the main dynamical parameters were calculated for all four curves. The calculations

Fig. 7.3 Parabolic curves of radial density distribution calculated by Eq. 7.73

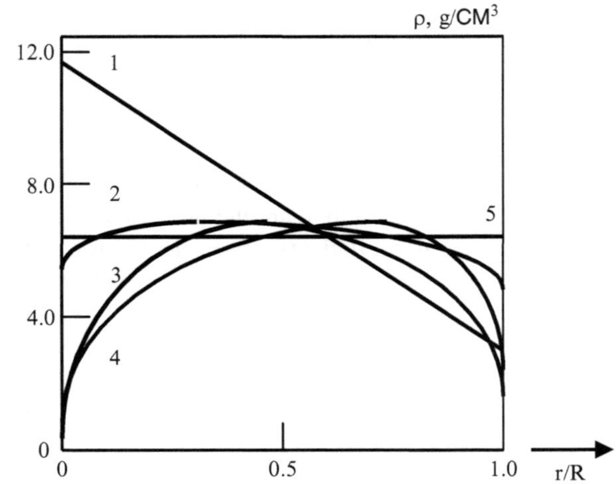

were done by the known formulae of the theory of attraction (Duboshin 1975) and taking into account the relations of (7.48) and (7.50) obtained in Sect. 7.2. These calculations are presented below for equation (4), as an example.

The potential energy of the non-uniform sphere with the density distribution law $\rho(r)$ is found from the equation:

$$U = 4\pi G \int_0^R r\rho(r)m(r)dr, \tag{7.74}$$

where

$$\rho(r) = \rho_0\left(a\frac{r^2}{R^2} + b\frac{r}{R} + c\right), \; a = -5.24; \; в = 5.132; \; c = 0.295;$$

$$m(r) = 4\pi \int_0^r r^2\rho(r)dr = 4\,\pi \int_0^r r^2\rho_0\left(a\frac{r^2}{R^2} + b\frac{r}{R} + c\right)dr$$

$$= \frac{4}{3}\,\pi r^3\left(\frac{3}{5}a\frac{r^2}{R^2} + \frac{3}{4}b\frac{r}{R} + c\right)$$

Then

$$U = 4\pi G \int_0^R r\rho_0\left(a\frac{r^2}{R^2} + b\frac{r}{R} + c\right)\frac{4}{3}\pi r^3\left(\frac{3}{5}a\frac{r^2}{R^2} + \frac{3}{4}b\frac{r}{R} + c\right)dr$$

$$= \left(\frac{4}{3}\pi\rho_0\right)^2 GR^5\frac{R}{R}\left(\frac{1}{5}a^2 + \frac{81}{160}ab + \frac{9}{28}b^2 + \frac{24}{35}ac + \frac{7}{8}bc + \frac{3}{5}c^2\right)$$

$$= 0.660143\frac{GM^2}{R}$$

The form-factor of the potential energy is $\alpha^2 = r_g{}^2/R^2 = 0.660143$, and the reduced radius of gravity is $r_g = \sqrt{0.660143R^2} = 0.8124918R$.

In accordance with Eq. 7.48, the potential energy of the non-uniform sphere is expanded into the components

$$U = U_0 + U_t + U_\gamma. \tag{7.75}$$

The potential energy of the uniform sphere is equal to

$$U_0 = \frac{3}{5}\frac{GM^2}{R}. \tag{7.76}$$

The form-factors of the potential and kinetic energy are equal to $\alpha_0^2 = 0.6$ and $\beta_0^2 = 0.6$ accordingly.

In accordance with the second term of the right–hand side of Eq. 7.48, the tangential component of the non-uniform sphere is written as

$$U_t = -\frac{1}{2}4\pi G\int_0^R r\rho_t(r)m_0(r)dr, \tag{7.77}$$

where

$$\rho_t(r) = \rho(r) - \rho_0 = \rho_0\left(a\frac{r^2}{R^2} + b\frac{r}{R} + c\right) - \rho_0 = \rho_0\left(a\frac{r^2}{R^2} + b\frac{r}{R} + c - 1\right);$$

$$m_0(r) = 4\pi\int_0^r r^2\rho_0 dr = \frac{4}{3}\pi\rho_0 r^3.$$

The coefficient ½ in (7.77) is taken as the ratio of the second terms of the right–hand side of Eqs. 7.48 and 7.50, as in this particular case the tangential component of the potential energy is determined through the tangential component of the kinetic energy and is equal to half its value. Then

$$U_t = -\frac{1}{2}4\frac{4}{3}(\pi\rho_0)^2 G\int_0^R r^4\left(a\frac{r^2}{R^2} + b\frac{r}{R} + c - 1\right)dr$$

$$= -\frac{1}{2}\frac{GM^2}{R}\left(\frac{3}{7}a + \frac{1}{2}b + \frac{3}{5}c - \frac{3}{5}\right) = 0.0513571\frac{GM^2}{R}$$

The form-factors of the tangential components of the potential and kinetic energy are equal to $\alpha_t^2 = 0.051357$ and $\beta_t^2 = 2 \cdot 0.051357 = 0.102714$ accordingly.

In accordance with the third term in the right–hand side of Eq. 7.48, the dissipative component of the potential energy of the non-uniform sphere is

$$U_\gamma = 4\pi G \int_0^R r\rho_t(r)m_t(r)dr, \tag{7.78}$$

where

$$\rho_t(r) = \rho(r) - \rho_0 = \rho_0\left(a\frac{r^2}{R^2} + b\frac{r}{R} + c - 1\right);$$

$$m_0(r) = 4\pi \int_0^r r^2\rho_t(r)dr = 4\pi \int_0^r r^2\rho_0\left(a\frac{r^2}{R^2} + b\frac{r}{R} + c - 1\right)dr$$

$$= \frac{4}{3}\pi\rho_0 r^3\left(\frac{3}{5}a\frac{r^2}{R^2} + \frac{3}{4}b\frac{r}{R} + c - 1\right).$$

Then

$$U_\gamma = 4\frac{4}{3}(\pi\rho_0)^2 G \int_0^R r^4\left(a\frac{r^2}{R^2} + b\frac{r}{R} + c - 1\right)\left(\frac{3}{5}a\frac{r^2}{R^2} + \frac{3}{4}b\frac{r}{R} + c - 1\right)dr$$

$$= \frac{GM^2}{R}\left(\frac{1}{5}a^2 + \frac{81}{160}ab + \frac{24}{35}ac - \frac{24}{35}a + \frac{9}{28}b^2 + \frac{7}{8}bc - \frac{7}{8}b + \frac{3}{5}c^2 - \frac{5}{6}c + \frac{3}{5}\right)$$

$$= 0.008786\frac{GM^2}{R} \tag{7.79}$$

So, the value of the form-factor of the dissipative component is $\alpha_\gamma^2 = 0.008786$.

The radial distribution of the potential energy for interaction of a test mass point with the non-uniform sphere is

$$U(r) = \frac{4\pi G}{r}\int_0^r r^2\rho(r)dr + 4\pi G\int_r^R r\rho(r)dr = \frac{4\pi G}{r}\int_0^r r^2\rho_0\left(a\frac{r^2}{R^2} + b\frac{r}{R} + c\right)dr$$

$$+4\pi G\int_r^R r\rho_0\left(a\frac{r^2}{R^2} + b\frac{r}{R} + c\right)dr = \frac{GMm_1}{R}\left(-\frac{3}{20}a\frac{r^4}{R^4} - \frac{1}{4}b\frac{r^3}{R^3} - \frac{1}{2}c\frac{r^2}{R^2} + \frac{3}{4a} + b + \frac{3}{2}c\right)$$

$$= \frac{GMm_1}{R}\left(0.786\frac{r^4}{R^4} - 1.283\frac{r^3}{R^3} - 0.1475\frac{r^2}{R^2} + 1.6445\right). \tag{7.80}$$

At r/R = 0, then $\alpha_u^2(r) = 1.6445$; and at r/R = 1 then $\alpha_u^2(r) = 1$.

The radial distribution of the interaction force of the test mass point with the non-uniform sphere is

$$q(r) = -\frac{4\pi G}{r^2}\int_0^r r^2\rho(r)dr = -\frac{4\pi G}{r^2}\int_0^r r^2\rho_0\left(a\frac{r^2}{R^2} + b\frac{r}{R} + c\right)dr$$

$$= -\frac{GMm_1}{R^2}\left(\frac{3}{5}a\frac{r^3}{R^3} + \frac{3}{4}b\frac{r^2}{R^2} + c\frac{r}{R}\right)$$

$$= -\frac{GMm_1}{R^2}\left(-3.144\frac{r^3}{R^3} + 3.849\frac{r^2}{R^2} + 0.295\frac{r}{R}\right). \quad (7.81)$$

At r/R = 0 then $\alpha_q^2(r) = 0$; and at r/R = 1 then $\alpha_q^2(r) = 1$.

Table 7.1 demonstrates the results of the calculated dynamical parameters for all the density curves (7.73) and Fig. 7.4 shows the curves of radial distribution of the potential energy and gravity force for the test mass point.

We wish to evaluate all four curves of mass density distribution in order to recognize which one is closer to the real Earth. In this case we keep in mind that the observed density jumps can be obtained for any curve by approximation of its continuous section with the mean value for each shell.

Figure 7.4 shows that the radial density values are substantially different for each curve. It refers, first of all, to the surface and centre of the body. At the same time Table 7.1 demonstrates the complete identity of the dynamical parameters of all the non-uniform spheres. It means that a fixed value of the polar moment of inertia permits us to have a multiplicity of curves of the radial density distribution with identical dynamical parameters of the body. The found property of the non-

Table 7.1 Physical and dynamical parameters of the Earth for the density distribution presented by Eq. 7.73

Equation no	1	2	3	4
ρ_s, g/cm^3	2.76	2.08	1.65	1.03224
ρ_c, g/cm^3	13.8	10.455	6.315	1.6284
ρ_{max}, g/cm^3/km	13.8/0	10.455/0	8.26/2096	8.57/3122
$\beta^2_\perp$	0.3315	0.3315	0.3315	0.3315238
β^2	0.49725	0.49725	0.49725	0.49725858
β_t^2	0.10275	0.10275	0.102752	0.102714
α^2	0.660737	0.660737	0.660737	0.660143
α_t^2	0.051371	0.051371	0.0513714	0.0513571
α_γ^2	0.009366	0.009366	0.009366	0.0087859
r_g, km	5178.6	5178.7	5178.6	5176.4
r_m, km	4492.6	4492.6	4492.6	4492.7

ρ_s, ρ_c, ρ_{max} are the density on the sphere's surface, in the centre, and maximal accordingly
$\beta^2_\perp$, β^2, β_t^2 are the form-factors of the axial, polar, and tangential components of the radius of inertia accordingly
α^2, α_t^2, α_γ^2 are the form-factors of the radial, tangential, and dissipative components of the force function accordingly
r_g, r_m are the radiuses of the gravity and inertia

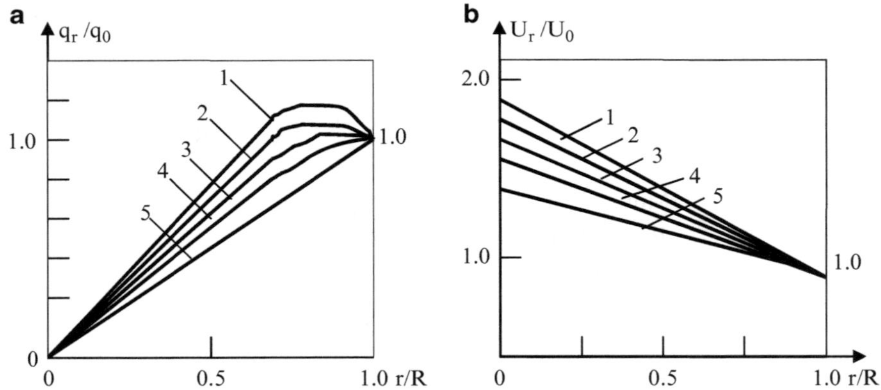

Fig. 7.4 The curves of the radial distribution of the potential energy (**a**) and gravity force (**b**) for the mass point test done by Eqs. 7.74 and 7.81

uniform self-gravitating sphere proves the rigor of the discovered functional relationship between the potential (kinetic) energy and the polar moment of inertia of the sphere. This property, in turn, is explained by the energy conservation law of a body during its motion and evolution in the form of the dynamical equilibrium equation or generalized virial theorem.

If we accept the conditions of the mass density separation presented in Sects. 7.5–7.8, then the range of curves of the density distribution gives a principal picture of its evolutionary redistribution and can be applied for reconstruction of the Earth's history. It follows from Eq. 7.71 that the density value of each overlying shell of the created Earth should be higher than the mean density of the inner mass. Otherwise, such a shell can not be retained and should be dispersed by the tidal forces. It follows from this that the planet's formation process should be strictly operated by the dynamical laws of motion in the form of the virial oscillations and accompanied by differentiation of the non-uniform shells. The model of a cyclonic vortex which was proposed by Descartes is the most acceptable from the point of view of the considered ideas of planets' and satellites' creation from a common nebula. This problem needs a separate consideration. We only note here that from the presented curves of radial density distribution the parabola (4) more closely reflects the present-day planet's evolution as fixed by observations. In this case location of the Earth's reduced inertia radius falls on the lower mantle and the reduced gravity radius – on the upper mantle. The density maximum falls also on the lower mantle. Its value is found by ordinary means, namely, by taking the derivative from the density distribution law as equated to zero. From here $\rho_{max} = 8.57$ g/cm^3 is found to be at a distance of $r = 3{,}122$ km. It means that the density maximum comes close to the border of the outer core where, as seismic observations show, the main density jump occurs. Curve (4) corrects the values of the radial density distribution in the mantle and changes its earlier interpretation in the outer and inner core. Because of zero values of the transverse velocities the matter of the inner core has

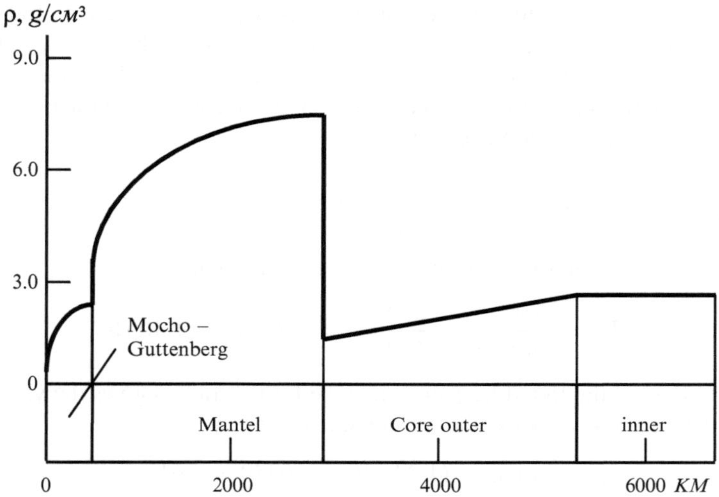

Fig. 7.5 Radial density distribution of the Earth by the authors' interpretation

a uniform density structure and, from the point of view of the equilibrium state, seems to be in a gaseous state at a pressure of 1–2 atmospheres. Taking into account the location of the maximum density value, there is a reason to assume that the outer core matter stays in the liquid or supercritical gaseous stage. In any case, the density and pressure of the inner and outer core are much lower and should have values corresponding to the seismic wave velocities. On the basis of the equation of mass density differentiation (7.71) we interpret the density jumps observed (by seismic data) near-by the Mohorovičić-Guttenberg and at the outer core borders as the borders of the shell's dynamical equilibrium. A shell which is found over that border appears in a suspended state due to action of the radial component of the gravitational pressure developed by the denser underlying shell. While the thickness of the suspended shell is growing it acquires its own equilibrium pressure (iceberg effect). The extremely high pressures in the Earth's interior, which follow from the hydrostatic equilibrium conditions, are impossible in its own force field.

The concept discussed above in relation to the Earth's density distribution is illustrated in Fig. 7.5 .

7.10 Oscillation Frequency and Angular Velocity of the Earth's Shell Rotation

In order to determine numerical values of frequency of the virial oscillations and the angular velocities, which are the main dynamical parameters of the Earth's shells, we accept equation (4) of the density distribution (7.73) as the first approximation. All further relevant calculations can be made by applying this equation.

We know the mean values of the planet's density $\rho_0 = 5.519$ g/cm^3 and angular velocity of the upper shell $\omega_t = 7.29 \cdot 10^{-5}$ s^{-1}. Applying these values, the frequency and period of the virial oscillations, and the coefficient k_e of the tangential component of the inner forces, can be found. In accordance with Eq. 7.67 the frequency of the upper shell is equal to

$$\omega_0(r) = \sqrt{\frac{4}{3}\pi G\rho_0(r)} = \sqrt{\frac{4}{3} \cdot 3.14 \cdot 6.67 \cdot 10^{-8} \cdot 5.519} = 1.24 \cdot 10^{-3}\, s^{-1}.$$

The period of oscillation is found from the expression

$$T_\omega = \frac{2\pi}{\omega_0(r)} = \frac{6.28}{1.24 \cdot 10^{-3}} = 5060.4\, s = 1.405\, h.$$

The product of the found frequency and the Earth's radius gives the value of the planet's first cosmic velocity, the mean value of which is

$$v_1 = \omega_0(r) r_e = (1.24 \cdot 10^{-3}) \cdot 6370 = 7.9\, km/s.$$

Unlike the usual expression for the first cosmic velocity in the form of $v_1 = \sqrt{GM/r}$, we used here the physical condition of the dynamical equilibrium at the Earth's surface between the inner gravitational pressure of interacting masses and the outer background pressure including atmospheric pressure.

Given below our own observation data on the near-surface atmospheric pressure and temperature oscillations at the near-surface layer and the results of the spectral analysis prove the above theoretical calculations of the planet's frequency of virial oscillations.

Now, applying the known mean value of the Earth's angular velocity $\omega_t = 7.29 \cdot 10^{-5}$ s^{-1} and the known value of the frequency of virial oscillations for the upper shell $\omega_o = 1.24 \cdot 10^{-3}$ s^{-1} by Eq. 7.69 the coefficient k_e can be found

$$k_e = \frac{\omega_t^2}{\omega_0^2} = \frac{(7.29 \cdot 10^{-5})^2}{(1.24 \cdot 10^{-3})^2} = \frac{1}{289.33} = 0.003456.$$

The coefficient k_e is known in geodynamics as a parameter that shows the ratio between the centrifugal force at the Earth's equator and the acceleration of the gravity force there equal to $k_e = 1/289.37$ (Melchior 1972). The parameter is used to study the Earth's figure based on the Clairaut hydrostatic theory.

7.10.1 Thickness of the Upper Earth's Rotating Shell

It is known that the value of the mean linear velocity of the upper planet's shell is $v_e = 0.465$ km/c. We can find the thickness h_e at which the velocity v_e corresponds to the found frequency of radial oscillations of the shell $\omega_o = 1.24 \cdot 10^{-3}$ s^{-1}

$$h_e = \frac{v}{\omega_o(r)} = \frac{0.465}{1.24 \cdot 10^{-3}} = 375\ Km. \tag{7.82}$$

Such is the thickness of the upper shell of the Earth which is rotating by forces in its own force field. It is assumed that the shell is found in the solid state. In reality it is known that the rigid shell has a thickness less than 50 km. The remaining more than 300 km-thick part of the shell has a viscous-plastic consistency, the density of which increases with depth. The border of the shell has a decreased density because of the melted substance due to high friction and saturation by a gaseous component. The border plays a role of some sort of spherical hinge. Because the density of the Earth's crust is lower than that of the underlying matter, then it occurs in the suspended state. During the oscillating motion the crust shells are affected by the alternating-sign acceleration and the inertial isostatic effects.

7.10.2 Oscillation of the Earth's Shells

Let us obtain the expression of virial oscillations for the Earth's other shells by applying expression (4) of (7.73) for the radial density distribution. Write Eq. 7.67

$$\omega_0(r) = \sqrt{\frac{4}{3}\pi G\rho_0(r)},$$

Where

$$\rho_0(\mathrm{r}) = \frac{\mathrm{m}_0(\mathrm{r})}{\frac{4}{3}\pi \mathrm{r}^3} = \frac{4\pi\int_0^r \mathrm{r}^2\rho(\mathrm{r})\mathrm{dr}}{\frac{4}{3}\pi \mathrm{r}^3} = \frac{\frac{4}{3}\pi \mathrm{r}^3\rho_0\left(\frac{3}{5}\mathrm{a}\frac{\mathrm{r}^2}{\mathrm{R}^2} + \frac{3}{4}\mathrm{b}\frac{\mathrm{r}}{\mathrm{R}} + \mathrm{c}\right)}{\frac{4}{3}\pi \mathrm{r}^3}$$
$$= \rho_0\left(\frac{3}{5}\mathrm{a}\frac{\mathrm{r}^2}{\mathrm{R}^2} + \frac{3}{4}\mathrm{b}\frac{\mathrm{r}}{\mathrm{R}} + \mathrm{c}\right).$$

Then

$$\omega_0(\mathrm{r}) = \sqrt{\frac{4}{3}\pi\,\mathrm{G}\rho_0(\mathrm{r})} = \sqrt{\frac{4}{3}\pi\,\mathrm{G}\rho_0\left(\frac{3}{5}\mathrm{a}\frac{\mathrm{r}^2}{\mathrm{R}^2} + \frac{3}{4}\mathrm{b}\frac{\mathrm{r}}{\mathrm{R}} + \mathrm{c}\right)}$$
$$= 1.24\cdot 10^{-3}\sqrt{\left(-3.144\frac{\mathrm{r}^2}{\mathrm{R}^2} + 3.849\frac{\mathrm{r}}{\mathrm{R}} + 0.295\right)}. \quad (7.83)$$

At $r/R = 0$ then $\omega_o(r) = 0.6743 \cdot 10^{-3}\ s^{-1}$; at $r/R = 1$ then $\omega_o(r) = 1.24 \cdot 10^{-3}\ s^{-1}$; at $\rho_{max} = 8.57\ g/cm^3$ $\omega_o(r) = 1.486 \cdot 10^{-3}\ s^{-1}$, where $r/R = 0.49$.

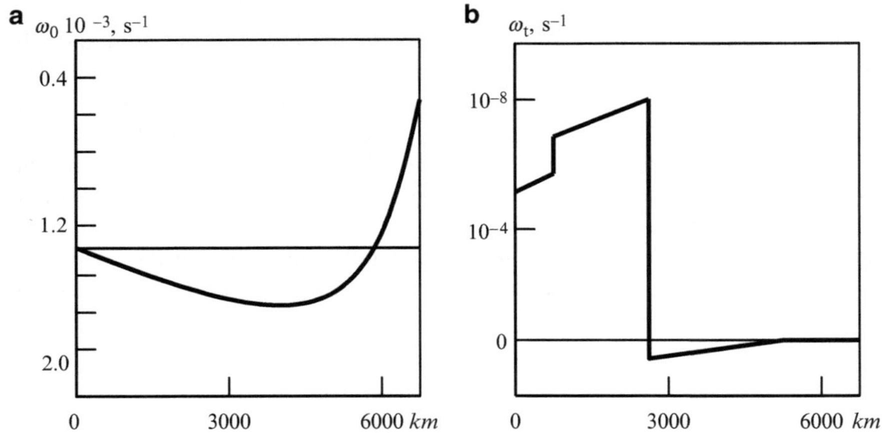

Fig. 7.6 Radial change in virial oscillation frequencies (**a**) and angular velocity of rotation (**b**) according to Eqs. 7.83 and 7.84

Figure 7.6a shows changes in the virial oscillation frequencies of the Earth's shells.

7.10.3 Angular Velocity of Shell Rotation

Angular velocity of the Earth's shell rotations is determined from Eq. 7.68

$$\omega_t(r) = \sqrt{\frac{4}{3}\pi G\rho_t(r)} = \sqrt{\frac{4}{3}\pi G\rho_0(r)\left(\frac{3}{5}a\frac{r^2}{R^2}+\frac{3}{4}b\frac{r}{R}+c\right)k_e(r)}$$

$$= \omega_0(r)\sqrt{\left(\frac{3}{5}\,a\,\frac{r^2}{R^2} + \frac{3}{4}\,b\,\frac{r}{R} + c\right)k_e(r)}$$

$$= \omega_0(r)\sqrt{\left(-3.144\,\frac{r^2}{R^2} + 3.8475\,\frac{r}{R} + 0.295\right)k_e(r)} \qquad (7.84)$$

where $\omega_t(r)$ is the angular velocity of the shell rotation; $\omega_o(r)$ is the shell oscillation frequency which is determined by Eq. 7.83.

The geodynamic parameter $k_e(r)$, which expresses the ratio of the tangential component of the force field and the gravity force acceleration for the upper shell, is approximated as

$$k_e(r) = \frac{\omega_t^2(r)}{\omega_0^2(r)} = \frac{h_c^2}{R^2}$$

where h_c is the distance between the sphere's surface and the density jump; R is the sphere's radius.

At $r/R = 1$ $k_e(r) = 0.003456$; at $r/R = 0$ $k_e(r) = 1$, $\omega_t(0) = \omega_o(0)$, i.e. the virial oscillation frequency corresponds to the gravity pressure of the uniform density masses. In this particular case we are interested in changes of the angular velocity of rotation of the upper (1,000 km) and lower (up to the core border) mantle (2,900 km) shells. Figure 7.6b shows the radial change of the angular velocity of rotation calculated by Eq. 7.84. It is seen that the angular velocity at the lower mantle – outer core is close to zero but changes its direction.

We emphasize once more that Eqs. 7.83 and 7.84 express the third Kepler law which determines radial distribution of both the virial oscillation frequencies and the angular velocities of rotation. Numerical values of these parameters are determined by the radial density distribution law. It also determines the density jumps which mark the effect of the shell's isostatic equilibrium.

7.11 Perturbation Effects Studied on Basis of Dynamic Equilibrium

The most noteworthy effects of dynamics of the Earth are the interrelated phenomena of the precession and nutation of the axis of rotation, tidal effects of the oceans, and atmosphere, the axial obliquity and declination of the plumb-line and the gravity change at each point of the planet's outer force field. The present-day ideas about the nature of these phenomena were formed on the basis of the Earth's hydrostatic equilibrium and since old times were considered as effects of perturbation from the Sun, the Moon and other planets. All the above phenomena represent periodic processes and many observational and analytical works were done for their understanding and description. The present-day studies of these processes are still continuing to be specified and corrected. This is because such topical problems as correct time, ocean dynamics, short and long-term weather and climate changes and other environmental changes are important for every-day human life.

Now, after it was found that the conditions of the hydrostatic equilibrium are not acceptable for study of the Earth's dynamics, we reconsider the nature of the phenomena by applying the concept of the planet's dynamical equilibrium and developing a novel approach to solving the problem.

7.11.1 The Nature of Perturbation for Orbital Motion of a Body

To begin, let us consider physical meaning of the gravitational perturbation for interacting volumetric (but not point) body masses. To the contrary of hydrostatics, where the measure of perturbation in the precession-nutation and the tidal phenomena is the perturbing force, in the dynamic approach that measure of perturbation is power pressure. In Chap. 2 we concluded that the mass points and the vector forces as a physical and mathematical instrument in the problem solution of dynamics of the Earth in its own force field are not applicable. This is because the outer vector central force field of the interacting volumetric masses incorrectly expresses dynamical effects of their interaction. As a result, the kinetic effect of interaction of the mass particles, namely, the kinetic energy of their oscillation, is lost. And also the geometric center of a body is accepted as the gravity center and center of the inertia (reaction). In dynamics it leads to wrong results and conclusions. In this connection we found that in dynamics of the Earth as a self-gravitating body the effect of gravitational interaction of mass particles should be considered as the power pressure. In addition, in this case we are free in our choice of a reference system. Our conclusion does not contradict to Newton's physical ideas which are presented in Book I of his *Principia* where he says: "I approach to state a theory about the motion of bodies tending to each other with centripetal forces, although to express that physically it should be called more correct as pressure. But we are dealing now with mathematics and in order to be understandable for mathematicians let us leave aside physical discussion and apply the force as its usual name".

Accepting the power pressure as an effect of gravitational interaction, we come to understanding that, in the considered problem of the mutual perturbations between the Earth, the Moon, and the Sun, the interaction results not between the body centers or shells along straight lines, but between the outer force fields of the bodies and between their inner force fields of the shells. Satellite observations show that the outer force field, induced by the Earth's mass, has 4π-outward direction of propagation and acquires a wave nature. We consider this outer wave force field as a physical media by which the bodies transmit their energy. Thus, the Earth and other planets are held and move on the orbits by the power of the outer wave field of the Sun. This power, in terms of its normal and equal to its tangential components, remains valid since the planets separation (see above Sects. 7.5, 7.6 and Chap. 8). In order to demonstrate validity of the above conception let us calculate the mean values of velocity of the Earth and the Moon orbital motion from the frequencies of oscillation of the respective outer wave fields of their parents.

In accordance with Eq. 7.67, the frequency ω_s of oscillation of the self-gravitating Sun's outer force field at the mean distance $R_e$of the Earth's orbit is

$$\omega_s = \sqrt{\frac{GM_s}{R_{es}^3}} = \sqrt{\frac{6.67 \cdot 10^{-8} \cdot 1.99 \cdot 10^{33}}{\left(1.496 \cdot 10^{13}\right)^3}} = 1.9931 \cdot 10^{-7}\,s^{-1},$$

where M_s is the Sun's mass; R_{es} is the mean distance between the Earth and the Sun.

In accordance with wave mechanics, the mean value of the Earth's orbital velocity is

$$v_e = \omega_s R_{es} = 1.9931 \cdot 10^{-7} \cdot 1.496 \cdot 10^{13} = 2.98 \cdot 10^6\,cm/s = 29.9\,km/s$$

If we extend the outer force fields of the Sun and the Earth up to equality of their reduced densities, then in accordance with the same Eq. 7.67 the border between the two interacting fields can be found. By calculation the mean (nodal) value of the Earth's field border is found to extend up to $2.128 \cdot 10^9$ m and the Sun's border – to $1.478 \cdot 10^{11}$ m.

Applying the same procedure, the orbital frequency oscillation $\omega_e$of the Earth's outer force field at the Moon distance R_{me}has value

$$\omega_e = \sqrt{\frac{GM_e}{R_{me}^3}} = \sqrt{\frac{6.67 \cdot 10^{-8} \cdot 5.976 \cdot 10^{27}}{(3.844 \cdot 10^{10})^3}} = 2.64907 \cdot 10^{-6}\,s^{-1}, \tag{7.85}$$

where M_e is the Earth's mass; $R_{me}$is the mean distance between the Moon and the Earth.

Then the Moon's orbital velocity is

$$\bar{v}_m = \omega_e R_m = 2.64907 \cdot 10^{-6} \cdot 3.844 \cdot 10^{10} = 1.0183\ km/s. \tag{7.86}$$

The Moon's border of the outer force field in the nodal plane extends to the Earth up to $0.72 \cdot 10^8$ m and the Earth's border – to $3.724 \cdot 10^8$ m.

The obtained values of velocities, as well as the values for the pericenters and apocenters, are exactly the same as known from observation. It means that the observed ecliptic inclination relative to the equatorial plane of the Sun and inclination of the Moon's orbit relative to the ecliptic reflect asymmetric distribution of the solar and the planet's masses. It also means that the observed inclination of the Moon's orbit plane and the ecliptic are governed by asymmetric distribution of the Earth's and the Sun's force fields. The force fields of the Earth and the Moon, together with the bodies themselves being local "secondary" inclusions in the powerful force fields of the Sun and the Earth, are obliged to adjust their positions in order to be in dynamic equilibrium. The observed parameters of the orbits and their inclination relative to the plane diameters of the Sun, the Earth and the Moon give a general view of the asymmetric distribution of the body's masses. In particular, the northern hemisphere of the Earth is more massive than the southern one. In the perihelion the northern hemisphere is turned to the less massive hemisphere of the Sun. So that, the polar oblateness of each body controls the location of its pericenter and apocenter, and the equatorial oblateness of each body responds to location of its nodes. Thus, the body motion in the outer force field of its parent occurs under strict conditions of dynamic equilibrium which is also the main condition of its separation. It follows from the condition of dynamic equilibrium that the orbital motion of the Earth and the Moon reflects asymmetry in mass density distribution of the Sun, the Earth, and the Moon and asymmetry in the potential of the outer wave field distribution. Only the structure of the Sun's outer wave field controls the Earth's trajectory at the orbital motion and the Earth's force field manages the orbital motion of the Moon, but not vice versa or somehow else.

7.11.2 Change of the Outer Force Field and the Nature of Precession and Nutation

At the right time of motion of the bodies with the outer wave fields, their mutual perturbations are transferred not directly from each body to the other one or from their shells, but through the outer fields by means of the corresponding active and reactive wave pressure of the interacting fields. There is an important dynamic effect of all the perturbations. This is the continuous change in the outer wave field of each body which proceeds from its non-uniform radial distribution of the mass density. As it was earlier shown, the non-uniform radial distribution of mass density initiates the differential rotation of the body shells. And, in accordance with Eqs. 7.67 and 7.68 expressing the third Kepler's law, the reduced body shells' perturbing effects are transferred to the other body by means of the outer wave field. So that the Sun, for instance, through its outer wave field, continuously transfers to the Earth all the perturbations resulting during rotation of the interacting masses of the shells. The Earth, in the framework of the energy conservation law, demonstrates all the perturbations by changes in its orbit turns around the Sun (see below Fig. 7.7).

Earlier it was shown that in the case of non-uniform distribution of mass density the body's potential and kinetic energies have radial and tangential components which induce oscillation and rotation of the shells. It was defined by Eq. 7.82 that the observed daily rotation of the Earth concerns only the upper shell with thickness of ~375 km and reaches the near-by Mohorovičić-Guttenberg discontinuity. By the

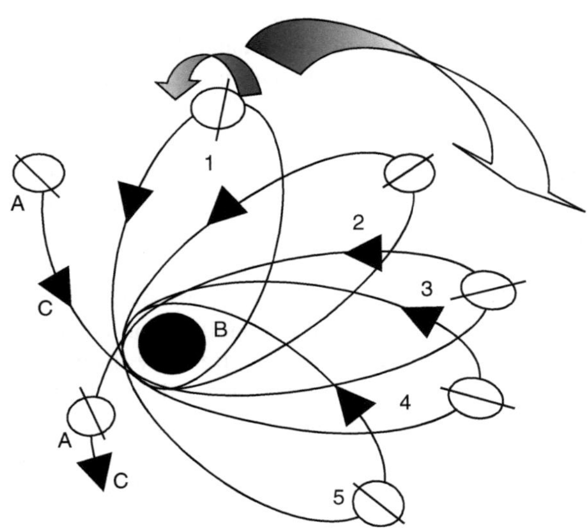

Fig. 7.7 Real picture of motion of a body A in the force field of a body B. Digits identify succession of turns of the body A moving around body B along the open orbit C

same reasoning it is not difficult to find the thickness of the upper shells for the Sun and the Moon correspondingly equal to:

$$h_s = \frac{v_s}{\omega_{0s}(R_s)} \approx \frac{2}{6.28 \cdot 10^{-4}} \approx 1600 \text{ km}, \quad (7.87)$$

$$h_m = \frac{v_m}{\omega_{0m}(R_m)} \approx \frac{4.56 \cdot 10^{-3}}{9.66 \cdot 10^{-4}} \approx 4.72 \text{ km}. \quad (7.88)$$

We do not know real values and angular velocities for the inner shells of the three bodies. These velocities have a direct interrelation with the observed changes in parameters of the orbital motion of the Earth and the Moon including the retrograde motion of the orbital nodes and the apsidal line. In this connection let us try to understand first of all the nature of precession and nutation of the bodies from the point of view of the dynamic approach.

It was noted above that, in accordance with the hydrostatic approach, precession of the equinoxes of the Earth is an effect of the net torque of the Moon and the Sun on the equatorial "bulge" aroused from gravitational attraction. The torque aspires to diminish inclination of the equatorial belt with surplus mass relative to the ecliptic and induce the retrograde motion of the nodal line. In addition, because the ratio of distance between the interacted bodies is changed, then the relationship between the forces is also changed. In this connection the precession is accompanied by nutation (wobbling) motion of the axes of rotation.

Analysis of orbits of the artificial satellite motion around the Earth shows that, in spite of absence of the equatorial "bulge" of mass, the apparatus demonstrates the precession effect. Its orbital plane has a clockwise rotation with retrograde motion of the nodal line. But a new explanation of the phenomenon is given. It appears that the retrograde motion of the nodal line associates with the Earth equatorial and polar oblateness. The amplitude of the nodal line shift depends on the satellite orbit inclination to the Earth's equatorial plane. In the case of the poles' orbital plane the nodal line shift is completely absent. This is because the pole motion excludes both the polar and the equatorial oblatenesses of the Earth. The direction of motion of the apsidal line depends on the satellite's orbit inclination and is determined by the Lentz law.

It is also known that for the other free-of-satellite planets the retrograde motion of the nodal line is also a characteristic phenomenon called the "secular perihelion shift". It was found from observation of Mercury, Venus, Earth and Mars that their secular perihelion shifts are decreased from ~ 40″ through ~ 8.5″, ~ 5″ to ~ 1.5″ accordingly [Chebotarev 1974].

All these facts imply that the explanation given for the satellites' precession depending on their orbital inclination to the ecliptic is correct. But the nature of this unique phenomenon, characteristic for all celestial bodies, are inconsistent with the hydrostatic approach and should be reconsidered, taking also into account the satellite observations.

7.11.3 Observed Picture of a Body Precession

The precession of the Earth, the Moon and the artificial satellites in the form of motion of an orbital plane toward the backward direction of the body's motion should be considered as a virtual explanation of the phenomenon. In fact, the orbit's plane is a geometric shape traced by the body. And there is no reason to consider its movement without the body itself. There is no difficulty to present the real body motion in space in two opposite directions synchronously. In particular, the actual picture of the Earth, the Moon and the satellite motion in counterclockwise direction and retrograde movement of the nodal line is shown in Fig. 7.7.

Here the satellite is moving in the counterclockwise direction along the unlocked elliptic orbit 1 in the continuously changing (perturbed by oblatenesses) planet's force field. Because of the counterclockwise rotation of the Earth's mass, the satellite in perigee started to move on the orbit 2 and makes a shift in retrograde direction in the ascending and descending nodes. At the same time the eccentricity of the orbit 2 changes by a proper value. Analogously the body passes on orbit 3, 4, 5 and so on. The theory of dynamic equilibrium of the Earth explains the physics of the observed phenomenon as follows.

7.11.4 The Nature of Precession and Nutation

The dynamic equilibrium theory assumes that the Earth is a self-gravitating body, the interacting mass particles of which induce the inner and outer force fields. Separation of the planet's asymmetric shells results by the inner force field and depends on the law of the radial mass density distribution. The normal component of the body's power pressure provides oscillation, and the tangential component induces rotation of the shells having a different angular velocity. At the same time the mantle shells A and the outer shell of the core B may have the same (Fig. 7.8a) or opposite direction (Fig. 7.8b) of rotation depending on the radial mass density distribution.

The seismic data show that the inner core C has a uniform density distribution. Because of this, it does not rotate and its potential energy is realized in the form of oscillation of the interacting particles. The potential E of the outer force field is controlled by integral effect of the interacted masses of all the shells and presented by the reduced shell D having continuously changing power.

The energy of the Earth's outer force field is changed from the body surface in accordance with the 1/r law and at every r is continuously varied because of differences in the angular velocity of rotation of the shell's masses. This force field controls the direction and angular velocity of orbital motion of a satellite. Taking into account the non-uniform and asymmetric distribution of the masses of rotating shells, the change in the trajectory of the body motion is accompanied by a corresponding change in eccentricity of the orbit both at each and subsequent turns.

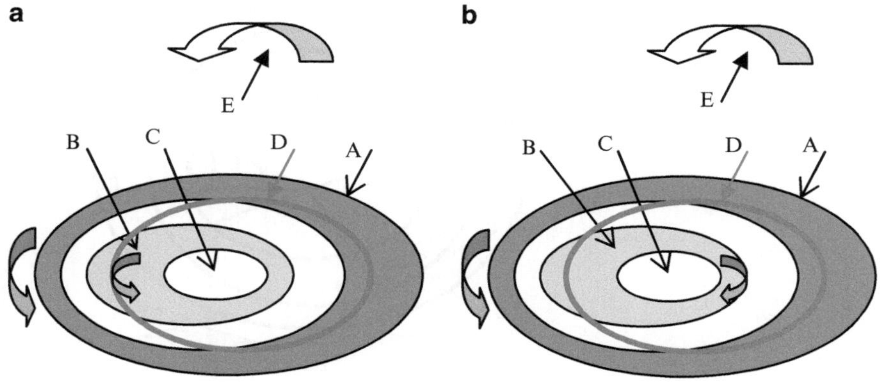

Fig. 7.8 Sketch of rotation of the Earth's shells by action of the inner force field: *A* is the mantle shells; *B* is the outer core; *C* is the inner core; *E* is the outer force field; *D* is the reduced shell of the inner force field of the planet

Its maximum value is reached when the non-uniformities of the rotating masses coincide and the minimal value appears at the opposite position.

It is worth noting that the effect of retrograde motion of the nodal line of the Earth, the Moon and artificial satellites appears to be a common phenomenon because the induced by the Sun and the Earth outer force fields are changing with a finite velocity. The conclusion follows from here that the Sun has the same effects in its shell structure and motion. It is obvious that the other planets with their satellites have the same character of structure and motion.

If one takes into account the effect of a planet's orbital plane inclination to the equatorial plane of the Sun, then the above changes are found to follow the law of 1/r This observable fact proves our conclusion that the changes in the outer force field of a body are controlled by rotation of its reduced inner force shell (see the force shell D on Fig. 7.8). It explains why Mercury has maximal value of the "secular perihelion shift" between the other planets.

Thus, the Earth's orbital motion and retrograde movement of its nodal line are controlled by the Sun's dynamics of the masses through the outer force field. The Earth plays the same role for the Moon and the artificial satellites. As to the nutation motion, then its nature is related to the same peculiarities in the structure and motion of the bodies but the effects of their perturbations are fixed by the axis wobbling.

7.11.5 The Nature of Possible Clockwise Rotation of the Outer Core of the Earth

The question arises why the outer planet's core may have a clockwise rotation. It was shown in Section 2.6 that the law of radial density distribution determines the direction of a body's shell rotation (Fig. 2.2).

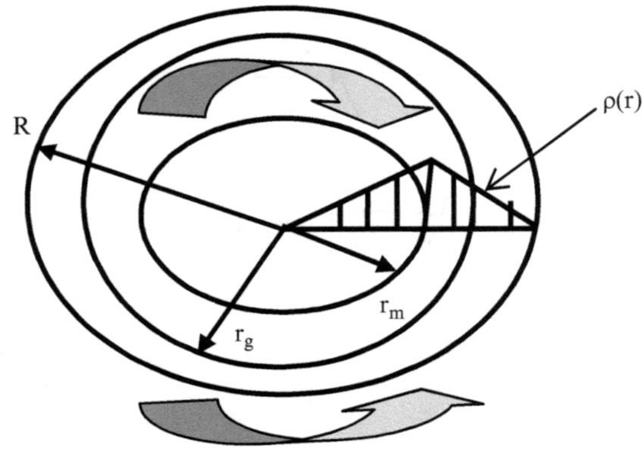

Fig. 7.9 Dependence of the parabolic law of radial density distribution on the shell rotation for the Earth. Here r_m and r_g are the reduced radiuses of inertia and gravitation

It was found that in the case of uniform mass density distribution all energy of the mass interaction is realized in the form of oscillation of the interacting particles (Fig. 2.2a). If the density increases from the body's surface to the center, then there are oscillations and counterclockwise rotation of shells (Fig. 2.2b). Increase of mass density from center to surface leads to oscillation and clockwise rotation with different angular velocities of the body shells (Fig. 2.2c). Finally, the parabolic law of radial density distribution (Fig. 7.9), where the density increases from the surface and then it decreases, leads to oscillation and reverse directions of rotation. Namely, the upper shells have a counterclockwise and the central shells – clockwise rotation. The case demonstrated on Fig. 7.9, obviously, is characteristic for a self-gravitating body.

Note that direction of the body rotation depends on radial density distribution and corresponds with the Lenz right-hand or right-screw rule, well known in electrodynamics. Taking into account the observed effect of the retrograde motion of the satellite nodal line, the gravitational induction of the inner and outer force fields of the Earth has a common nature with electromagnetic induction noted earlier. Just Fig. 7.9 may explain the nature of the retrograde motion of the nodal line of a satellite orbit related to the finite velocity in the potential changes of the outer Earth's force field induced by the interacted mass particles. The continuous and opposite-directed movement of the asymmetric mass density distribution of the mantle and the outer core (Fig. 7.8) seems to be the physical cause of precession, nutation and variation of the inner and outer force fields observed by satellites. This idea is proved by the satellite data about the retrograde motion of the nodal line depending on inclination of its orbital plane with respect to the planet's equatorial plane

It is worth recalling, from the literature, that the idea of dynamical effects of the probably liquid core of the Earth has been discussed among geophysicists for a long time. (Melchior 1972).

7.11.6 The Nature of the Force Field Potential Change

It follows from the above discussion that in the frame of the considered dynamical approach, the variation of potential of the inner and outer force field relates to the non-uniform distribution of mass density of a self-gravitating body. Rotation of the outer and inner shells together with the induction finite velocity delay leads to the observed effects of precession, nutation and variation of their own force fields.

7.11.7 The Nature of the Earth's Orbit Plane Obliquity

Celestial mechanics does not discuss the problem of obliquity of the planet's and satellite's orbit planes and accepts it as an observable fact. The theory of dynamic equilibrium explains this phenomenon and the nature of apocenters and pericenters by asymmetric distribution of masses and by effect of rotation of asymmetric shells of self-gravitating bodies (see Fig. 7.9). In fact, if the mass of the Sun's shells has an asymmetric distribution, then the potential of the outer force field has the same asymmetry. This asymmetry determines inclination of the Earth's orbit plane relative to the plane of the Sun's rotation. Each point of the orbit reflects a condition of dynamical equilibrium of the interacting outer force fields of the planet and the Sun. The position of the Earth's aphelion and perihelion reflects the position of the reduced maximal and minimal concentration of the Sun's mass density in the shells. Because the Sun's asymmetric shells have different angular velocities of rotation, then amplitude of the nodal line will decrease with increasing distance between the mass anomalies and vise versa. The effect of variation of the nodal line is proved by observation. So, the present-day angle of ecliptic inclination to the plane of rotation of the Sun equal to ~7°15 expresses the relation between maximum and minimum concentrations of the reduced mass density of the Sun's shells. An analogous effect is shown by inclination of the Moon's orbital plane to the plane of rotation of the Earth.

7.11.8 The Nature of Chandler's Effect of the Earth Pole Wobbling

As it was noticed, changes in the planet's inner force field are observed in the form of nutation or wobbling of the axis of rotation. The axis itself reflects the dynamics of the upper planet's shell, the thickness of which, by our estimate, is about 375 km. The Moon is rotating about the Sun in the force field of the Earth which is perturbed by its natural satellite. Its maximum yearly perturbation should be the Chandler effect. The Moon's yearly cycle seems to be the ratio of the Earth's to the Moon's month (in days). Then this cycle is $365(30.5/27) \approx 410$ days.

7.11.9 Change in Climate as an Effect of Rotation of the Earth's Shells

The above analysis of dynamical effects of the Earth's shells is based first of all on the data of satellite's orbit changes and measurements of the planet's force field. Unfortunately, a specific feature of an artificial satellite orbital motion is its artificial velocity which is ~16 times higher than the angular velocity of the upper Earth's shell. In this connection all its parameters of satellite motion are unnatural. So, we cannot directly divide the natural component of its nodal retrograde shift in order to get the total picture of perturbations which propagate the Earth's inner shells. This is an experimental problem.

But there are also long term astronomical observations of the Earth's dynamics relative to the far stars, the results of which correspond to the presented ones. In addition, periodicity in rotation of asymmetric inner shells of the Sun can be fixed by climatic changes on the Earth over a long period of time. Such changes were being studied, for instance, by data of the oxygen isotopic composition in mollusk shells over a number of years. Figure 7.10 demonstrates the results of Emiliani (1978) who studied the core obtained during deep sea drilling in the Caribbean basin.

The author obtained the picture of climate change in the Pleistocene era over 730,000 years. It is seen that the periods of climate change vary from 50 to 120,000 years. It means that the pure period of rotation of the asymmetric mass shells of the Sun is absent and the orbital trajectory has not been locked into place during the studied time.

7.11.10 The Nature of Obliquity of the Earth's Equatorial Plane to the Ecliptic

It is obvious that the obliquity of the planet's equatorial plane is related to the polar and equatorial oblateness of the Earth's masses. It follows from Eq. 7.68 that the obliquity, in turn, is determined by the tangential component of the inner force pressure generated by the non-uniform radial mass density distribution.

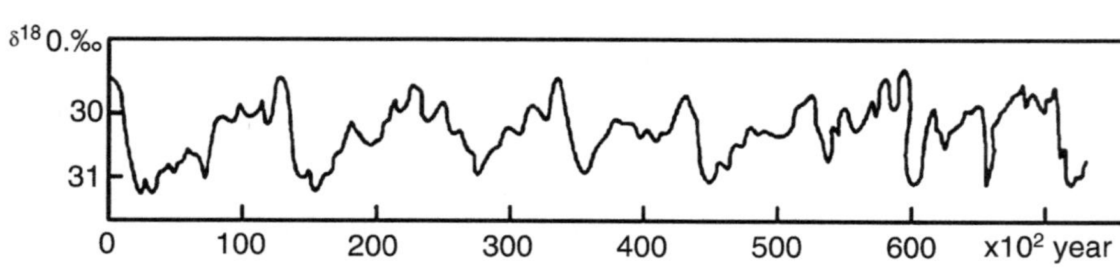

Fig. 7.10 Isotopic composition of oxygen in shells of mollusk *Globigerinoides Sacculifera* within time period 0–730,000 year [Emiliani 1978]

This tangential component of the inner force field induces the inner field of the rotary moments, the energy of which was discussed in Section 6.9 and presented in Table 4. The obliquity value can be obtained from the ratio of the potential energy of the uniform U_o and non-uniform U_t body of the same mass. Accepting this physical idea and the data of Table 4, we write and obtain:

$$\cos\Theta = \frac{U_o}{U_t} = \frac{\alpha_o^2}{\alpha_t{}^2} = \frac{0.6}{0.66} = 0.909, \quad \Theta = 24.5^o, \tag{7.89}$$

where α_o^2 and $\alpha_t^2$are the structural form-factors taken from Table 4.

The error obtained in calculation of obliquity by formula (6.89) equal to about 1° or $\Delta\alpha_t^2 = 0.006$ – can be explained by the accepted law of the continuous radial distribution of the planet's mass density.

Equation 7.89 expresses the integral effect of the obliquity of the planet's equatorial plane which is observed on the surface of the upper rotating shell. It was shown earlier that the observed obliquity is really an integral dynamical effect of the Earth's mass including the upper part of the Gutenberg shell. But being in a suspended state, relative to the other parts of the body, the upper shell is able to wobble as if on a hinge joint by perturbation from the Sun and the Moon. This effect of the upper shell wobbling gives an impression of the axial wobbling.

By the same cause the obliquity of the ecliptic with respect to the solar equator is determined by the Sun's polar and equatorial oblateness. The trajectory of the Earth's orbital motion at each point is controlled by the outer asymmetric solar force field in accordance with the dynamic equilibrium conditions. And only in the nodes, which are common points for equatorial oblateness of the Sun and the Earth, is the Huygens' effect of the innate initial conditions fixed by the third Kepler's law.

7.11.11 Tidal Interaction of Two Bodies

Let us consider the mechanism and effects of interaction of the outer force pressure of two bodies being in dynamic equilibrium. Come back to the mechanism and conditions of separation of a body mass with respect to its density when a shell with light density is extruded to the surface. Rewrite Eq. 6.71 for acceleration of the gravity force in points A and B of the two body shells (Fig. 7.7b) and their densities ρ_M and ρ_m

$$q_{AB} = 4\pi G r\left(\frac{2}{3}\rho_M - \rho_m\right), \tag{7.90}$$

After the shell with density ρ_m appears on the outer surface of the body, the condition of its separation by Eq. 7.90 will be:

$$\rho_M > 2/3\ \rho_m \tag{7.91}$$

The gravitational pressure will replace the shell up to the radius A + δA, where the condition of its equilibrium reaches $\rho_M = \rho_m$. This condition is kept on the new border line between the body and its upper shell. Taking into account that the shell in any case has a thickness, then, by the Archimedes law, the body will be subject to its hydrostatic pressure. If the separated shell is non-uniform with respect to density, then a component of the tangential force pressure appears in it, and the secondary self-gravitating body (satellite) is formed. The new body will be kept on the orbit by the normal and equal tangential components of the outer force pressure. In this case the reaction of the normal gravitational pressure will be local and non-uniform. If the upper shell is uniform with respect to density, then the reaction of the normal gravitational pressure along the whole surface of the body and the shell remains uniform. In this case the separated shell remains in the form of a uniform ring.

The above schematic description of the physical picture of the separation and creation of a secondary body can be used for construction of a mechanism of the tidal phenomena in the oceans, the atmosphere and the upper solid shell at interaction between the Earth and the Moon. The outer gravitational pressure of the Moon, due to which it maintains itself in equilibrium on the orbit, at the same time renders hydrostatic pressure on the Earth's atmosphere, oceans and upper solid shell through its outer force field. This effect determines the tidal wave in the oceans and takes active part in formation and motion of cyclonic and anti-cyclonic vortexes. In accordance with the Pascal law, the reaction of the Moon's hydrostatic pressure is propagated within the total mass of the ocean water and forms two tidal bulges. Because the upper shell of the Earth is faster-moving relative to motion of the Moon, the front tidal bulge appears ahead of the moving planet. Our perception of the ocean tides as an effect of attraction of the Moon appears to be speculative.

7.12 Dynamics of the Earth's Atmosphere and Ocean

The atmosphere and the oceans (collectively) are both upper shells of the Earth. The first of them occurs in a gaseous and the second one in a liquid phase. These shells exist in the solid Earth's outer force field but the atmosphere exists in dynamical equilibrium, and the oceans are in a weighted transition to a hydrostatic state. The atmosphere as a gaseous shell is totally "dissolved" in the Earth's outer force field in the form of atomic and molecular sub-layers, differentiated with respect to density, and these self-gravitating masses appear in a dynamical equilibrium state. Relatively homogeneous water masses of the oceans have too low a density (approximately 2/3) in comparison with the mineral crust to be an independent, with respect to its dynamics, self-gravitating shell. Therefore, it appears to be suspended in a semi-hydrostatic equilibrium relative to the crust's shell. Its inner gravitational pressure on the shell's surface is equilibrated with the atmospheric and outer gravitational pressure. The surface water is practically found to be in a limiting hydrostatic equilibrium. The small portion of the continuously pumping solar energy varies within 7% in the annual cycle and leads to the dynamical process of water transfer from its liquid to vapor, and vice verse phase.

Dynamics of the main nitrogen and oxygen components with close densities (atomic weights) of the atmospheric masses in a non-perturbed state could be represented by its own virial oscillation and rotation about the solid Earth. But the water vapor which is continuously injected into the gaseous shell, being a component of very large-capacity energy (1 cm^3 of water generates more than 1,000 cm^3 water vapor) appears to be the cause of the stormy dynamical perturbation within the near-surface shell of the atmosphere. The cyclonic activity of the atmospheric vapor has scientific, but mainly practical, interest. The weather and climate change by variation of the solar energy flux through the water vapor dynamics have a negative affect on the biosphere.

In turn, the oceans being in a balanced hydrostatic state, are continuously perturbed both by the inner gravity pressure of the planet related to the density differentiation of the shells and by perturbation from the Sun's and the Moon's outer force fields. All the above dynamical processes seem to be of interest for consideration from the point of view of dynamical equilibrium theory.

In this chapter we search for a solution of Jacobi's virial equation for the non-perturbed atmosphere as the Earth's shell which is affected by both inner and outer perturbations. In order to justify applicability of the virial equation for the study of dynamics of the atmosphere in the framework of a model of a continuous medium, we derive this equation from the Euler equations.

7.12.1 Derivation of the Virial Equation for the Earth's Atmosphere

We consider the problem of global oscillations of the Earth's atmosphere as its shell. We accept that the atmosphere is found to be in dynamical equilibrium both in its own force field and in the outer force field of the Earth. Derivation of the virial equation is done in the framework of the continuous medium model. We accept also that the dynamics of the atmosphere is described by Euler's equations and the medium is composed of an ideal gas.

The Euler equations are written in the form (Landau and Lifshitz 1954)

$$\rho \frac{\partial \mathrm{v}}{\partial \mathrm{t}} + \rho(\mathrm{v}\nabla)\mathrm{v} = -\mathrm{grad}\ \mathrm{p} + \rho \mathrm{F}, \tag{7.92}$$

where ρ is the gas density; p is the gas pressure; $\partial \mathrm{v}/\partial \mathrm{t}$is the rate of velocity change at a fixed point of space; F is the density of mass forces.

In addition, the continuity equation

$$\frac{\partial \mathrm{p}}{\partial \mathrm{t}} + \mathrm{div}\ \rho \mathrm{v} = 0 \tag{7.93}$$

holds for a gas medium.

Multiplying Eq. 7.93 by v and summing the product with Eq. 7.1, we obtain

$$\frac{\partial \rho}{\partial \mathrm{t}}\mathrm{v} = -\rho(\mathrm{v}\nabla)\mathrm{v} - \mathrm{v}\mathrm{div}\,\rho\mathrm{v} - \mathrm{grad}\,\mathrm{p} + \rho\mathrm{F}. \tag{7.94}$$

We take the divergence of Eq. 7.94 and note that

$$\frac{\partial}{\partial t}[\mathrm{div}\,\rho\mathrm{v}] = \left[\frac{\partial}{\partial \mathrm{t}}(\mathrm{div}\rho\mathrm{v})\right] = -\frac{\partial^2 \rho}{\partial \mathrm{t}^2}.$$

Then Eq. 7.94 can be rewritten in the form

$$\frac{\partial^2 \rho}{\partial \mathrm{t}^2} = \mathrm{div}[\rho(\mathrm{v}\nabla)\mathrm{v} + \mathrm{v}\mathrm{div}(\rho\mathrm{v}) + \mathrm{gradp} - \rho\mathrm{F}] \tag{7.95}$$

Multiplying Eq. 7.95 by $r^2/2$ and integrating the obtained expression over the whole volume of the atmosphere, we have

$$\int\limits_{(\mathrm{V})} \frac{\mathrm{r}^2}{2}\frac{\partial^2 \rho}{\partial \mathrm{t}^2}\mathrm{dV} = \int\limits_{(\mathrm{V})} \frac{\mathrm{r}^2}{2}\,\mathrm{div}[\rho(\mathrm{v}\nabla)\mathrm{v} + \mathrm{v}\mathrm{div}(\rho\mathrm{v}) + \mathrm{grad}\,\mathrm{p} - \rho\mathrm{F}]\mathrm{dV}. \tag{7.96}$$

The left-hand side of Eq. 7.96 can be rewritten in the form

$$\int\limits_{(\mathrm{V})} \frac{\mathrm{r}^2}{2}\frac{\partial^2 \rho}{\partial \mathrm{t}^2}\mathrm{dV} = \frac{\partial^2}{\partial \mathrm{t}^2}\left(\frac{1}{2}\int\limits_{(\mathrm{V})} \mathrm{r}^2\rho\mathrm{dV}\right) = \ddot{\Phi},$$

where Φ is the Jacobi function of the atmosphere.

Thus, we obtain Jacobi's equation for the Earth's atmosphere derived from the Eulerian equation (7.92) and the continuity equation (7.93) as follows:

$$\ddot{\Phi} = \int\limits_{(\mathrm{V})} \frac{\mathrm{r}^2}{2}\,\mathrm{div}[\rho(\mathrm{v}\nabla)\mathrm{v} + \mathrm{v}\mathrm{div}(\rho\mathrm{v}) + \mathrm{grad}\,\mathrm{p} - \rho\mathrm{F}]\mathrm{dV}. \tag{7.97}$$

We assume that the Earth is a rigid spherical body with mass M and radius R and that the mass of the atmosphere is negligible in comparison with the Earth's mass. We can transform the right-hand side of Eq. 7.97 as follows:

$$\int\limits_{(\mathrm{V})} \frac{\mathrm{r}^2}{2}\mathrm{div}[\rho(\mathrm{v}\nabla)\mathrm{v} + \mathrm{v}\mathrm{div}(\rho\mathrm{v}) + \mathrm{grad}\,\mathrm{p} - \rho\mathrm{F}]\mathrm{dV}$$

$$= 2\mathrm{T} + \mathrm{U} + 3\int\limits_{(\mathrm{V})} \mathrm{p}\mathrm{dV} - 4\pi\mathrm{r}^3\mathrm{p}\Big|_{\mathrm{R}}^{\mathrm{R}_{\max}},$$

where

$$T = \frac{1}{2} \int_{(V)} \rho v^2 dV$$

is the kinetic energy of the gas of the atmosphere in the Earth's gravitational field;

$$U = GM \int_{(V)} \frac{\rho dV}{r}$$

is the potential energy of the atmosphere in the Earth's field;

$$3 \int_{(V)} p dV$$

is the internal energy of the gas atmosphere;

$$-2\pi r^3 p|_{Re}^{Rmax}$$

is the energy of the outer surface force pressure effecting the atmosphere.

In fact, we can write an expression for the mass forces, taking into account the spherical symmetry of the system considered:
$F = -GM\frac{r}{r^3}$.
Then

$$-\int_{(V)} \frac{r^2}{2} \operatorname{div}(\rho F) dV = \int_{(V)} \frac{r^2}{2} \operatorname{div}\left(\rho GM \frac{r}{r^3}\right) dV = \int_{R}^{Rmax} \frac{r^2}{2} \frac{1}{r^2} \frac{\partial}{\partial r}\left(\frac{r^2 \rho GM}{r^2}\right) 4\pi^2 dr$$

$$= \frac{4\pi r^2}{2} \rho GM|_{R}^{Rmax} - 4\pi GM \int_{R}^{Rmax} \rho r dr = U + \frac{4\pi r^2}{2} \rho GM|_{R}^{Rmax}.$$

Analogously, we have

$$\int_{(V)} \frac{r^2}{2} \operatorname{div}(\operatorname{grad} p) dV = 4\pi \int_{R}^{Rmax} \frac{r^2}{2} \frac{1}{r^2} \frac{\partial}{\partial r}\left(r^2 \frac{\partial p}{\Im r}\right) r^2 dr$$

$$= \frac{4\pi r^2}{2} \frac{\partial p}{\partial r}\bigg|_{R}^{Rmax} - 4\pi r^3 p|_{R}^{Rmax} + 3 \cdot 4\pi \int_{R}^{Rmax} r^2 p dr$$

$$= 3 \int_{(V)} p dV - 4\pi r^3|_{R}^{Rmax} + \frac{4\pi r^4}{2} \frac{\partial p}{\partial r}\bigg|_{R}^{Rmax}.$$

It is easy to show that

$$\int_{(V)} \frac{r^2}{2}\mathrm{div}[\rho(v\nabla)v + v\,\mathrm{div}(\rho v)]dV = \int_{(V)} v^2\rho dV = 2T.$$

Owing to the immobility of the rigid Earth and the atmospheric boundary as well as the condition of continuity for the gas, the equilibrium hydrostatic condition holds along the normal to this border, which can be written in the form

$$\rho\frac{GM}{r^2}|_R = -\frac{\partial p}{\partial r}|_R.$$

If we assume $\rho r^2|_{Rmax} = 0$, $r^4\frac{\partial p}{\partial r}|_{Rmax} = 0$,then

$$\frac{4\pi r^2}{2}\rho GM|_R^{Rmax} + \frac{4\pi r^4}{2}\frac{\partial p}{\partial r}|_R^{Rmax} = 0$$

It follows from this that Jacobi's virial equation (7.97) takes the form

$$\ddot{\Phi} = 2T + U - 4\pi r^3 p|_R + 3\int_{(V)} pdV. \tag{7.98}$$

Finally, using the energy conservation law for a continuum (Sedov 1970) and the conservativity of a system, following from the accepted model (the Earth is a solid body and atmospheric gas is ideal), Eq. 7.98 can be written in the form of the standard Jacobi virial equation

$$\ddot{\Phi} = 2E - U, \tag{7.99}$$

where the total energy of the atmosphere is conserved and equal to

$$E = T + U - 2\pi r^3 p|_R + \frac{3}{2}\int_{(V)} pdV. \tag{7.100}$$

Let us make some important notes concerning the study of the atmosphere by a conventional approach based on use of the virial theorem.

The standard solution of the problem using the same assumption at zero approximation (barometric height formula) expresses the equilibrium condition between the atmospheric gravity and the gas pressure. In our case this condition is satisfied if Jacobi's virial equation (7.99) is averaged with respect to time, i.e. it is reduced to the condition of the hydrostatic equilibrium. But in that case the kinetic energy of the particle interaction is excluded. Assuming that the motion of the gas particles is finite, and after time averaging Eq. 7.99 over a sufficiently large time interval, the validity of the virial theorem is easily shown to be

$$E = U/2. \tag{7.101}$$

At known parameters of the atmosphere and the Earth, the potential and total energies of the atmosphere can be estimated as

$$U = -\frac{GMm}{R} \approx -3.2 \cdot 10^{33}\ \text{erg},$$

$$E \approx -1.6 \cdot 10^{33}\ \text{erg}.$$

7.12.2 Non-perturbed Oscillation of the Atmosphere

Let us now consider the solution of Eq. 7.99 for the spherical model of the atmosphere and find the dependence of its Jacobi function Φ (polar moment of inertia) on time in explicit form.

In the previous chapter it was shown that Eq. 7.99 is resolved both for the uniform medium and for the medium with radial density distribution by some law. In the last case the polar moment of inertia and potential energy are expanded on the uniform and tangential components and instead of Eq. 7.99 two Eqs. 7.61 and 7.62 are written.

Let us consider a solution for the uniform component of the atmosphere whose radial density distribution changes by the barometric equation. The uniform component of Eq. 7.99 has a solution when there is a relationship between the Jacobi function and the potential energy of the system in the form

$$|U|\sqrt{\Phi} = \text{const}. \tag{7.102}$$

Assume that the Earth has a spherical shape with mass M and radius R and be enveloped by a uniform atmosphere with mass m which has thickness Δ. Then the potential energy of the shell U_a, which is in the gravitational field of the sphere, is

$$U_a = -GM\rho_a \int\limits_R^{R+\Delta} \frac{4\pi r^2}{r} dr = -2\pi GM\rho_a \left(2R\Delta + \Delta^2\right),$$

where ρ_a is the mass density of the shell.

The Jacobi function of the atmosphere is

$$\Phi_a = 4\pi\rho_a \frac{1}{2}\int\limits_{R}^{R+\Delta} r^4 = \frac{4\pi\rho_a}{2.5}[5R^4\Delta + 10R^3\Delta^2 + 10R^2\Delta^3 + 5R\Delta^4 + \Delta^5]. \quad (7.103)$$

Expressing the gas density ρ_a through its mass, we can write the relationship (7.102) in the form

$$B = |U|_a\sqrt{\Phi_a} = \frac{GMm}{\sqrt{2}}\frac{3(2+\lambda)}{2(3+3\lambda+\lambda^2)}\sqrt{\frac{3(5+10\lambda+10\lambda^2+5\lambda^3+\lambda^4)}{5(3+3\lambda+\lambda^4)}}, \quad (7.104)$$

where $\lambda = \Delta/R$.

Note that Eq. 7.104 depends only on the ratio of thickness of the shell to the radius of a central body and varies over limited ranges, while λ varies from 0 to ∞. At $\lambda = 0$

$$|U_a|\sqrt{\Phi_a} \to \frac{GMm^{3/2}}{R},$$

and at $\lambda \to \infty$

$$|U_a|\sqrt{\Phi_a} \to \frac{GMm^{3/2}}{R}\left(\frac{27}{20}\right)^{1/2}.$$

On the other hand, the Jacobi function (7.102) expressed through the mass of the shell, is written in the form

$$\Phi_a = \frac{3m}{10}R^2\frac{3(5+10\lambda+10\lambda^2+5\lambda^3+\lambda^4)}{5(3+3\lambda+\lambda^4)}. \quad (7.105)$$

It follows from (7.14) that the Jacobi function of the atmosphere does not depend only on the value λ but also on the radius of the body R. Moreover, the value Φ_a varies over unlimited ranges when λ runs from 0 to ∞. The same can be said about the potential energy of the atmosphere:

$$U_a = \frac{GMm}{R}\frac{3}{2}\frac{2+\lambda}{2(3+3\lambda+\lambda^2)}$$

Accepting the mass of the atmosphere m $= 10^{21}$ kg, we can find the value B as

$$|U_a|\sqrt{\Phi_a} = \frac{GMm}{R}\left(\frac{mR^2}{2}\right)^{1/2} = 1.0374 \cdot 10^{53}\ [g^{3/2}\,cm^3\,s^{-2}]. \qquad (7.106)$$

Taking (7.106) into account, Eq. 7.99 can be written in the form of an equation of virial oscillations of the atmosphere (subscript a at Φ is farther drop):

$$\ddot{\Phi} = -A + \frac{B}{\sqrt{\Phi}}, \qquad (7.107)$$

where $A = -2E$; $E = -1.6 \cdot 10^{33}$ erg.

As shown in Chap. 4, at A = const and B = const, Eq. 7.107 has two first integrals, as follows:

$$C = -2A\Phi + 4B\sqrt{\Phi} - \dot{\Phi}^2, \qquad (7.108)$$

$$-\arccos\frac{(A/B)\sqrt{\Phi}-1}{\sqrt{1-AC/2B^2}} - \sqrt{1-\frac{AC}{2B^2}}\sqrt{1-\left[\frac{(A/B\sqrt{\Phi}-1)}{\sqrt{1-AC/2B^2}}\right]^2} = \frac{(2A)^{3/2}}{4B}(t-t_0), \qquad (7.109)$$

where C and t_o are integration constants. The constant C has a dimension of square angular momentum.

The integrals (7.108) and (7.109) are solutions of the Eq. 7.107. Introducing new variables

$$E'' = -\arccos\frac{(A/B\sqrt{\Phi}-1)}{\sqrt{1-AC/B^2}}, \quad e'' = \sqrt{1-\frac{AC}{2B^2}},$$

and rewriting (7.109) in the form of Kepler's equation

$$E'' - e''\sin E'' = \frac{(2A)^{3/2}}{4B}(t-t_0) = M'',$$

expressions can be obtained for Φ(t) and Φ(t) from the first integrals (7.108) and (7.109) in explicit form, using Lagrange's series (Duboshin 1975):

$$\Phi(t) = \frac{B^2}{A^2}\left[1 + 2e''\cos M'' - e''^2\left(1-\frac{3}{2}\cos^2 M''\right) - \frac{5}{2}e''^3(1-\cos^2 M'')\cos M'' + \ldots\right] \qquad (7.110)$$

$$\Phi(t) = \sqrt{\frac{2}{A}}\varepsilon B\left[\sin M + \frac{1}{2}\varepsilon \sin 2M + \frac{\varepsilon^2}{2}\sin M(2\cos^2 M - \sin^2 M + \ldots\right],$$

where $M'' = (2A)^{3/2}(t - t''_0)/n(t - t''_0)$; n'' is the frequency of the atmospheres own virial oscillations; t_o is the time moment when Φ acquires its maximal value.

At $0 < C < 2B^2/A$ the Jacobi function changes in time in accordance with Eq. 7.110. At $C = 2B^2/A$ the Jacobi function is equal to $\Phi = B^2/A^2$, which corresponds to the hydrostatic equilibrium of the system or to the virial theorem.

The solution represents the non-linear periodic pulsation of the Jacobi function of the atmosphere as a whole with period T_v. Using the numerical values of constants $B = 1.03 \cdot 10^{53}$ and $A = -U = 3.2 . 10^{33}$ erg, the period of unperturbed virial oscillations of the Earth atmosphere is equal to

$$T_v = \frac{8\pi B}{(2A)^{3/2}} = 2\pi\sqrt{\frac{R^3}{Gm}} = 5060.7\,\text{s} = 1.4057\,\text{h}. \tag{7.111}$$

Note that the expression (7.111) for the period of virial oscillations T_ν includes three fundamental constants: the body M, radius R and gravity constant G. The simplest combination of these three constants, which gives the dimension of time, coincides with (7.111). In this case the nature of the virial oscillations of the atmosphere can be explained by the change in time of the gravitational potential of the solid Earth. The period (7.111) coincides with the period of revolution of a satellite along a circular orbit with first cosmic velocity $v = 7.9$ km/s, radius of the Earth and with a mathematical pendulum the length of whose filament is equal to the radius of the Earth. This is because the parameters considered are defined by the same constants g, m, and R.

7.12.3 Perturbed Oscillations

We now consider a general approach to solving the problem of perturbed virial oscillations of the atmosphere, taking as an example the perturbations caused by the variation throughout the year of the solar energy flux owing to the ellipticity of the Earth's orbit. We assume that all the dissipative processes that occur during the interaction between the atmosphere and the hydrosphere and in the atmosphere itself are compensated by solar energy. We note, however, that the value of the flux is evidently dependent on time. Assuming also that the eccentricity of the Earth's orbit and the mean total energy of the atmosphere are known and remain unchanged in time, the total energy of the Earth's atmosphere is proportional to the power of the solar energy flux L(x) which reaches the atmosphere at a given point of the orbit.

Then

$$L(t) = L_0 \frac{a^2}{r^2}, \tag{7.112}$$

where Lo is the mean energy flux reaching the Earth's atmosphere; r is the radius of the Earth's orbit; a is the semi-major axis of the Earth's orbit.

Using the property of the elliptical motion, we obtain

$$L(t) = L_0 \left(1 - 2e' \cos E' + e'^2 \cos^2 E'\right)^{-1}, \tag{7.113}$$

where e' is the eccentricity of the Earth's orbit; E' is the eccentric anomaly that characterizes the location of the Earth on the orbit and is linked with time by the Keplerian equation

$$E' - e' \sin E' = n'(t - t_0) = M'_0, \tag{7.114}$$

where $n = 2\pi/\tau$ is the cyclic frequency of the Earth's revolution round the Sun; τ is the period of revolution and is equal to 1 year; t_o is the moment of time required by the Earth to pass through the orbit's perihelion; M' is the mean of the anomaly.

If the eccentricity $e' \leq \bar{e} = 0.6627\ldots$, which is the Laplacian limit, then, using the Lagrangian series, we can obtain expressions for E' and $\varphi(E')$ in the form of an absolute convergent infinite series expanded by entire positive powers of e'.

Note that in order to obtain the expression for $(1 - e' \cos E')$ we can write the equality which follows from the Keplerian equation:

$$\frac{1}{(1 - e' \cos E')^2} = \left(\frac{dE'}{dM'}\right)^2.$$

Expanding the eccentric anomaly E' in the Lagrangian series by the power of eccentricity e' [Duboshin 1975], we obtain

$$E' = \sum_{k=0}^{\infty} \frac{e'^k}{k!} \frac{d^{k-1}\left(\sin^k M'\right)}{dM'^{k-1}}. \tag{7.115}$$

This can be rewritten in a more convenient form:

$$E' = \sum_{k=0}^{\infty} e'^k E_k(M'),$$

where

$$E_k(M') = \sum_{k=0}^{\infty} \frac{1}{k!} \frac{d^{k-1}\left(\sin^k M'\right)}{dM'^{k-1}}$$

As the absolute convergent series can be differentiated term by term, we obtain

$$\frac{a}{r} = \frac{dE'}{dM'} = \sum_{k=0}^{\infty} e'^k \bar{R}_k(M'), \tag{7.116}$$

where

$$\bar{R}_k(M') = \frac{dE_k(M')}{dM'}.$$

Multiplying the series (7.116) by itself, we obtain

$$\frac{a^2}{r^2} = \sum_{k=0}^{\infty} e'^k \bar{R}_k^{(2)}(M'), \tag{7.117}$$

where

$$\bar{R}_k^{(2)}(M') = \sum_{k=0}^{k} \bar{R}_s(M')\bar{R}_{k-s}(M')$$

Then the expression (7.112) for the solar energy flux reaching the atmosphere can be rewritten:

$$L(t) = L_0 \sum_{k=0}^{\infty} e'^k \bar{R}_k^{(2)}(M'). \tag{7.118}$$

In agreement with (7.118), the change of the total energy of the Earth's atmosphere is proportional to the change of the solar energy flux $L(t)-L_o$. Then the expression for the total energy of the atmosphere is

$$E(t) = E + k[L(t) - L_0],$$

where k is a proportionality factor.

Thus, in our problem of virial oscillations of the atmosphere perturbed by the solar energy flux varying during the motion of the Earth along the orbit, the equation can be written as

$$\ddot{\Phi} = -A + \frac{B}{\sqrt{\Phi}} + X(e', M'), \tag{7.119}$$

where $X(e', M')$ is the perturbation function, which has the form

$$X(e', M') = 2k[L(t) - L_0]\left[\sum_{k=0}^{\infty} e'^k R_K^{(2)}(M') - 1\right]. \tag{7.120}$$

To estimate the geophysical effect of the variation of the solar energy within a time period of 1 year, we introduce the perturbation function (7.120) into Eq. 7.119 of perturbed oscillations, to an accuracy of squared eccentricity, i.e.,

$$X(M') = kL_0[2e' \cos M' + \frac{5}{2}e'^2 \cos 2M], \tag{7.121}$$

where $kL_o \approx 3 \cdot 10^{31}$ erg.

In this case, the expressions for the Jacobi function of the atmosphere and its first derivative to an accuracy of *e* have the form

$$\Phi(M) = \Phi_0(1 - 2e\cos M), \tag{7.122}$$

$$\dot{\Phi}(M) = 2\Phi_0 n \sin M, \tag{7.123}$$

Then, differentiating the expression for the eccentricity of the perturbed oscillations $e = (1-AC/2B^2)^{1/2}$ with respect to time and using Lagrange's method of varying arbitrary constants, we obtain

$$\frac{de}{dt} = \frac{A\dot{\Phi}(M)X(M')}{2B^2e} = \frac{A\Phi_0 kL_0 ne'}{B^2}[\sin(M + M') + \sin(M - M')$$

$$+\frac{5}{2}e' \sin(M + 2M') + \frac{5}{2}e' \sin(M - 2M')\Big]. \tag{7.124}$$

Integrating (7.124) with respect to time, we obtain the law of variation of the virial oscillation eccentricity as a first approximation of the perturbation theory:

$$e(t) = \bar{e} - \frac{A\Phi_0 ne'}{B^2}\left[\frac{\cos(M + M')}{n + n'} + \frac{\cos(M - M')}{n - n'}\right.$$

$$\left.+\frac{5}{2}e'\frac{\cos(M + 2M')}{n + 2n'} + \frac{5}{2}e'\frac{\cos(M - 2M')}{n - 2n'}\right] \tag{7.125}$$

Finally, putting the expression for the eccentricity of the perturbed oscillations into (7.122), we obtain

$$\Phi(t) = \Phi_0 + \frac{1}{2}\alpha_1[\cos(2M + M') + \cos M'] + \frac{1}{2}\alpha_2[\cos(2M - M') + \cos M']$$

$$+\frac{1}{2}\alpha_3[\cos(2M + 2M') + \cos 2M'] + \frac{1}{2}\alpha_4[\cos(2M - 2M') + \cos(2M'], \tag{7.126}$$

where

$$\alpha_1 = \frac{2A\Phi_0^2 kL_0 ne'}{B^2(n+n')}, \qquad \alpha_2 = \frac{2A\Phi_0^2 kL_0 ne'^2}{B^2(n-n')},$$

$$\alpha_3 = \frac{5A\Phi_0^2 kL_0 ne'^2}{B^2(n+2n')}, \qquad \alpha_4 = \frac{5A\Phi_0^2 kL_0 ne'^2}{B^2(n-2n')},$$

$$M = n(t - t_0), \qquad M' = n'(t - t'_0)$$

$$n = 1.2 \cdot 10^{-3}\, s^{-1}, \qquad n = 1.2 \cdot 10^{-7}\, s^{-1}$$

It follows from (7.125) that a contribution of the long-periodic part of the first-order perturbations to the variation of the Jacobi function of the Earth's atmosphere is

$$\Phi_a(t) = \text{const} + 1.44 \cdot 10^{-3}\Phi_0 \cos M' + 1.44 \cdot 10^{-3} e'\Phi_0 \cos 2M', \tag{7.127}$$

where $\Phi_o = 1.06 \cdot 10^{39}$ g $\cdot$ cm^2; $\Phi_a/I_a = 3/4$.

Assuming in first approximation that the rotations of the Earth and the atmosphere are synchronous, and using the law of conservation of angular momentum for the Earth-atmosphere system as $(I_\oplus + I_a)\omega_\oplus = \text{const}$, we obtain

$$\frac{\Delta I_a}{I_\oplus} = -\frac{\Delta\omega}{\omega_\oplus} = \frac{\Delta T}{T_0}, \tag{7.128}$$

where ω_o is the angular velocity of the Earth's daily rotation; $T = 8.64 \cdot 10^4$s; $I_\oplus = 8.04 \cdot 10^{44}$ g cm^2.

It is easy to show that the Earth's rate of rotation has annual variations with daily amplitude of variation of about 2 ms duration and can be approximated in our estimate by the sum of two harmonics with a period of 1 year and half a year respectively. This estimate is in good agreement with the observed data of the seasonal variation of the Earth's angular velocity.

7.12.4 Resonance Oscillation

We now consider the solution for identification of the resonance frequencies of the perturbed oscillations of the atmosphere due to the change of solar energy flux during the Earth's motion along an elliptic orbit.

Let us assume that the Earth's atmosphere satisfies all the conditions needed for writing the Eq. 7.107 of unperturbed virial oscillations. We can solve this equation because its two first integrals of motion (7.108) and (7.109) are known. We accept from the solution of Eq. 7.107 that the Jacobi function Φ changes in time with the period $\tau'' = 5{,}060.7\ s^{-1}$ and the frequency $n'' = 2\pi/\tau'' = 0.00124\ s^{-1}$. Assuming also that the perturbation of the Earth's atmosphere is affected only by the change of power of the solar radiation flux during the year owing to the ellipticity of the Earth's orbit, then the equation of the perturbed virial oscillations should have the form of (7.28) and the perturbation function $X(e', M')$, represented by (7.120), should be a periodic function of time with period $\tau' = 31{,}556{,}929.9747$ s and frequency $n' = 2\pi/\tau' = 1.9910638\ s^{-1}$.

We use the Picard method to obtain the solution of Eq. 7.119, which in this case is written as

$$\ddot{\Phi} = -A + \frac{B}{\sqrt{\Phi}} + 2kL_0\left[\sum_{k=0}^{\infty} e'^2 R_k^{(2)}(M') - 1\right], \tag{7.129}$$

where $A = -2E_o$; $B = |U_a|\sqrt{\Phi_a}$; $e' = 0.014$ is the eccentricity of the Earth's orbit; $M' = n'(t - t_o) = E' - e' \sin E'$ is the mean anomaly determined from the Kepler equation; E' is the eccentric anomaly.

Using Lagrange's method of variation of the arbitrary constants C and t, which determine the solution of Eq. 7.129 in the form

$$\Phi = \Phi(C, t''_0, t), \tag{7.130}$$

$$\dot{\Phi} = \dot{\Phi}(C, t''_0, t), \tag{7.131}$$

we can write the system of differential equations (5.19) and (5.20) (see Chap. 5) determining the change of C и t_o'' in time:

$$\frac{dC}{dt} = -2\dot{\Phi}X(M'), \tag{7.132}$$

$$\frac{dt''_0}{dt} = -2\dot{\Phi}X(M')\Psi(\Phi, C), \tag{7.133}$$

where the function Ψ(Φ, C) was defined earlier, and X(M′) is a periodic function of the argument M′ with the period 2π.

Substituting the expressions for Φ and $\dot{\Phi}$ from (7.130) and (7.131) into (7.132) и (7.133), we obtain the system of two differential equations:

$$\frac{dC}{dt} = F_1(M', M'', C, t''_0), \tag{7.134}$$

$$\frac{dt''_0}{dt} = F_2(M', M'', C, t''_0), \tag{7.135}$$

where F_1 and F_2 are periodic functions of the arguments M′ and M″ with the period 2π.

Owing to the periodicity of the right-hand side of the system of Eqs. 7.134 and 7.44, they can be expanded into a double Fourier series. In this case the system (7.134) and (7.135) can be written:

$$\frac{dC}{dt} = \left\{ A_{00} + \sum_{k',k''=-\infty}^{\infty} \left[A_{k',k''} \cos(k'M' + k''M'') + B_{k',k''} \sin(k'M' + k''M'') \right] \right\}, \tag{7.136}$$

$$\frac{dC}{dt} = \left\{ a_{00} + \sum_{k',k''=-\infty}^{\infty} \left[a_{k',k''} \cos(k'M' + k''M'') + b_{k',k''} \sin(k'M' + k''M'') \right] \right\}. \tag{7.137}$$

Here the coefficients A_{oo} , $A_{k'k''}$, $B_{k',k''}$, a_{oo}, $a_{k'k''}$, $b_{k',k''}$ do not depend on M' and M'', but are functions of the unknown quantities C and t_o''.

Using the Picard procedure, we determine $C^{(1)}$ and $t_o''^{(1)}$ in the first approximation by substituting the constant values $C^{(o)}$ and $t_o''^{(o)}$ into the expressions for F_1 and F_2. The values $C^{(o)}$ and $t_o''^{(o)}$ could be found through the initial conditions of $\Phi_0$and $\dot{\Phi}_0$using corresponding formulas (7.108) and (7.109) which describe the unperturbed virial oscillations.

After integration of the system (7.136) and (7.137) with respect to time, we have

$$\begin{aligned} C^{(1)} = C^{(0)} + A_{00}^{(0)}(t - t_0) + \sum_{k',k''\to-\infty}^{\infty} \frac{1}{k'n' + k''n''} \Big\{ A_{k',k''}^{(0)} [\cos(k'M' + k''M'') \\ - \cos(k'M'_0 + k''M''_0)] - B_{k',k''}^{(0)} [\sin(k'M' + k''M'') \\ - \sin\left(k'M'_0 + k''M''_0)\right] \Big\}, \end{aligned} \tag{7.138}$$

$$\begin{aligned} t''^{(1)}_0 = t''^{(0)}_0 + a_{00}^{(0)}(t - t_0) + \sum_{k',k''\to-\infty}^{\infty} \frac{1}{k'n' + k''n''} \Big\{ a_{k',k''}^{(0)} [\cos(k'M' + k''M'') \\ - \cos(k'M'_0 + k''M''_0)] - b_{k',k''}^{(0)} [\sin(k'M' + k''M'') \\ - \sin\left(k'M'_0 + k''M''_0)\right] \Big\}, \end{aligned} \tag{7.139}$$

where $C^{(1)}$ and $t_o''^{(1)}$ are the arbitrary constants which determine solution of the Eq. 7.28; $A_{k',k''}^{(0)}$, $B_{k',k''}^{(0)}$, $a_{00}^{(0)}$, $a_{k',k''}^{(0)}$, $b_{k',k''}^{(0)}$ are corresponding coefficients of the system (7.136) and (7.137) after replacing C и t_o'' by $C^{(o)}$ and $t_o''^{(o)}$.

Thus we have obtained the analytical structure of the solutions (7.138) and (7.139) known in general perturbation theory which have three classes of terms: constant, periodic and secular. Of the periodic terms, the most important are the resonance terms, i.e. those quantities $n_s = k'n' + k''n''$ which are substantially less then both n" and n'. These terms give a series of long periodic inequalities (their number is infinite) and they allow prediction of the development of the natural processes within relatively long intervals of time.

Let us calculate, as an example, such lower resonance frequencies which have climatic significance:

$$n_{2:12471} = (2.12415 - 12471 \cdot 1.9910638)10^{-7} = 0.556 \cdot 10^{-7}s^{-1},$$

$$n_{3:18706} = (3.12415 - 18706 \cdot 19910638)10^{-7} = 0.161 \cdot 10^{-7}s^{-1},$$

$$n_{11:68589} = (11.12415 - 68589 \cdot 1.9910638)10^{-7} = 0.07 \cdot 10^{-7}s^{-1}$$

and so on, and corresponding to those frequencies the periods are:

$$\tau_{2:12471} = 3.6 \text{ years}; \quad \tau_{3:18706} = 12.36 \text{ years}, \quad \tau_{11:68598} \approx 28 \text{ years}.$$

It should, however be kept in mind that the first approximation obtained in the framework of perturbation theory is in good agreement with observations within short (not cosmogenic) intervals of time.

Note that one can find analogously the resonance frequencies and the periods of virial oscillations which occur owing to other perturbations, such as, for example, diurnal perturbations, because of the rotation of the Earth around its axis and its latitudinal perturbations, etc. As is well known, the mean solar day has period $\tau = 8.64{\cdot}10^4$ s and frequency $n_s = 2\pi/\tau = 7.27 \cdot 10^{-5}\,s^{-1}$. Then calculating the resonance frequency $n_s = k'n' \vec{+} k''n'' << n, n''$, we obtain:

$$n_{1:17} = (1 \cdot 124.17 - 17 \cdot 7.27)10^{-5} = 0.58 \cdot 10^{-5}\,s^{-1},$$

$$n_{2:24} = (2 \cdot 124.17 - 24 \cdot 7.27)10^{-5} = 1.16 \cdot 10^{-5}\,s^{-1},$$

and so on, and corresponding periods:

$$\tau_{1:17} = 12.5 \text{ d}; \quad \tau_{2:24} = 6.24 \text{ d andsoon.}$$

For the monthly Moon perturbations, when $\tau = 2.352672{\cdot}10^6$ s and $n_s = 2.67{\cdot}10^{-6}\,s^{-1}$, we have

$$n_{1:465} = (1 \cdot 1241.7 - 2.67 \cdot 465)10^{-6} = 0.15 \cdot 10^{-6}\,s^{-1},$$

$$n_{2:930} = (2 \cdot 1241.7 - 2.67 \cdot 930)10^{-6} = 0.3 \cdot 10^{-6}\,s^{-1}$$

and so on, and corresponding periods:

$$\tau_{1:465} = 1.3 \text{ years}; \quad \tau_{2:930} = 0.66 \text{ years}.$$

7.12.5 Observation of the Virial Eigenoscillations of the Earth's Atmosphere

It was predicted in Sect. 7.12.2, by means of the solution of Jacobi's virial equation, that the eigenoscillations of the Earth's atmosphere with a period of $T_v = 1^h.4$ and frequency $\omega = 1.24{\cdot}10^{-3}\,s^{-1}$ exists. This solution describes the periodic change of the Jacobi function of the Earth's atmosphere in time and can be expressed in the form of a trigonometric Fourier series expanded by entire multiple values of argument M related to t as

$$M = \frac{2\pi}{T_v}(t - t'_0),$$

where t'_o is the moment of time defining the phase of the virial oscillations.

The first four terms of this series are

$$\Phi(t) = \Phi_0\left[1 + \frac{3}{2}e^2 + \left(-2e + \frac{e^3}{4}\right)\cos M - \frac{e^2}{2}\cos 2M - \frac{e^3}{3}\cos 3M + \ldots\right], \quad (7.140)$$

where Φ_o is the mean value of the Jacobi function, determined by the virial theorem; e is the parameter of the virial oscillations of the atmosphere which characterizes the amplitude of the Jacobi function change; $e \ll 1$.

It has been shown theoretically that the period of chance of the Jacobi function of the atmosphere depends on the value of its total energy, and in the case of non-perturbed atmosphere is equal to $1^h.4$. In the framework of the model considered with spherically symmetric atmosphere, the change of Jacobi function takes place owing to the change of the radial mass density distribution of the atmosphere having the same period. Direct experimental test of this statement is difficult because the value of the Jacobi function cannot be measured directly to prove the expression (7.140). But, as it was shown in Sections 2.2, permanent changes in the Earth's Jacobi function (polar moment of inertia) is fixed by artificial satellites and earthquake measurements. Moreover, the process of virial oscillations is accompanied by synchronous changes of pressure, temperature, air moisture, magnetic field intensity and other measurable geophysical parameters at the Earth's surface. In addition, from the condition of a global scale of the virial oscillations, it follows that all the geophysical parameters are pulsating with the same period and are coherent both within the considerable interval of time and in space over all the Earth's surface as well as vertically. The expression for the virial pulsations of the atmospheric pressure p(t) and temperature T(t) can be expanded into a Fourier series (7.140) as follows:

$$p(t) = p_0\left[1 + \frac{1}{2}e^2 + \left(2e + \frac{3}{4}e^3\right)\cos M + \frac{5}{2}e^2\cos 2M + \ldots\right], \quad (7.141)$$

$$T(t) = T_0\left[1 + \frac{1}{2}e^2 + \left(2e + \frac{3}{4}e^3\right)\cos M + \frac{5}{2}e^2\cos 2M + \ldots\right], \quad (7.142)$$

where p_o and T_o are the mean values of the atmospheric pressure and temperature averaged over all the Earth's surface and through the mass of the atmosphere respectively.

We shall use for this expansion the analysis and interpretation of the experimental data. As is well known, regular observation of the atmospheric pressure and temperature at various points of the Earth prevents the discovery of a rigorous periodicity in changes of atmospheric parameters, especially for short periods. There are a number of reasons for this, including their variability which defines the dynamics of air masses at any point of observation. The parameters recognized and studied until now are seasonal, diurnal and semidiurnal periodicity, in addition to variation of the atmospheric parameters connected with the motion of the planet along the elliptical orbit around the Sun, with the obliquity of the axis of rotation to the ecliptic, and with perturbations caused by the Moon.

In our case, in order to prove that the predicted oscillations of various geophysical parameters with period 1h.4 exist, we used the applicable spectral analysis of experimental data.

Oscillation of the temperature. We now describe the results of spectral analysis of temperature data recorded by our colleagues in the Central Hydrometeorological Observatory at the Ostankino TV Tower, Moscow. Let us consider two sets X_1 and Y_1 representing regular records of the air temperature variation of 34 h duration, each of which were obtained simultaneously in July 1971 at heights of 503 and 83 m (Fig. 7.11). The discreteness interval of the numerical record of the temperature was 120 s, the interval of the gauge was ~60 s, and the sensitivity was около 0.1°C. The sets X_1 and Y_1 contain 1,024 discrete values of recorded temperature starting at 10 h 34 min on 17 July through 21 h 34min on 18 July. Spectral analysis of the data received was carried out by computer using the method of quick Fourier transformation with the program developed by A.B. Leybo and V.Yu. Semenov.

Figure 7.11a shows the recorded power spectra S_{xx} and S_{yy} of the sets X_1 and Y_1. With the help of this initial information, we calculated the function of the mutual spectral density $S_{xx} = S'_{xy} + iS''_{xy}$. Then the function of the mutual coherence was found to be

$$Co^2 = \frac{|S_{xy}|^2}{|S_{xx}||S_{yy}|} \tag{7.143}$$

and the function of the phase difference was defined as

$$\Delta\varphi = \text{atctg}\frac{S''_{xy}}{S'_{xy}}. \tag{7.144}$$

They are both plotted in Figs. 7.11b, c.

It is known that the range of the confidence intervals for estimating the phase difference $\Delta\varphi$ tends to zero as Co^2 runs to unity. Hence the higher the value of the function of the mutual coherence, the higher the stability of the phase difference of the harmonics of the two processes X and Y. Therefore, the relationship between the harmonics is probable for those frequencies where the value of Co^2 is close to

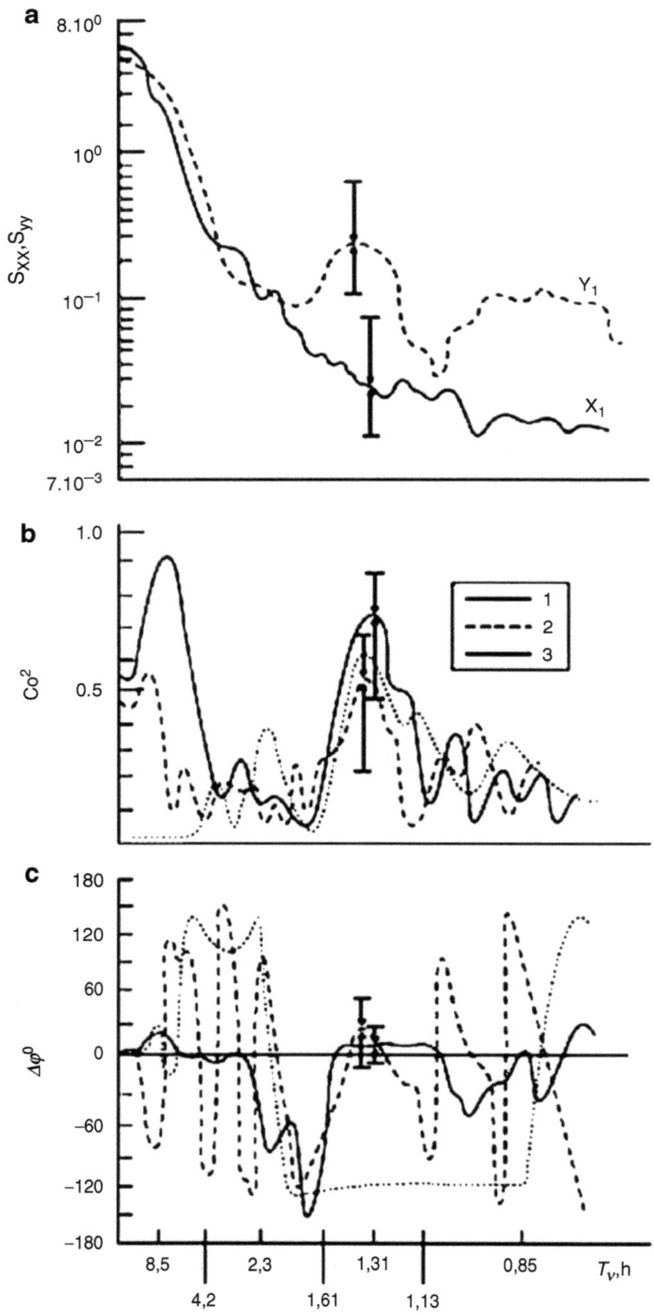

Fig. 7.11 Power spectra of sets of temperature variation at heights of 503 and 83 m (**a**); functions of mutual coherence of the sets X_1 and Y_1 (*1*), X_2 and Y_2 (*2*), X_3 and Y_3 (*3*) (**b**); functions of phase difference of the sets X_1 and Y_1 (*1*), X_2 and Y_2 (*2*), X_3 and Y (*3*) (**c**). (*Arrows* show the range of 95% of the confidence intervals)

unity. At the same values of Co^2 the meaning of the phase difference is lost because of the wide range of the confidence interval.

We can see from Fig. 7.11b that in the vicinity of the period $T_v = 1^h.4$, predicted by theory, the function of the mutual coherence of the sets X_1 and Y_1 is significant, i.e. has high probability of not being equal to zero, and sometimes of being even greater then 0.7. Note also that the phase difference function in the neighborhood of the period of oscillations T_v is equal to zero, which indicates that the harmonic constituents of the temperature variations with the period T_v within the considered interval of time are coherent at heights of 503 and 83 m.

Twice as much extended time for sets X_1 and Y_1 does not change the general character of the discovered regularity. Figures 7.11b, c demonstrate values of the mutual coherence function Co^2 and the phase difference $\Delta\varphi$ of two sets X_2 and Y_2 each containing 2,048 experimental points and recorded synchronously with the same discreteness interval. Recording was continued for 3 days starting at 10 h 34 min on 17 July through 6 h 54 min on 20 July 1971. However, reduction of time for the sets leads to an increase in values of the mutual coherence function on the low frequencies, but the values of the frequencies corresponding to the period T_v do not increase. This proves the theoretically predicted conclusion concerning the existence of a coherent harmonic with the period T_v.

Figure 7.11b plots the mutual coherence function for two sets of temperature variations recorded at 17 h 2 min in each case. These sets were recorded at a height of 503 m with the same discreteness interval as discussed above, but in different years. Set X_3 covers a time interval from 10 h 34 min on 17 July until 3 h 36 min on 18 July 1971, and Y_3 was recorded from 9 h 14 min 30 July to 2 h 16 min on 31 July 1971. We can see from the plot that in this case the mutual coherence function acquires a value equal to 0.6 at $T = 1$ h.31, which proves the theory of the steady state virial oscillations within sufficiently long intervals of time.

Let us estimate the amplitude of temperature virial oscillations in the neat surface layer of the atmosphere. For this purpose we recall that the value of the power spectrum of a process at a given value of the frequency is proportional to the square of the amplitude value of the harmonics with the same frequency in Fourier analysis of the process. We plot the power spectrum of a sinusoid of some known amplitude, for example, equal to 1°C with period $1^h.4$ and a given discreteness $\tau \approx 120$ s. The value of the power spectrum of the sinusoid for the frequency related to period $T_v = 1^h.4$ was found equal to 3.4 grad^2 per hertz, and the value of the power spectrum of temperature micro-fluctuation at heights of 83–503 m in July 1971, according to the Observatory recording, was 0.3 grad^2 per hertz. It is easy to obtain the estimated value of the amplitude of the temperature virial oscillations, which is 0.3°C.

Oscillation of the pressure. We now consider our results of the spectrum analysis of the atmospheric pressure records made by means of a microbarograph designed by V.N. Bobrov in the Institute of Earth Magnetism and Radio Wave Propagation, Russian Academy of Sciences. Analysing the records, we took the pressure ordinates correct to 1 mm at amplitude of pressure oscillation within several cm (at microbarograph sensitivity equal to 0.02 mb/mm). The spectral analysis was done using the methodology described above.

Figure 7.12a shows the power spectrum of two sets X_4 and Y_4 containing 1,024 of the pressure ordinates taken with discreteness interval 76.5 s covering the time

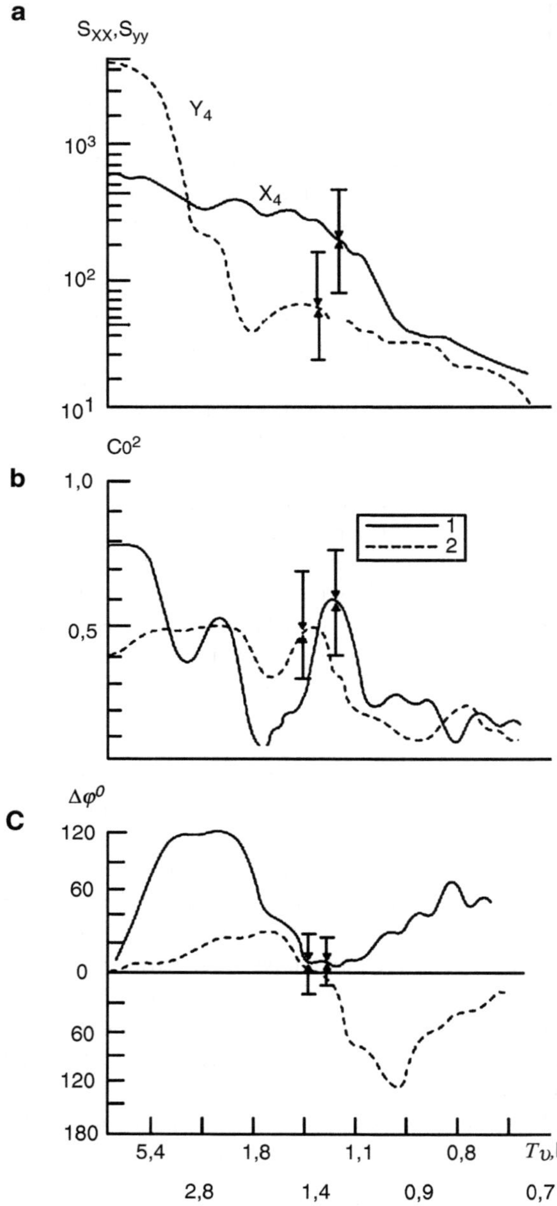

Fig. 7.12 Power spectra of the sets X_4 and Y_4 of atmospheric pressure micro-fluctuations northeast of the coastal area of the Caspian Sea (**a**); functions of mutual coherence of the sets X_4 and Y_4 (*1*) and X_5 and Y_5 (*2*) (**b**); functions of the phase difference of the sets X_4 and Y_4 (*1*) and X_5 and Y_5 (*2*) (**c**). (*Arrows* show the range of 95% of the confidence intervals)

interval from 18 h on 31 August through 15 h 30 min of 1 September 1977 and from 15 h 00 min on 1 September through 12 h 30 min on 2 September 1977.

The experimental data of micro-fluctuations of the atmospheric pressure were obtained at a near surface layer of the northeast coastal area of the Caspian Sea by our colleagues from the Institute of Earth Magnetism and Radio Wave Propagation. Using the function of mutual spectral density S_{xx}, we calculated the functions of mutual coherence Co^2 and phase difference $\Delta\varphi$, shown in Fig. 7.12b, c.

The spectral densities were analyzed by averaging the values of seven experimental points (the number of degrees of freedom equal to 14). The graph of spectral densities was plotted in relative units and on logarithmic scale (in order to have the same range of confidence intervals for any value of the spectral density).

Figure 7.12b shows that in the vicinity of the frequency value corresponding to the period $T_v = 1^h.4$, the coherence coefficient is significant and is equal to $Co^2 = 0.6$. It is also important to note that, for the same frequency, the phase difference is stable and close to zero (see Fig 7.12c). These facts prove the theory of the existence of coherent oscillations of the Earth's atmosphere with period close to T_v.

Figures 7.12b, c also show the curves of mutual coherence and phase difference for the other two sets X_5 and Y_5 plotted on a base of 512 experimental points of atmospheric pressure recorded within time intervals starting at 16 h 56 min on 28 August and at 2 h 45 min on 29 August and at 18 h 00 min 31 August through 3 h 49 min on 1 September 1977, respectively. The mutual coherence function has a value equal to 0.5 in the vicinity of T_V, and the function of the phase difference has a value close to zero. This also proves the conclusion discussed above.

We can also show that virial pulsation of the Earth's atmosphere with period $T_v = 1^h.4$ is observed in both mid and low latitudes. For this purpose we studied records of micro-fluctuations of the atmospheric pressure obtained by the same researchers during their expedition to Cuba.

Let us analyze two sets of experimental data X_6 and Y_6 taken from their records of micro-fluctuations of the atmospheric pressure with discreteness interval equal to 180 s. The process X_6 covers a time interval starting at 19 h 00 min on 7 May through 20 h 36 min on 8 May 1976 and the process Y_6 covers an interval from 21 h 54 min on 7 May through 23 h 30 min on 9 May 1976. Figure 7.13a shows the power spectra of the processes X_6 and Y_6. The functions of mutual coherence and phase difference which, in the vicinity of the period T_v, have values $Co^2 = 0.56$ and $\Delta\varphi = 0°$ respectively, are shown in Fig. 7.13b, c. The same figures also show functions of mutual coherence and phase difference for two other sets, X_7 and Y_7, representing 256 experimental points each and analyzed with the same discreteness intervals. They cover time intervals starting at 21 h 37 min on 25 April through 10 h 23 min on 26 April and on 20 h 00 min on 28 April on 9 h 14 min on 29 April 1976. The value of function $Co^2 = 0.6$ and $\Delta\varphi = 0°$ relative to frequencies with period $T_v = 1^h.4$.

Our estimate of the amplitude of virial oscillations of the atmospheric pressure made by using the above procedure gives the value of 0.1 mbar.

Figure 7.13b shows the function of mutual coherence of two sets of atmospheric pressure X_8 and Y_8 taken with discreteness interval $\Delta t = 6$ min and recorded in various years and at different points of the globe: (a) in Cuba within a time interval

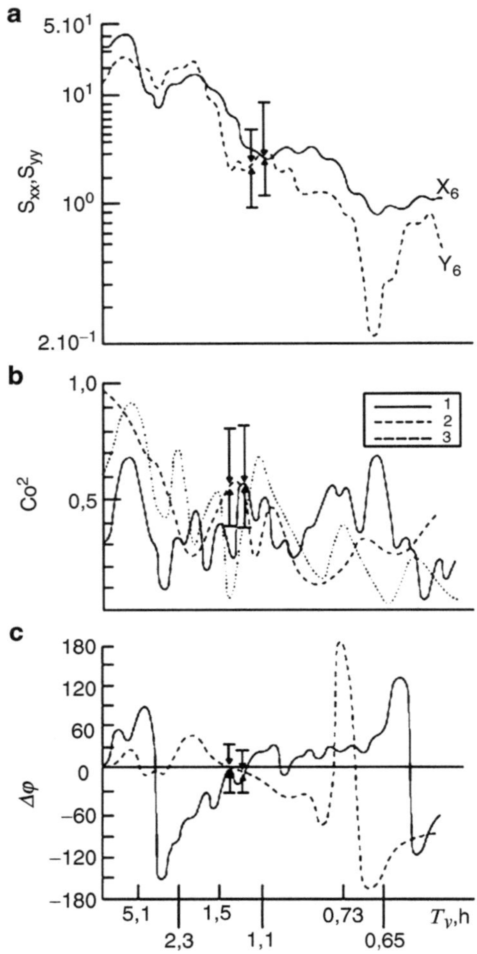

Fig. 7.13 Power spectra of the sets X_6 and Y_6 of atmospheric pressure variations in Cuba (**a**); functions of mutual coherence of the sets X_6 and Y_6 (*1*) and X_7 and Y_7 (*2*) and X_8 and Y_8 (*3*) (**b**); functions of the phase difference of the sets X_6 and Y_6 (*1*) and X_7 and Y_7 (*2*) (**c**). (*Arrows* show the range of 95% of the confidence intervals)

starting at 19 h 00 min on 7 May through 2 h 30 min 8 May 1976; (b) on northeastern shore of the Caspian Sea at 19 h 00 min on 4 September through 20 h 30 min on 5 September 1977. We also observed here that the function Co^2 takes a value of about 0.65 in the vicinity of the period $T_v = 1^h.4$. This also proves our hypothesis.

Analogous results were obtained on analysis of the experimental data of micro-fluctuation of the atmospheric pressure recorded at the Black Sea. The analyses were made with discreteness interval equal to 120 s.

We have shown the existence of harmonics with period close to $T_v = 1^h.4$, coherent within long time intervals for micro-fluctuations of atmospheric pressure and temperatures in the near surface layer of the atmosphere, obtained at various points of the globe and in different seasons of the year. The resulting data from analyses of experimental records of micro-fluctuations of pressure and temperature as well as the geomagnetic field of the Earth are given in Table 7.2.

7.12.6 Dynamics of the Oceans

It follows from Eq. 7.71 that the world ocean is found to be in a suspended state because the density value of its water is by far less (2/3) than the mean density of the solid Earth. Accepting this criterion, we may assume that in the earlier stage of creation of the planet its hydrosphere remained in the gaseous phase. Only after irradiation of the corresponding part of the potential energy has the water vapor condensed into the liquid phase (see Chap. 9). Taking into account the observed geographic distribution of the Atlantic and Pacific oceanic basins, we also may assume that their floors were formed as a consequence of the planet's equatorial oblateness during formation. Applying the same argument and geography of location, one may conclude that the Indian oceanic floor was formed on the same base at the Earth's polar flattening and due to the asymmetric deformation of the southern hemisphere (see Chap. 2).

The oceans have their own potential energy value which is equal to $U \approx 2{\cdot}10^{32}$ erg. This value is by four degrees less of their oscillating and rotating kinetic energy. It means that the oceans stay in hydrostatic equilibrium in the outer force field of the solid Earth which provides rotational motion of this shell

This suspended stay of equilibrium of the oceans determines their dynamics. According to Eq. 7.71, angular velocity of the rotating oceanic water, because of low density, is less than that of the solid Earth. Therefore, the oceanic water in its rotary motion has lower angular velocity with respect to the solid Earth. Encountering the continents on their way, the ocean waters form the latitudinal currents along both American and Asian continents. The observed multiply disordered regional and local linear and eddy currents are the consequences of different perturbations, which dominate over the regular currents due to the above energy effects.

The regular currents, which move along the east shores of the continents, play an important role as heating system in formation of weather and climate processes in the middle and high latitudes.

7.12.7 The Nature of the Weather and Climate Changes

The atmosphere and oceans comprise a common natural system which controls the weather and climate on the planet. Both shells, being uniform in density and staying in semi-hydrostatic and semi-dynamic equilibrium in the outer force field of the solid Earth, are affected by virial oscillations of the planet and rotate with a

Table 7.2 Experimental data after reciprocal spectral analysis

Place of observation and parameters	Data and hours of observation	Frequency points (h)	Discreteness (Δt, s)	Time (h)	$Co^2/\Delta\varphi^\circ$
Caspian Sea, pressure	X:31.08 (18-00)–01.09.1977 (05-05) Y:28.08 (17-00)–29.08.1977 (04-05)	512/11	78.5	1.39	0.5/13
Caspian Sea, pressure	X:31.08 (18-00)–01.09.1977 (15-38) Y:01.09 (15-38)–02.09.1977 (12-41)	1024/22	78.5	1.36	0.56/6
Caspian Sea, pressure	X:28.08 (16-45)–29.08.1977 (04-26) Y:09.09 (22-00)–10.08.1978 (09-30)	512/11	78.5	1.39	0.47/0
Black Sea, pressure	X:13.08 (22-30)–14.08.1979 (11-00) Y:14.08 (11-00)–14.08.1979 (23-30)	512/12,5	87.7	1.39	0.76/−1.4
Moscow, temper. (503 m)	X:24.01 (08-14)–25.01.1973 (01-16) Y:25.01 (02-30)–25.01.1973 (19-32)	512/17	120	1.42	0.60/0
Moscow, temper. (50, 83 m)	X:17.07 (10-34)–18.07.1971 (03-36) Y:17.07 (10-34)–18.07.1971 (03-36)	512/17	120	1.42	0.67/−30
Moscow, temper. (503, 83 m)	X:17.07 (10-34)–18.07.1971 (20-36) Y:17.07 (10-34)–18.07.1971 (03-36)	1024/17	120	1.26	0.72/4
Moscow, temper. (503, 83 m)	X:17.07 (10-34)–20.07.1971 (06-54)	2048/68	120	1.28	0.55/10

(continued)

Table 7.2 (continued)

Place of observation and parameters	Data and hours of observation	Frequency points (h)	Discreteness (Δt, s)	Time (h)	$Co^2/\Delta\varphi^\circ$
	Y:17.07 (10-34)–20.07.1971 (06-54)				
Moscow, temper. (503, 83 m)	X:17.07 (10-34)–18.07.1971 (03-36)	512/17	120	1.31	0.62/0
	Y:30.07 (09-14)–31.07.1977 (02-16)				
Cuba, pressure	X:25.04 (21-37)–26.04.1976 (10-23)	256/13	180	1.42	0.57/10
	Y:28.04 (20-00)–29.04.1976 (09-14)				
Cuba, pressure	X:07.05 (19-00)–08.05.1976 (20-36)	512/26	180	1.28	0.59/1.7
	Y:08.05 (21-54)–09.00.1976 (23-30)				
Cuba, pressure	X:16.11 (12-00)–18.11.1976 (15-12)	1024/51	180	1.42	0.50/−17
	Y:07.05 (19-00)–09.05.1976 (22-15)				
Cuba, magnetic field	X:16.02 (12-00)–18.02.1976 (15-12)	1024/51	180	1.65	0.88/120
Cuba, pressure	Y:16.02 (12-00)–18.02.1976 (15-12)			1.38	0.50/160
Cuba, pressure	X:07.05 (19-00)–08.05.1976 (20-30)	256/26	360	1.1	0.65/−40
Caspian Sea, pressure	Y:14.08 (11-00)–14.08.1979 (23-30)				

corresponding angular velocity. However, irregular seasonal pumping of solar energy, because of ellipticity of the Earth's orbit and the precession and nutation perturbation of the upper solid shell, leads to continuous redistribution of solar energy in the latitude and longitude directions. Those perturbations cause permanent change in balance of the evaporated water from the ocean surface and lead to changes in baric topography and in trajectories of the cyclonic vortexes which carry clouds of moisture. From point of view of the considered theory, those are the perturbations that appear to be the cause of weather and climate change.

The principles of perturbation and resonance oscillation of the atmosphere presented in this chapter could be used as a basis for development of analytical solution of the problem.

7.13 Lyapunov Stability of Motion in Jacobi Dynamics

It is important to note that in the dynamic approach to the solution of problem of the system's dynamics, the integral characteristics of a system (Jacobi function and total energy), presents in Jacobi's equation, are immanent to their own integrals. Estimating the Lyapunov stability of motion of a system, they play the role of Lyapunov functions.

Studying Lyapunov stability of the virial oscillations of celestial bodies, we use the Duboshin criterion, which is applicable when permanent perturbations are present. For conservative systems, the potential energy of the system plays the role of such a perturbation. Thus, the nature of the virial oscillations can be understood as an effect of non-linear resonance between the kinetic and the potential energies.

7.13.1 Lyapunov Stability of Motion of a System Described in Terms of Co-ordinates and Integral Characteristics

Let us recall the definition of stability of motion according to Lyapunov (Duboshin 1952). If for any $\varepsilon > 0$ there exist $\delta > 0$ such that for any perturbations of the initial data x_{oj}, satisfying the condition $| x_{oj}| < \delta$, the inequalities $|x_{oj}| < \varepsilon$ ($j = 1, 2, \ldots, n$) holds, then the perturbed motion is stable; otherwise it is unstable.

Studying the stability of motion according to Lyapunov, we shell use his theorem (Duboshin 1952) which follows:

If for differential equations of a perturbed motion the function V of fixed sign (positive defined or negative defined) can be found (Lyapunov function), whose derivative by vitrue of these equations is the function of the constant sign opposite to the sign of V or identically equal to zero, then the unperturbed motion is stable.

The theory of Lyapunov stability considers variations in the initial data. Thus the equations for the perturbed and unperturbed motions ate the same, and one can therefore write Lyapunov's equation for the variation of perturbations (sequential perturbations) in the form

$$\dot{x}_j = X_j(x_1, \ldots, x_j \ldots, x_n, t) \quad j = 1, 2, \ldots, n \tag{7.145}$$

where

$$X_j(x_1, \ldots, x_j \ldots, x_n, t) = P_j(f_i + x_1, \ldots, f_j + x_j, , \ldots, f_n + x_n) - P_j(f_i, \ldots, f_j, , \ldots, f_n t)$$

is equal to the difference between the two particular solutions of one and the same equation of motion of the system which was obtained by subtracting the two equations. The right-hand sides x_j (...) of Eq. 7.145 are the functions equal to zero at the stationary point $x_j = 0$ ($j = 1, 2, \ldots, n$), i.e. in the absence of initial data perturbations. Thus, according to Lyapunov, the problem of stability can be reduced to the problem of stability of the zero (trivial) solution of the perturbation equations, having on the right-hand side at the origin of the new co-ordinates (x_j) the function x_j (...) identically equal to zero.

Investigating the Lyapunov stability of the system whose equations of motion are written in co-ordinates, the difficulty is to find the Lyapunov function V. Progress in this field has been achieved owing to the development of Chetaev's method (Chetaev 1962), which we shall employ.

The method is as follows. If the first m integrals $F_1(x_1,\ldots, x_n) = h_1,\ldots, F_m(x_1,\ldots, x_n) = h_m$ of the perturbed motion are known, then the Lyapunov function can be sought in the form

$$V = \lambda_1[F_1 - F_1(0)] + \ldots + \lambda_m[F_m - F_m(0)] + \mu_1\left[F_1^2 - F_1^2(0)\right] + \ldots \\ + \left[F_m^2 - F_m^2(0)\right]. \tag{7.146}$$

In writing the equations for the system's dynamics in terms of integral characteristics it appears that the latter (Jacobi function and total energy) are immanent to their own integrals. Let us illustrate this statement by well-known examples.

Example 1. Stability of stationary motion in the two-body problem.

Let r_o be the radius of circular motion of mass m in the two-body problem; ω is its angular velocity. Then according to Kepler's law we write

$$\omega^2 r_0^3 = Gm = \mu = \text{const}. \tag{7.147}$$

If r, ψ, θ are co-ordinates of the elliptic motion, then the integrals of their perturbed motion are as follows:

$$T + U = \frac{m}{2}\left[(\dot{r})^2 + r^2(\theta)^2 + r^2\cos^2\theta\left(\dot{\psi}\right)^2\right] - \mu\frac{m}{r} = \frac{m}{2}h_1, \tag{7.148}$$

$$\frac{\partial T}{\partial \dot{\psi}} = mr^2\cos\theta\dot{\psi} = mh_2. \tag{7.149}$$

Introducing the perturbations x_j,

$$r = r_o + x_1; \quad \dot{r} = 0 + x_2,$$

$$\theta = x_3; \quad \dot{\theta} = 0 = x_4,$$

$$\dot{\psi} = \omega + x_5$$

Then integrals (7.148) and (7.149) of the perturbed motion can be rewritten in the form

$$F_1 = x_2^2 + (r_0 + x_1)^2 x_4^2 + (r_0 + x_1)^2 \cos^2 x_3 (\omega + x_5)^2 - \frac{2\mu}{r_0 + x_1} = h_1, \quad (7.150)$$

$$F_2 = (r_0 + x_1)^2 \cos_2 x_3 (\omega + x_5) = h_2. \quad (7.151)$$

Let us write the Lyapunov function according to Chetaev's method as follows:

$$V = F_1 - F_1(0) + \lambda[F_2 - F_2(0)] + k\left[F_2^2 - F_2^2(0)\right]$$

The function V is of fixed sign when $2\omega + \lambda + 2kr_0^2\omega = 0$. It is a positive definite function when the condition $k > 3r_0^2$ is satisfied.

All the conditions needed for application of the stability theorem mentioned earlier are satisfied at $k, > 3r_0^2$ and the motion in the accepted co-ordinates is therefore stable.

To estimate the stability of Keplerian motion in terms of integral (volumetric) characteristics, only one integral of motion is needed, namely the angular momentum M_o:

$$M_0 \propto \Phi\omega. \quad (7.152)$$

It should be pointed out that these integral characteristics (Φ and ω) form a canonical pair since their product is proportional to the action of the system

$$h \propto \Phi\omega \quad (7.153)$$

and the Jacobi function Φ is the integral characteristic of motion. It is easy to verify that Keplerian motion is unstable at $M_o = 0$, i.e. when linear motion takes place.

Example 2. Stability of a pendulum at vertical oscillation of its fixed point of suspension (Kapitza 1951).

In contrast to the case considered above, this example is related to the problems of stability of the system at permanent perturbations (e.g. due to dissipation of feedback of energy). Here the equations of unperturbed motion (without dissipation of energy) and perturbed motion (when dissipation of energy takes place) are different. The procedure of subtraction of equations does not lead to the equation with the right-hand side identically equal to zero in the stationary point (for trivial solution).

Such systems can be studied with the help of an approach developed by Duboshin (1952). Duboshin's criterion of stability of motion at permanent perturbations is as follows (1978): Unperturbed motion is stable et permanent perturbations $R_s(t/x_\sigma)$ if for any given arbitrary positive $\varepsilon \leq A$ there exist such positive numbers $\lambda \leq \varepsilon$ and r depending on ε that at any initial data x_s^o satisfying the conditions $\left|X_s^o\right| \leqslant \lambda$ and at any functions F_s satisfying at $|x_s| \leq \varepsilon$ at $t > t_o$ the inequalities

$$|R_s(t/x_\sigma)| < r \tag{7.154}$$

all the solutions $x_s(t)$ of the system of differential equations for $t > t_o$ satisfy the conditions $|x_s(t)| < \varepsilon$. In other cases the motion is called unstable at permanent perturbations, whatever their amplitude r is.

We shell employ Duboshin's criterion while studying the stability of a pendulum's oscillations and the stability of the virial oscillations.

The equation of motion of a pendulum near the stable (in the absence of oscillations of the fixed point) lowest equilibrium state is written

$$\ddot{x} + \left(\frac{g}{l} + d\frac{\omega^2}{l}\cos\omega t\right)x = 0, \tag{7.155}$$

where x is the angle measured from the equilibrium state; ℓ is the length of the pendulum; d is the amplitude of vertical oscillations of the fixed point; and ω is the frequency of oscillations of the fixed point.

For a pendulum oscillating near the lowest stable equilibrium state, the loss of stability (the parametric resonance) appears at the frequency of oscillations

$$\omega = \frac{2}{n}\sqrt{\frac{g}{\ell}} \tag{7.156}$$

where n is a natural number.

For oscillations of a pendulum near the unstable upper state, the condition of stability of the motion is expressed as

$$\omega > \frac{\sqrt{2g\ell}}{d}. \tag{7.157}$$

Thus variation of the system's parameters affects the stability of the motion. It should be noted that Eq. 7.155 itself (even if the dependence of amplitude d on time is taken into account) is written in linearized form relative to x, although the amplitudes of swing can be large and even rotation can be considered. Therefore, the more correct form of the equation will be, without linearization:

$$\ddot{x} + \left(\frac{g}{\ell} + d\frac{\omega^2}{\ell}\cos\omega t\right)\sin x = 0. \tag{7.158}$$

In this connection, let us consider the case when $d = 0$ and the effect of potential energy of the free rotation of the system is taken into account.

$$\ddot{x} = 0, \quad \dot{x} = \Omega = \text{const},$$

where Ω is the angular velocity.

The energy of rotation here is equal to the kinetic energy

$$T = \frac{1}{2} m\ell\Omega^2$$

The phase of revolution of a pendulum is written as $x = \Omega t + x_o$, where x_o is the initial phase .The potential energy of the system is $U = (mg/\ell)\sin x$, where U is proportional to sinx.

In this case Eq. 7.158 of the perturbed motion of a pendulum can be considered as a non-linear resonance of the potential and kinetic energies.

Unperturbed motion (rotation) of the system has angular velocity depending on its energy $T = 1/2m\ell^2\Omega$ and therefore

$$\Omega = \sqrt{\frac{2T}{m\ell^2}}$$

i.e.

$$\Omega \propto T^{1/2} \tag{7.159}$$

Thus, the unperturbed motion is non-linear although harmonic. Non-linearity is developed here in non-isochronous oscillations.

The perturbation function, which is the potential energy, oscillates with the same frequency as the perturbed system itself but has a phase shift. The curves indicating the variation of the phase of revolution (unperturbed motion) and the potential energy (perturbation function) are plotted in Fig. 7.14.

In this figure τ denotes the period of revolution of the pendulum. Curve 1 represents the phase of revolution, which is a function of time: $x = \Omega t$. Curve 2 is parallel to curve 1 but is shifted in the phase by 2π. Curve 3 represents the phase of the potential energy, which is the following function of time:

$$U = \frac{mg}{\ell} \sin x\alpha\left(x - \frac{x^3}{3} + \ldots\right)$$

In a linear approximation we have $U \propto x$ and thus the phase $x = \Omega t$ coincides with the phase of revolution. It follows from Fig. 7.14 that if the value of the coefficient mg grows, then curve 3 intersects curve 2. The differences of phase of revolution and perturbation reaches the value 2π and the character of the motion changes. Therefore, if the potential energy is considered as a permanently acting perturbation, the existence of the bifurcation point of the system with respect to permanent mg is evident.

Let us write all the forgoing in analytical form. The equation of motion is

$$\ddot{x} + \frac{g}{\ell}\cos x = 0. \tag{7.160}$$

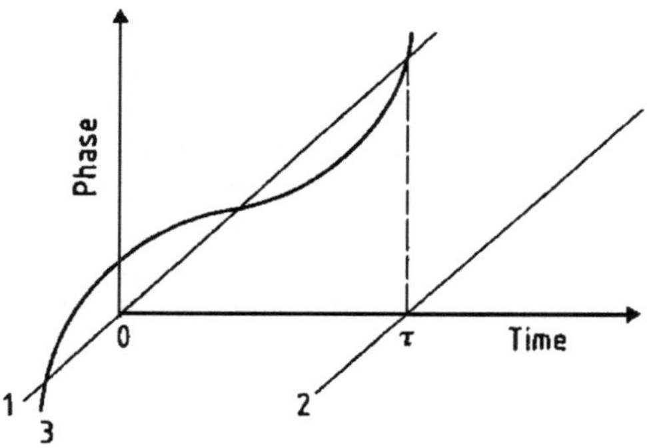

Fig. 7.14 Variation of the phase of revolution (unperturbed motion) and the potential energy (perturbation function) for a pendulum of vertical oscillations of its fixed point of suspension

Multiplying (7.160) by 2 ẋ and integrating over t, one obtains

$$(\dot{x})^2 + \frac{2g}{\ell}\sin x + b = 0 \tag{7.161}$$

or

$$(\dot{x})^2 = -b - +\frac{2g}{\ell}\sin x. \tag{7.162}$$

The left-hand side of the last equation obtains non-negative values. Therefore there exist two qualitatively different types of solution depending on the value of parameter b. If b has negative values ($-b > 0$) and $|b|>2g/\ell$, then any value of x is possible ($|\sin x|\leqslant\ell$and thus . If $|b|<2g/\ell$, then only those values of x are possible at which the inequality $|\sin x|\leqslant|b|/(2g/\ell)<1$ holds.

Thus at $g/\ell = |b|/2$ we obtain bifurcation of the system by the parameter g/ℓ; at $g/\ell\leqslant|b|/2$ we obtain rotation of the system; and at $g/\ell>|b|/2$ we obtain oscillations of the system. It can be seen that the kinetic energy T of the system oscillations is periodically overpumping from the system into potential energy (perturbation energy V) and back, but at any time the total energy remains, $E = T + U = \text{const}$. The amplitude of swing of a system is the difference between the phase of unperturbed motion (rotation) and that of perturbation (the potential energy), which is responsible for the existence of the non-linear resonance (note that unperturbed motion is non-linear and non-isochronous, and the difference between the phase of rotation and perturbation is less than 2π.

There is no contradiction here or for estimates of stability of motion according to Lyapunov, since in Lyapunov's theory sufficiently small initial perturbations can be taken into consideration. Here we are studying the whole range of perturbation magnitudes. At the same time, it is obvious how Duboshin's criterion (7.154) works. If the stability of rotation of the system is studied at permanently acting perturbations whose role is played by the potential energy, then the value of perturbation is bound from above by a constant value until the bifurcation point $g/\ell = |b|/2$ and the rotation of he system occurs.

Now the general case of stability of motion can be easily considered, when $d \neq 0$ in Eqs. 7.155–7.158. The term (d cosωt)sinx = v in Eq. 7.158 can be considered as an external perturbation of the free (rotating) system and thus the system becomes unstable when it passes the bifurcation point (the parameter d has the same dimension as the term mg, i.e. it is proportional to energy). Therefore, the energy of the new perturbation should be greater than or equal to the energy of previous perturbations. The energy of the initial perturbations is equal to $mg\ell$ and that of the new perturbation to $mv^2/2$, where v is the velocity of movement of the fixed point of a pendulum equal to dω. The energy of the new perturbation is

$$\frac{mv^2}{2} = \frac{md^2\omega^2}{2}$$

The following inequalities should hold:

$$\frac{mv^2}{2} > mg\ell$$

$$\frac{d^2\omega^2}{2} > g\ell \quad \text{or} \quad d > \sqrt{\frac{2g\ell}{\omega}}$$

Thus, we have shown that by using Duboshin's criterion in studying the stability of motion of a pendulum with vertical oscillations of its fixed point of suspension in the integral characteristics, the result can be obtained directly. The analogous result (7.157), following from Mathieu's theory when studying the dynamics of the system in co-ordinates, requires more complicated solutions.

Lyapunov's theory provides the possibility of studying the stability of motion relative to different generalized co-ordinates and associated with their momenta. But the choice of suitable variables for the particular system cannot be made if the physics of the studied phenomenon is ignored.

Keplerian motion of a particle in an elliptical orbit can serve as an example of such a case, being unstable if determined relative to the radius-vector of a system. But this motion will be stable if it is defined relative to the variable $z = r - p/\ell + e\cos\psi$, where p and e are the eccentricity parameters of an ellipse, and r and ψ are its polar co-ordinates. The number of generalized co-ordinates should be equal to the number of degrees of freedom of a system (where co-ordinates are linearly independent). If the integral characteristics are used for the description of motion of a system, the number of characteristics must be complemented up to the number of degrees of freedom, at the expense of unimportant variables.

In the case of the two-body problem (taking the law of conservation of momentum into account) we should have the two canonical pairs. They are action-angle (usually mentioned) and Jacobi function-frequency (there is no example of application of the last pair in other works).

Duboshin's criterion allows the stability of a system to be studied when its potential energy is considered (partially or as a whole) as perturbation. Such an approach to the use of potential energy is employed in quantum mechanics, where all the potential energy is considered as perturbation (Landau and Lifshitz 1963). In this connection the virial oscillations of celestial bodies from the physical viewpoint should be considered as an effect of non-linear resonance between the kinetic and potential energies.

Comparing the approaches to estimations of the system's stability in a system of co-ordinates and in integral characteristics, let us point out the following. The method, which uses a full co-ordinate-momenta description of the system's dynamics, is too detailed, and classification of the studied system with respect to stability is therefore impossible. There are an infinite number of cases of systems based on it which can be chosen.

Using the integral approach, the following classification is possible. For example, in Keplerian motion, relative to the ratio of the integrals (energy and angular momentum), all possible types of orbit can be subdivided into four classes: two are of stable motion (elliptical and hyperbolic) and two are unstable (circular and parabolic).

A system's motion described by co-ordinates and momenta can be classified as unstable only in a formal way that can be accounted for by the choice of co-ordinate system. In fact, since we do not know the solution of the problem beforehand, the wrong co-ordinate system might be chosen.

In order to study the Lyapunov stability of a system, the Lyapunov function must be known. However, there is no standard rule prescribing how to find it; it is, rather, an art.

Considerable progress has been made in this field in the development of Chetaev's method. In this connection let us point out two facts:

(a) Chetaev's method for obtaining the Lyapunov function uses the combination of integrals of motion of the system. It is obvious that this way of choosing the Lyapunov function has considerable advantages.
(b) The integrals (and not their combination) are not positively defined functions of co-ordinates and momenta, which is why they are not used in such an analysis. They cannot be co used since arbitrary (and not their own) co-ordinate-momenta ate strange to these integrals. In practice, for many soluble problems, combination of a number of integrals of motion of a system corrects this deficiency. However, this can be done only for systems with small numbers of degrees of freedom which resemble a more general description of a system.

In describing a system in terms of integral characteristics, the characteristics themselves are immanent to their own integrals of a system and therefore they play the role of the Lyapunov function. In this case it is not even necessary to carry out formal mathematical analysis to study the stability of a system.

7.13.2 Stability of Virial Oscillations According to Lyapunov

Let us first estimate the stability of motion which is described by the classical virial theorem, i.e. when

$$\ddot{\Phi}_{\text{virial}} = 0. \tag{7.163}$$

The equation of the perturbed oscillations is in this case Jacobi's equation with the same value of the constant $\alpha^2\beta$:

$$\ddot{\Phi} = -A + \frac{B}{\sqrt{\Phi}}$$

The perturbation function in this case is

$$\eta = \Phi - \Phi_{virial}. \tag{7.164}$$

This function satisfies the equation

$$\eta = \ddot{\Phi} - \ddot{\Phi}_{virial} = -A + \frac{B}{\sqrt{\eta + \Phi_{virial}}} = x(\eta), \tag{7.165}$$

where $x(\eta)|_{r=0} = 0$

Here the Jacobi function plays the role of the co-ordinate, and, since $x(0) = 0$, we obtain the classical problem of stability of the trivial solution according to Lyapunov. Using the form of $x(\eta)$ and taking into account that, when the condition of the virial theorem holds at any moment in time, Φ_{virial} is a constant and A and B are such that $-A + (B/\Phi) = 0$. It follows that $x(\eta)$ is not a function of the defined sign. Therefore, the stability of the virial state does not follow automatically; it follows only on the condition that, in a virial state, the function Φ_{virial} assumes minimal values among all the possible configurations of the system. But the last condition is even more speculative then our hypothesis of constancy of the $\alpha^2\beta$ product.

Thus, the general, and the main, case remains the one when Φ does not remains constant. The case described by Φ_{virial} can be considered as a smooth state relative to the main state, i.e. it is related to the condition when averaging over the period of oscillations has been carried out.

Now let us consider the stability of virial oscillations of a conservative system according to Lyapunov described by

$$\ddot{\Phi} = -A + \frac{B}{\sqrt{\Phi}} \tag{7.166}$$

when the product of the form factors $\alpha^2\beta$ is not a constant, i.e. variation of the parameter B is allowed.

When there are no variations of B, Eq. 7.166 is written

$$\ddot{\Phi}_1 = -A + \frac{B_1}{\sqrt{\Phi_1}} \tag{7.167}$$

and when variations of the parameter B occur, the equation is written

$$\ddot{\Phi}_2 = -A + \frac{B_2}{\sqrt{\Phi_2}}, \tag{7.168}$$

where A is a constant in the absence of energy dissipation.

The perturbation function η is written

$$\eta = \Phi_2 - \Phi_1. \tag{7.169}$$

Then the perturbation equation is written

$$\ddot{\eta} = \frac{B_2}{\sqrt{\Phi_1 - \eta}} - \frac{B_1}{\sqrt{\Phi_1}} = x(\eta). \tag{7.170}$$

If η is a continuous function, we have

$$x(\eta)|_{r=0} = 0 \tag{7.171}$$

If η is a discontinuous function, then

$$x(\eta)|_{r=0} \neq 0 \tag{7.172}$$

If η is continuous function < i.e. variation of $\alpha^2\beta$ is continuous, we obtain the classical problem of Lyapunov stability of the trivial solution of the perturbed equation, and all the comments are ordinary, i.e. the motion is stable (the necessary and sufficient conditions hold).

At discontinuous variations of $\alpha^2\beta$ (isolated singular points), the stability problem can be studied using Duboshin's criterion. As was shown in Chap. 5, the equation for dissipative system can be reduced to a system of linear equations with constant coefficients. When the ratio of frequencies for each of these equations is a rational number, the solution can be given by a closed algebraic curve (i.e. a set of isolated points). If the ratio of frequencies is an irrational number, the curve corresponding to the solution fills the whole set of possible values of the Jacobi equation. Here the last possibility corresponds to continuous variation of the function η.

It is interesting to point out that the well-known three-body problem theorem by Poincaré studies this case. There, instead of the mass of one body, in the course of the proof of the theorem, the continuity is employed.

But as Wintner (1941) has pointed out, the result obtained by Poincaré corresponds to the case when the masses of the bodies are not fixed. This note made by Wintner is rather general and can be related to the product of the form factors $\alpha^2\beta$ which should not be considered as a continuous function but rather as a set of isolated points.

On the whole, study of the stability of virial oscillations gives us evidence that the proof of the constancy of the product $|U|\Phi$ for existing celestial bodies is a sufficient but not a necessary condition. The necessary condition for the proof of the above relationship follows from the direct derivation of the equation of virial oscillations of celestial bodies from Einstein's equation and also by showing that Schwartzschild's solution in the framework of the general relativity theory is the solution of Jacobi's equation when $\Phi = 0$.

7.13.3 Stability of Virial Oscillations of Celestial Bodies as Dissipative Systems

Let us now estimate the stability of oscillations of celectial bodies emitting electromagnetic energy. This is the case for systems existing in nature.

For a conservative system, Jacobi's equation is written

$$\ddot{\Phi}_c = 2E_c - U \tag{7.173}$$

where the subscript c indicates that the corresponding characteristics of the system relate to a conservative system, and the subscript d, below, denotes a dissipative system.

Let us study the stability using Eq. 7.173, assuming dissipation of energy, which can be written

$$\ddot{\Phi}_d = 2E_d - U$$

A variation of the solution of Eq. 7.173 is written $\eta = \Phi_d - \Phi_c$.

Then

$$\eta = 2(E_d - E_c) = 2E_\gamma = At,$$

where t is the time elapsed from the beginning of evolution.

According to Duboshin's criterion, the system (7.173) which emits energy will be stable at any finite time interval, ranging from $t = t_o$ to $t = \tau$, where τ is the bifurcation time of the system (see Chap. 8).

Chapter 8
Creation and Evolution of the Solar System Bodies

The study of creation and evolution of celestial bodies and their systems is one of the main subjects in astrophysics, cosmogony and cosmology. It is known from observation that the matter of the Solar System bodies is very identical with respect to its substantial and chemical content and from this point of view all the bodies are of common origin. But the attempts to find general mechanism of creation of the bodies encounter irresistible contradiction. The point is that the planets having only ~0.015% of the system's mass possess 98% of the orbital angular momentum. At the same time ~99.85% of the Sun's masses produce no sooner than 2% of the moment of momentum which, in both cases, is accepted to be a conservative parameter. Also, the specific (for unit of the mass) orbital angular momentum is increased with the distance from the Sun. These results follow from the problem solution based on the hydrostatic approach.

We study the problem in the framework of the Jacobi dynamics and with the assumption that all evolutionary processes are developed as a consequence of the energy loss through radiation. And also we consider virial oscillations of celestial bodies to be a mechanism of generation and release of the energy emitted.

Our study appeared to be successful, since, applying dynamic approach, we resolve the above discrepancy. It was found that the orbital planetary moment of momentum and the angular momentum of the axial rotating body are kinetic effects which originate from different components of the potential energy. It was shown in Sect. 7.2 of Chap. 7, that axial rotation is a dynamical effect which is provided by tangential component of the body's potential energy. Rotating energy comprises very small portion of the total potential energy. It equals to $\sim 3 \cdot 10-^{2}$ for the Earth and $\sim 10^{-4}$ for the Sun from their total energy . Rotating energy is the function of the radial density distribution law and appears as an effect of interaction of the uniform and non-uniform mass particles with respect to their density. For the uniform body its rotary energy is equal to zero. This is physics of the body's angular momentum.

Quite different origin has orbital moment of momentum. Interaction (collision and scattering) of the body's mass particles is accompanied by continuous redistribution of the mass density. According to the Roche's tidal dynamics (see Sect. 7.5 of Chap. 7), redistribution of the density leads to the shell separation. As it follows

V.I. Ferronsky et al., *Jacobi Dynamics*, Astrophysics and Space Science Library 369, DOI 10.1007/978-94-007-0498-5_8, © Springer Science+Business Media B.V. 2011

from Eq. 7.71 of Chap. 7, when the body's upper shell density reaches value more than 2/3 of the mean density of all others shells, then the upper shell becomes weightless. From physical point of view it means that the own force field of the upper shell reaches dynamical equilibrium with respect to the parent's outer force field. If the density of such a shell around the body has non-uniform distribution, then its tangential component of the potential energy converts it into secondary body with mean density according to Eq. 7.70. In the case of its uniform density, the shell is preserved as a parent's ring. In general case the shell is decomposed on parts of different masses. The orbital motion of a body now is resulted from the parental body by its outer force field induced by the normal component of the body's potential energy. In this case the orbital velocity of the newly created body should be equal to the first cosmic velocity of the parental body on its surface and the direction of motion is determined by the Lenz law. Just this energy (and corresponding velocity) is the function of the orbital moment of momentum of the secondary body.

So, it is clear from the above that the induced outer force field, which is formed by the normal component of the solar potential energy, operates the orbital motion of a secondary body. Doing so, the secondary body conserves the value of the parental energy at the time of its creation. To the contrary, the body's own shell rotation is provided by the tangential component of the potential energy. From here, the energy and the orbital moment of momentum are conservative parameters. The body's total energy is also conservative parameter, but the tangential component of energy and the corresponding angular momentum are not conservative values (see minus sign in Eq. 7.50 of Chap. 7).

8.1 The Third Kepler's Law as a Kinematics Basis for the Problem Solution of Creation of the Solar System Bodies

Using the above physics, we discovered a very interesting phenomenon, which opens the way for solving the problem. It appears that the mean orbital velocities and periods of revolution of all the planets and asteroids are equal to the first cosmic velocity and corresponding period of the contracting Protosun, having its radius equal to the semi-major axes of each planet's orbit. The same was happened with the planets' satellites. The subsequent expansion of the space has not broken the above regularity.

Tables 8.1 and 8.2 demonstrate the observable and calculated values of the orbital periods of revolution of the planets, asteroids (small planets) and satellites obtained by applying first cosmic velocities of the Protosun and the protoplanets which prove the above said.

Table 8.1 Observable orbital periods of revolution of the planets around the Sun and calculated periods of oscillation of its corresponding outer shell

Planets	Orbital radius, $R \times 10^{11}$ (m)	Observable period of revolution T (year)	Calculating period of oscillation T_1 (year)
Mercury	0.579	0.24	0.2408
Venus	1.082	0.62	0.6153
Earth	1.496	1.0	1.00
Mars	2.28	1.88	1.8823
Vesta	3.53	3.63	3.7594
Juno	3.997	4.37	4.3733
Ceres	4.13	4.6	4.598
Themis	4.68	5.539	5.5397
Jupiter	7.784	11.86	11.8781
Saturn	14,271	29.48	29.4802
Uranus	28.708	84.01	84.1951
Neptune	44.969	164.8	164.9185
Pluto	59.466	248.09	250.8882

Table 8.2 Observable orbital periods of revolution of the satellites around the planets and calculated periods of oscillation of their corresponding outer shells

Planets	Satellites	Orbital radius $R \times 10^6$, m	Observable period of revolution T, year	Calculated period of revolution T_1, year
Earth	Moon	384.4	27.32	27.4103
Mars	Phobos	9.4	0.319	0.3208
	Deimos	23.5	1.262	1.2604
Jupiter	V	181	0.498	0.4973
	Io	422	1.769	1.7706
	Europa	671	3.551	3.5508
	Ganymede	1,070	7.155	7.154
	Callisto	1,880	16.69	16.6709
	XIII	11,100	240.92	239.0960
	VII	11,750	259.14	259.5899
	XII	21,000	620.77	660.7744
	1X	23,700	758.90	745.1833
Satutn	Janus	151.5	0.7	0.6956
	Mimas	185.6	0.94	0.9431
	Enceladus	238.1	1.37	1.3704
	Tethys	294.7	1.89	1.8869
	Dione	377.4	2.74	2.7366
	Titan	1,212.9	15.95	15.7548
	Iapetus	3,560.8	79.33	79.2494
	Phoebe	12,944	548.2	549.2722
Uranus	Cordelia	49.751	0.3350	0.3348
	Cupid	74.8	0.618	0.6172
	Miranda	129.39	1.4135	1.4043
	Ariel	191.02	2.5204	2.5189
	Umbriel	266.3	4.1442	4.1463
	Titania	435.91	8.7058	8.6840
	Oberon	583.52	13.4632	13.4503

(continued)

Table 8.2 (continued)

Planets	Satellites	Orbital radius $R \times 10^6$, m	Observable period of revolution T, year	Calculated period of revolution T_1, year
Neptune	Triton	354.8	5.877	5.8523
	Nereid	5513.4	360.14	359.8227
Pluto	Charon	19.571	6.387	9.5065
	Nix	48.675	24.856	37.2873
	Hydra	64.780	38.206	54.2482

The first cosmic velocity v_1 of the Protosun's and protoplanetary bodies and the period of oscillation of the corresponding outer shell T_1 of the created bodies were calculated by the formulae from which, in fact, the third Kepler's law follows:

$$v_1 = \omega R = R\sqrt{\frac{Gm}{R^3}} = \sqrt{\frac{Gm}{R}}, \quad T1 = \frac{2\pi}{\omega} = \frac{2\pi R}{v_1}, \quad \frac{(2\pi)^2}{T_1^2} = \frac{Gm}{R^3},$$

where m is the body's mass; G is the gravity constant; R is the semi-major axis; $\omega = v_1/R$ is the frequency of virial oscillation of the outer shell, which appears to be equal to the angular velocity of the orbital motion. Note, the frequency of virial oscillation of the outer weighty shell does not equal to its angular velocity because the frequency is the parameter of the force field.

For example, when the Protosun's radius R extended up to the present day Earth's orbit ($m = 1.99 \cdot 10^{30}$ kg, $R = 1.496 \cdot 10^{11}$ m), then its first cosmic velocity was equal to

$$v_1 = \omega R = \sqrt{\frac{Gm_s}{R}} = \sqrt{\frac{6,67 \cdot 10^{11} \cdot 1,99 \cdot 10^{30}}{1,496 \cdot 10^{11}}} = 29786.786\,\mathrm{m/s}$$
$$= 29.786786\ \mathrm{km/s}.$$

This value corresponds to the observed mean orbital velocity of the Earth.

The period of oscillation of the interacted mass particles of the Protosun's outer shell ($R = 1.496 \cdot 10^{11}$ m, $v_1 = 29786.786$ m/s) was equal to

$$T_1 = \frac{2\pi R}{v_1} = \frac{6,28 \cdot 1.496 \cdot 10^{11}}{29786.786} = 3.1540428 \cdot 10^7 \mathrm{s} = 1\ \mathrm{year},$$

which is equal to the observed period of the planet's orbital revolution.

When the Protoearth's radius R extended up to the present day Moon's orbit ($m_e = 5.976 \cdot 10^{24}$ kg, $R = 3.844 \cdot 10^8$ m), then its first cosmic velocity was equal to

$$v_1 = \sqrt{\frac{Gm_e}{R}} = \sqrt{\frac{6,67 \cdot 10^{11} \cdot 5.976 \cdot 10^{24}}{3.844 \cdot 10^8}} = 1018.3018\ \mathrm{m} = 1.0183918\ \mathrm{km/s},$$

which is the present day Moon's mean orbital velocity.

The period of oscillation of the interacted mass particles of the Protoearth's outer shell (R $= 3.844\ 10^8$ m, $v_1 = 1018.3018$ m/s) was equal to

$$T_1 = \frac{2\pi R}{v_1} = \frac{2 \cdot 3.14 \cdot 3,844 \cdot 10^8}{1018.3018} = 23.706449 \cdot 10^5 \text{s} = 27.438019 \text{ days},$$

which corresponds to the present day Moon's period of orbital revolution.

The obtained results mean that all the planets and satellites were launched by first cosmic velocity of the self-gravitating Protosun and protoplanets after their outer shells acquired weightlessness. As it was said above, the process of evolutionary loss of energy by emission led to redistribution and differentiation of the body's mass density: increase it in the inner shells and decrease in the outer one by the light components dilution. In general, due to this process and contraction in the form of separation of the matter, the shell separation of a body with respect to density was developed up to the state of weightlessness and self-gravitation of the outer shell's matter and creation of the body.

The discovered regularity of the Solar system's planets and satellites creation seems to be valid for the process of separation of the Protosun itself and the other ptotostars from the protogalaxy Milky Way. If we accept for the Galaxy's known astrometric data (mass $m_g = 2.5 \cdot 10^{41}$ kg, distance of the Sun from the Galaxy center $R_s = 2.5 \cdot 10^{20}$ m), then it is not difficult to calculate that the first cosmic velocity of the proto-Galaxy, which size was limited by the Sun's semi-major orbital axes, is equal to 230 km/s, and the orbital period of revolution is $220 \cdot 10^6$ year. The values are close to those found by observation, namely: mean orbital velocity of the Sun is called as (230–250) km/s, and the orbital period of revolution $T_s = (220\text{–}250) \cdot 10^6$ year.

The observed today picture of the Milky Way, consisting of a bar-shaped core surrounded by a disc of gaseous matter and stars, which create two major and four or more smaller logarithmic spiral arms, prove the generally common mechanism of creation of the galactic system. This picture demonstrates that the huge of mass and size of the Protogalaxy rotating body was subjected during evolution by the polar and the equatorial oblateness. Due to redistribution of mass density and after reaching a state of their weightlessness the stars were separated from the outer shell in different surface regions. It appears that the main parts of the mass were separated in two regions (in the pericenter and the apocenter) and less in all others. But because the first cosmic velocity is the function of radius of the body, the regions of separation in the space formed the logarithmic spirals of the moving stars in accordance with the third Kepler's law. The same logarithmic spirals were formed also by the planets and the satellites in the Solar System.

The following initial values of density ρ_i and radius R_i of the Protosun and protoplanets can be obtained on the basis of their dynamic equilibrium state.

The Protosolar cloud has separated from the Protogalaxy body when its outer shell in the equatorial domain has reached the value of the first cosmic velocity. In fact, the gaseous cloud should represent chemically non-homogeneous rotating body. As it follows from Roche's dynamics (see Chap. 7, Eq. 7.71), the mean density of the gaseous Protogalaxy outer shell should be $\rho_s = 2/3\rho_g$. The condition $\rho_s = 2/3\rho_g$ is the starting point of separation and creation of the Protosun from the outer Protogalaxy shell. Accepting the above described mechanism of formation of the secondary body, we can find the mean density of the Protogalaxy at the Protosun separation as

$$\rho_g = \frac{m_g}{\frac{4}{3}\pi R^3} = \frac{2.5 \cdot 10^{41}}{\frac{4}{3} \cdot 3.14 \cdot \left(2.5 \cdot 10^{20}\right)^3} = 1.67 \cdot 10^{-21} \mathrm{kg/m^3} = 1.67 \cdot 10^{-24} \mathrm{g/cm^3}.$$

The mean density of the separated proto-Galaxy shell is

$$\rho_s = 2/3\rho_g = 2/3 \cdot 1.67 \cdot 10^{-24} = 1.11 \cdot 10^{-24} \mathrm{g/cm^3}.$$

In accordance with Eq. 6.70, the mean density and radius of the initially created Protosun body should be

$$\rho_s = 2\rho_g = 2 \cdot 1.67 \cdot 10^{-24} = 3.34 \cdot 10^{-24} \mathrm{g/cm^3};$$

$$R_s = \sqrt[3]{\frac{2 \cdot 10^{33}}{\frac{4}{3} 3{,}34 \cdot 10^{-24}}} = 7.5 \cdot 10^{18} \mathrm{cm} = 7.5 \cdot 10^{16} \mathrm{m}$$

The mean density and the radius of the initially created Protojupiter, Protoearth and Protomoon are as follows:

$$\text{the ProtoJupiter}: \quad \rho_j = 2 \cdot 10^{-9} \mathrm{g/cm^3}, \quad R_j = 6.2 \cdot 10^{13} \mathrm{cm} = 6.2 \cdot 10^{11} \mathrm{m};$$

$$\text{the ProtoEarth}: \quad \rho_e = 2.85 \cdot 10^{-7} \mathrm{g/cm^3}, \quad R_e = 1.9 \cdot 10^{11} \mathrm{cm} = 1.19 \cdot 10^{9} \mathrm{m};$$

$$\text{the Protomoon}: \quad \rho_m = 5 \cdot 10^{-4} \mathrm{g/cm^3}, \quad R_m = 1.1 \cdot 10^{9} \mathrm{cm} = 1.1 \cdot 10^{7} \mathrm{m};$$

Analogous unified process was repeated for all the planets and their satellites.

Creation of the other small bodies like comets, meteors and meteorites are also found their explanation within the considered mechanism and physics. In fact, the only condition for separation of outer body's shell is its weightlessness (its corresponding mean density relative to the body's mean density), but not a limit of some amount of mass. In this connection any volume and amount of mass could probability be separated at any time. For example, we found by calculation that the short-periodic Encke's Comet (1970 I, T = 3.302 year) has semi-major orbital axis $R \approx 1.5 \cdot 10^{11}$ m and has separated from the Proto-Sun after small planet Vesta and before the Mars. The short-periodic Halley's Comet (1910 II, T = 76.1 year) has semi-major orbital axis $R = 2.7 \cdot 10^{12}$ m and has separated from the Proto-Sun after Saturn and before Jupiter. The long-periodic Ikeya-Seki's Comet (1965 III, T = 874 year) has semi-major orbital axis $R = 1.35 \cdot 10^{14}$ m and has separated from the Proto-Sun before the Pluto. Like the asteroid belt between the Jupiter and Mars, the comet belts should definitely exist between the orbits of all the Jupiter group planets. As to the meteors and meteorites, they all should be separated from the planets by the same way. From point of view of dynamical equilibrium of their orbital motion the orbits of all the small bodies (comets, meteors and meteorites) should have large eccentricities and dip angles of inclination to the equator of their central bodies. This is because of probable oblateness of the proto-Sun body, where its polar regions should have higher values of the first cosmic velocity. Those small bodies and meteorites, which have not reached or have later on lost dynamical equilibrium, fell down on the planet's or satellite's surface.

As it was shown in the Table 8.1, the small planets of the asteroid belt separated from the Protosun by the same mechanism. From point of view of the orbital motion and first cosmic velocities there are no any features of their separation from a broken planet.

The above consideration takes off an old misunderstanding about the difference in the orbital planet's and the Sun's moment of momentum. The secondary body conserves the creation energy and orbital moment of momentum in accordance with the third Kepler's law. As to direction of body's axial rotation and orbital revolution, then these parameters enter by the inner and outer force field, like in electrodynamics, in accordance with the Lenz law. As to the specific (for unit of the mass) orbital moment momentum of the planets and satellites which increases along with distance from the central body, the explanation of this gives the increasing radius from the central body

The revealed physics and kinematics of creation and separation of the Solar System bodies prove the Huygens' law of motion on semi-cubic parabola of his watch pendulum, which synchronously follows the Earth motion. This curve was called evolute. And the curve perpendicular to the series of tangents to the evolute is called evolvent. The relationship between the evolute and evolvent represents the relationship between function and its derivative or between function and its integral. These relations exist not locally like in mathematical analysis but in integral form and geometrically visible. While plotting a series of the evolvents with fixed lengths of the pendulum a peculiarity of the same type is appeared in each point of the initial curve of evolvent. The peculiarity is the semi-cubic parabola of type $x^2 = y^3$ or $x = y^{3/2}$. This is just universal law of a body motion in the nature, which is the consequence of the simple fact. The degree 3/2 is the ration between the body's mass volume, which generates the energy, and the body's surface, which emit this energy. And also, in any task of motion we always have some initial conditions, which the moving object is inherited. In the case of the Huygens' oscillating pendulum the suspension filament starts unrolling in a fixed point. In the case of a celestial body, creation of satellite starts in a fixed point of its parental body where the initial conditions are transferred by the third Kepler's law.

This is because just in a fixed point of dynamical equilibrium between the generated volumetric energy (cubic degree of radius) and the irradiated from outer surface energy (square degree of radius) is broke.

First cosmic velocity was practically applied by man only in twentieth century. The nature seems to use it perpetually as the main instrument of the Universe evolution. Our Universe seems to be a pulsating system and its basic infinitesimall particle is ~10^{-36} g in weight (see calculation later on in Sect. 8.3) is responsible for the system's equilibrium. Because of the matter evolution and energy conservation law is the process, continuing infinitely long time. Some details related to the evolution of a gaseous sphere we discuss below.

8.2 Evolution by Radiation and Gravitational Contraction of a Gaseous Sphere

We consider here several problems in the gravitational evolution of a gaseous sphere based on the generalized virial theorem and the relationship between the potential energy and the moment of inertia of the sphere in the form

$$-U\sqrt{I} = \alpha^2 \frac{Gm^2}{R} \sqrt{m(\beta R)^2} = \alpha^2 \beta G m^{5/2}, \tag{8.1}$$

where U is the potential energy of the sphere; I is the polar moment of inertia; G is the gravitational constant; m is the body mass; R is the sphere radius; and α^2 and β^2 are dimensionless structural parameters depending on the radial mass density distribution of the spherical body.

From (8.1), and taking into account Eqs. 2.43, 2.45, 7.54 and 7.55, we have the following relationships between the structural form factors:

$$\alpha^2 = \frac{r_g^2}{R^2} \quad \text{and} \quad \beta^2 = \frac{r_m^2}{R^2}, \tag{8.2}$$

$$\alpha^2 \beta = a = \text{const}, \tag{8.3}$$

where $\alpha^2 = \alpha_0^2 + \alpha_t^2 + \alpha_\gamma^2$; $\beta^2 = \beta_0^2 - \beta_t^2$; $\alpha_0^2 = \beta_0^2 = 0.6$; $2\alpha_t^2 = \beta_t^2$; $\alpha_0^2\beta_0 = a_0 =$ const; $\alpha_t^2\beta_t = a_t =$ const; r_g and r_m are the reduced radius of gravity and radius of inertia; α_o, β_o, α_t, α_γ, β_t, are form factors of the normal, tangential and dissipative components of the energy for non-uniform mass density distribution of a system.

In Chaps. 6 and 7 we found that the constancy of the form factors product (8.3) is independent of the body mass, radius and radial mass density distribution for spherical and elliptic bodies. Eq. 8.3 is therefore a key expression in our further consideration.

8.2.1 Equilibrium Boundary Conditions for a Gravitating Gaseous Sphere

It is well known that polytropic models require the boundary mass density of a gravitating body to be rigorously equal to zero. Hence this condition gives us no opportunity to consider any physical processes during evolution.

If Eq. 8.1 for the spherical and elliptical gravitating system is valid, it allows us to consider convenient boundary conditions which can be used in the study of evolutionary problems.

In deriving the physical boundary conditions for a self-gravitating and rotating gaseous sphere, we consider its rotation as an effect of the tangential component of energy generated by the interacted non-uniform particles. As it was shown in Chap. 7, the ellipticity of the body is formed not as a result of its rotation but because of its self-gravitation. The key relationship (8.3) used here as the basis of our consideration prevents any possible errors. When we have to introduce the moment of inertia, the rotating sphere boundary at the equator will be defined by Kepler's law.

The fact that gaseous sphere boundary equilibrium conditions differ from those of the interior explains the difference between a free molecular boundary particle movement and an internal chaotic one. It is a consequence of the discrete matter structure dominant at the boundary (Jeans 1919).

Let us now consider the thermodynamic boundary conditions. Surely, we can define the boundary temperature only in the case of its real existence which, in turn, depends on the existence of the thermodynamic equilibrium between matter and radiation. Otherwise, it cannot be considered as black body radiation, and the Stefan-Boltzmann equation is inapplicable.

Thermodynamic equilibrium at the boundary can be reached only when the energy and momentum carried away by the radiation flow is greater than that carried away by the flow of particles from the sphere surface per unit time. Such a surface cannot increase further without disturbing the thermodynamic equilibrium.

We shall consider the evolutionary process of the gaseous sphere to be a successive series of hydrodynamic states in equilibrium. We shell also assume that the radiation energy loss causes the sphere to contract during the time periods between the equilibrium states.

Taking these ideas into account, we can express the hydrodynamic equilibrium at the boundary either by an expression representing particle flow 'locking' by the gravitational force, or, equivalently, by an equation showing the absence of particle dissipation from the boundary surface, which can be written in the form

$$\frac{\mathrm{Gm}\mu}{\mathrm{R}^2} = \frac{\mu\bar{\mathrm{v}}^2}{\mathrm{R}}, \tag{8.4}$$

where μ is the mass of the particle, and $\bar{v}$ is the velocity of the particle heat movement at the sphere boundary of the pole (more precisely it is the velocity of a particle running from the gravitational field).

For gravitational contraction between any two equilibrium states, Eq. 8.4 must be written as

$$\frac{Gm\mu}{R^2} > \frac{\mu\bar{v}^2}{R}. \tag{8.5}$$

Let us prove that the expression (8.4) for the gaseous spherical body boundary satisfies the virial relations.

First we consider one particle at the sphere boundary surface with mass μ and moving in the volumetric central field of the body with mass m and radius R. Then it is easy to see that

$$\left(\frac{\mu \ddot{R}^2}{2}\right) = \mu\left(\overline{\ddot{R}R} + \left(\dot{R}\right)^2\right). \tag{8.6}$$

The kinetic energy K_p of the particle is

$$\mu\left(\dot{R}\right)^2 = \frac{2\mu v^2}{2} = 2K_p. \tag{8.7}$$

From Newton's law we have

$$\ddot{R} = -\frac{Gm}{R^3}R. \tag{8.8}$$

The potential energy U_p of the particle in the gravitational field of the body is

$$\mu\overline{\ddot{R}R} = -\frac{Gm\mu}{R^3}\left(\overline{RR}\right) = -\frac{Gm\mu}{R} = U_p. \tag{8.9}$$

Therefore

$$\frac{d^2}{dt^2}\left(\frac{\mu R^2}{2}\right) = U_p + 2K_p. \tag{8.10}$$

Summing over all particles at the boundary layer and neglecting their interaction energy, we obtain

$$\frac{d^2}{dt^2}\left(\frac{m_s R^2}{2}\right) = U_s + 2K_s, \tag{8.11}$$

where m_s is the mass of the boundary spherical layer.

Or finally

$$\frac{3}{4}\ddot{I}_s = U_s + 2K_s,$$

$$\ddot{\Phi}_s = U_s + 2K_s \tag{8.12}$$

which represents Jacobi's virial equation for a spherical gaseous layer.

The exchange of particles between the gaseous body and its boundary layer takes place at the same radius R and lasts for a short time while the total mass of the layer remains constant. So Eq. 8.12 is rigorous.

The solution of Eq. 8.12 will be exactly the same as that obtained in Chap. 4 for a gravitating sphere, except that corresponding parameters of the sphere must be replaced by those of the boundary layer.

If one time-averages over time intervals which are longer than the period of boundary-layer oscillations, then the left-hand side of Eq. 8.12 tends to zero (i.e. the layer enters into the outer force field) and a quasi-equilibrium boundary state is obtained determined by the generalized classical virial relation between the potential and kinetic energies:

$$\dot{\Phi}_s = U_s + 2K_s. \tag{8.13}$$

Thus, we have proved that Eq. 8.4 written for the gaseous sphere boundary is a virial relation. We shell use this expression further in solving the problem of contraction velocity for gravitating gaseous sphere.

8.2.2 *Velocity of Gravitational Contraction of a Gaseous Sphere*

In considering the evolution of a gaseous sphere, one does not usually take into account its rotation because the total kinetic energy exceeds the rotational energy. Other authors who accepted the rotation of the gaseous sphere could not manage with the angular momentum accepted as conservative value during contraction (Zeldovich and Novikov 1967; Spitzer 1968; Alfvén and Arrhenius 1970).

It was shown in Chaps. 2 and 7 that the main part of kinetic energy of a celestial body is represented by oscillatory energy of the interacting elementary particles. The rotational part is much smaller of oscillatory energy and appears to be an indication of degree of the body matter non-homogeneity. Slow rotating bodies like the Sun, Mercury, Venus, and Moon have more homogeneous density distribution. Their part of rotational energy from the total kinetic one is ~$1/10^4$. For the other planets of the Solar System this figure is ~1/300. It follows from (7.50) of Chap. 7 that the value of oscillatory energy for a body as a whole is a conservative parameter. The value of rotary energy is a changeable parameter.

The solution of the virial equation obtained earlier enables us to propose the following mechanism for gravitational contraction of a gaseous sphere. During each period of the sphere's oscillation, a certain amount of energy is lost through radiation. Hence, the contraction amplitude is larger than the expansion amplitude. The difference between the two amplitudes is the value of the gaseous sphere contraction averaged over one period of oscillation. Taking into account the adiabatic invariant relation (Landau and Lifshitz 1973a), we shell consider the problem of the gravitational contraction of a gaseous sphere using the virial relations and the key relationships (8.1) and (8.3). Note that we consider here the process of evolution without loss of body equilibrium.

Since we consider the evolution process of a gaseous sphere as a successive moment from one equilibrium state to another, it is natural that the minimum time interval for averaging varying parameters should be a little larger than that required for establishing the hydrodynamic equilibrium. So it is not difficult to control the variations of parameters during evolution which are not in contradiction with the equilibrium. (Later, we shell consider these restrictions to be nonexistent).

It is convenient for our purpose to write the generalized virial theorem in the form

$$-\mathrm{U} = -2(\mathrm{E} - \mathrm{E}_\gamma) \equiv 2(\mathrm{E}_\gamma - \mathrm{E}), \tag{8.14}$$

where $\mathrm{E} = \mathrm{U} + \mathrm{K}$ is the total energy of the gaseous sphere which is a constant over time; E_γ is the electromagnetic energy radiated up to the considered moment of time; K is kinetic energy which includes the energy of rotation and oscillation of the interacted mass particles; E and U are negative parameters.

The time derivative of E_γ is the gaseous sphere luminosity L which is a function of the sphere radius R and the boundary surface temperature $\mathrm{T_o}$:

$$\frac{\mathrm{d}}{\mathrm{dt}}(\mathrm{E}_\gamma) = \mathrm{L} = 4\pi\sigma\mathrm{R}^2\mathrm{T}_0^4, \tag{8.15}$$

where σ is the Stefan-Boltzmann constant.

From Eq. 8.14 it follows that

$$\frac{\mathrm{d}}{\mathrm{dt}}(\mathrm{E}_\gamma) \equiv \frac{\mathrm{d}}{\mathrm{dt}}(\mathrm{E}_\gamma - \mathrm{E}) + \frac{1}{2}\frac{\mathrm{d}}{\mathrm{dt}}(-\mathrm{U}).$$

The potential energy is in turn a function of the radius R:

$$-\mathrm{U} = \alpha^2\frac{\mathrm{Gm}^2}{\mathrm{R}}.$$

The time derivative of $(-\mathrm{U})$ is

$$\frac{\mathrm{d}}{\mathrm{dt}}(-\mathrm{U}) = \mathrm{v_c}\frac{\mathrm{d}}{\mathrm{dR}}(-\mathrm{U}),$$

where $v_c = dR/dt$ is the gaseous sphere contraction velocity. To find this velocity we write

$$\frac{1}{2} v_c \frac{d}{dR}\left(\alpha^2 \frac{Gm^2}{R}\right) = \frac{dE_\gamma}{dt} = L$$

and finally, with the help of Eq. 8.3, we obtain

$$v_c = \frac{8\pi\sigma}{Gm^2} \frac{R^2 T_0^4}{(d/dR)(\alpha^2/R)}. \quad (8.16)$$

From Eq. 8.16 it is easy to see that v_c contains two unknown functions: $\alpha^2 = \alpha^2(R)$ and $T_o = T_o(R)$.

As was found in Chap. 2, the structural form factor α^2, as well as β^2, is the function of radial mass density distribution of the sphere. In Chap. 7 we considered this function presented by (8.2) and (8.3). It was found that the contraction velocity of the gaseous sphere depends on the mass density redistribution which determines kinetic energy of the body and its shells. So, the function $\beta = \beta(R)$ can be found from the condition of kinetic energy conservation of the body's upper shell after its separation.

It follows from (8.3) that during the gravitational contraction of the gaseous sphere its radius $R \to R_1$ and $\beta^2 \to 1$ (where R_1 is the orbital radius of separation). If $R \to R_1$ then velocity of rotation $v \to v_1$ (v_1 is the first cosmic velocity).

The kinetic energy of the body's upper shell before K_b and after K_a shell is written as

$$K_b = I\omega^2 = \beta^2 m\omega^2 R^2, \quad (8.17)$$

$$K_a = mv_1^2 = m_s\omega^2 R_1^2, \quad (8.18)$$

where I is the polar moment of inertia of the body; ω is the frequency of the radial oscillations; m and m_s are the body and its upper shell mass; $R - R_1$ is the thickness of the upper shell or the contraction value.

From Eqs. 8.17 and 8.18 we can write

$$\beta^2 = \frac{m_s\omega^2 R_1^2}{m\omega^2 R^2} = \kappa \frac{R_1^2}{R^2},$$

$$\beta = \sqrt{\kappa}\frac{R_1}{R},$$

$$\alpha^2 = \frac{a}{\beta} = \frac{a}{\sqrt{\kappa}} \frac{R}{R_1}, \quad (8.19)$$

where κ is the ratio of the Protosun's mass to the mass of a separated body.

Thus, we obtained an expression for α as a function of R, which is valid when the kinetic energy of the upper body's shell conserves in the orbital motion of the separated creature.

Let us now try to obtain the relationship between the gaseous sphere boundary temperature T_o and the radius R. We introduced the virial equilibrium boundary conditions by Eq. 8.4. This equilibrium was defined as particle flow 'locking' by the gravitational force, or, equivalently, by an equation showing the absence of particle dissipation from the boundary surface. Let us now rewrite it:

$$\frac{Gm\mu}{R^2} = \frac{\mu\bar{v}^2}{R}. \tag{8.20}$$

The heat velocity $\bar{v}^2$ depends on the boundary temperature T_o as

$$\mu\bar{v}^2 = 3kT_0, \tag{8.21}$$

where k is the Boltzmann constant.

Therefore we can rewrite the condition for particle flow 'locking' (8.20) with the help of Eq. 8.21 as

$$\frac{Gm\mu}{3k} = T_0R. \tag{8.22}$$

From the law of equal energy distribution over the degrees of freedom for the case of a gas particle mixture in equilibrium, it follows that

$$\mu_1\bar{v}_1^2 = \mu_2\bar{v}_2. \tag{8.23}$$

It is easy to see from (8.22) that the equilibrium radius of a gaseous sphere depends on the chemical composition of the gas. This conclusion follows from Eqs. 8.22 and Eq. 7.71 of Chap. 7, where the mechanical equilibrium condition of a body's upper shell is considered. Those results explains the effect of the particle flow 'locking' on the pole by the gravitational force which is based on the concept of mass and radiation equilibrium. Care must therefore be taken when the gas mixture is analyzed, i.e. if there are a small number of particles with light masses, the mixture will dissipate easily and the particles flow 'locking' will take place in the case of the heavier particles of the gas mixture.

When the quantities of the various mass particles are approximately equal, the particle flow 'locking' condition can be found only by a numerical solution. The gaseous sphere radius can be determined only after the equilibrium equation is solved, and to solve it we must consider all the given types and concentrations of particles in the flow. Formally, we can apply the effective particle mass μ which depends on a value averaged over all the particle masses. The problem can also be solved by numerical methods for a gas mixture consisting of many particles, and especially when the processes of ionization and recombination and chemical reactions occur.

Another interesting phenomenon, which we shell discuss, arises from the fact that electromagnetic forces are much stronger then gravitational forces. When some electrons escape the gravitating body it becomes positive by charges that create huge forces which tend to stop the process of electron dissipation. That is why it is necessary to use the proton mass μ_p when the gaseous cloud consists of neutral hydrogen partly ionized at the gaseous sphere boundary surface (the position of the boundary shell is specified by the radius R). The flow of electrons will be ‘locked’ by the extra forces appearing as a result of their primary dissipation. In addition, this uncompensated positive charge should have a drift at the boundary surface and small flow of cold plasma should be observed.

In the course of contraction of the gaseous sphere and the increase of its average temperature, the process of gas ionization should also increase. When the flow of electrons is large enough, and the limiting equilibrium between the gravitational forces and the charged protons is achieved, the protons should also start to run off the body’s gravitational field. In this case, the increasing electron flux has to be ‘locked’ by electrostatic forces. The boundary equilibrium change from the proton ‘locking’ to electron ‘locking’ should start at this moment.

Thus, we come to the conclusion that at least two phases of gaseous sphere evolutions should exist: that of the proton and that of the electron, with a transitional domain between them which can be calculated by numerical methods in each specific case.

Figure 8.1 illustrates all that we have said. The process of gravitational contraction of the gaseous sphere is represented by the curve ABCD. Within the AB range, the body equilibrium is kept by the gravitational field ‘locking’ of the proton flow (the proton phase). Within the same range of sphere contraction, the radius R decreases while the temperature T_o increases. Point B is the critical one; here the transformation of equilibrium boundary conditions from proton ‘locking’ to electron ‘locking’ begins. The process spreads up to point C. While the sphere radius decreases in the range BC, the boundary temperature remains constant.

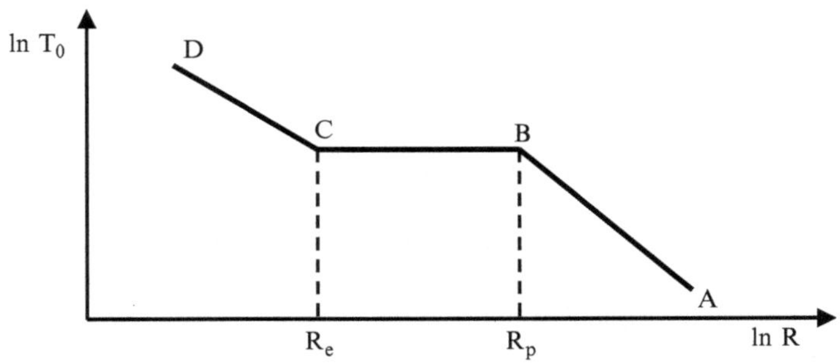

Fig. 8.1 Proton AB and electron CD equilibrium phases of the boundary shell of contracting gaseous sphere

In the electron equilibrium phase in the range CD, we can see that during the contraction process the boundary temperature increases again.

Let us check the derived expression (8.22) and the conclusion concerning the existence of two boundary equilibrium phases on the observed Sun data.

First, we calculate the numerical value of T_oR in the CGS system with the help of Eq. 8.22. Assuming numerical values for proton and electron masses, we obtain

$$T_pR_p = A_p = \frac{Gm\mu_p}{3k} = \frac{6.67 \cdot 10^{-8} \cdot 2 \cdot 10^{33} \cdot 1.67 \cdot 10^{-24}}{3 \cdot 1.38 \cdot 10^{-16}} = 5 \cdot 10^{17} \text{cm} \cdot \text{K},$$

$$T_eR_e = A_e = \frac{Gm\mu_e}{3k} = \frac{6.67 \cdot 10^{-8} \cdot 2 \cdot 10^{33} \cdot 9.1 \cdot 10^{-28}}{3 \cdot 1.38 \cdot 10^{-16}} = 2.73 \cdot 10^{14} \text{cm} \cdot \text{K}.$$

For the contemporary Sun we know that $R = 7 \cdot 10^{10}$ cm and $T_o = 5{,}000$ K so that $T_oR = 3.5 \cdot 10^{14}$ cm · K.

As at $T_o = 5{,}000$ K, where gas ionization must be fairly complete, we have a very good coincidence of the calculated and the observed data for the products T_oR and T_eR_e. For this temperature the proton radius of the Sun R_p is equal to 10^{14} T_oR cm, which corresponds to the orbit radius of Jupiter.

Thus, we have found α, β and T_o as functions of the radius R. We can now obtain the gaseous sphere contraction velocity. We rewrite Eq. 8.16:

$$v_c = \frac{8\pi\sigma}{Gm^2} \frac{(RT_0)^4}{R^2(d/dR)(\alpha^2 R)} \tag{8.24}$$

and, using (8.19), we can evaluate the denominator as

$$R^2 \left| \frac{d}{dR}\left(\frac{\alpha^2}{R}\right) \right| = R^2 \left| \frac{d}{dR}\left[\frac{a}{\sqrt{\kappa}} \frac{1}{R} \frac{R}{R_1}\right] \right| = \frac{a}{2\sqrt{\kappa}} \frac{R}{R_1}.$$

Finally, we write contraction velocity v_c as

$$v_c = \frac{16\pi\sigma}{Gm^2} \frac{A^4}{a} \sqrt{\kappa} \frac{R_1}{R}, \tag{8.25}$$

where $A = A_e = R_eT_e$ and $A = A_p = R_pT_p$ are for the electron and the proton phases of the gaseous sphere evolution, respectively.

Let us use Eq. 8.25 to obtain the contraction velocity and the time of contraction of the Protosun during the proton and the electron phase of the gaseous sphere evolution using its corresponding parameters.

If we take for the proton phase of the Protosun, after its separation from the Protoalaxy, $A_p = 5 \cdot 10^{17}$ cm · K, initial radius $R = 7.5 \cdot 10^{18}$ cm, final radius of the proton phase evolution (at the asteroid belt, after separation of the Protojupiter),

$R_1 = 4.2 \cdot 10^{13}$ cm, a = 0.46, and $\sigma = 5.76 \cdot 10^{-5}$ erg · cm^{-2} · s · (K)^4 as initial, we obtain

$$\overline{v}_{csp} = \frac{16 \cdot 3.14 \cdot 5.76 \cdot 10^{-5}\left(5 \cdot 10^{17}\right)^4}{6.67 \cdot 10^{-8}\left(2 \cdot 10^{33}\right)^2 \cdot 0.46} \cdot \sqrt{\frac{2 \cdot 10^{33}}{2.65 \cdot 10^{30}}} \cdot \frac{4.2 \cdot 10^{13}}{7.5 \cdot 10^{18}}$$

$$= 2.23 \cdot 10^5 \mathrm{cm} \cdot \mathrm{s}^{-1},$$

$$t_{sp} = \frac{7.5 \cdot 10^{18}}{2.23 \cdot 10^5} = 3.36 \cdot 10^{13}\mathrm{s} = 1.06 \cdot 10^6 \mathrm{years}.$$

We can find now the contraction velocity and the time of contraction of the Protosun during the electron phase of the gaseous sphere evolution. We take now for the electron phase $A_e = 2.73 \cdot 10^{14}$ cm·K, initial radius of the Protossun, after separation of the Protojupiter, $R = 4.2 \cdot 10^{13}$ m, final radius of the electron phase let be present day value $R_1 = 7 \cdot 10^{10}$ m, a = 0.46, and $\sigma = 5.76 \cdot 10^{-5}$ erg · cm^{-2} · s · (K)^4. Then we obtain

$$v_{cse} = \frac{16 \cdot 3.14 \cdot 5.76 \cdot 10^{-5}\left(2.73 \cdot 10^{14}\right)^4}{6.67 \cdot 10^{-8}\left(2 \cdot 10^{33}\right)^2 \cdot 0.46} \cdot \sqrt{\frac{2 \cdot 10^{33}}{1.18 \cdot 10^{28}}} \cdot \frac{7 \cdot 10^{10}}{4.2 \cdot 10^{13}}$$

$$= 8.96 \cdot 10^{-5} \mathrm{cm} \cdot \mathrm{s}^{-1},$$

$$t_{se} = \frac{4.2 \cdot 10^{13}}{8.96 \cdot 10^{-5}} = 4.7 \cdot 10^{17}\mathrm{s} = 14.9 \cdot 10^9 \mathrm{years}.$$

The found values show that the contemporary Solar System has formed during the proton phase (the Jupiter group of planets) within one million years and during the electron phase (the Earth group of planets) within next 15 billion years. Here we have not taken into account the effects of chemistry of the gaseous sphere on the equilibrium boundary conditions of the evolutionary process. But the obtained figures of evolution time show that our calculations give good approximation to the reality.

8.2.3 The Luminosity–Mass Relationship

To obtain the luminosity-mass we again consider the gaseous sphere evolution plot given in Fig. 8.1. It follows from (8.15) that in proton (AB) and electron (CD) evolutionary phases, the gaseous sphere luminosity is proportional to $1/R^2$. The boundary surface temperature T_o remains practically constant during the transition period (BC) when the equilibrium transformation from the proton to the electron phase takes place. But the gaseous sphere luminosity will decrease sharply. One can see that the luminosity decrease here is proportional to

$$L \propto \frac{\mu_p^2}{\mu_e^2}, \tag{8.26}$$

i.e. it is proportional to the ratio of the proton and electron mass squared as the gaseous sphere surface decreases proportionally to R^2. Thus, while going from point B to point C of the plot, the luminosity of the contracting body decreases by six orders of magnitude. We can suppose that the observed variations of variable star brightness are related to their virial energy pulsations, when stars are at the stage of evolution being considered.

As shown in the previous section, the most continuous period of proton or electron phase evolution is on the right end of the plot intercept (AB) and (CD). For these principal evolution time intervals we can write

$$L = 4\pi\sigma R^2 T_o^4 = \frac{(RT_o)^4}{R^2} \propto m^4. \tag{8.27}$$

This expression, derived from our theoretical considerations, is in good agreement with the well-known luminosity-mass relation which follows from observations. That is why Eq. 8.22 can be considered as an additional relation between the luminosity, the radius and the boundary surface temperature.

Let us take one more example. In Campbell's work (1962), 13 elliptical galaxies from the Virgo Cluster are considered and an analysis of the radius-mass relation for the observed data is given. To interpret these data, Jeans' relation (Jeans 1919) is used:

$$Gm\mu = \frac{3}{2}kT_oR \quad \text{or} \quad \frac{m}{R} = \frac{3}{2}\frac{3kT_o}{2G\mu}, \tag{8.28}$$

where μ is the proton mass.

On the plot presented in this work reflecting the mass-radius dependence, all the points are found to lie on a straight line with slope corresponding to $T_o \approx 1.5 \cdot 10^7$ K. Campbell concludes from this that the Jeans condition of self-gravitational instability is valid.

We note that Jeans' formula was derived on the assumption of low gas temperature and that all the kinetic energy of the gas is used for particle heat movement. The radiation energy was not taken into account.

Because of absence of direct temperature measurements, the theoretically found high temperature values at very steep line slopes need other explanations. We mast stress that in the observational data presented, the distance to the objects (in relative units) have been found with high degree of precision so that the experimentally derived constancy of the line slope should be trusted.

We interpret Campbell's data on the basis of our expression (8.22) where we consider the mass-radius relation to be dependent on electron temperature. That is why, contrary to Jeans, we write

$$\frac{m}{R_e} = \frac{3kT_e}{G\mu_e}.$$

Now the value of the boundary surface temperature of Campbell's galaxies is $T_o \approx 4{,}000$ K. This value corresponds to the usual boundary temperatures of celestial bodies whose evolution goes according to the electron phase of the equilibrium.

Hence, the experimental data presented by Campbell in his paper confirms once more the validity of Eq. 8.22 and the assumption of the existence of two evolutionary phases for celestial bodies.

In connection with the interpretation of Campbel's data, it is possible to use Eq. 8.22 to obtain the limiting temperature which should be reached by a gaseous sphere in its evolution. We write (8.22) as

$$\frac{Gm}{c^2}\frac{1}{R} = \frac{3kT_e}{\mu c^2} \quad \text{or} \quad \frac{R_g}{R} = \frac{3kT_o}{\mu c^2}. \tag{8.29}$$

Hence, during the evolution of a gaseous sphere through the electron phase of equilibrium, when $R \to R_g < T_o \to \mu_e c^2/3$ k or, equally,

$$3kT_0 \to \mu_e c^2 \approx 0.5 \text{ MeV},$$

$$T \approx 5 \cdot 10^9 \text{ K}.$$

This means that the temperature of the bodies approaches the electron temperature.

8.2.4 *Bifurcation of a Dissipative System*

In Chap. 5 we considered the dynamics of a dissipative system assuming that its evolution is a consequence of the loss of energy due to its radiation. Let us consider the problem in some details.

Jacobi's virial equation for a system was written as

$$\ddot{\Phi} = -A_0[1 + q(t)] + \frac{B}{\sqrt{\Phi}}, \tag{8.30}$$

where the function $A_o[1 + q(t)] = E - E_\gamma$ increases monotonically, reflecting the change of the total energy of a system as a function of time, and E_γ is the energy radiated up to time $t[E_\gamma > 0]$.

The solution of Eq. 8.30 was found to be

$$\begin{aligned} &-\arccos W + \arccos W_0 - \sqrt{1 - \frac{A_0[1+q(t)]C}{2B^2}} = \sqrt{1-W^2} \\ &+\sqrt{1-\frac{A_0 C}{2B^2}}\sqrt{1-W_0^2} = \pm\frac{[2A_9(1+q(t))]^{3/2}}{4B}(t-t_0). \end{aligned} \tag{8.31}$$

Equations of the discriminant curves which bound oscillations of the moment of inertia (Jacobi function) (see Fig. 5.1) are

$$\sqrt{I_{1,2}} = \frac{2B}{A_0[1+q(t)]}\left\{1 \pm \sqrt{1 - \frac{A_0[1+q(t)]C}{2B^2}}\right\}. \tag{8.32}$$

From analysis of the solution of Eq. 8.30 it follows that the dissipative system during its evolution must inevitable reach the state when its stability breaks; that moment (see Fig. 5.1) can be defined by the point O_b which is the physical bifurcation point. The position of the point can be defined by Eq. 8.32 as

$$\frac{2B^2}{A_0[1+q(t_b)]} = C, \tag{8.33}$$

Where $q(t_b)$ is parameter of the bifurcation point which can be found from condition (8.33)

$$q(t_b) = \frac{2B^2}{A_0 C} - 1. \tag{8.34}$$

The moment of inertia (Jacobi function) of the system corresponding to the bifurcation point, where the discriminant lines coincide, is

$$I_b = \frac{B}{A_0\left(1 + \frac{2B^2}{A_0 C} - 1\right)} = \frac{C^2}{4B^2}. \tag{8.35}$$

To find the moment of time of t_b where the system reaches its bifurcation point, one must know the law of energy radiation of the body q(t) or $E_\gamma(t)$, entering Eq. 8.30.

We give below our model solution for $E_\gamma(t)$.

The solution for the energy $E_\gamma(t)$ radiation up to t is based on the assumed existence of the proton and the electron phases of evolution for celestial bodies proposed in this chapter. On this basis, we have found a relationship between the body luminosity L and its radius R. During 'smooth' intervals of the body evolution, when $E_\gamma(t)$ is a continuous and monotonic function of time, the following relation holds:

$$\frac{Gm\mu_p}{3k} = RT_0, \tag{8.36}$$

where μ_p is the mass of the particle (proton or electron) which provides the boundary heat equilibrium of the body; k is the Boltzmann constant; and T_o is the gaseous sphere boundary temperature.

Let us write down the expression for the body luminosity L in relation to the time derivative of E_γ:

$$\frac{dE_\gamma}{dt} = L = 4\pi\sigma R^2 T_0^4, \tag{8.37}$$

where σ is the Stefan-Boltzmann constant.

Now we shell find an explicit expression for $E_\gamma(t)$ with the initial condition

$$E_\gamma(t_0)|_{t_0=0} = 0.$$

Equation (8.37) between the limits 0 and t can be integrated with the help of (8.36):

$$E_\gamma(t) = \int_0^t 4\pi\sigma R^2 T_0^4 dt = \int_0^t \frac{4\pi\sigma R^4 T_0^4}{R^2} dt = \int_0^t \frac{4\pi\sigma(Gm\mu_p)^4}{(3k)^4}\frac{1}{R^2}dt = \int_0^t \frac{K}{R^2}dt, \tag{8.38}$$

Where $K = 4\pi\sigma(Gm\mu_p)^4(3k)^4$.

Now let us make use of the expression (8.25) for the velocity of the gravitational contraction of the gaseous sphere v_c, which we had found earlier in this chapter:

$$v_c = \frac{dR}{dt} = \frac{32}{3}\frac{\pi\sigma}{Gm^2}\left(\frac{Gm\mu_p}{3k}\right)^4 \frac{\sqrt{\kappa}}{a}\sqrt[4]{\frac{R_1}{R}}, \tag{8.39}$$

Integrating this equation,

$$\int_0^R R^{1/4} dR = -\frac{32}{3}\frac{\pi\sigma}{Gm^2}\left(\frac{Gm\mu_p}{3k}\right)^4 \frac{1}{a}\sqrt[4]{\kappa^2 R_1}\int_0^t dt,$$

we obtain

$$\frac{4}{5}R^{5/4} - \frac{4}{5}R_0^{5/4} = -\frac{32}{3}\frac{\pi\sigma}{Gm^2}\left(\frac{Gm\mu_p}{3k}\right)^4 \frac{1}{a}\left(\sqrt[4]{\kappa^2 R_1}\right)t. \tag{8.40}$$

Then

$$R = \left(-Dt + R_0^{5/4}\right)^{4/5},$$

where

$$D = \frac{40\pi\sigma}{3Gm^2}\left(\frac{Gm\mu_p}{3k}\right)^4 \frac{1}{a}\sqrt[4]{\kappa^2 R_1}.$$

Finally, substituting the found expression for (8.40) into (8.38), we have

$$\begin{aligned} E_\gamma(t) &= \int_0^t \frac{Kdt}{\left(-Dt + R_0^{5/4}\right)^{8/5}} = \frac{5K}{3D}\left[\left(R_0^{5/4} - Dt\right)^{-3/5} - R_0^{3/4}\right] \\ &= \frac{5}{3}\frac{K}{D}\left[\frac{1}{\left(R_0^{5/4} - Dt\right)} - \frac{1}{R_0^{3/4}}\right] \end{aligned} \tag{8.41}$$

Thus we have obtained an expression in explicit form which we can use to calculate the energy loss by radiation during the time intervals of 'smooth' evolution of celestial bodies and hence find the parameters of the bifurcation point of a dissipative system.

8.3 Cosmo–Chemical Effects

From the analysis of the solution of Eq. 8.30 for a dissipative system, we found that, during the period of energy dissipation, the primary celestial body reaches a bifurcation point, characterized by separation of its outer shell which angular frequency coincides with frequency of virial oscillations. According to our theory of bifurcational creation of secondary bodies (in Alfvén's definition), some portion of the mass of the rotating primordial cloud reaches equilibrium relative to the tidal forces of the whole cloud at the bifurcation point, and moves further in a Kepler's orbit. As a result, during the subsequent dissipation of energy, the primary body continues its contraction by means of redistribution of the mass density without a separated secondary body. This secondary body conserves the corresponding angular moment $M_1 = mv_1R_1 = mv_1^2/\omega$ which in fact is the kinetic energy divided by frequency of the interacted mass particles. In accordance with (8.3), the value of this tangential component of the kinetic energy equal to doubled potential energy ($2\beta_t = \alpha_t$) at the moment of a secondary body separation.

It is commonly known that when both the gravitational and electromagnetic interactions are taken into account, the condition to attain an equilibrium state by some portion of the mass (secondary body) can be written in the form suggested by Chandrasekhar and Fermi (1953):

$$\int\limits_{(V)} \left[\rho \bar{v}^2 + 3p + \frac{H^2 + E^2}{8\pi} - \frac{(\nabla U)^2}{8\pi G} \right] dV = 0, \tag{8.42}$$

where ρ is the density of the substance of the secondary body; v the mean velocity; p the internal pressure; H and E are the components of the electromagnetic field; G the gravitational constant; V the volume of the system; and ∇U the gradient of the gravitational field.

Since the bifurcational point of a system is characterized by the zero amplitude of the virial oscillations, the kinetic terms in Eq. 8.42 are small compared to the mass terms. In this case, Eq. 8.42 can be rewritten as

$$\int\limits_{(V)} \left[3p - \frac{(\nabla U)^2}{8\pi G} \right] dV \approx 0,$$

or

$$\int\limits_{(V)} 3p dV \approx 0.1 \frac{Gm}{R}, \tag{8.43}$$

where the coefficient 0.1 represents the electromagnetic component in expansion of the potential energy (8.43) found by astronomical observation of the equilibrium nebulae (Ferronsky et al. 1996).

The left-hand side of (8.43) is proportional to the energy of the Coulomb interactions of the charged particles (electrons, protons, ionized atoms and molecules). The right-hand side of this expression is proportional to the energy of the gravitational interaction of the particles.

Thus, assuming the separated secondary body to have mass m, radius R and the average mass of its constituent particles to be μ, expression (8.43) can be rewritten in the form of an equality of the energies of the gravitational and Coulomb interactions

$$\frac{m}{\mu} \frac{e^2}{R\sqrt[3]{\mu/m}} = 0.1 \frac{Gm^2}{R}, \tag{8.44}$$

where $e = 4.8 \cdot 10^{-10}$ e.s.u. is the electron charge.

Expression (8.44) is the equivalent of

$$m\mu \propto \frac{e^3}{G^{3/2}}. \tag{8.45}$$

Table 8.3 Critical and averaged masses of the constituent particle for the Planets

Planets	m_c (g)	μ_a (g)	μ_a (aum)
Mercury	$0.33 \cdot 10^{27}$	$0.78 \cdot 10^{-21}$	469
Earth	$5.97 \cdot 10^{27}$	$0.18 \cdot 10^{-21}$	114
Jupiter	$2 \cdot 10^{30}$	$0.00 \cdot 10^{-23}$	6.02
Saturn	$0.57 \cdot 10^{30}$	$1.87 \cdot 10^{-23}$	11.3
Uranus	$0.087 \cdot 10^{30}$	$4.79 \cdot 10^{-23}$	28.8

The last expression relates the critical mass m_c of the separated secondary body to the averaged mass μ_a of its constituent particles (electron, proton, molecules), responsible for the hydrodynamic equilibrium of the body, as

$$m_c\mu_a^2 \propto \left(\frac{e^2}{G^{3/2}}\right) = \text{const} = 2 \cdot 10^{-16} g^3. \tag{8.46}$$

To illustrate this relationship, we determined the average values for the masses of the individual particles constituting the planets, stars and galaxes.

Planets: Table 8.3 shows critical masses of the constituent particles for the planets of the Solar System.

Thus, assuming that the bifurcation theory describes the formation of the Solar System correctly, the particles determining the hydrodynamic gas pressure in the case of the considered planet at the moment of their separation from the proto-solar cloud could have been composed of such elements as H, He, O, Si, Mn, Fe in atomic or molecular form. The average masses of the particles obtained can be used as a criterion in the development of cosmo-chemical models of planets with a complicated chemical composition at the moment of their separation from the proto-solar cloud and also for the construction of their chemical evolution models.

Stars: From (8.46) can be found the boundary values for all stellar critical masses, corresponding to the masses of the proton and the electron – particles which can be responsible for the hydrodynamic pressure inside the stellar cloud at the moment of separation at the bifurcation point of the proto-galactic cloud.

For the mass of the proton $\mu_p = 1.6 \cdot 10^{-24} g, \quad m_c = 10^{32} g.$
For the mass of the electron $\mu_e = 0.9 \cdot 10^{-27} g, \quad m_c = 2 \cdot 10^{38} g.$
In the case of $\mu_a = \sqrt{\mu_p \mu_a} = 0,4 \cdot 10^{-25} g, \quad m_c = 10^{35} g.$

Therefore, considering a typical stellar mass to be ~ 10^{33} g, we obtain in the framework of the bifurcation theory of formation of celestial bodies that the hydrodynamic equilibrium of the gas at the moment of separation of the proto-stellar cloud is supported both by electron and proton.

Galaxes: The presence of the factor $(e^2/G)^{3/2}$ in the right-hand side of (8.46) allows us to carry out the following transformations:

$$m_c\mu_a^2 = \left(\frac{e^2}{\hbar c}\right)^{3/2} \left(\frac{\hbar c}{G}\right) = \left(\frac{1}{137}\right)^{3/2} m_p^3, \tag{8.47}$$

where ħ is Planck's constant; c the velocity of light; and m_p the Planck's mass.

Thus, in the right-hand side of (8.47) there are two fundamental constants: the Planck mass m_p (2.2 · 10^{-5} g) and the fine-structure constant $\alpha = 1/137$. The presence of the constant α in the right-hand side of (8.47), being the universal constant of the weak and electromagnetic interactions, shows that this relation is applicable not only to electromagnetic but also to weak interactions. Then, putting the experimentally found values for the neutrino mass $\mu_\nu = 10^{-30}$ g (Shirkov 1980) into (8.44), we obtain

$$m_c = \frac{2 \cdot 10^{-16}}{\left(10^{-30}\right)^2} = 2 \cdot 10^{44} \mathrm{g}. \tag{8.48}$$

This mass, following from (8.48), is a typical mass of galaxies. Therefore, in the framework of the bifurcation theory of formation of celestial bodies, the hydrodynamic equilibrium (8.41) of the substances of galaxies at the moment of their formation can be provided by the pressure of neutrinos.

Universe: In the framework of the virial oscillation theory, the evolution of the Universe can be described by a pulsating model (for c = constant) of the system of material elementary particles. Such a system indefinitely long time exists. The mass of the particle responsible for hydrodynamic equilibrium of the Universe at the moment of its maximal compression (singularity stage) can be obtained from the same expression (8.46). Assuming $m_c \approx 10^{56}$ g we obtain

$$\mu_a \approx 10^{36} \mathrm{g}. \tag{8.49}$$

In the bifurcation theory the maximal average mass of particles in cosmic space can be determined from the condition $\mu_a = m_c$. Then,

$$\mu_{max} = 6 \cdot 10^{-6} \mathrm{g}.$$

This value is close to the Planck mass.

8.4 Direct Derivation of the Equation of Virial Oscillation from Einstein's Equation

Weinberg (1972) reduced Einstein's equation for homogeneous isotropic space, with the help of the Robertson-Walker metric, to the following scalar form:

$$3\ddot{R} = -4G(\rho + 3p)R, \tag{8.50}$$

$$\ddot{R}R + 2\left(\dot{R}\right)^2 + 2k = 4\pi G(\rho - p)R^2, \tag{8.51}$$

where R is the radius of the Universe; p radiation pressure (mass defect); and ρ the density of matter without mass defect.

Multiplying Eq. 8.50 by R/3 and summing it with (8.51), we obtain:

$$\left(\ddot{R}^2\right) + 2k = 8\pi G R^2\left(\frac{1}{3}\rho - p\right). \tag{8.52}$$

When $\rho < < p$ and $\rho R^3 = \text{const}$ (dust cloud), and taking into account that for curved space (Landau and Lifshitz 1973b)

$$\rho R^3 = \frac{m}{2\pi^2}, \tag{8.53}$$

where m is the total mass of the particles constituting the cloud, expression (8.53) is transformed into

$$\left(\ddot{R}^2\right) + 2k = \frac{8\pi}{3} G \frac{m}{2\pi^2} \frac{1}{R}. \tag{8.54}$$

Since from the Jacobi function we have $\Phi = mR^2/2$, Eq. 8.54 can be rewritten as

$$\ddot{\Phi} + km = \frac{2}{3\pi} Gm^2 \sqrt{\frac{m}{2}} \frac{1}{\sqrt{\Phi}} \tag{8.55}$$

or

$$\ddot{\Phi} + km = \frac{\sqrt{2}}{3\pi} \sqrt{G^2 m^5} \frac{1}{\sqrt{\Phi}}. \tag{8.56}$$

Finally, the equation of virial oscillations can be easily obtained in the known form

$$\ddot{\Phi} = -A + \frac{B}{\sqrt{\Phi}}, \tag{8.57}$$

where $A = km = E$ is the total energy, and B is a constant equal to $Gm^{5/2}$ multiplied by a factor which depends on the realization of the mass defect and on the period of the $\alpha^2\beta$ form factors (equal to $1/\sqrt{2}$).

When $p = \rho/3$, the equation of virial oscillations for radiation can be obtained from Eq. 8.52:

$$\ddot{\Phi} = -A.$$

Equations 8.50 and 8.51 are valid for all natural systems which exhibit a central symmetry of mass distribution. For celestial bodies Eq. 8.52 is written as

$$\rho R^3 = \frac{3m}{4\pi}.$$

Then, from (8.52) it follows that

$$(\ddot{R})^2 + 2k = \frac{8\pi}{3} G R^2 \left(1 - \frac{3p}{\rho}\right) = \frac{2Gm}{R}\left(1 - \frac{3p}{\rho}\right).$$

Now, Eq. 8.57 becomes

$$\ddot{\Phi} = -A + \frac{B}{\sqrt{\Phi}}\left(1 - \frac{3p}{\rho}\right). \tag{8.58}$$

As Weinberg (1972) pointed out, the inequality $0 < 3p \leq \rho$ holds for celestial bodies, and in the most general case we can write

$$3 = (\gamma - 1)(\rho - n\mu)$$

where n is the density of particles and μ is the mass of a particle.

Therefore, $(\rho - n\mu)$ is the mass defect and γ is the politropic index, which for stable system ranges from 0 to 5/3 for non-relativistic objects, and $\gamma \geq 4/3$ for ultra-relaticvistic objects. For $\gamma > 5/3$, the body expands indefinitely, and at $\gamma \leq 4/3$ collapse of the body occurs.

For actually existing celestial bodies, where the absence of heat equilibrium is taken into account (in the case of a discrete system), pressure is defined as (Weinberg 1972).

$$p = \frac{1}{3}[\rho + f(\rho, n)],$$

where $f(\rho, n) = T_\alpha^o$ is a function of the energy density ρ and the density n (number of particles per unit volume). This function is equal to zero in the ultra-relativistic limit and in the non-relativistic limit it is equal to $[-n\mu + (\rho - n\mu)] = -2n\mu + \rho$.

In both limiting cases, pressure p is

$$p = \frac{1}{3}\rho \quad \text{and} \quad p = \frac{2}{3}(\rho - n\mu).$$

Hence, in Eq. 8.58 the undetermined factor in B is equal to zero and $[(2\mu n/\rho) - 1)$ or $(1 - (2\Delta/\rho)]$, where $\Delta = \mu n - \rho$ is the mass defect.

Finally, taking into account the mass defect in Eq. 8.58 shows that the constant $B = B_o D$, where B_o is of Newtonean nature ($aGm^{5/2}$) and D, a relativistic correction, is smaller than 1.

Now let us estimate this correction D in the case of the white dwarf and the neutron star models according to Weinberg.

The equation determining the density of particles of particles when Fermi-Dirac statistics hold can be written as

$$n = \frac{k_F^3}{3\pi\hbar^2},$$

where n is the number of particles in the volume; k_F the radius of the Fermi sphere; and ħ is Planck's constant.

The density of matter of a star is written as

$$\rho = n\mu_p n_p,$$

where μ_p is the mass of a proton and n_p the average number of protons in a nuclei.

The critical density of matter in a star is

$$\rho_{cr} = \frac{\mu_p n_p {\mu_e}^3}{3\pi\hbar^3},$$

where μ_e is the electron mass.

Introducing the new variables $Z_1 = \rho/\rho_{cr}$ and $Z_2 = \rho/\rho_{cr}$, the equation of state for white dwarfs can be rewritten as follows:

$$Z_1 = \frac{3\mu_e}{\mu_p} F_1(Z_1),$$

$$Z_2 = \frac{3\mu_e}{\mu_p} F_2(Z_2),$$

where F_1 and F_2 are some transcendental functions.

For neutron stars the critical density is

$$\rho_{cr} = \frac{\mu_p^4}{3\pi\hbar^3}$$

and the equations of state are written

$$Z_1 = 3F_1(Z_1),$$

$$Z_2 = 3F_2(Z_2).$$

Solving the equations of state for the two limiting cases when $\rho < < \rho_{cr}$ (i.e. when the polytropic indexes are 5/3 and 4/3 respectively), we obtain for white dwarfs, respectively,

$$\rho_e = \frac{3}{2}p \quad \text{and} \quad \rho_e = 3p.$$

For neutron stars, in the limiting cases ($\rho << \rho_{cr}$) and ($\rho >> \rho_{cr}$), we have the same form of relations:

$$\rho = \frac{3}{2}p \quad \text{and} \quad \rho = 3p,$$

where ρ is the total density of matter.

In the ultra-relativistic limit, the relativistic correction will have very large values ($D = 0$), which means that the total collapse of the star (Oppenheimer-Volkoff limit) is leading to the formation of a black hole.

Note that besides the gravitational interaction, there are only two types of known interactions: the unified electroweak and the strong. Thus it follows that there cannot be any other types of collapse, since the collapse of white dwarfs corresponds to the first type of interaction and the formation of neutron stars to the second.

Thus, we have obtained the equation of virial oscillations (8.58) directly in the most general case and without having to assume the constancy of the form factor product $\alpha^2\beta$. Since the same equation follows from Jacobi's equation with the use of the hypothesis, we conclude that the relation $\alpha^2\beta = \text{const}$ was proven.

We should also note that modern astrophysical studies of the oscillation of celestial bodies in the non-relativistic approximation are based on the supposition that these movements have a homologous structure (Misner et al. 1973; Weinberg 1972; Frank-Kamenetsky 1959; Zeldovich and Novikov 1967). It can easily be verified that the supposition of homology is a sufficient condition to prove the constancy of the form factor product $\alpha^2\beta$ which is the main point in the derivation of the equation of virial oscillations from Jacobi's equation.

The mathematical formulation of the homologous motion of matter in the course of oscillation of a celestial body is written as follows:

$$r(t) = t(0) \cdot f(t),$$

where $r(t)$ is the radius of a given layer-shell of the body and $f(t)$ is an arbitrary function of time.

Let us introduce the Lagrange co-ordinates, where m is the mass inside the sphere of radius r, and dm is the mass of shell of radius r and thickness dr. According to the property of Lagrange co-ordinates, they are independent of time.

Then, the Jacobi function and the potential energy are written as:

$$\Phi = \frac{1}{2}\int_0^m r^2 dm,$$

$$|U| = G\int_0^m \frac{m dm}{r}.$$

Using the assumption that the motion is homologous, these expressions can be rewritten:

$$\Phi = \frac{1}{2} f^2(t) \int_0^m r^2(0) dm,$$

$$U = \frac{G}{f(t)} G \int_0^m \frac{mdm}{r(0)}.$$

Integrals on the right-hand side of these expressions do not depend on time and are therefore constants. Thus, the product $U^2\Phi$ does not depend on time and is also a constant.

Note that in the works of the authors mentioned above, the formula for the pulsation frequency of celestial bodies has been obtained assuming small amplitudes and the validity of the harmonic law of pulsations. Our approach allows the same frequency of pulsations to be obtained without the above restricting assumptions. Moreover, by comparing the two expressions which give equivalent results, it is possible to obtain the politropic index which enters into the astrophysical formula for the frequency of pulsations.

Chapter 9
The Nature of Electromagnetic Field of a Celestial Body and Mechanism of Its Energy Generation

The hydro-magnetic dynamo, action of which is provided by the planet's liquid metal core or the solar gas plasma, is the most popular idea for explanation of a body electromagnetic field generation. Its essence is in the motion of the conducting liquid core where self-excitation of the electric and magnetic poloidal (meridional) and toroidal (parallel) fields are happened. During rotation of the inner planet's shells with different angular velocities, in the case of asymmetric thermal convection of the shell mass, the intensity of the fields is increased. This condition, for example, for the Earth is achieved because the rotation and magnetic axes are not coincided and the thermal convection supposedly takes place. But physically justified theory of the observed planet's and solar phenomenon of electromagnetic field is absent. There is no explanation of mechanism of generation of the energy of this field except of general physical principle of the mass and charge interaction. Also the ideas or hypotheses about source of refilling of the planets energy which is spent for the gravitational and thermal irradiation are absent. The only source of the solar and star irradiated energy is accepted to be the interior nuclear fusion.

In order to find solution of the problem, in this Chapter we discuss a novel idea based on the innate capacity of body's energy for performing motion. As it was shown in Chaps. 2 and 3, the energy is the measure of the motion and interaction of particles of any kind of body's matter. The various forms of energy are interconvertible and its sum for a system remains constant. The above unique properties of the energy, with its oscillating mode of the motion in our dynamics, make it possible to consider the nature of the electromagnetic and gravitational effects of celestial bodies as interconnected events.

It was shown in Chap. 7 that the body's gravitational (potential) energy results in the body's matter volumetric pulsations, having oscillating regime, frequencies of which depend on the mass density. In our consideration the planets and stars are accepted as self-gravitating bodies. Their dynamics is based on the own internal force field and the potential and kinetic energies are controlled by the energy of oscillation of the polar moment of inertia, i.e. by interaction of the body's elementary particles.

V.I. Ferronsky et al., *Jacobi Dynamics*, Astrophysics and Space Science Library 369, DOI 10.1007/978-94-007-0498-5_9, © Springer Science+Business Media B.V. 2011

Applying the dynamical approach and the results obtained, we show below that the nature of creation of the electromagnetic field and mechanism of its energy generation appears to be the effect of the volumetric gravitational oscillation of the body's masses. This effect is also characteristic for any celestial body.

9.1 Electromagnetic Component of the Interacted Masses

It was shown in Sect. 7.2 that the electromagnetic energy is a component of the expanded analytical expression of the potential energy. The expansion was done by means of the auxiliary function of the density variation relative to its mean value. The expression of the body's potential gravitational energy in the expanded form (7.48) was found as

$$U = \alpha^2 \frac{GM^2}{R} = \left[\frac{3}{5} + 3\int_0^1 \psi x dx + \frac{9}{2}\int_0^1 \left(\frac{\psi}{x}\right)^2 dx\right] \frac{GM^2}{R}, \tag{9.1}$$

where U is the potential energy of the gravitational interaction; α^2 is the form-factor of the force function; G is gravity constant; M is the body mass; R is its radius; $\Psi(s)$ is the auxiliary function of radial density distribution relative to its mean value.

We have considered and applied the two first right-hand side terms of Eq. 9.1. The third term in dimensionless form represents an additive part of the potential energy of the interaction of the non-uniformities between themselves, which was written as

$$\frac{9}{2}\lambda = \frac{9}{2}\int_0^1 \left(\frac{\psi}{x}\right)^2 dx \equiv \frac{9}{2}\int_0^1 \left(\frac{\psi}{x^2}\right)^2 x^2 dx. \tag{9.2}$$

where $\lambda = \int_0^1 \left(\frac{\psi}{x}\right)^2 dx \geq 0$.

The non-uniformities are determined as the difference between the given density of a spherical layer and the mean density of the body within the radius of the considered layer. For interpretation of the third term we apply the analogy of electrodynamics (Ferronsky et al. 1996). Each particle there generates an external field, which determines its energy. The energies of some other interacted particles and their own charges are determined by this field. As far as the potential of the field is expressed by means of the Poisson equation through the density of charge in the same point, then the total energy can be presented in additive form through the application of the squared field potential. If the body mass is considered as a moving system, then the Maxwell radiation field applies.

In our solution the dimensionless third term of the field energy is written as

$$\frac{9}{2}\lambda = \frac{9}{2}\int_0^1 \left(\frac{\psi}{x}\right)^2 dx \equiv \frac{9}{2}\int_0^1 \left(\frac{\psi}{x^2}\right)^2 x^2 dx \equiv \frac{9}{2}\int_0^1 E^2 dV, \tag{9.3}$$

where $E = \Psi/x^2$ is dimensionless form of the electromagnetic field potential which is a part of the gravitational potential; Ψ plays role of the charge; $dV = x^2dx$ is the volume element in dimensionless form.

In order to determine the numerical value of λ, the calculations for a sphere with different laws of radial density distribution including the politropic model were done (Ferronsky et al. 1996). These models were used in our earlier numerical calculations of the form factors α^2 and β^2. The results, presented in Tables 9.1 and 9.2, show that for the density distributions which have physical meaning (Dirac's envelopes, Gaussian and exponential distributions) and also for the politrops with index 1.5, the parameter λ has the same constant value. We interpret this fact for a steady-state dynamical system as evidence of the existence of equilibrium radiation between a celestial body and the external flow. The numerical value of the parameter λ is equal to 0.022. There is also an observational confirmation of this conclusion. Spitzer (1968) demonstrates observational results of nebulae of different mass and size in Table 3.2 of his book, which we reproduce here in Table 9.3.

One can see that for masses of solar order and up to huge size the value of m/R^2 remains constant. This fact proves the statement of the physical meaning of the expression (9.2) which is the equilibrium radiation of a celestial body.

Table 9.1 Numerical values of the parameter λ for a sphere with different laws of density distribution

Law of radial density distribution	α^2	β	$Q = \alpha^2\beta$	9λ
$\rho = \rho_o$	0.6	0.77	0.46	0
$\rho = \rho_o\,(1- r/R)$	0.74	0.63	0.47	0.086
$\rho = \rho_o\,[1-(r/R)^2]$	0.71	0.65	0.47	0.060
$\rho = \rho_o \exp(-k\, r/R)$	0.16 k	3.45/k	0.53	0.19
$\rho = \rho_o \exp[-(k\, r/R)^2$	$\sqrt{k/2\,\pi}$	$1.8\sqrt{1/k}$	0.49	0.19
$\rho = \rho_o\, \delta(1- r/R)$	0.5	1.0	0.5	0.20

Table 9.2 Numerical values of the parameter λ for politropic models of a sphere

Index of politrope	α^2	β	$Q = \alpha^2\beta$	9λ
0	0.6	0.77	0.46	0
1	0.75	0.62	0.465	0.08
1.5	0.87	0.55	0.475	0.24
2	1.0	0.48	0.482	0.43
3	1.5	0.34	0.5	1.31
3.5	2.0	0.26	0.52	2.26

Table 9.3 Observational parameters of equilibrium nebulae

	Visible dark nebulae			
Parameters	Small globula	Large globula	Intermediate cloud	Large cloud
m/m_{Sun}	>0.1	3	$8 \cdot 10^2$	$1.8 \cdot 10^4$
R (pc)	0.03	0.25	100	20
п (п/cm^3)	$>4 \cdot 10^4$	$1.6 \cdot 10^3$	100	20
$m/\pi R^2$ (g/cm^2)	$>10^{-2}$	$3 \cdot 10^{-3}$	$3 \cdot 10^{-3}$	$3 \cdot 10^{-3}$

Thus, the virial approach to the problem solution of the Earth global dynamics gives a novel idea about the nature of the planet's electromagnetic field. The energy of this field appears to be the component of the potential energy of the interacted masses. The question arises about the mechanism of the body's energy generation, which provides radiation in wide range of the wave spectrum from radio through thermal and optical to x and γ rays.

9.2 Potential Energy of the Coulomb Interaction of Mass Particles

In Sect. 6.4.4 of Chap. 6 with the help of model solution, we showed that for the Coulomb interactions of the charged particles, constituting a celestial body, the relationship between the potential energy of a self-gravitating system and its Jacobi function holds.

Considering a one-component, ionized, quasi-neutral and gravitating gaseous cloud with a spherical symmetrical mass distribution, we found that the form-factor entering the expression for the potential energy of the Coulomb interaction acquires the same physical meaning what it has in the expression for the potential energy of the gravitational interaction of the masses. It represents the effective shell to which the charges in the sphere are reduced.

The considered task about the potential energy of the Coulomb interactions of the charged particles proves the legitimacy of solution of the virial equation of dynamical equilibrium for study electromagnetic effects of a celestial body.

9.3 Emission of Electromagnetic Energy by a Celestial Body as an Electric Dipole

In Chap. 5 we considered the solution of the virial equation of dynamical equilibrium for dissipative systems written in the form

$$\ddot{\Phi} = -A_0[1 + q(t)] + \frac{B}{\sqrt{\Phi}}. \tag{9.4}$$

Here the function of the energy emission [1−q(t)] was accepted on the basis of the Stefan-Boltzmann law without explanation of the nature of the radiation process. Now, after the analysis of the relationship between the potential energy and the polar moment of inertia, considered in the previous section, and taking into account the observed relationship by artificial satellites, we try to obtain the same relation for the celestial body as an oscillating electric dipole (Ferronsky et al. 1987).

Equation 9.4 for a celestial body as a dissipative system can be rewritten as

$$\ddot{\Phi} = -A_0 + \frac{B}{\sqrt{\Phi}} + X(t - t_0),$$

where $X(t - t_o)$ is the perturbation function sought, expressing the electromagnetic energy radiation of the body as

$$X(t - t_0) = E_\gamma(t - t_0).$$

The electromagnetic field formed by the body is described by Maxwell's equations, which can be derived from Einstein's equations written for the energy-momentum tensor of electromagnetic energy. In this case only the general property of the curvature tensor in the form of Bianchi's contracted identity is used. We recall briefly this derivation sketch (Misner et al. 1975).

Let us write Einstein's equation in geometric form:

$$G = 8\pi T, \tag{9.5}$$

where G is an Einstein tensor and T is an energy-momentum tensor.

In the absence of mass, the energy-momentum tensor of the electromagnetic field can be written in arbitrary co-ordinates in the

$$4\pi T^{\mu\nu} = F^{\mu\alpha}F^{\nu\beta}g_{\alpha\beta} - \frac{1}{4}g^{\mu\nu}F_{\sigma\tau}F^{\sigma\tau}, \tag{9.6}$$

where $g_{\alpha\beta}$ is the metric tensor in co-ordinates, and $F^{\mu\nu}$ the tensor of the electromagnetic field.

From Bianchi's identity

$$\nabla G \equiv 0, \tag{9.7}$$

where ∇ is a covariant 4–delta, follows the equation expressing the energy-momentum conservation law:

$$\nabla T \equiv 0. \tag{9.8}$$

In the component form, the equation is

$$F^{\mu\alpha}{}_{;\sigma}g_{\alpha\tau}F^{\sigma\tau} + F^{\mu\alpha}{}_{;\tau}g_{\alpha\sigma}F^{\tau\alpha} = g^{\mu\nu}(F_{\nu\tau;\sigma} + F_{\sigma\nu;\tau})F^{\sigma\tau}. \tag{9.9}$$

After a series of simple transformations, we finally have

$$F^{\beta v}{}_{;v} = 0. \tag{9.10}$$

here and above, the symbol ' ; ' defines covariant differentiation.

To obtain the total power of the electromagnetic energy emitted by the body, Maxwell's equations should be solved, taking into account the motion of the charges constituting the body. In the general case, the expressions for the scalar and vector potentials are

$$4\pi\phi = \int\limits_{(V)} \frac{[\rho]dV}{R}, \tag{9.11}$$

$$4\pi\overline{A} = \int\limits_{(V)} \frac{[j]dV}{R}, \tag{9.12}$$

where ρ and j are densities of charge and current; [j] denotes the retarding effect (i.e. the value of function j at the time moment t – R/c); R is the distance between the point of integration and that of observation, and c the velocity of light.

In this case, however, it seems more convenient to use the Hertz vector Z of the retarded dipole p(t – R/c) (Tamm 1976). The Hertz vector is defined as

$$4\pi Z = \frac{1}{R}\rho\left(t - \frac{R}{c}\right). \tag{9.13}$$

Electromagnetic field potentials of the Hertz dipole can be determined from the expressions

$$\phi = -\mathrm{div}Z, \tag{9.14}$$

$$\overline{A} = \frac{1}{c}\frac{dZ}{dt}. \tag{9.15}$$

Moreover, the Hertz vector satisfies the equation

$$\Box Z \equiv \left(\nabla^2 - \frac{1}{c^2}\frac{\partial^2}{\partial t^2}\right)Z = 0, \tag{9.16}$$

where □ is the d'Alembertian operator.

The intensities of the electric and magnetic fields E and H are expressed in terms $\overline{Z}$ by means of the equations

$$\overline{H} = \mathrm{rot}\dot{\overline{Z}}, \tag{9.17}$$

$$\overline{E} = \text{grad div}\overline{Z} - \frac{1}{c}\ddot{\overline{Z}}. \tag{9.18}$$

The radiation of the system can be described with the help of the Hertz vector of the dipole p = qr, where q is the charge and r the distance of the vector from the charge (+q) to (–q).

From the sense of the retardation of the dipole p(t – R/c) we can write the following relations:

$$\frac{d\bar{p}}{\partial R} = -\frac{1}{c}\dot{\bar{p}}; \quad \frac{d^2\bar{p}}{dR^2} = \frac{1}{c}\ddot{\bar{p}}.$$

Then the components of the fields E and H of the dipole are as follows:

$$H_\varphi = \frac{\sin\theta}{c^2 R}\ddot{p}\left(t - \frac{R}{c}\right), \tag{9.19}$$

$$E_\theta = \frac{\sin\theta}{c^2 R}\ddot{p}\left(t - \frac{R}{c}\right), \tag{9.20}$$

where θ is the angle between p and $\bar{R}$; $H_\varphi \perp E_\theta$ and $\perp R$; the other components of E and H in the wave zone are tending to zero quicker than 1/R in the limit $R \to \infty$.

The flax of energy (per unit area) is equal to

$$S = \frac{c}{4\pi}E_\theta H_\varphi = \frac{1}{4\pi c^2}\frac{\sin\theta}{R^2}\left(\ddot{\bar{p}}\right)^2. \tag{9.21}$$

The total energy radiated per unit time is given by

$$\oiint S d\sigma = \frac{2}{3c^3}\left(\ddot{\bar{p}}\right)^2 \tag{9.22}$$

Thus, transforming the dissipative system to an electric dipole by means of the Hertz vector, we have reduced the task of a celestial body model construction to the determination of the dipole charges +Q and –Q through the effective parameters of the body.

This problem can be solved by equating expression (9.22) for the total radiation of a celestial body as an oscillating electric dipole. In addition, the relation for the black body radiation expressing through effective parameters was presented below in Sect. 8.2.3.

The expression (9.22) for the total rate of the electromagnetic radiation *J* of the electric dipole can be written in the form (Landau and Lifshitz 1973b)

$$J = \frac{2}{3}\frac{Q^2}{c^3}\left(\ddot{\bar{r}}\right), \tag{9.23}$$

where Q is the absolute value of each of the dipole charges, and r is the vector distance between the polar charges of the dipole. Its length in our case is equal to the effective radius of the body.

In our elliptic motion model of the two equal masses the vector r satisfies the equation

$$\ddot{\bar{r}} = -Gm\frac{\bar{r}}{r^3}. \tag{9.24}$$

Thus, the total rate of the electromagnetic radiation of the dipole is

$$J = \frac{2}{3}\frac{Q^2}{c^3}\frac{(Gm)^2}{r^4}. \tag{9.25}$$

In order to obtain the average flux of electromagnetic energy radiation, the value of the factor $1/r^4$ should be calculated averaged during the time period of one oscillation. Using the angular momentum conservation law, we can replace the time-averaging by angular averaging, taking into consideration that

$$dt = \frac{mr^2}{2M}d\varphi, \tag{9.26}$$

where M is angular momentum, and φ is the polar angle.

The equation of the elliptical motion is

$$\frac{1}{r} = \frac{1}{a(1-\varepsilon^2)}(1 + \varepsilon\cos\varphi), \tag{9.27}$$

where *a* is the semi-major axis, and ε is the eccentricity of the elliptical orbit.

The value of $1/r^4$ can be found by integration. In our case of small eccentricities, we neglect the value of ε^2 and write

$$\overline{\left(\frac{1}{r^4}\right)} = \frac{1}{a^4}. \tag{9.28}$$

Finally we obtain

$$\bar{J} = \frac{2}{3}\frac{Q^2}{c^3}\frac{Gm^2}{a^4}. \tag{9.29}$$

Earlier it was shown (Ferronsky et al. 1987) that

$$\bar{J} = 4\pi\sigma\frac{1}{a^2}A_c^4, \tag{9.30}$$

where σ is the Stefan-Boltzmann constant; $A_e = Gm\mu_e/3k$ is the electron branch constant; μ_e is the electron mass; and k is the Boltzmann constant.

Equating relations (9.29) and (9.30), we find the expression for the effective charge Q as follows:

$$Q = \sqrt{6\pi\sigma}\frac{A_e^2}{cr_g}, \tag{9.31}$$

where $r_g = Gm/c^2$ is the gravitation radius of the body.

We have thus demonstrated that it is possible to construct a simple model of the radiation emitted by a celestial body, using only the effective radius and the charge of the body. Moreover, it was shown a practical method of determining the effective charge using the body temperature from observed data.

The logical question is raised what is mechanism of the energy generation of the bodies which they emitted in the wide range of oscillating frequencies spectrum. Let us consider this important question at least in first approximation.

9.4 Quantum Effects of Generated Electromagnetic Energy

The problem of the energy generation technology for human practical use has been solved far ago. In the beginning it was understood how to transfer the wind and fair energy into the energy of translational and rotary motion. Later on people have learned about production of the electric and atomic energy. Technology of the thermo-nuclear fusion energy generation is the next step. It is assumed that the Sun replenishes its emitted energy by the thermo-nuclear fusion of hydrogen, helium and carbon. The Earth thermal energy loss is considered to be filled up by convection of the masses and thermal conductivity. But the source of energy for convection of the masses is not known.

The obtained solution of the problem of volumetric pulsations for a self-gravitating body based on their dynamical equilibrium creates real physical basis to formulate and solve the problem. In fact, if a body performs gravitational pulsations (mechanical motions of masses) with strict parameters of contraction and expansion of any as much as desired small volume of the mass, then such a body, like a quantum generator, should generate electromagnetic energy by means of its transformation from mechanical form through the forced energy levels transitions and their inversion on both the atomic and nuclear levels. In short, the considered process represents transfer of mechanical energy of the mass pulsation to the energy of electromagnetic field (Fig. 9.1).

An interpretation of the process can be presented as follows. While pulsating and acting in regime of the quantum generator, the body should generate and emit coherent electromagnetic radiation. Its intensity and wave spectrum should depend on the body mass, its radial density distribution and chemical (atomic) content.

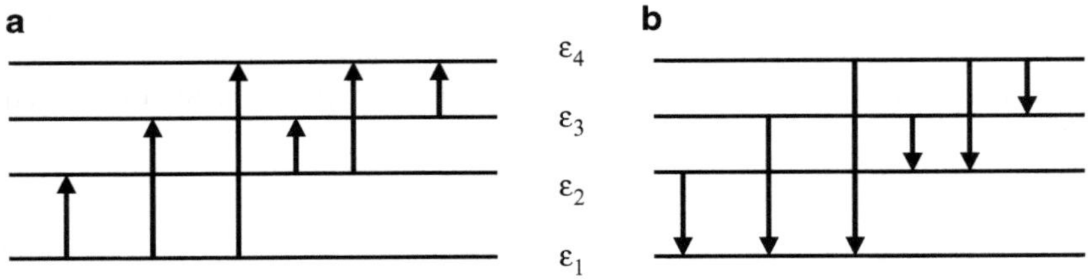

Fig. 9.1 Quantum transition of energy levels at contraction phase of the body mass (**a**) and inversion at the phase of its expansion (**b**); $\varepsilon_1, \varepsilon_2, \varepsilon_3, \varepsilon_4$ are levels of energy

As it was shown in Sect. 7.4, the body with uniform density and atomic content provides pulsations of uniform frequency within the entire volume. In this case, the energy generated during the contraction phase will be completely absorbed at the expansion phase. The radiation appeared at the body's boundary surface must be in equilibrium with the outer flux of radiation. The phenomenon like this seems to be characteristic for the equilibrated galaxy nebulae and for the Earth water vapor in anti-cyclonic atmosphere.

The pulsation frequencies of the shell-structured bodies are different but steady for each shell density. In the case of density increase to the body center, the radiation generating at the contraction phase will be partially absorbed by an overlying stratum at the expansion phase. The other part of radiation will be summed up and transferred to the body surface. That radiation forms an outer electromagnetic field and is equilibrated by interaction with the outer radiation flax. The rest of the non-equilibrated and more energetic in the spectrum of radiation moves to the space. The coherent radiation which reaches the boundary surface has a pertinent potential and wave spectrum depending on mass and atomic content of the interacted shells in accordance with Moseley law. The Earth emits infrared thermal radiation in an optical short wave range of spectrum. The Sun and other stars cover the spectrum of electromagnetic radiation from radio- through optical, *x* and gamma ray of wave ranges. The observed spectra of star radiation show that total mass of a body takes part in generation and formation of surface radiation. According to the accepted parabolic low of density distribution of the Earth it has maximum density value near the lower mantel boundary. The value of the outer core density has jump-like fall and the inner core density seems to be uniform up to the body center. The discussed mechanism of the energy generation is justified by the observed seismic data of density distribution. It is assumed, that the excess of generated electromagnetic energy from the outer core comes to the inner core and keeps there the pressure of dynamical equilibrium at the body pulsation during the entire time of the evolution. The parabolic distribution of density seems to be characteristic for most of the celestial bodies.

In connection with the discussed problem it is worth to consider the equilibrium conditions between radiation and matter on the body boundary surface.

9.5 The Nature of the Star Emitted Radiation Spectrum

We assume that the Novas and Supernovas after explosion and collapse pass into neutron stars, white dwarfs, quasars, black holes and other exotic creatures which emit electromagnetic radiation in different ranges of the wave spectrum. The discussed in the book effects based on dynamical equilibrium evolution of self-gravitating celestial bodies allow the exotic stars to be interpreted from a new position. We consider the observed explosions of stars as a natural logical step of evolution related to their mass differentiation with respect to the density. The process is completed by separation of the upper 'light' shell. At the same time the wave parameters of the generated energy of the star after shell separation are changed because of changes in density and atomic contents. As a result, the frequency intensity and spectrum of the coherent electromagnetic radiation on the boundary surface are changing. For example, instead of radiation in optical range the coherent emission in x or gamma ray range takes place. But the body's dynamical equilibrium should remain during all the time of evolution. The loss of the upper body shell leads to decrease of the angular velocity and increase of the oscillation frequency. The idea of the star gravitational collapse seems to be an effect of the hydrostatic equilibrium theory.

As to the high temperature on the body surface, the order of which from Rayleigh -Jeans' equation is 10^7 K and more, then in our interpretation as applying Eq. 8.33 for evolution of a star of solar mass at the electron phase (Fig. 8.1), the limiting temperature $T_0 \rightarrow \mu_e c^2/3k$ or (Ferronsky et al. 1996)

$$3kT_0 \rightarrow \mu_e c^2 \approx 0.5 \text{ MeV},$$

$$T \approx 5 \cdot 10^9 \text{ K}.$$

This means that on the body surface the gas approaches to the electron temperature because the velocity of its oscillating motion runs to c.

The energy is a quantitative measure of interaction and motion of all the forms of the matter. In accordance with the low of conservation the energy does not disappear and does not appear itself. It only passes from one form to another. For a self-gravitating body the energy of mechanical oscillations, induced by the gravitational interactions, passes to electromagnetic energy of the radiation emission and vise versa. The process results by the induced quantum transition of the energy levels and their inversion. Here transition of the gravitational energy into electromagnetic and vice versa results in the self-oscillating regime. In the outer space of the body's border the emitted radiation energy forms the equilibrium electromagnetic field. The non-equilibrium part of the energy in corresponding wave range of the spectrum is irradiated to the outer space. The irreversible loss of the emitted energy is compensated by means of the binding energy (mass defect) at the fission and fusion of molecules, atoms and nuclei. The body works in the regime of a quantum generator. Those are conclusions followed from the theory based on the body dynamical equilibrium.

9.6 Temperature of the Relict Radiation

The conclusion about the constancy of the equilibrium radiation parameter λ in (9.2) enables to obtain the numerical value of its temperature. For this we equate the expression of the black radiation energy and the volumetric density of that part of the gravitational energy which corresponds to the radiation energy in the Maxwell theory, that is

$$\sigma T^4 = \lambda \frac{Gm^2}{R} \cdot \frac{1}{4/3\pi R^3}, \tag{9.32}$$

where σ is the Stefan-Boltzmann's constant and T is absolute temperature.

From here

$$T = \sqrt[4]{\frac{3}{4\pi} \frac{\lambda G}{\sigma}} \cdot \sqrt{\frac{m}{R^2}}. \tag{9.33}$$

In accordance with Spitzer (1968), the value of $m/\pi R^2$ for globulae of solar to galactic mass is equal to $3 \cdot 10^{-3}$ g/cm^2 ($m/R^2 = 10^{-2}$) Our data shows that $\lambda = 0.022$. Then from Eq. 9.33 the value of $T \approx 4$ K. This is the observed relict radiation..

Let us now consider expression (9.32) from the point of view of the solar system. The expression can be written in the form

$$(RT)^4 = \frac{\lambda G m^2}{4/3\pi\sigma}.$$

From here with the solar mass

$$RT = \sqrt{m} \cdot \sqrt[4]{\frac{\lambda G}{4/3\pi\sigma}} = 5 \cdot 10^{17} \text{cm} \cdot {}^{\circ}\text{K}.$$

This numerical value of RT was obtained in our work (Ferronsky et al. 1979a; 1987) as the value of the proton branch of the Sun evolution. This theoretical solution has been proved by observational data of Spitzer. The electron branch of the evolution was also found there. Its RT value is 2,000 times less than that of the proton branch in accordance with the ratio of their masses. Hence, the radius of the Protosun cloud is $\sim 10^{18}$ cm for the proton branch of the evolution and $\sim 10^{13}$ cm for the electron branch. This is possible explanation for the division of the solar system planets in two groups.

The planetary masses are by 10^3–10^7 times less than the solar mass and their radii are 10^{-2}–10^{-4} solar radius. This estimation follows from the recalculation of the proton and electron branches of evolution. Therefore, the ratio m/R^2 for the planets is only 10–100 times more than that of a protostar obtained by Spitzer. The radiation of the planets and satellites should be in equilibrium with the other flux which is the internal solar flax. In this connection the equilibrium temperature of the planets and satellites should be 3–10 times more than the star's temperature of 10–100°K. The existing direct measurements give evidence of two peaks in the relict radiation spectrum (4 K and 20 K). It means that our physics and numerical estimates are reasonable.

In conclusion we wish to stress that the discovered by the artificial satellites relationship between the gravitational field (potential energy) and the polar moment of inertia of the Earth leads to understanding the nature an mechanism of the planet's energy generation as the force function of all the dynamical processes release in the form of oscillation and rotation of the matter. Through the energy nature we understand the unity of forms of the gravitational and electromagnetic interactions which, in fact, are the two sides of the same natural effect.

Chapter 10
Conclusions

It might appear surprising that the integral approach to description of dynamics of natural systems, which has a number of obvious advantages, has been developed far less well than the differential hydrostatic one. However, if we consider the development of the apparatus of mathematical physics from this viewpoint, the picture changes completely.

In fact, as soon as the concept of the field was formulated – even through at first this concept was a purely mathematical one (e.g. of the electrostatic and magnetic fields) – Gauss' theorem relating to the flux of a field vector through a closed surface was put forward. This integral characteristic of a field enclosed within a surface is an invariant of the field. In the case of electrostatics it is charge which give rise to the field.

The concept of vector flux through a closed surface has been generalized and developed. For example, such a generalization is Stokes' theorem relating to the circulation of a vector around a closed circuit, which can be used to identify vortex sources in vector fields. These theorems, which by their very nature are distinctly integral ones, have served as the basis for the whole mathematical theory of continuum mechanics, the electromagnetic theory of Maxwell and Poisson's theory of Newtonian gravitation.

Thus, the development of the mathematical apparatus of physics has taken the course of the integral approach to the description of natural phenomena. The concepts of divergence and the rotor introduced in this connection have served as instruments for finding the sources and sinks of a field and its vortices.

However, the idea of the continuity of a field, which gave rise to these concepts itself placed a limit on them, because the size of the region in which the charge was enclosed by a surface had to tend to zero. The Gaussian surface integral was thus replaced by divergence as a differential operation.

Circulation was similarly replaced by the rotor as a differential operation. It is these operations which are used in the Maxwellian field theory. This is because of the erroneous idea that the electric charges giving rise to the field are themselves continuous quantities distributed over the volume and also over the surface of dielectrics and conductors. The theorems of Gauss and Stokes are therefore limited to volumes shrinking to nil, and the theory became a purely differential one. This situation was later improved by Lorentz, who introduced into the field discrete

V.I. Ferronsky et al., *Jacobi Dynamics*, Astrophysics and Space Science Library 369, DOI 10.1007/978-94-007-0498-5_10, © Springer Science+Business Media B.V. 2011

charge points of finite magnitude scattered in empty space. According to his theory, Maxwell's equations remain applicable in the empty space between the small regions enclosing point singularities. On the closed surfaces surrounding these regions containing field singularities, the solutions to the field equations satisfy integral conditions. The flax of the field vector through these surfaces is equal to the sum of discrete charges enclosed by the total surface,

With the solution averaged over space, Lorentz' theory led to Maxwell's theory, which was in fact his objective. This is how the integral approach to the description of natural phenomena came into being.

The same approach was used by Einstein in the interpretation of his general theory of relativity and for deriving the equations of motion of matter in accordance with Newton's theory from his own equations.

It is, of course, well known that Einstein constructed his general theory of relativity as a relativistic theory of gravitation. For this, he first wrote Newton's equations in the form of field equations using Poisson's equation, and then gave the latter a relativistic, generalized character.

Einstein went further and abandoned inertial counting system, which had been accorded a position of privilege. Thus, the invariance was no longer assumed to be Lorenzian but universal in relation to any improper continuous transformation. Here, use was also made of Lorentz' idea, which we have mentioned earlier, of the discrete nature of the distribution of matter. Matter is concentrated in point singularities of a field, and between them there is empty space, for which Einstein's field equations hold true. The equations are not satisfied at singular points, which must be surrounded by closed surfaces. For the latter, the integral relations of Gauss in tern hold true, i.e. the flux of the field through these surfaces is equal to the charges found inside them. It should be emphasized once again that the actual fields inside these regions need not satisfy the conditions of the Einstein's equations.

Einstein's theory is, therefore, by its very nature and because of the basis on which it is constructed, an integral one. This fact is not usually realized, which is why we draw attention to it. It is by this condition, which in mathematical terms amounts to the requirement that the divergence of the original tensor should become exactly nil, that the nature of Einstein's tensor is uniquely defined. Such a tensor is one, the divergence of which is twice the contracted Bianchi identity for the Riemann curvature tensor.

If all the singularities of a field are surrounded by small spheres, in the space between them the field will everywhere be regular and its equations can be expanded in descending series in terms of the reciprocals of the velocities of light. Equating the coefficients in terms of the same powers, we obtain a series of equations. Every such system contains new quantities not found in the previous systems and is easily solved.

The motion of singularities (i.e. of particles) is determined by virtue of the fact that the left-hand sides of the systems of equations being solved satisfy four identities. The right-hand side of these equations must therefore also satisfy these identities or, with the singularities taken into account, the integral conditions. In the absence of singularities these conditions are automatically satisfied and provide nothing new. But if there are present, they determine the equations of motion. Einstein followed all calculations through and obtained Newton's equations. This method can also be used when gravitational and magnetic fields exist simultaneously, and the result of the calculation is positive. In this way, Einstein showed that even the classical interaction of mass points is caused by the non-linearity of the field equations. This fact is usually emphasized, but the role of integral conditions tends not to be mentioned.

Einstein's equations therefore contain Newton's equations and thus also their solutions and combinations.

Jacobi's virial equation is derived from Newton's equations and consequently, must itself be contained in Einstein's equations. However, it is not immediately apparent whether Newton's or Jacobi's equation is the more fundamental. Newton's equations were obtained by Einstein from his second-order equations by approximation. Jacobi's equation was obtained from Einstein's by the method of oscillation moments, also in second-order but by an exact method. This makes Jacobi's equation the more fundamental one; moreover, unlike Newton's equation, it remains integral and dynamical in nature.

As we have mentioned, the way in which the whole problem is formulated gives Jacobi's moment equation an exact, closed from which in fact solves the problem itself. In the case of the universe the problem is also one of its non-steady-state nature. A clever solution to this problem was found earlier by Friedmann. His solution is a solution to Jacobi's equation or to the smoothed Einstein equation. This is an analogue of Maxwell's equation in the form of a smoothed Lorentzian equation for charge points.

For the empty space between point singularities an anisotropic solution to Einstein's equation has been found (also by indirect means). This solution is Kasner's metric. Analysis of this metric shows that the empty space being considered pulsates. It is compressed on two axes, expands on one, and vice versa. Since this solution has been obtained for the case of space without matter, i.e. without its interaction, so that the law of interaction is without significance, the oscillatory nature of processes in nature is universal. The solution, however, is a formal one and its physical significance needs to be elucidated.

In fact, in Newton's well known law of gravitation for two masses it is assumed that these are mass points. Otherwise, the inverse-square law ceases to apply to their interaction. This in turn contravenes the law of remote screening mentioned in Chap. 1, which makes it impossible for approximately isolated (conservative) systems to exist.

The law of gravitation thus permits the existence of infinitely small radii of curvature and thereby of an infinitely large curvature of space time, i.e. of singularities. There are other examples of motion towards or away from a singularity, such as the formation of stars and planets, the expansion of the universe etc. Newton's law of gravitation therefore non-explicitly reflects the conditions for the existence of singularities, and the generalization of his theory by Einstein retains and, on the basis of the principle of equivalence, clearly demonstrates these singularities.

Singularities are therefore an empirical fact. So what are they?

In accordance with Einstein's theory, curvature is produced by mass. Consequently, empty-space time is not abstract emptiness but a physical vacuum with its own structure and also an analogue of mass, which in fact reflects Kasner's solution to Einstein's equation. This view is now widely held. In most models a vacuum is considered to be a quantum-mechanical system of virtual particles and to behave in a way similar to an elastic medium. Belinsky et al. (1970) studied the behavior of Einstein's equation for non-empty space-time but near a singularity. They showed that with increasing proximity to (distance from) the singularity a moment is approached at which the vacuum curvature exceeds the curvature from matter and the solution to Einstein's equation again becomes Kasner's solution.

Its solution, however, is a case of uniform – although anisotropic – space-time. Belinsky, Lifshitz and Khalatnikov also examined the case of inhomogeneous space-time and came to conclusion that the nature of the solution was the same but that the Kasner parameters were dependent on the co-ordinates and time.

In the case of further evolution of the Kasner solution with expansion of space away from the singularity, the original anisotropic space in gradually converted into isotropic space, i.e. into the Friedmann model, which is a solution to the second-order virial equation.

The oscillatory law of the dynamics of natural processes is thus a universal law of nature. It should, however, be noted that into all the approaches mentioned above the concept of finite time and of a beginning of time counting has been introduced. In some models there is also the concept of the end of the world. Only in one of them (in which the average density of matter for the space being considered is strictly determined) do the periodically alternating processes of expansion and contraction infinitely. It is this mode which is determined by the solution to Jacobi's virial equation.

A special feature of the Kasner solution for the general anisotropic case of space-time is the appearance in it of dependence of metric coefficients of time in accordance with the $|t|^{2/3}$ law, where t is a time interval. This law was found for the most general case in which there is no external symmetry, i.e. no symmetry which is not associated only with the internal arrangement of singularities.

The sources of the important relation $|t|^{2/3}$ go back to Kepler, who found experimentally the law in accordance with which the squares of the periods of rotation of bodies of the solar system are the cubes of the semi-axes of the ellipses in which they undergo motion.

It was pointed out in Chap. 6 that in Newton's theory about the attraction of mass points such a law is also found to be asymptotic for the case in which *n* bodies collide simultaneously. It was also shown there that within this asymptotic limit the simultaneous collision of *n* bodies leads to a homologous configuration. And for it in turn the condition of the applicability of Jacobi's general virial equation with two functions holds true. Thus, using a solution of the Kasner type, the applicability of Jacobi's virial equation within the asymptotic limit of simultaneous collision between *n* bodies which was found earlier for Newton's theory is extended to the case of the solution of Einstein's general equation. This indicates the universal nature of Jacobi's virial equation in dynamics.

Let us note further important aspect of the solutions under consideration, which relates to the change of Kasner epochs. Their number is infinitely independent of whether the world has a beginning and end. This occurs as a result of a decrease in the duration of an individual epoch as a singularity is approached.

Let us now consider yet another aspect of the fundamental nature of Jacobi's virial equation. As we have already pointed out, Newton's law of gravitation permits the existence of a curvature in space-time, which is derived from Einstein's theory. However, there is one fundamental difference between the two theories. According to Newton, the gravitational interaction is a long-range one, corresponding to an infinite velocity of propagation of the interaction. Einstein assumes a short-range interaction. It is propagated at finite velocity (at the velocity of light). Consequently, Newton's theory is formulated in terms of Euclidian geometry. Nevertheless, with both theories space-time is distorted.

Newton's theory is constructed on the basis of a simple empirical law of Kepler's and does not make use of another empirical law, namely the principle of equivalence derived from the experiments of Eötvös.

So what common ground is there between the theories?

The fact is that Newton's theory is constructed as Newtonean mechanics plus his own law of gravitation. In Newtonean mechanics there are three axioms, but the type of interaction is not determined; this is done experimentally. In generalized Newton's theory, it is the mechanics that should have been generalized and not the type of interaction.

With Einstein the type of interaction is replaced by the principle of equivalence. The mechanics, on the other hand, is generalized in accordance with the principle of the invariance of equations. Long-range interaction is thus not involved here, and the type of interaction makes no difference.

Jacobi's virial equation, which was obtained from Newton's equations, also does not so much generalize the type of interaction law, in the way that this was done in his (Jacobi's) conclusions, as take into account the mass defect (potential energy). It is therefore linked with the principle of equivalence. The mass defect, in turn, is determined by a system that has already been formed and, consequently, does not depend on the type of interaction during the process of formation (long-range or short-range).

As was thought by Wintner, Jacobi's virial equation therefore reflects the type of interaction law only integrally over the whole period of time in which the mass defect is formed. Also, if there is no delay, as in the case of Newton's long-range law of gravitation, it will be simultaneously a specific and instantaneous type of interaction, as pointed out by Wintner.

If a delay does take place, for example in accordance with Einstein's short-range interaction law, instead of a specific, instantaneous type of interaction, the equation will include an expression which has been strongly averaged over time, and the dependence on the type of interaction will cease to be of significance. It will be replaced by an assertion about the dependence on instantaneous mass or on the mass defect which has built up over a long time.

This is the answer to the question posed. At the same time, the strength of the Jacobi equation is evident. Since in the general theory of relativity the usual problems in the framework of a short time interval – and even the classical two-body problem – are not solved, the enormous practical significance of solving Jacobi's virial equation becomes obvious. The fact that there are oscillations even in empty space-time indicates the exclusively fundamental nature of this equation.

Moreover, it has now become obvious that Jacobi's virial equation, which was obtained from Newton's equations, is a particular case of more general virial equation derived from Einstein's equations. This equation will thus be studied from the most general global points of view, namely that of empty space-time, which will not be called a vacuum, and that of models of an evolving universe. It should be noted here that the models that have so far been developed from Jacobi's equation of an open, a closed and a pulsating universe have been obtained automatically as its natural solutions as a function of the source data – the quantities of total moment and mass defect. In this case, all possible types of solution are encompassed, and the question of the completeness of the set of possible models of the universe is thereby solved.

Let us now consider an example which demonstrates the use of the integral approach for constructing a complete closed theory based on Hooke's law. The theory concerned is the theory of elasticity.

In this theory, for any volume of a continuum, only quantities and parameters which ate integral from the point of view of an external observer are considered, namely deformation, stress and modulus of elasticity. The elements of the volume interact through their surface. A quantitative measure of their interaction is provided by strains, and a quantitative measure of the results of interaction by relative changes in the external dimensions of elements, in other words, their deformation. The internal structure of the material is demonstrated quantitatively by means of integral parameters, namely the mass density, the modulus of elasticity and the Poisson coefficient.

The interaction between the element of interest of a body and the external world takes place through external surface and volume forces. The external surface forces act only on the surfaces of the whole body and not on that of any of its elements. External volume forces amount to the application of surface forces to the surfaces of any element, and thereby to tensions. Here, external surface forces do not come into the equilibrium equations but into the boundary conditions of the problem and are thus excluded as forces.

It is important to stress this point. It was mentioned earlier by Hertz, who set himself the problem of constructing a system of mechanics without forces. The fact that he was relatively unsuccessful is because in his days Minkowski's idea about the unity of space-time was as yet unknown. The link between static and dynamics was not as clear as it would be after Minkowski.

Exposition of the theory of elasticity usually begins with the formulation of Hooke's law in the form of the relation between deformation and force in the context of the tensions or compression of a uniform beam. The elongation or compression of the beam is in linear proportion to the force applied to its ends.

This relation is then established more precisely by taking into account the dimension of the area of the beam's cross-section, the concept of tension as a ratio of force to the cross-section and the elongation of the beam as a function of the initial length. The deformation is thus the ratio of relative elongation to the total length. The intrinsic properties of the material of the beam are taken into account by introducing the concept of Young's modulus, which has the dimensionality of the volume density of energy. Only then is it established that, with longitudinal tension and compression, the beam is simultaneously undergoing transverse compression and tension. The relative deformations about the three axes will be in proportion to one another. The link between the relative deformations is determined by the Poisson coefficient, which is also a parameter characterizing the material of the beam. Usually, this aspect is not stressed. For our problem, however, it is of particular interest.

First, the beam has dimensions (length, height, width, surface area), which are determined only from the position of an external observer; secondly, all forces are also fixed from the beginning; thirdly, the internal arrangement of the beam is fixed by integral parameters (the modulus of elasticity and the Poisson coefficient); fourthly, when external forces are applied, the beam reacts as a unified whole and simultaneously in three dimensions. Since here the relative deformations are in proportion to one another with a constant Poisson coefficient, the principle of the superimposition of deformations is applicable. The whole theory is therefore found to be strictly linear.

Finally, it should be noted that two states of the beam are always considered – the initial and the final states before and after the application of the forces. One of them is generally the equilibrium state. If these two states of one and the same system occur at different times, displacement deformations are replaced by velocity deformations. In this case the approach followed takes the form of the theory of viscous or liquid media of gases. As far as Hooke's law itself is concerned, for its purposes deformations and stresses are characterized by tensors of second order and the set of coefficients linking them in a linear fashion is an elasticity tensor of fourth order.

For a fluid, Hooke's law is written in the form of Pascal's law. This way of writing it expresses the condition of equilibrium of the medium, where the stresses on the main axes are equal to the pressure of the fluid. Another condition of equilibrium for a fluid is the law of the conservation of matter.

If in the context of Hooke's law we move to the point of view of Minkowskian unified space-time, and effect a Lorentzian transformation from a stationary system of co-ordinates to a moving one, the equilibrium conditions in accordance with Pascal's law or in the form of any other Hookean tension law can be expressed as Euler equations and as an equation of the continuity of the medium. Here it is important to note that, when deriving the Euler equations of motion, it is not obligatory to use Newton's second law of mechanics and that a Hookean system equilibrium equation can be used.

Nor are any dynamic laws used to justify the Minkowski approach, which is based directly on experimental values and is considered to be valid.

It should be noted that, in the context of Hooke's law, a rigorous solution can be found to Jacobi's virial equation for conservative system. In this case Hooke's law determines the constancy of the product of potential energy and of the Jacobi function; this constancy is written in the form$|U|\sqrt{\Phi} = aGm^{5/2}$.

In this relation the coefficient $a = \alpha^2\beta$ (which stands for the product of form factors included in the expressions for the potential energy and the Jacobi function) acts as a modulus of the dynamic elasticity of the system. It remains a constant and reflects the constancy of the law of mass density distribution of the system within the limits of its elastic deformations with virial oscillations. The deformation of the system is characterized by its integral parameter the Jacobi function – and the stresses are determined by the term $Gm^{5/2}/U$. As a result, the virial pulses of the system will be strictly periodic, and the deformations will be found to be elastic and therefore reversible.

On this basis it was shown in Chap. 7 that the parameters of the virial oscillations of the Earth, which are detected, can be used as if the Earth were an elastic body for determining its potential energy. This option remains open for when natural systems are being examined in the framework of other models of continuous and discrete media.

We have mentioned a number of aspects of the universality of Jacobi dynamics in the examination of natural systems. We shell now consider the prospects for solving a number of practical problems in the context of the dynamic approach.

One of these problems is the dynamics of the Solar System, of its evolution and its origin. In Chap. 8 we made first step in this direction and obtained the basic common solution on creation of celestial bodies and their systems. It appears that any rank of new celestial body (from galaxy to meteorite and even to molecule and atom) is born by self-gravitating parent as consequence of loss of its energy by radiation. It means that the stage of self-gravitation and separation mast be changed by the stage of gravitation and joining of the matter. Thus, the present day stage of expansion of the Universe, after total separation of the matter, should come to the stage of its contraction and gathering. Generally saying, our Universe is a closed pulsating and perpetual system. New more detailed solutions in this direction are desirable.

In the context of the dynamic approach a new problem of dynamics of the self-gravitating Earth and its interaction with the Sun and the Moon were considered in Chap. 7. The found normal and tangential components of the potential and kinetic energies of a self-gravitating body made it possible to understand mechanism of separation of the body's shells, their oscillation and rotation by the inner force field. It was understood that the induced outer force field, which has all the properties of the electromagnetic field, acquires the property to conserve the irradiated energy and potential in the orbital motion of its secondary body. But because of limited velocity of propagation of the changing potential the orbital trajectory is found to be open. This fact is proved both by the artificial satellites and the observed precession of all the planets and the moons. The found important effect makes it possible to interpret inner structure of the Sun, the Earth, the Moon and other celestial bodies.

And also it rises the problem of improving of the Kepler's approximation of the Earth's and other body's orbits which are found to be too rough.

In our opinion, the dynamics of the microcosm is a very interesting application of Jacobi dynamics. This book takes only the first step in this direction. It is shown that Jacobi dynamics is also applicable for the solution of this type of problem. An attractive idea is to use the dynamic approach for studying the physics of molecules, atoms and nuclei as dissipative systems, which might lead to discovery of many interesting effects.

References

Alfven, H., Arrhenius, G.: Origin and evolution of the solar system. Astrophys. Space Sci. **9**, 3–33 (1970)

Belinsky, V.A., Lifshitz, E.M., Khalatnikov, I.M.: Usp. Fis. Nauk **102**, 463 (1970)

Bogolubov, N.N., Mitropolsky, Y.A.: Asymptotic Methods in the Theory of Non-Linear Oscillations. Nauka, Moskow (1974)

Bullen, R.E.: Introduction to the theory of seismology. Cambridge Univ. Press, London (1974)

Campbell, P.M.: Proc. National Acad. Sci., **48**, 12, 1993 (1962)

Chandrasekhar, S.: An introduction to study of stellar structure. Univ. Chicago Press (1939)

Chandrasekhar, S.: Ellipsoidal Figures of Equilibrium (trans: from Engl.). Mir, Moscow (1969)

Chandrasekhar, S., Fermi, E.: Problems of gravitational stability in the presence of a magnetic field. Astrophys. J. **118**, 113 (1953)

Chebotarev, G.A.: Celestial mechanics. In: Great Soviet Encyclopedia, 3rd edn., pp. 386–388, Soviet Encyclopedia, Moscow (1974)

Chetaev, N.G.: The Stability of Motion. Works on Analytical Mechanics, USSR Academy of Sciences, Moscow (1962)

Dive, P.: Bull. Soc. Math. Fr. **59**, 128 (1931)

Duboshin, G.N.: Foundations of the stability of motion. Moscow State Univ., Moscow (1952)

Duboshin, G.N.: Celestial Mechanics: The Main Problems and the Methods. Nauka, Moskow (1975)

Duboshin, G.N.: Celestial Mechanics: Analytical and Qualitative Methods. Nauka, Moscow (1978)

Duboshin, G.N., Rybakov, A.I., Kalinina, E.N., Kholopov, P.N.: Reports of Sternberg Astronomical Institute, Moscow State University Publications, Moscow (1971)

Emiliani, C.: The cause of the ice ages. Earth Planet. Sci. Lett. **37**, 349–352 (1978)

Ferronsky, V.I., Denisik, S.A. and Ferronsky, S.V.: Virial oscillations of celestial bodies: II. General approach to the solution of perturbed oscillations problem and electromagnetic effects. Celest. Mech. **27**, 285–304 (1982)

Ferronsky, S.V.: Observation of coherent pulsations in the Earth atmosphere with period about one and half hour, Phys. Atmos. Ocean, **20**, 922–928 (1984)

Ferronsky, V.I., Denisik, S.A., Ferronsky, S.V.: Virial oscillations of celestial bodies: III. The solution of the evolutionary problem in the non-Newtonian time scale. Celest. Mech. **32**, 173–183 (1984)

Ferronsky, V.I., Denisik, S.A. and Ferronsky, S.V.: Virial oscillations of celestial bodies: IV. The Lyapunov stability of motion. Celest. Mech. **35,** 23–43. (1985)

Ferronsky, V.I., Ferronsky, S.V.: Dynamics of the Earth. Scientific World, Moscow (2007)

Ferronsky, V.I.: Virial approach to solve the problem of global dynamics of the Earth. Investigated in Russia, pp. 1207–1228 http://zhurnal.ape.relarn/articles/2005/120.pdf (2005)

Ferronsky, V.I., Denisik, S.A., Ferronsky, S.V.: The solution of Jacobi's virial equation for celestial bodies. Celest. Mech. **18**, 113–140 (1978)

V.I. Ferronsky et al., *Jacobi Dynamics*, Astrophysics and Space Science Library 369, DOI 10.1007/978-94-007-0498-5,

Ferronsky, V.I., Denisik, S.A., Ferronsky, S.V.: The virial-based solution for the velocity of gaseous sphere gravitational contraction. Celest. Mech. **19**, 173–201 (1979a)
Ferronsky, V.I., Denisik, S.A., Ferronsky, S.V.: The asymptotic limit of the form- factor α and β product for celestial bodies. Celest. Mech. **20**, 69–81 (1979b)
Ferronsky, V.I., Denisik, S.A., Ferronsky, S.V.: The solution of Jacobi's virial equation for nonconservative system and analysis of its dependence on parameters. Celest. Mech. **20**, 143–172 (1979c)
Ferronsky, V.I., Denisik, S.A., Ferronsky, S.V.: Virial oscillations of celestial bodies: I. The effect of electromagnetic interactions. Celest. Mech. **23**, 243–267 (1981)
Ferronsky, V.I., Denisik, S.A., Ferronsky, S.V.: Jacobi Dynamics. Reidel, Dordrecht (1987)
Ferronsky, V.I., Denisik, S.A., Ferronsky, S.V.: Virial oscillations of celestial bodies: V. The structure of the potential and kinetic energies of a celestial body as a record of its creation history. Celest. Mech. Dyn. Astron. **64**, 167–183 (1996)
Flügge, S.: Practical Quantum Mechanics. Springer, Berlin (1971)
Frank-Kamenetsky, L.A.: Physical Processes in the Interior of Stars. Fizmatgiz, Moscow (1959)
Garcia, L.D., Mosconi, M.B., Sersic, J.L.: A global model for violent relaxation. Astrophys. Space Sci. **113**, 89–98 (1985)
Goldstein, H.: Classical Mechanics, 2nd edn. Addison-Wesley, Reading (1980)
Grushinsky, N.P.: Theory of the Earth Figure. Nauka, Moscow (1976)
Jacobi, C.G.J.: Vorlesungen über Dynamik. Klebsch, Berlin (1884)
Janke, E., Emde, F., Lösch, F.: Tafeln höherer funktionen, 6 Aufgabe. BG Teubner, Stuttgart (1960)
Jeans, J.H.: Problems of Cosmogony and Stellar Dynamics. Cambridge University Press, Cambridge (1919)
Jeffreys, H.: The Earth: Its Origin, History and Physical Constitution, 5th edn. Cambridge University Press, Cambridge (1970)
Kapitza, P.L.: Journal of Experimental and Theoretical Physics, **21**, 5 (1951)
Kittel, C., Knight, W.D., Ruderman, M.W.: Mechanics, Berkeley Physics Course, vol. 1. McGraw Hill, New York (1965)
Klein, F., Sommerfeld, A.: Theorie des Kreisels, Heft III. Teubner, Leipzig (1903)
Landau, L.D., Lifshitz, E.M.: Mechanics of Continuous Media. Gostechizdat, Moscow (1954)
Landau, L.D., Lifshitz, E.M.: Quantum Mechanics. Nauka, Moscow (1963)
Landau, L.D., Lifshitz, E.M.: Mechanics, Elektrodynamics. Nauka, Moscow (1969)
Landau, L.D., Lifshitz, E.M.: Mechanics. Nauka, Moscow (1973a)
Landau, L.D., Lifshitz, E.M.: Field Theory. Nauka, Moscow (1973b)
Melchior, P.: Physique et Dynamique Planetaires. Vander-Editeur, Bruxelles (1972)
Misner, C.W., Thorne, K.S., Wheeler, J.A.: Gravitation, Freeman, San Francisco (1973)
Misner, C.W., Thorne, K.S., Wheeler, J.A.: Gravitation. Freeman, San Francisco (1975)
Poincare, H.: Lectures on Celestial Mechanics. Nauka, Moscow (1965)
Sedov, L.I.: Mechanics of Continuous Media, vol. 2. Nauka, Moscow (1970)
Shirkov, D.V. (ed.): Microcosmos Physics, Small Encyclopedy. Nauka, Moscow (1980)
Spitzer Jr., L.: Diffuse Matter in Space. Interscience, New York (1968)
Tamm, I.E.: Fundamentals on Theory of Electricity. Nauka, Moscow (1976)
Tolman, R.C.: Relativity, Thermodynamics and Cosmology. Clarendon Press, Oxford (1969)
Weinberg, S.: Gravitation and Cosmology. Wiley, New York (1972)
Whittaker, E.T.: A Treatise on the Analytical Dynamics of Particles and Rigid Bodies. Cambridge University Press, Cambridge (1937)
Wintner, A.: The Analytical Foundations of Celestial Mechanics. Princeton University Press, Princeton (1941)
Zeldovich, Y.B., Novikov, I.D.: Pelativistic Astrophysics. Nauka, Moscow (1967)
Zharkov, V.N.: Inner Structure of the Earth and Planets. Nauka, Moscow (1978)

Index

MIX
Papier aus verantwortungsvollen Quellen
Paper from responsible sources
FSC® C105338

If you have any concerns about our products,
you can contact us on
ProductSafety@springernature.com

In case Publisher is established outside the EU,
the EU authorized representative is:
Springer Nature Customer Service Center GmbH
Europaplatz 3, 69115 Heidelberg, Germany

Printed by Libri Plureos GmbH
in Hamburg, Germany

Media and Digital Modernism

Sefer Kalaman and Bilal Süslü (eds.)

Media and Digital Modernism

New Communication Environments

In collaboration with
Serhat Bolat and Ayşe Nesil Demir

PETER LANG

Bibliographic Information published by the Deutsche Nationalbibliothek
The Deutsche Nationalbibliothek lists this publication in the Deutsche Nationalbibliografie; detailed bibliographic data is available online at http://dnb.d-nb.de.

Library of Congress Cataloging-in-Publication Data
A CIP catalog record for this book has been applied for at the Library of Congress.

Cover illustration: Gerd Altmann

ISBN 978-3-631-76485-5 (Print)
E-ISBN 978-3-631-76495-4 (E-PDF)
E-ISBN 978-3-631-76496-1 (EPUB)
E-ISBN 978-3-631-76497-8 (MOBI)
DOI 10.3726/b14540

Peter Lang – Berlin · Bern · Bruxelles · New York · Oxford · Warszawa · Wien

This publication has been peer reviewed.

www.peterlang.com

Foreword

Teaching the concepts, complexities and possibilities of digital media has always been a challenge for a non-digital native like me. I feel lucky that I have been the ultimate witness to the transformation of the reader/user in this vast mediated environment. Starting from the mid-1990s, the definitions of time and space changed radically. We have been trapped between "real-time" and "virtuality" trying to make sense of this new communication environment. My students in college had no problem in adjusting themselves though. They were either born into this digital world or had no problem adjusting themselves to new digital realities.

There came new approaches, debates and concepts to define the gray areas to explain the existing binaries where the qualities of modernism were placed over against the characteristics of postmodernism. Digimodernism (digital modernism) did not choose modernism over postmodernism or vice versa. As a new paradigm, it aimed to shed light on the intersection of real-time and virtuality by making them the twin axes of digimodernism. This term, which was coined by British cultural critic Alan Kirby, brought a fresh look into the debates on the new cultural landscape shaken up by digitization. Kirby, in addition to his work on explaining the impacts of computer technology on cultural products from personal communication to blockbuster movies, from popular television to radio programming, offers a bold new lens at the roles of the author and reader/viewer. He goes beyond the limitations of postmodernism by emphasizing the increasing irrelevance of postmodernism and the need of a new theory to underpin our current digital culture. He traces a path through an author-led textuality characterized by fixed meaning, through post-structuralism's death of the author. He even writes "You click, you punch the keys, you are 'involved', engulfed, deciding. You are the text, there is no-one else, no 'author'; there is nowhere else, no other time or place. You are free: you are the text: the text is superseded".

Since millennials are the generations of digimodernism, it is interesting for an educator to observe their relations with textuality on social media. The most thought-provoking discussions in my opinion come with the images' ephemerality on Snapchat. Snapchat has a strong connection to digimodernism and captures the change in today's young demographic. The social media platform is perhaps one of the best examples of evanescence, defined as: "text is difficult to capture and archive, has no interest in being reproducible". Snapchat pictures are taken and sent to the viewer who can then open and view the image for any time from 1 to 10 seconds. Users use this to their advantage and can change the view time based on how much they want the viewer to be able to read, or how long they want the viewer to be able to see the picture. Pictures are taken and sent very quickly without the intention of being reproduced. This allure of disposability and the millennials' disinterest for collective memory, capturing moments as trophies can be placed at the core of the digimodernism discussions.

This book edited by Sefer Kalaman and Bilal Süslü is an effort to give a comprehensive overview on different aspects of digimodernism, which will be extremely useful for both learners and educators. The breadth and extensive coverage of Media and Digital Modernism tackle a wide range of topics and debates varying from corporate identity to cyber identity, from popular culture and literary narratives to blockbuster movies and self-representation, and from digital voyeurism to digital vigilantism and visual culture. I truly believe that Media and Digital Modernism will be a humble step for a better understanding of how digitalization has been reshaping our cultural life.

Aslı Tunç
Istanbul Bilgi University, Turkey

Aslı Tunç is a professor at the Media and Communication Systems Department at Istanbul Bilgi University, Turkey. She has a BA in communication sciences from Istanbul University and an MA in film and television studies from Anadolu University. She received her PhD in media and communications at Temple University in Philadelphia in 2000. She has written numerous publications on the issues of democracy and media, social impacts of new media technologies, and media ownership structure in Turkey. She is the co-author of a book in Turkish (Blogdan Al Haberi, Haber Blogları, Demokrasi ve Gazeteciliğin Geleceği Üzerine) on the impact of political news blogs on the future of journalism. She is a regular contributor to P24: Platform for Independent Journalism website (platform24.org) writing on media issues.

Contents

Sefer Kalaman

Digimodern Era, New Communication Technologies, and Sociocultural Change

Introduction

People have communicated with each other through different methods and techniques since their existence. Communication has been one of the indispensable elements in human life that they have narrated their expressions to other individuals or societies with pictures, texts, music, dances, photographs, voice recordings, video recordings, or electronic messages. Communication, which is individually realized, has evolved into a mass structure with the increase of technological developments after the invention of writing. The emergence of newspapers, cinema, radio, and television, and their integration into human life initiate mass communication and create media concept.

These mass media, which are now called the traditional media, have an important role in human life with many functions such as informing, training, and entertaining. The rapid development of technology has paved the way for a new communication environment after the second half of the 20th century. The Internet emerges due to the invention of the computer and this causes a radical change in the field of communication. With the development of internet technology and its accession by a large mass of people in a short period of time, traditional media tools have gradually begun to integrate into Internet. In addition to traditional media tools, the Internet has constituted several communication environments and created a new communication environment called new media.

Both Internet technology and other technological developments have integrated into human life at the end of the 20th century and enabled many fields/environments to evolve from analog to digital.

These technological developments, especially Internet, indicate that a new era has begun according to many theorists or writers. Digimodernism is probably the best name to describe this new era that is expressed by the author or the theoreticians with different names. Alan Kirby (2009: 1), describing this era dominated by digital technology, has handled all environments and situations related to human life such as social, cultural, political, economic aspects from the viewpoint of digitality.

Digital modernism, briefly digimodernism, argues that digital technologies have penetrated almost all of the routine of human life, digitalized to transform many activities of life such as education, shopping, socialization, banking, sports, and entertainment. The people born in the Digimodern era easily adapt this period and its principles, that is, they are grown up to integrate into these digital technologies. People who are born in the postmodern era and continue to live in digimodern era have later adapted to the digital technologies of this era.

Digital technologies, which enable digimodernism to emerge, have radically transformed the different fields of human life as well as the media field. Digimodernism has caused major and radical changes in traditional media tools namely newspapers, magazines, photographs, cinema, radio, and television, and created its own media (new media). Online games, social networks, online internet forums, etc., are some of these new communication environments.

The newspapers, one of the oldest media tools/environments, have gradually begun to decrease their circulation in the 21st century. While some newspapers reduce the number of print newspapers, some completely abandon the production of print newspapers and continue to publish on the internet. The newspapers have adapted to internet technology in a short period of time with the digimodern era and they have maintained their activities both in print and digital journalism. Digital newspapers, reaching their readers via the Internet, have begun to be read and followed by large masses for reasons such as having an interactive structure, being easy or accessible at any time, and being free.

On the other hand, the camera, which was invented in 1826 by French Nicephore Niepce, is one of the most common communication tools today, and the first photograph taken with this machine turn into completely different things. The camera was once an analog machine; the photographs are totally different now. In this digimodern era, digital cameras are used by a large mass; the photographs are digital and kept or displayed in digital media.

The television, for which a great interest was shown by the society at the time of its invention and watched by a large mass in the following years, integrates itself into internet technology in this digimodern era. With the interactive structure via the Internet, it has become more efficient in terms of access and watching. Televisions have begun to be watched both on computers and the internet via smartphones. In addition to this, television broadcasting, which costs more money to broadcast, has evolved into another dimension with personal television channels (web TVs) created in the virtual environment and costs less.

The cinema, which is another tool of mass media and entered human life before television technology, has also had radical changes from the late 19th century to the present. In this period, the interest shown by people to the cinema has gradually continued to increase. The film-makers, who shoot films to increase the box-office gross and not to lose this interest, always use technological possibilities to increase the quality of the films and keep the interest of the audiences alive. Developments in light and sound technology, effect technology and camera technology have always been used in the film industry. Just like the development of digital technologies and the interest of the cinema about these developments result these technologies to be used in the films. Many technological improvements such as the creation of the place, the physical appearance of the actors, the lighting system of the place, the sounds of people or objects, the effects, to the creation of objects or living beings (animations), are naturally involved in the film industry. The films have now reached a completely different dimension due to the digital technologies of the digital era. It is possible to say that this development/change will evolve to an inconceivable dimension.

The new media, which is included in human life after the traditional media, is integrated with the previous mass media and transforms into a composite structure. Besides, traditional media tools have adapted to this new mass media environment in a short time and continued to develop. New media is an integrated virtual environment that includes different mass media. This integrated mass media, the media of the Digimodern era, is used by almost half of the world's population (wearesocial, 2017: 5). With its versatile, interactive, inclusive, entertaining, and informative structure, this virtual environment has made communication as easy, fast, and inclusive as ever before. People use different environments of the new media to practice different activities. The most popular examples of these environments are obviously social networks and digital online games.

Social networks are clearly the most important mass media environment. These networks, which are used by about one-third of people (wearesocial, 2017: 5), have brought socialization, entertainment, information, and news to different dimensions. There are perhaps a few technological developments or environments in the history of communication that has changed the structure of communication. With social networks, internet technology is one of these environments.

Digital games, on the other hand, are one of the mass media that are played by 2.2 billion people and have a turnover of hundreds of billions of dollars (McDonald, 2017). Two of the most important factors why digital games are accepted as mass media: messages are sent to the game players via game characters, environments, game scenarios, etc., and the other one is that game players in these digital games can communicate with other players and share information.

All these traditional and new media are tools/environments that enable communication to exist and spread. The emergence, development, and future of old and new communication tools are discussed within the scope of the study. In particular, it has been tried to examine what kind of structures the communication tools have in Digimodern era and will evolve in the future.

1 Development Process of Communication Technologies

Communication starts with people: it is a necessary condition for the existence of man and society. Without communication, it is impossible for people to maintain their social existence. People use natural and technological tools created by them in activities that they try to produce themselves and society. The organization, implementation, development, and modification of this usage can only occur via communication. The relations and communication established by artificial tools are called relations and communication with technologies. Communication is a necessary condition for the existence and implementation of this relation. Thus, the relation is not communication but the existence of the relation depends on communication. In other words, communication is not the relation itself, but it determines the existence of the relation (Erdoğan and Alemdar, 2005: 15). As Işık (2017: 11) quotes from Aristotle, man is a social entity. One consequence of this is that socialization occurs for people who live collectively and relate to the others. In other words, in order to maintain their daily life, people should be in contact with others and nature. Since it is not possible to meet all kinds of human needs on their own, it is essential to establish relations and communication on the axis of work sharing and specialization. To sum up, there is communication in every moment and everywhere in the flow of life such as shopping, education, communication. People are in communication at every moment of daily life. They constantly communicate at home, bus stop, office, on the street and at every moment and everywhere.

According to Oskay (1992: 15), a group or society formed by people who share similar experiences, problems, and similar feelings with one another, informing each other about changes in objects, events or attitudes, judgments, thoughts, and emotions expressed in community life is called communication. According to Andersen (1959), communication is a process that allows us to understand others and to be understood by others. This is a dynamic and constantly changing process depending on the situation.

Communication has become an important tool in transforming the lives of people and society into systematic structures. People have had the opportunity to communicate and to contact each other, to record the developments related to life, and to transfer their routine to other generations in a systemic communicative structure. Every one of the communication tools such as text, picture, paper, printing, newspaper, telephone, telegraph, radio, cinema, television, and computer has made an important contribution to the era in which each of them has developed, and has been effective in taking humanity a step further. In addition to that, some communication tools such as printing and information technology have caused humanity to open a brand new period (Güngör, 2011: 36). As communication technologies evolve and diversify, human relations and social practices are also affected.

While the communication pattern, which occurs in a narrower environment, provides community relations via face-to-face communication in its initial stages, it has been complicated and has expanded during the developing periods. People have gained the ability and the opportunity to communicate more complexly and quickly in wider environments among communities and societies, thanks to the developing communication technologies (Önür, 2002: 63). As social relations gradually get more and more complicated, there is more to be needed than the common memory of the group to remember important things. What is needed is a memory outside the body, sometimes called extrasomatic memory. Thus, the increase in the "communication" enables the emergence of "communication tools" and the development of tools that store and restore the increased data if necessary (Crowley and Heyer, 2011: 18). This process has continued from notches on the bone in the early ages to murals, stone inscriptions, printed books, analog technology, and communication tools that is capable of digital recording.

When the communication process is evaluated from the general perspective, it is possible to analyze the history of communication technologies in four stages (Vural, 2002: 5-6): The first stage is the period when physical and mechanical power is dominant. The Morse alphabet, the messenger pigeons, the communication with the smoke signals are at this stage. The second one involves the development of telephone, telegraph, radio, and television due to the invention of electricity and electromechanical power. The third one is the period of computer, chip, transistor, semi-conductor discoveries, thanks to the development of electronic possibilities. At this stage, telecommunication has begun to integrate with computer technologies. The fourth one is the transition from analog to digital networks, and the rise of electronic product consumption.

It does not take long to understand that electricity is an ideal technique in the transmission of knowledge in these technological developments because the use of electricity in telecommunication today is based on a history of about 150 years. Within this process, a significant progress has been accomplished in the field of communication every two decades: telegram by the 1850s, telephone between 1850–1880, transmission by electromagnetic waves by the 1990s, radio between

1920–1930, television technology between 1950–1960 and finally, internet technology has begun to be used since 1970 (Balle and Eymery, 1995: 18).

The invention and development of the computer has a different importance within these technological developments. Computers were initially used in large organizations, regulatory administrative agencies, scientific research, and in the military. After the development of microchip technology in the 1970s, computer circuits have spread to every part of the industry and become an essential part of household goods, from heaters to television receivers and games. With the emergence of personal computers, it has now become possible for people working at home to have access to data processing capabilities that is previously used only by large institutions. Today, it is seen that the global internet, the networks of local corporations, and the networks with World Wide Web rapidly spread to be used in a wide range of fields, including administrative use, research, public expression, art, social events, and business (Crowley and Heyer, 2011: 438).

In addition to the new communication tools/environments, the society has followed and attached importance to traditional media tools, and these tools enables radical changes in human life. No technological advancement has ever made people addicted in their life but Internet. (Dijk, 2006: 1). This virtual network, which is important due to the fact that it penetrates to almost every part of life, has a large number of users in a short time. Apart from personal use, this virtual network is integrated with several fields from banks to companies, public institutions to restaurants, from nurseries to factories in a short period of time.

As the Gutenberg press transformed the world (Firat and Dholakia, 2004: 140), moving cloistered knowledge into public spaces, and as the introduction of the telephone altered the spatial and geographical framework of interpersonal communication, so the internet has had similar effects today. It has surpassed our wildest imaginations (not to mention Newsweek's expectations), giving us access to more information that we could ever use in a lifetime, information we can access virtually (no pun intended) anywhere and anytime. It has changed the way we live our lives, the way we work, socialize, shop, meet people and very simply, communicate (Mulady, 2010: 48).

The internet, which changed the daily life activities of people, has caused the beginning of a new era. The internet, which has caused the daily life routines or activities to change and evolve, also allows for a fundamental change, renewal, and development of the media industry. This new technology called internet has brought new communication technologies/environments together with a new era called Digimodernism.

2 Digital Modernism and New Communication Technologies

After the completion of its platform since the second half of the 20th Century, the computer enters the daily life of the people and prepares the basis for the internet technology to be formed soon. The discovery of the Internet and its use by a large mass has resulted in direct or indirect changes in almost every aspect of people's lives. The Internet has penetrated and transformed almost every aspect such as accessing information, shopping, eating and drinking habits, entertainment, socialization, business, and this transformation has increasingly continued.

One of the most important platforms in which these changes are experienced is obviously the media. The integration of traditional media tools with the internet, which provides the requirements of the era, and the advancement of new communication tools have resulted in the emergence of a new formation/environment called new media.

The new media, which have been used in many studies in the same sense as the new communication environments, is a term coined by scientists in the 1970s who conduct social, psychological, economic, political, and cultural studies based on information and communication. However, the meaning mentioned in the 1970s has expanded and reached different dimensions with the computer and internet technology, which gain great momentum in the 1990s. New communication environments that are capable of gathering all known different media such as "multi-media" emerged with forms such as text, static image, moving image, sound (Kara, 2013: 11).

With this digital revolution, new media has penetrated to all aspects of our lives and brought to a new dimension in mass communication as well as interpersonal communication. In other words, the new media, which has rapidly increased its development and expansion with this technological progress, has led to a change in face-to-face and mass communication from quantitative and qualitative viewpoints.

This digital evolution and digital technology have already significantly changed the ways in which we create, spread, use, and understand information, including the information that has traditionally been contained in books (Konsa and Hermann, 2014: 37). This new communication environment has provided the basis for a radical change in the way people make friends, meet, experience the behavior norms and create/discover their identity and caused a new way of communication (Barnes, 2006).

According to Castells (2013: 21), this fundamental change in the field of communication in recent years is due to mass self-communication, that is, the use of Internet and wireless networks as digital communication platforms. This digital communication is also mass communication with the intention of reaching a large number of people and processing the messages which are likely to be connected to the infinite networks that transmit digital information, and it also becomes self-communication because the sender autonomously decides on the content of the message and determines the receiver.

Though it's not just the internet, technology has become an integral part of our lives. From the way we communicate with each other, to the way we shop, and from the way we consume media, to what kinds of media we have to consume, technology has had fundamental effects on all of these. The way we view television is radically different; not only is it now digital, but we can fast forward through adverts, pause live broadcasts, and record without using any kind of disc or 'tape'. We use our remote controls to 'vote', or to access further information (BBC); in other words, we interact with our televisions, or with the programmes or adverts. But why watch programmes on television when you can watch them online? Where you can interact with adverts to such an extent, you can create your own customised ending? If we're going that far, why don't you become the star of the advert? Or be the creative force behind the advert? Surely we cannot believe us to be in the same period of postmodernism as we were back in 1995? (Mulady, 2010: 14).

The new media, which includes many virtual environments such as forums, blogs, news sites, online media broadcasts (radio and television broadcasts), internet journalism, cinema, and social networks, has directly integrated the passive mass (audience/listener) into the communication process and made them active, and therefore, let them produce content and feedbacks.

Social networks, the most popular and widely used platforms of the new media, have become much more effective than other digital media at the point where people participate directly in the communication and play an active role and produce content. Besides, people who have the opportunity of establishing the desired social relations in these networks have been leading the virtual environment to reach out to the huge number of users in a short period of time.

Social sharing networks, typical examples of applications called Web 2.0, are web-based platforms including different media, information and communication technologies that allow users to explain/create their own contact lists, communicate with each other, show their contact lists, and create profiles with their information (Fuchs, 2011: 141).

The Internet, and therefore digital systems, has completely changed the rules of the game, the players and the game itself (Stoddart, 2014: 33). Since its first appearance in the second half of the 1990s under the impetus of new technologies, digimodernism has decisively displaced postmodernism to establish itself as the twenty-first century's new cultural paradigm. It owes its emergence and preeminence to the computerization of text, which yields a new form of textuality characterized in its purest instances by onwardness, haphazardness, evanescence, and anonymous, social and multiple authorship. These in turn become the hallmarks of a group of texts in new and established modes that also manifest the digimodernist traits of infantilism, earnestness, endlessness, and apparent reality. Digimodernist texts are found across contemporary culture, ranging from "reality TV" to Hollywood fantasy blockbusters, from Web 2.0 platforms to the most sophisticated videogames, and from certain kinds of radio shows to crossover fiction. In its pure form, the digimodernist text permits the reader or viewer to intervene textually, physically to make text, to add visible content or tangibly shape narrative development. Hence "digimodernism," properly understood as a contraction of "digital modernism," is a pun: it's where digital technology meets textuality and text is (re) formulated by fingers and thumbs (the digits) clicking and keying and pressing in the positive act of partial or obscurely collective textual elaboration (Kirby, 2009: 1).

The spread of the personal computer in the 1980s brought with it a new associated vocabulary, some of which, like "interfacing" or going "online," have been absorbed permanently into the language. If the emergence of the digimodernist text has had a comparable effect you might point to the discourse of "interactivity" as an example. Videogames, reality TV, YouTube, and the rest of Web 2.0 are all supposed to offer an "interactive" textual experience by virtue of the fact that the individual is given and may carry out manual or digital actions while engaging with them (Kirby, 2009: 60).

This era shaped by digital technology marks a turning point where people interact superficially in the virtual environment, including clicks, likes, chooses, and downloads mentioned by Kirby (Firkins, 2015).

There are various ways of defining digimodernism. It is the impact on cultural forms of computerization (inventing some, altering others). It is a set of aesthetic characteristics consequent on that process and gaining a unique cast from their new context. It's a cultural shift, a communicative revolution, a social organization. The most immediate way, however, of describing digimodernism is this: it's a new form of textuality (Kirby, 2009: 50).

There are various relations between digimodernism and postmodernism as mentioned. Postmodernism remains in the shadow of digimodernism at the point of expressing the dominant cultural, technological, and political elements of the period after the mid-1990s. Briefly, digimodernism has begun to take the place of postmodernism after this period (Kirby, 2009: 1–2). According to

Mulady (2010: 14–15), postmodernism is not completely over but has begun to evolve into a new cultural paradigm called digimodernism.

Postmodernism is dead or at best dying and has been replaced by a new cultural paradigm claim many philosophers, writers, and art critics. Though there are many philosophies and theories on the nature of this new cultural paradigm, it appears that whatever is coming after postmodernism is largely being defined by global advances in technology and the increased interaction of people via digital media (Conner, 2013: 1).

In addition to Kirby's ideas, there have been other writers and theoreticians who claim that postmodernism has ended or begun to evolve into a new cultural paradigm. Nicolas Bourriaud argues that the postmodern era ends and the economic, political, and cultural aspects of the era can be notably defined with the new paradigm called "Altermodernism". Raoul Eshelman meets on a common ground with digimodernism, thanks to the term "Performatism". Robert Samuel and his friend Gilles Lipovetsky, one of the writers whom Alan Kirby uses as a source for the concept of digimodernism, argue that the postmodern era is over and a new era called "hypermodernity" has begun (Mulady, 2010: 36–37).

In addition to that, different terms have been used by other writers, critics, or theorists to describe this new cultural paradigm after the postmodern era. "Transmodernism", "Avantpop", "New Sincerity", "Pseudomodernism", and "Automodernism" are a few of them (Bunnell, 2015: 2–3). The definitions of this era change but the consensus of these writers or theorists is that a new era has begun instead of postmodernism. However, this paradigm, which is defined as digimodernism or other terms, is neither a beginning nor an end.

While digimodernism begins instead of postmodernism in the course of time, the new paradigm in the future will begin instead of digimodernism in the same way (Mulady, 2010: 52). The new technological developments in the future will create their own era, cultures, and mass. Consequently, this has already happened now. The increasing popularity of the Internet and the adaptation of virtual environments have affected the society. This digitalization, which surrounds the whole society, has formed its own mass and the generation born in this era has completely connected to it.

All of them are "Digital Natives." They were all born after 1980, when social digital technologies, such as Usenet and bulletin board systems, came online. They all have access to networked digital technologies. And they all have the skills to use those technologies. There is one thing you know for sure: These kids are different. They study, work, write, and interact with each other in ways that are very different from the ways that you did growing up. They read blogs rather than newspapers. They often meet each other online before they meet in person.

They probably don't even know what a library card looks like, much less have one; and if they do, they've probably never used it. They get their music online often for free, illegally rather than buying it in record stores. They're more likely to send an instant message (IM) than to pick up the telephone to arrange a date later in the afternoon. They adopt and pal around with virtual Neopets online instead of pound puppies. And they're connected to one another by a common culture. Major aspects of their lives – social interactions, friendships, civic activities – are mediated by digital technologies. And they've never known any other way of life (Palfrey and Gasser, 2008: 1–2).

The reason for the use of the term "digital natives" for this generation with the eldest being 30 years old is that they are born into an era where the Internet has become widespread, so they grow up in a computer culture and are thought to be a "native" of this life. This also means that they have had a different socialization process than previous generations (Cross, 2011: 119).

Unlike the other people, Digital Natives live much of their lives online, without distinguishing between the online and the offline. Instead of thinking of their digital identity and their real-space identity as separate things, they just have an identity (with representations in two, or three, or more different spaces). They are joined by a set of common practices, including the amount of time they spend using digital technologies, their tendency to multitask, their tendency to express themselves and relate to one another in ways mediated by digital technologies, and their pattern of using the technologies to access and use information and create new knowledge and art forms. For these young people, new digital technologies – computers, cell phones, Sidekicks – are primary mediators of human-to-human connections. They have created a 24/7 network that blends the human with the technical to a degree we haven't experienced before, and it is transforming human relationships in fundamental ways. They feel as comfortable in online spaces as they do in offline ones. They don't think of their hybrid lives as anything remarkable. Digital Natives haven't known anything but a life connected to one another, and to the world of bits, in this manner (Palfrey and Gasser, 2008: 4–5).

Digital Natives are tremendously creative. It is impossible to say whether they are more or less creative than prior generations, but one thing is certain: They express themselves creatively in ways that are very different from the ways their parents did at their age. Many Digital Natives perceive information to be malleable; it is something they can control and reshape in new and interesting ways. That might mean editing a profile on MySpace or encyclopedia entries on Wikipedia, making a movie or online video, or downloading a hot music track – whether lawfully or not. Whether or not they realize it, they have come to have a degree of control over their cultural environment that is unprecedented. Digital Natives can learn how to use a new software program in a snap. They seemingly can take, upload, and edit pictures to share with friends online in their sleep. Digital Natives, at their most creative, are creating parallel worlds on sites like Second Life. And after they do, they record parts of that world and post a video of it on YouTube (if they live in California) or Daily Motion (if they live in Cannes) in a new art form called "machinima." Digital Natives can rework media, using off-the-shelf computer programs, in ways that would have seemed impossible a few short decades ago (Palfrey and Gasser, 2008: 6). The social networks, which these virtual communities (digital natives) come together and establish social relations, has become the temples or castles of them. The digital selves, who come together in this temple, display many of their information, photographs, or videos to each other's surveillance and present themselves in this way.

These digital natives have begun to learn and adopt all of the original aspects of the digital period since their birth. These natives, who closely follow digitalization and shape their lives according to this, have created their own languages and cultures apart from the generations born before the digimodern era. However, this created language and culture have resulted that the privacy, one of the most important characteristics of people, becomes public. The people, who adapt every aspect of their life to this virtual environment with the desire to socialize and share their personal information, have display their privacy to the other users, commercial corporations and the state, and allow their data to be surveilled.

People born in the digital era and affected by digital culture carry out their social relations, studies, communication, shopping, and even relationships through this virtual network. In addition to this generation called digital natives, other people also use the internet as a routine of a daily life. The needs related to several fields such as banking, hospital and office procedures, shopping, travelling, information exchange, education, and communication are carried out via this virtual network.

Digital immigrants, who are not born in the digital world, influenced by new technologies in their lives and benefit the advantages of it, have lived in the Internet era and the previous one, and also experienced the communication, socialization, studies and other practices in these periods. For this reason, digital immigrants can criticize and analyze Internet and social networks from a different perspective. As they are older and more experienced, skeptical and conscious of their privacy than the digital natives, this causes them to be more aware of being surveilled. Thanks to being a part of the Internet, it is not possible for digital natives to question or examine from another point of view to realize the damages and the future ones.

In digimodernist societies, in fact, "publics" are not simply publics, audiences, target groups, stake- or share-holders. Instead, they become active players, often unrecognizable leaders of social interactions and of social construction of reality and meanings around reality (Valentini et al., 2012: 875).

At the time when the concept of a public has changed and social, political, economic, and cultural developments are affected by the increasing use – and misuse – of communication technology that encourages the ideology of globalized market economies and of self-expression, the concept of relationship becomes even more central for 21st Century public relations theory. Not only do people become progressively more involved in the production and consumption of cultural artificants, but their relationships can be formed and dissolved more quickly, depending on their interests and concerns. Understanding the role of relationships in relationship dynamics can help public relations practitioners navigate in multiple communities, physical and virtual, and to overcome ambiguities and uncertainties that characterize digimodernist societies. Communication technology has created opportunities and challenges both for organizations and for people, requiring an increased need to reconsider and redefine the Dynamics of relationships within organizations and in society at large (Valentini and Kruckeberg, 2011; Valentini et al., 2012: 876).

As mentioned, the digital practices of the Digimodern era have dominated and transformed the human life. Digimodernism, the internet, digital natives, and digital immigrants have played an essential role in evolving or maintaining the existence of the media of Digital era and the traditional media tools/environments.

Digimodernism has primarily transformed traditional media tools with its digital technologies. Mass media (cinema, radio, television, newspaper etc.) have been forced to permanently change their structures under the dominance of digital technologies. However, cinema is probably the most significant environment that benefits the advantages of digital technologies in this virtual modern era.

If the internet and its use define and dominate pseudo-modernism, the new era has also seen the revamping of older forms along its lines. Cinema in the pseudo-modern age looks more and more like a computer game. Its images, which once came from the 'real' world – framed, lit, soundtracked and edited together by ingenious directors to guide the viewer's thoughts or emotions – are now increasingly created through a computer. And they look it. Where once special effects were supposed to make the impossible appear credible, CGI frequently [inadvertently] works to make the possible look artificial, as in much of The Lord of the Rings or Gladiator. Battles involving thousands of individuals have really happened; pseudo-modern cinema makes them look as if they have only ever happened in cyberspace. And so cinema has given cultural ground not merely to the computer as a generator of its images, but to the computer game as the model of its relationship with the viewer (Kirby, 2006: 35).

In addition to cinema, traditional mass media such as radio, television, newspapers, and magazines have also tried to preserve their target groups and aimed to expand them. After the adaption of mass media to internet technology, they make their virtual existence more prominent from their physical existence and also begin to invest in this direction. The Internet channels/pages for each channel or newspaper are created and they reach the audience via these channels/pages. Besides, advertisements, which form the basis of all the financial resources of all private TVs or newspapers, have also begun to be shown in this virtual platform. Thanks to the interactive structure of the internet, it is also possible for audience to respond, comment, like, or criticize. Audiences can participate in the live programs of TV channels by making comments via internet. Besides, it is also possible for audiences to follow these programs in the platforms based on their will due to virtualized broadcasts. In addition to that, these virtual broadcasts also facilitate the establishment of the radio and TV channel. Broadcasting in the platforms such as web TVs, YouTube channels, have become individualized.

In the new media and social networks, one of its most important platforms, things are more developed and different. People use these virtual communication environments almost every day and these have become routine of the daily life. As a result of this, people and their communication have gradually got more virtual and the communication completely become digital. Online multi-player digital games or personal digital games are also examples of this virtual environment. People can communicate, entertain and meet their different sociological-psychological needs in this environment. The avatars/virtual characters created in digital games are just like the characters created in social media. Although these virtual characters are not physically similar to each other, they have similar characteristics in terms of content or service. The commitment to these virtual characters/identities is the sign of virtual reality. In the course of time, people may experience deterioration in their perception of reality.

In the Digimodern era, it is possible to say that the communication and communication technologies will evolve into a much more complex and virtual structure in the future, thanks to new technological developments.

3 Future of Communication

While digital innovations and communication technologies influencing the whole society seem to reach the upper level in the digimodern era, it will be much more likely that digital/virtual technologies seem to obtain humanity and all of their activities in the future. Because new digital communication technologies are invented every single day and offered to use. Besides, the number of these developments is so high that people cannot follow them. Thus, those who are still adapting and using these technological developments, instantly face with other developments. Digital natives are eager to adapt to these technologies whereas a significant percentage of digital migrants usually fail to follow or even use them.

New communication technologies/environments have changed the rules of the game. Reading and writing rules are changing and new communication languages are derived. New media, which shapes the senses and provides radical changes, has continued to change the lifestyles, communication processes and many other dynamics of people (Yengin, 2014: 175). Scientists, studying the human brain's thinking system, have found that the human brain in the 20th century where the old media is at the peak and the human brain in the 21st century where the new media of the digital environment is dominant work quite differently (Turgut, 2013: 31). This explains the difference between digital immigrants and digital natives in the adaptation to digital technologies. In other words, new media and new communication tools have caused the working principle of the human brain to change. The human brain has learned to act with the internet and solve problems with new methods in order to meet the practices that people need. Briefly, it has adapted to this digital era.

It has also been argued that the heavy use of digital media actually restructures the human brain and makes it more suited to superficial skimming and scanning than the in-depth concentrated reading associated with traditional printed books (Carr, 2010).

In addition to all these developments and progresses, communication technology and other technologies have gradually entered human life with developing, renewing, and increasing processes. In particular, the developments in web technology are following one another and virtuality reaches different dimensions. Experts are designing Web 4.0 (intelligent web) technology, which will be the future communication and interaction environment after Web 1, Web 2, and Web 3 (semantic web) that is currently being developed and prepared to serve, used in human life.

In this regard, several different opinions have begun to emerge. First, it is stated that there is a world where people have avatars and can travel through these virtual environments with the introduction of the term "Augmented Reality" to the web. In order to have more information about a friend on Facebook, it is in a virtual world where people visit the created virtual house of their friends with their virtual characters and gather the information uploaded about the virtual characters of their friends without being online. Another Web 4.0 prediction is the structure that the operating systems called Web OS are also in the cloud environments and the transition from the Semantic Web to the Intelligent Web based on the concept of customer belonging will be planned (Esgin, 2015: 44).

Another technological development, which is designing and enters human life in the future, is the information transfer. Thanks to new technologies, the news flow will be gathered in one spot and it will be translated into the desired language perfectly. In addition to the news, people, who want to communicate and interact in different ways, will be able to exchange information and interact with each other through devices designed to be integrated into their bodies. The new communication technology that Google has designed is one of the best-known examples of this. With its new technology, Google makes simultaneous translations in 40 languages and it can transfer the translated speech of people to you in your desired language. Face-to-face communication will reach a quite different dimension with Google's new technology.

Alongside the new Pixel 2 smartphones Google unveiled in 5 October 2017, the company also launched a set of Bluetooth earbuds called the Pixel Buds with one standout feature: instant translation between 40 different languages using a Pixel smartphone. In a live demo on stage, the Pixel Buds were shown translating short phrases back and forth between English and Swedish using Google Translate running on a Pixel 2 smartphone. In many ways, the machine-learning and artificial intelligence expansion have pushed what we thought was possible out of the realms of science fiction into reality. None of the currently available technologies and gadgets are quite the true Babel fish in your ear, but with the Pixel Buds and the Bragi Dash Pro, the Hitchhiker's Guide to the Galaxy dream of a seamless removal of language barriers is closer than ever (Gibbs, 2017).

Another technology that will accelerate communication and information transfer is Li-Fi technology, which is planned to enter our life in the future. Wi-Fi technology, which has been used for many years under the name of wireless internet, is predicted to evolve into a system called Li-Fi. Li-Fi technology provides 100 times faster internet access compared to Wi-Fi. In addition to that, Li-Fi is much faster than Wi-Fi in transferring photos, videos, and texts. Li-Fi known as the new internet connection technology uses visible light spectrum to transmit data instead of using radio waves. Li-Fi basically known as "LIGHT FIDELITY" is an outcome of 21st century. The basic idea behind this technology is that the data can be transmitted through LED light whose intensity varies even faster than the human eye. As the transmission of the data takes place through the light emitting diodes (LEDs), the amount is comparatively small. In modern times, it is called as the optimized version of WI-FI. The advantageous thing is the wireless communication which decreases the cost enormously (Poonam and Siddiqui, 2014: 8031).

With the evolution of technology to micro structures, it is foreseen that there will be many revolutionary changes in communication technologies and environments. Professions such as cameramen and film editor will disappear and everybody will be able to record in any environment, thanks to the camera integrated into his body. In addition to this, it is also foreseen that hologram technologies will become popular among people and three-dimensional view will become a routine of daily life instead of two-dimensional recording or two-dimensional image view.

Beyond that, it can be said that the physical structure of the communication and the media tools/environments will almost disappear in the future. Communication, recording, and storage will not occur through large devices (desktop computer, laptop, voice recorder, telephone, camera, external disk, etc.) but micro devices integrated into the human body. For example, images can be recorded with the lens attached to the human eye (the eye will function as a camera), the functions such as recording, stopping, zooming can be done with blink of an eye, and recorded images will be stored with the microchips on the bodies. In other words, the human body will evolve into a structure integrated with technological tools. With the developing digital technologies, the communication tools in the micro structure will be preferred in terms of being suitable for daily use, easy to use and having more functional structures.

Technology is cheaper, more diverse and easier to access than the old one. In the future, it will become cheaper, more diverse and easier to access. Besides, the ability to use these technologies will increase. This will also increase the dependence of daily life on the Internet and digital technologies. At the same time, the percentage of using these technologies to do daily need/works will rise. According to a Cisco report prepared by Evans (2011), while the number of things connected to the internet in 2015 are 25 billion, this number will reach 50 billion in 2020. The number of physical objects that are currently connected to the Internet is less in general. In the foreseeable future, the things, which will connect on the internet, will include many different fields such as agriculture, environment, and education. By installing small sensors in the devices, many things have started to be controlled electronically. Smart houses and trash cans, bridges and buildings use these technologies to enable information exchange between the internet and objects. In the future, this kind of use is thought to include the whole society.

In the light of all these, it has been tried to understand what kinds of structures the new communication technologies are in the Digimodern era and how they will evolve in the future. Present communication technologies/digital technologies offer many possibilities/convenience to people. It is an unquestionable fact that these technologies shaped human life will appear in future life.

Conclusion

Influential social, political, economic, or technological events or developments have emerged in every period of history. The invention of writing, the French revolution, the industrial revolution, the invention of the printing press, the world wars, the birth of modern mass media and the internet are a few of the most important advancements.

These developments, which took place in pre-modern, modern, or post-modern eras, deeply affected societies and caused radical changes. Mass media has always been at the center of these developments since their invention. As mass media such as newspapers, radio, photography, and cinema have the characteristics to direct societies and wars, the states have given the necessary attention to these communication tools and these mass media have constantly been used to manipulate people.

However, the rapid development of communication technologies since the second half of the 20th century has a different importance than the previous eras. This era has a different structure from those of the periods when the mass media are merely propaganda-oriented or in slow moving process. Since this period, mass media have begun to be managed and run by professional people. At the same time, mass media have become cheaper and are being followed by the vast majority of the population. Moreover, these communication tools have evolved into a digital structure, thanks to the technological developments in that era. These developments eventually form a basis for the transformation of the life practices of people.

The most popular and important of these technologies is obviously the internet. It affects the society in terms of social, cultural, and economic aspects. People have started to practise most of their daily activities in this virtual network. The Internet has caused many radical changes in the field of communication as well as many other fields. News sites, blogs, social networks, web TVs, and several new communication environments have emerged. Today, these new communication environments are called new media.

The new media, which allows users to create and share content, includes socilization, cooperation and bi-directional interaction. While this makes it easier for users to communicate with each other, it promotes sharing and encourages people to build communities. Today, users can communicate, share, and socialize with many different social media platforms. Communication has become more and more virtual, thanks to these digital communication environments. This virtual environment begins to differentiate with the media or era called traditional or old in a short period of time.

As a result of this, a new era emerges, thanks to the internet and new media, the most important communication environment of it. The digital modern era called digimodernism includes this era because the internet has developed, used by large mass and involved in almost every aspect of life. Digimodernism constitutes a new culture and therefore, this new culture has created the virtual identities.

The dominant culture of Digimodern era is called virtual culture or internet culture. This virtual culture has created virtual individuals, the digital natives, and influenced the digital immigrants. Digital natives born in the internet world and influenced by digital culture, conduct their social relations and studies, communication, shopping, and even relationships through this virtual network. In addition to this generation called digital natives, other people have also used the internet as a routine of daily life.

As mentioned in the above paragraphs, the internet has now penetrated almost all aspects of human life. One of the most important aspects is the mass media where people meet their needs such as reaching news and information, entertaining, spending time. The Internet has transformed the communication tools known as the traditional media, and integrated these tools into it, and it also constitutes the media called new media that includes new communication environments.

New media and other digital communication technologies, the media of the Digimodern era, are at the center of dominant communication tools. In the future, these digital communication technologies will have more distinct and advanced structures. Analyzing the evolution of digital communication technologies, it is possible to say that these technologies will integrate into the human body in the future and virtuality will cover almost the entire human life. Just as in the past and present, media and new communication environments will also constitute their own mass and era in the future.

References

Andersen, M. P. (1959). What is Communication. Journal of Communication, Vol 9, No 5, p. 183–191.

Balle, F., Eymery, G. (1995). Yeni Medyalar. İstanbul: İletişim Yayınları.

Barnes, S. B. (2006). A Privacy Paradox: Social Networking in the United States. First Monday, Vol 9, No 4, p. 1–12.

Bunnell, N. (2015). Oscillating from a Distance: A Study of Metamodernism In Theory and Practice. Undergraduate Journal of Humanistic Studies, Vol 1, p. 1–8.

Carr, N. G. (2010). The Shallows: What the Internet is Doing to Our Brains. New York: W. W. Norton.

Castells, M. (2013). İsyan ve Umut Ağları - İnternet Çağında Toplumsal Hareketler. İstanbul: Koç Üniversitesi Yayınları.

Conner, D. (2013). Communicating The #Gospel Message Within A Post-Postmodern Digitally Connected Culture. Global Missiology English, Vol 1, No 11, p. 1–28.

Cross, M. (2011). Bloggerati, Twitterati: How Blogs and Twitter Are Transforming Popular Culture. Oxford: Praeger.

Crowley D., Heyer P. (2011). İletişim Tarihi - Teknoloji, Kültür, Toplum. Ankara: Siyasal Kitabevi.

Dijk, V. D. (2006). The Network Society. London: Sage Publication.

Erdoğan, İ., Alemdar, K. (2005). Öteki Kuram, Kitle İletişim Kuram ve Araştırmalarının Tarihsel ve Eleştirel Bir Değerlendirmesi. Ankara: Erk Yayınevi.

Esgin, M. S. (2015). Web Teknolojileri Dünyasında Yeni Trendler. A. Baloğlu (Ed.). Sosyal Medya Madenciliği (s. 17–59). İstanbul: Beta Yayıncılık.

Evans, D. (2011). The Internet of Things How the Next Evolution of the Internet Is Changing Everything. https://www.cisco.com/c/dam/en_us/about/ac79/docs/innov/IoT_IBSG_0411FINAL.pdf, Access Date: 12.08.2017.

Firat, F. A., Dholakia, N. (2004). Theoretical and Philosophical Implications of Postmodern Debates: Some Challenges to Modern Marketing. Marketing Theory, Vol 6, No 2, p. 123–162.

Firkins, S. (2015). Curvism: Beyond Postmodernism. http://www.curvism.com/beyondpostmodernism-writen/, Access Date: 10.08.2016.

Fuchs, C. (2011). An Alternative View of Privacy on Facebook. Information, Vol 2, p. 140–165.

Gibbs, S. (2017). Google Pixel Buds: is Babel fish dream of in-ear translation now a reality? https://www.theguardian.com/technology/2017/oct/05/

google-pixel-buds-babel-fish-translation-in-ear-ai-wireless-language, Access Date: 12.10.2017.

Güngör, N. (2011). İletişim Kuramlar ve Yaklaşımlar. Ankara: Siyasal Kitabevi.

Işık, M. (2017). İletişim Bilimine Giriş. Konya: Eğitim Yayınevi.

Kara, T. (2013). Sosyal Medya Endüstrisi - İnsan, Toplum, Ekonomi. İstanbul: Beta Yayıncılık.

Kirby, A. (2006). The Death of Postmodernism and Beyond. Philosophy Now, Vol 58, p. 34–38.

Kirby, A. (2009). Digimodernism: How New Technologies Dismantle the Postmodern and Reconfigure Our Culture. London: The Continuum International Publishing Group.

Konsa, K., Hermann E. (2014). The Logic of Text and Structure of Books: Changing the Functions. M. Lauristin and P. Vihalemm (Ed.). Reading in Changing Society (p. 28–38). Tartu: University of Tartu Press.

McDonald, E. (2017). The Global Games Market Will Reach $108.9 Billion in 2017 With Mobile Taking 42%. https://newzoo.com/insights/articles/the-global-games-market-will-reach-108-9-billion-in-2017-with-mobile-taking-42/, Access Date: 23.09.2017.

Mulady, L. (2010). Digimodernism; the Future is Now! http://pure.au.dk/portal/files/13871/Final_Paper_for_Upload.pdf, Access Date: 03.12.2016.

Oskay, Ü. (1992). İletişimin Abc'si. İstanbul: Simavi Yayınları.

Önür, N. (2002). Küreselleşen Dünyada İletişim ve Toplum. Ankara: Alp Yayınları.

Palfrey, J., Gasser, U. (2008). Born Digital: Understanding the First Generation of Digital Natives. New York: Basic Books.

Poonam, P., Siddiqui, S. (2014). Li-Fi Technology. International Journal of Computer Science and Information Technologies, Vol 5, No 6, p. 8031–8032.

Stoddart, E. (2014). (In)visibility Before Privacy: A Theological Ethics of Surveillance as Social Sorting. Studies in Christian Ethics, Vol 27, No 1, p. 33–49.

Turgut, S. (2013). Yeni Medya. İstanbul: Destek Yayınevi.

Valentini, C., Kruckeberg, D. (2011). Public Relations and Trust in Contemporary Global Society: A Luhmannian Perspective of the Role of Public Relations in Enhancing Trust among Social Systems. Central European Journal of Communication, Vol 4, No 1, p. 91–107.

Valentini, C., Kruckeberg, D., Starck, K. (2012). Public Relations and Community: A Persistent Covenant. Public Relations Review, Vol 38, p. 873–879.

Vural, B. A. (2002). Information Communication Technologies and Change: Human Resources, Society and Organizational Perspective. İzmir: Ege Üniversitesi İletişim Fakültesi Yayınları.

Wearesocial (2017). Digital in 2017: Global Overview. https://wearesocial.com/special-reports/digital-in-2017-global-overview, Access Date: 03.10.2017.

Yengin, D. (2014). Yeni Medya ve Dokunmatik Toplum. İstanbul: Derin Yayınları.

Ceren Yegen

Digivoyeurism: Follower Monitoring via Applications

Introduction

Voyeurism refers to watch beyond one's knowledge (often during stripping or sexual act). Voyeurism, which is also called a kind of paraphilia in the literature (Ertekin et al., 2013), nowadays falls into a digital form that individuals/users follow each other via innovations in communication technologies (their actions, their sharing, location information, etc.) in new internet-based media environments. In this regard, the voyeurism, which is accompanied with exhibitionism especially in social networking sites (Facebook, Twitter, etc.) and various applications (Instagram, Snapchat, WhatsApp, etc.) is discussed in the new media literature in terms of privacy. Companies with their applications developed in the new media market like researchers, who work in the new media field, also contribute to the agenda of Voyeurism with various applications. For instance, Instagram – one of the applications that is used by individuals extensively, is one of the areas in which voyeurism is actively seen today and is also being discussed in terms of various applications that are developed for monitoring/tracking of the individuals by their followers. Therefore, the subject of this study is to understand the digitalized form of voyeurism via various applications. For this reason, it is aimed to understand the structure of follower monitoring via Reports + application, which allows follower monitoring and the changing structure of voyeurism in communication technologies. For this purpose, a field survey for Instagram users will be carried out by using a netnographic method in the study, the users will be interviewed and the research will be analyzed on the basis of the application possibilities and structure by Reports +. In this respect, 6 (six) Instagram users will be conducted by in-depth interviews. In the theoretical part, contributions are expected to be made to the studies on new media with the study, which is expected to be discussed on interactive technologies, new media, voyeurism, and exhibitionism.

Voyeurism

Voyeurism refers to the monitoring of someone else (İncesu, 2004; Çakan, 2017: 17). According to some approaches, voyeurism is a type of paraphilia and "repetitive affairs with fantasies and activities involving monitoring naked or undressed, dressed or sexually active people". Voyeurism may be accompanied with fetishism, which is another paraphilic condition and "intense, sexually stimulating fantasies about the use of inanimate objects by a person, sexual repulsions or behaviors appearing in a repetitive manner and claimed to be mostly seen by men" (Ertekin vd., 2013: 70–71). Events which are reflected in the justice by perverse actions such as exhibitionism, monitoring, fetishism, and transvestism in Turkey are assessed with the Turkish Penal Code (TCK) (Geyran and Uygur, 1996: 47) and constitute with crime (O'Donohue, 2014; Postmust, 2013).

Voyeurism practices can be of different types, and the actor who is monitoring may not be in direct interaction with the object of voyeurism. Objects usually do not know that they are under observation; they behave quite routinely and naturally (Sidi and Midin, 2008).

The etiology of voyeurism is not fully understood, and its causes are thought to be multifaceted. According to some approaches, particular causes appear to be a combination of biological, social, and cultural determinants (Green, 2017). Voyeurism, in short the observation of a person who is undressing or naked, may also be associated with obsessive-compulsive spectrum disorder (OCD). Voyeurism emerges in situations involving the observation of an innocent person who is naked or involved in sexual activity during a period of at least 6 months (Sidi and Midin, 2008).

Voyeurism is the result of a verb in French which is translated into English with -eur suffix. In voyeurism, the voyeur is generally a male, but at the same time many women are claimed to be enjoying voyeurism. Voyeurism can take place in some institutions like gyms and schools. In this respect, some measures have been taken from time to time for the usage of mobile phones in such environments and institutions. In recent days there has been a numerical increase in voyeurism due to the privacy problems caused by the innovations in communication technologies. In Muslim countries like Saudi Arabia, camera mobiles are a big problem in this respect, even introduce a ban together with (Sidi and Midin, 2008).

The results of a research on university students are important to determine whether "normal" people are engaged in voyeurism or not. Because the individuals have reported that they are willing to observe someone rather than their fear of being caught. That is to say, social, economic and cultural situation and statute in the related action are the sharp determinants or directors (Rye and Meaney, 2005).

Today, voyeuristic actions have various forms. For example, using a person's personal computer, or personal phone, monitoring and scratching is also a voyeuristic action. Because, it is a violation of privacy and is wrong (Holmes et al., 1998: 17–18) to access special things from the individual's personal information, credit card information, or personal photographs/videos.

New Media Geneology

Today, new media geneology (genealogy) remains on the agenda with many concepts. With the development of the internet, new concepts and approaches have emerged in the field almost every day, and most of the studies on new media are about related concepts and approaches. Since the importance of new media technologies, citizens have entered into a process that is both cultural and social as well as technological and become a part of the process; many new changes have been made through new models; media-based cultures and approaches have been formed (Burgess, 2007; Burri, 2012; Carrabine, 2014).

As Manovich states, the new media (Lazer et al., 2017: 1–7), which is accelerated in areas like Facebook and Google, but a democratic platform in which fake news and content is also visible, may not be synonymous with digital media, but they can often overlap. Because both concepts cover processes related to digital and new fields. In these areas interaction, speed, and concurrency are essential (Grieb, 2003).

Although the Internet has reached a level between 80 % and 90 % in the West, there is a serious digital divide (Friemel, 2014: 1). The Internet and new media environments, which cover most of the daily lives of the individuals, are now being discussed with negative aspects as well as the positive aspects. Today, especially digital natives (Marsh et al., 2005; Mills et al., 2015) have made the internet a part of their daily lives.

The Internet involves a lot of needs, especially for young people, from getting information to having fun. The search engine Google is today the most frequent and highly referenced consulting tool, and also differentiates traditional publishing (Bermejo, 2009; Frandsen, 2012). It is important to note that the report titled "Social Media Prison Internet Addiction Detention Camp" in the Show TV prime-time newscaster on November, 2017, pointed out that there is an environment for the rehabilitation of young people who are claimed to be Internet addicts in China. Because, this environment, which is interpreted as a "social media prison," takes the individuals, who cannot live without social media and games in a digital detox camp. In this environment where telephone, computer, and tablets are prohibited, a discipline like military camp and a uniform dominates. In this camp, which begins at 6 in the morning, sports and hard trainings are held, and the time for bed is at 21:30. In majority of the camps there are young people who are forced to camp by their families. This camp is an example of how the internet and new media environments have changed the individuals, their daily practices, their habits, and even their moods.

Recent changes in the media environment caused by digital media also affect media organizations in many ways. Journalism, one of the founding institutions of the media, is undergoing significant changes parallel to new conditions. Current research in journalism generally focuses on how journalism reacts to new media and how journalism challenges new conditions (Frandsen, 2012: 5–6).

Voyeurism and Exhibitionism in the New Media Age

The e-mail, internet and various digital accelerated over the last decade are important arguments for researchers studying on the new media field. Researchers use a variety of methodological strategies to discover social, political, and cultural events related to the growth of these innovations, and they have reached important data (Schneider and Foot, 2004). New media has created new environments, structures, and systems in many parts of the world (Morris et al., 2015; Schäfer and Taddicken, 2015).

The public space, namely the surface, appears as an exhibition area. The public space as an exhibition area is transforming a public space, while the private space defined by privacy refers to a place where there is slowly changing and difficult to reach (Fındıklı, 2016: 109). Nowadays, since the new media is also a public space, exhibitionism, which is a kind of voyeurism is practiced in this field (Çakan, 2017: 16–17). Because social network sites and many applications also allow exhibitionism like voyeurism, for this reason individuals share many things about their daily life, their experiences, feelings on social networks, and some others are monitoring them as well. Especially in the world age, Facebook with huge number of users defines an important area in this regard. A study indicates that voyeurism and exhibitionism on Facebook are carried out in order to satisfy the social and emotional needs of the individual. Facebook is interpreted as an important, easily accessible and usable source of motivation. Because Facebook allows individual the opportunity of introducing, reflecting, and expressing oneself online. However, we also see that Facebook users can browse other users' information based on their privacy settings. Users have the right to decide what level of information about themselves are accessible by others (Ju Su, 2012: 1–9). It is also noted that some users' profiles are public in order to be seen and recognized.

Like voyeurism, the etiology of exhibitionism is quite uncertain. Possible causes include situations like exposure of sexual arousal, stress, and anger. Again like voyeurism, there may not be any direct physical contact with the victim in exhibitionism (Green, 2017).

A research (Baruh, 2010) indicates that voyeurism is related to "having a curiosity secretly". However, it must be indicated that there are many claims about the media, especially television, which is one of its important tools, makes a voyeuristic call (Hammer, 2016: 89–90 cited from Baruh, 2010). Today, social media environments such as television are also calling for voyeurism and exhibitionism. Furthermore, celebrities or people phenomenon are exhibiting themselves on social media environments or various applications, and many individuals are monitoring by wondering about them. For example, the Kardashian family, who are always on the agenda by their practices, can be given as an example (Hammer, 2016: 95).

Follower Monitoring As a Voyeuristic Practice

The media is an important and effective measure of voyeurism and exhibitionism (Yesil, 2001). Today, new media together with digital environments, especially social media has been added to the approaches that claim television mediates voyeurism (Bagdasarov et al., 2010). Moreover, there are current approaches which claim that voyeurism is a privacy issue in the social media (Yurdigül and Elitaş, 2015). According to these approaches, the disclosure of a lot of things in the social media negatively affects privacy.

Social media today has developed a digital voyeurism that comes in different ways from television (in the context of television programs, series, etc.), which allows voyeurism and exhibitionism. Individuals socialize by watching/monitoring each other via social media and by exhibiting themselves (their lives, experiences, feelings, etc.). Relevant socialization today is a habit for social sharing sites like Facebook, Twitter, Instagram, and applications such as Snapchat, Periscope, and WhatsApp.

Voyeurism must be seen in the context of men and women (Soldin, 2015). When voyeurism is discussed in social media, it is not right to make a gender discrimination in terms of voyeur. Because both sexes have voyeuristic practice, especially via social network sites. However, it can be considered that voyeurism is a feeling, impulse presentation (Rudinow, 1979). Because individuals often follow each other on social media for these reasons like jealousy, getting intense and up-to-date information and news, and curiosity.

However, there are some matters of curiosity like who followed whom, which sharings are liked, and who unfollowed whom in social media and some applications. Especially Instagram is an important tool in order to get rid of these related needs. So this impulse met new terms to the field like “stalk”, “stalker”, which characterize someone who secretly observes someone and who is doing this action.

The practice of famous people and phenomena on Instagram has also led individuals to become intensive Instagram users. Indeed, individuals now buy followers or likes, and sellers of follower like and apps are derived from in order to be popular identities, Instagram phenomena or celebrities (Güz and Yegen, 2017).

Instagram and Digital Voyeurism

Instagram is a relatively new communication channel that allows users to share photos and videos. Since its introduction in October, 2010, the number of users of Instagram worldwide has grown rapidly (Alkhowaiter, 2016), and it is now more than a social-sharing platform in many places. Instagram is now a business, fashion, shopping, and social environment/tool. Social media, especially Instagram, which meets the needs of individuals to socialize, entertain, spend time, and acquire information, is highly preferred by individuals because of their freedom of self-expression (Kale, 2016: 121).

Instagram, like Snapchat, which is defined as a rapidly rising social network application, is being used as a public relation and marketing tool because of its popularity by businesses and marketers today. Celebrities or those, who want to be popular, also use both of these applications in particular (Lim, 2017: 1–15).

There are also a number of problems in Instagram, which is a very popular online photo sharing platform. Along with the problems of privacy, violation of privacy also becomes a serious concern on Instagram. Recent statistics show that between 9 % and 25 % of users are bullyragged on Instagram, which is even more common on Twitter and Facebook (Hao Li et al., 2016: 3952).

Individuals/users today are following, monitoring, exhibiting themselves or others, and introducing digital and new style of voyeurism and exhibitionism via Instagram. In this respect, Instagram, which has emerged as a photo sharing platform and getting popular day by day, can be interpreted as a form of digital voyeurism like Facebook and Twitter, etc.

An Application That Allows Digital Voyageism: Reports +

Intensive use and widespread adoption of Instagram has made it possible to put various applications into the market. One of these applications is Reports +, which will be examined as a sample in this study. Because the application allows individuals to have digital voyeuristic practices and has been developed for this purpose. However, there are different applications in Instagram such as *Followers Insight, who looked at my Instagram profile* and *Followers - Social Analytics For Instagram* that are developed on behalf of follower monitoring for Instagram. As mobile devices become the focus of users, publishers are trying to enter into their lives with different alternatives, *InstaTakip For Instagram*, for instance, is an application that offers follow-up options for Instagram that is the most popular social media photo sharing platform. The application is an average of three years old, free and promising to be a popular social environment for people to share their photos with friends and to give them an edge in popular racing by multiplying their surroundings and enjoyment by gaining extra follower. The application allows you to connect to the system by logging in username and password from your personal Instagram account. It is provided with the application that the number of followers is increased, and the shared photographs also receive a lot of likes (http://instatakip-for-instagram.indir.com/Android, 05.11.2017).

The application Reports +, which will be examined as a sample in this study (Fig. 1), is a Turkish initiative and developed for iOS.

App Store, which is a virtual marketplace of Apple that allows to download applications, MobileX Software Open Company Yaşar Utku Anıl Şimşek and his partner is known to be the developer of the application. The size of the continuously updated app is 27.2 MB. The age limit is 4+ and it is emphasized that the application requires iOS 9.0 or higher version by November 2017. Application languages are: Turkish, German, French, Portuguese, Russian, English, Spanish, and Italian. The application's privacy policy, update history, developer website, and developer applications are also easily accessible. The application is entered with Instagram username and password. When you enter the application you can see the users who unfollow and those who visit your profile. The application sends a notification to the user when he/she is unfollowed by another user (Fig. 2).

Reports + charts also allow the user to see the increase or decrease of followers systematically and quantitatively (Fig. 3).

The app has also been commented on by the users on the App Store. It is also observed that comments about the application are generally positive (Fig. 4).

A Netnographical Analysis Specific to Reports+. In this section, a field survey was conducted for Instagram users via a netnographical method. The netnographical method is used in new-media-based studies and allows online research, interviews, data collection (Cengiz and Başaran, 2016). In this regard, in-depth interviews with Instagram users were conducted in terms of follower monitoring and the data were recorded.

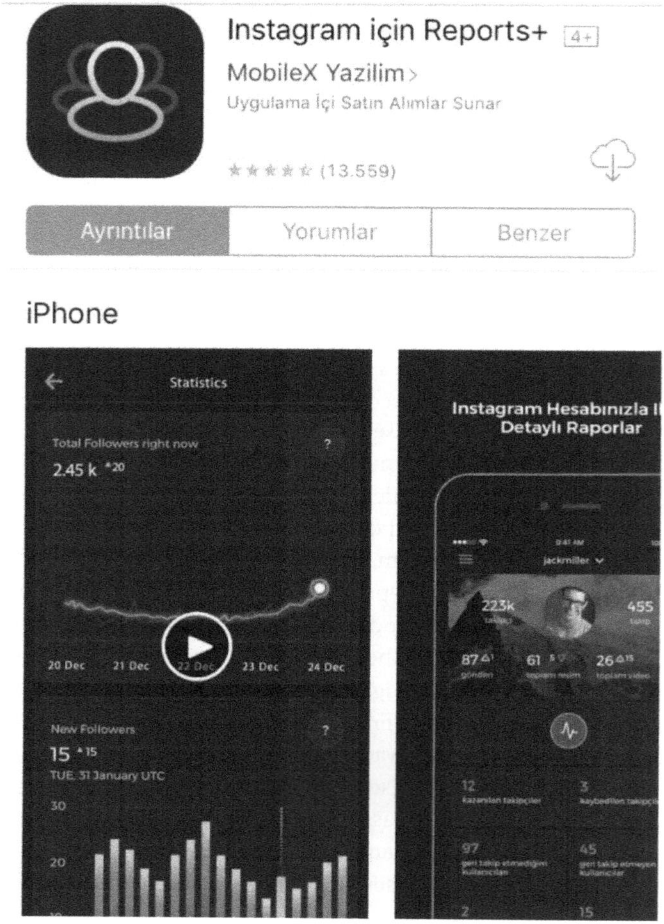

Fig. 1: App Store'da Reports +

Interviews were conducted face to face and through Instagram application. Users are selected with a simple random sample between the students who use Instagram study in various departments of Faculty of Communication in Gazi University. The sample was limited to 6 students in order to access limited data in the study.

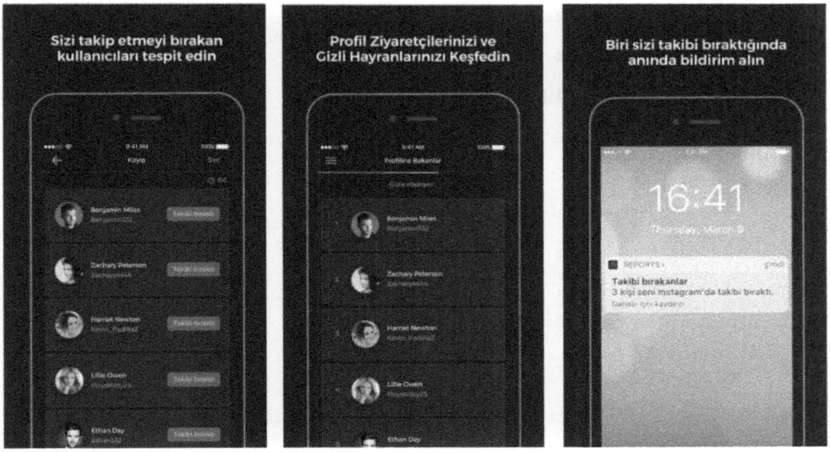

Fig. 2: *Reports*+ Possibilities

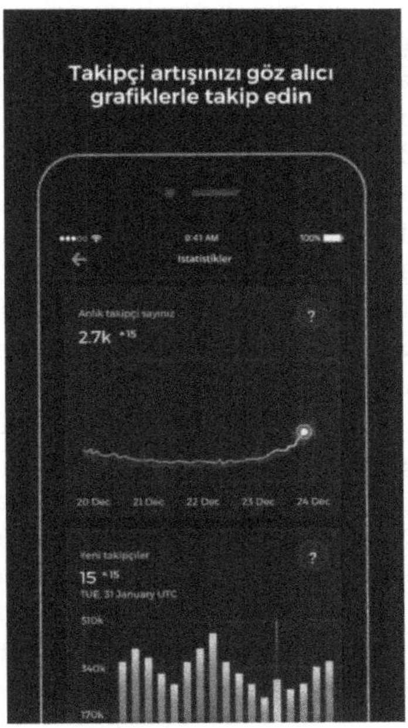

Fig. 3: Reports + Sample Graph

Fig. 4: Reports + Comments on the App Store

The questions asked in the interviews are as follows: How old are you? In which department are you studying? What is the monthly average income of your family? Do you use Instagram for what purpose/purposes? Does Instagram meet your social needs? Do you have any idea about follow-up Reports + Follower monitoring application? Do you use this application? What do you think about follower monitoring on Instagram? Do you wonder who is following you or unfollowing you? In this respect, do you often check your Instagram account? What are the reasons for this impulse?

The names and surnames of the users are hidden for the ethical requirements; the subjects, the Instagram users who are interviewed are: User 1, User 2, User 3, User 4, User 5, and User 6. Interviews are as follows:

User 1 reported that he was 20 years old and was studying in the Journalism Department. According to his declaration, the average monthly income of his family is between 3501 TL and 5000 TL. User 1 said that he used Instagram to share photos/videos and communicate with the social environment. According to him, Instagram partially meets the communicative social needs. User 1 has no idea about the Reports + follower monitoring program and User 1 does not use this application. According to User 1, follower monitoring on the Instagram is unnecessary and is caused by too much free time. User 1 reports that he does not wonder who follow or unfollow him, so he does not often check his account, followers, and his follows. According to User 1, the reason of this impulse is curiosity, but an unwarranted curiosity.

User 2 reports that he is 19 years old and studying in the Journalism Department. According to his statement, the average monthly income of his family is between 2001 TL and 3500 TL. User 2 reported that he uses Instagram to upload photos and display them. According to him, Instagram is an environment that meets many social needs. User 2 has no idea about the Reports + follower monitoring program and he is not using this application. According to User 2, monitoring the follower on Instagram is something interesting and the user does not realize why. User 2, who is not curious about who follows or unfollows him, and he thinks has to lean this is an interesting impulse to consider.

User 3 is 22 years old and reports that he is studying in the department of Public Relations and Publicity. According to the statement of User 3, the monthly income of the family is between 2001 TL and 3500 TL. He reported that he uses Instagram to communicate with his friends. According to him, Instagram meets some of the social needs of an individual. User 3 has no idea about the Reports + follower monitoring program and he does not use this application. According to him, follower monitoring on Instagram is an unnecessary action. User 3 reports that he is not curious about who follows or unfollows his account. Because "when someone sends you a message, when you send a message, when you like a share, or when you share a comment, you are already notified by the application and you become aware of them. User 3, states that he does not need this action, saying "accept it, reject it or ignore it, it is up to you." Therefore, he does not check his account frequently, and according to his statement, those who are involved in this action are usually adolescents with curiosity.

User 4 is 21 years old and reports that he is studying in the department of Radio, Television and Cinema. According to his statement, the monthly income of his family is between 3501 TL and 5000 TL. User 4 uses Instagram to maintain communication with his surroundings and to share things about his life. According to User 4, Instagram is a platform that meets many social needs. User 4 has no idea about the Reports + follower monitoring program and does not use this application. On the other hand, User 4 states that he is using another application that allows follower monitoring, and expresses the name of the application is *SOCIALTOOL*. According to User 4, it is important to know who follows or unfollows him. Thus, User 4, who states that he is using the program *SOCIALTOOL*, thinks that this action is increasingly directly proportional to the popularity of Instagram. According to him, it is important how people have positioned each other in this created area (Instagram account) during communication. Because according to him, if a person puts you on pedestal, he/she will follow you. If not, he/she will unfollow you, will not need to follow or interact with you. User 4 indicates that he frequently checks his account. According to him, the reason of this impulse is the desire to see value and curiosity.

User 5 is 22 years old and says that he is studying in the department of Radio, Television, and Cinema. The monthly income of his family is minimum wage (in Turkey the new minimum wage is determined as monthly gross salary 1.777,50 TL, monthly net salary as 1.404,06 TL, monthly gross is determined as 59,25 TL per day). User 5 uses Instagram to communicate with his surroundings and family, sometimes to share something and to be aware of the new media agenda. According to him, Instagram is an application that meets social and popular needs. User 5 has no idea about the Reports + follower monitoring program and does not use this application. According to User 5, it does not matter who follows or unfollows him. The user indicates that he sometimes wonders who follows him (he gets attention when he gets a notification) and he does not make any extra effort to see who unfollows. According to User 5, Instagram has become so popular in recent months that it is unlikely to try to stay away from things related to the application or to stay away from it. User 5 says: "Instagram has become one of the indispensables of daily life via Instagram phenomena with high number of followers. There are even people like them who want to have a lot of followers, even I have friends, who follows people he/she does not know."

User 6 reports that he is 20 years old and he is studying in the department of Journalism. According to his statement, the monthly income of the family is between 3501 TL and 5000 TL. User 6 reports that he uses Instagram to share photos and communicate with friends. According to User 6, Instagram meets a number of social needs of the individual. For example, people communicate with their friends who have not talked for many years so that they can fulfill their longing, make themselves happy by sharing and even maybe they are satisfying their egos. User 6 has no idea about the follower monitoring program called Reports + and User 6 does not use this application. According to him, it does not matter who follows or unfollows him. In this regard, he, who says that he does not check his account often, thinks that a person checks the followers or who unfollows in order to measure his own value in the eyes of the individual and this impulse is unreasonable to him.

Conclusion and Evaluation

It is possible to say that new media technologies digitize the world, life, habits day by day. The era in which we live is discussed no longer digital but post-digitally is also discussed on the basis of the involving of the new media densely. So now digital words have begun to be added at the beginning or ending of concepts.

Furthermore, people are to be a part of the new media process, which they are thinking about and problematizing locally (in the context of negative effects and professionalism towards the traditional media). In this study, this concept is also tried to be explained by the conceptualization of the digivoyeurism. Because in the study it is analyzed how the voyeurism and the exhibitionism is evolved in the process of new media and how digitalization is realized via social networks and various applications.

It is seen in the study that Facebook, Twitter, Instagram, etc., are the virtual media and applications in which individuals show themselves via sharing and monitoring somebodies. New media environments that offer numerous possibilities to users are the basis of current and digital format of voyeurism and exhibitionism.

During the in-depth interviews with Instagram users, it is observed that Instagram is generally used to share and to connect/communicate with the surroundings. Monthly income of their parents is generally between 3501 TL and 5000 TL, 3 of the students/users are in the department of journalism, 2 in Radio, Television and Cinema and 1 in Public Relations and Publicity. That is to say, the most participants are in the department of Journalism. The least participants are in the department of Public Relations and Publicity.

Most of the participants (total of 5 participants) do not have any idea about the Reports + follower monitoring program and most of the participants also state that they do not use any follower monitoring programs. On the other hand, one of the participants indicates that he uses a follower monitoring application (an application called *SOCIALTOOL*) for Instagram. All of the participants agree that Instagram meets social needs.

It is important that the majority of the participants are not interested in who follows or unfollows their Instagram accounts. That means the average of an Instagram user often uses Instagram appropriately for its main purpose and does not interpret the application for different purposes. So most of the users do not check whether they are followed or not in their Instagram accounts. The number of users who sometimes check their account is 1, while one of the participants checks his account in this regard. According to the majority of participants, the need to check whether they are followed or not is not very clear, and at this point the participant's common guess is curiosity and one's self-esteem. However, social networks are not the places where the value, emotion, or status of the individual can be determined and measured, and it is not clear to question the identity of the self, identity via social networks.

As a result, it is not known whether the reason is being followed of follow in the practice of Instagram, which become more popular day by day, it will be discussed later. As digital actions and impulses of the individuals to have identity, status, and value increases, innovations and developments will be added to the programs like Reports +, which are considered as a sample in this study and which allow followers to monitor. This action, which I have described in the form of Digivoyeurism, will probably evolve towards post digital areas. It should be noted that it is not Instagram, but the applications that enable the followers to take control of accounts, the likes and followers of accounts in the Instagram, that are digital voyeurs. However, at this point it is necessary to discuss Instagram as an application, in terms of exhibitionism/voyeurism.

References

Alkhowaiter, W. (2016). The Power of Instagram in Building Small Businesses Conference on e-Business, e-Services and e-Society, I3E 2016: Social Media: The Good, the Bad, and the Ugly pp. 59–64.

Apple (2017). *Reports+* Application.

Bagdasarov, Z.; Greene, K.; Banerjee, C. S.; Krcmar, M.; Yanovitzky, I. & Ruginyte, D. (2010). I Am What I Watch: Voyeurism, Sensation Seeking, and Television Viewing Patterns, *Journal of Broadcasting & Electronic Media*, 54 (2): 299–315.

Bermejo, F. (2009). Audience manufacture in historical perspective: from broadcasting to Google, *New Media Society*, 11 (1&2): 133–154.

Burgess, J. (2007). Vernacular Creativity and New Media, A Dissertation presented in fulfilment of requirements for the Degree of Doctor of Philosophy, Creative Industries Faculty Quennsland University of Technology.

Burri, M. (2012). Controlling New Media (without the Law) (November 19, 2011). *HANDBOOK OF MEDIA LAW AND POLICY*, Monroe Price & Stefaan Verhulst (Eds.), Routledge, pp. 327–342, NCCR Trade Regulation Working Paper No. 2011/71. Available at SSRN: https://ssrn.com/abstract=1961942.

Carrabine, E. (2014). Seeing things: Violence, voyeurism and the camera, *Theoretical Criminology*, 18 (2): 154–158.

Çakan, S. (2017). Avrupa İnsan Hakları Sözleşmesi Bağlamında Cinsel Suç Faillerinin Hadım Edilmesi, *Türkiye Barolar Birliği*, 129, pp. 11–36.

Cengiz, H. & Başaran, S. (2016). Cep Telefonu Kullanıcılarının Tüketici Tecrübelerinin Değerlendirilmesi: Online Tüketici Yorumları Üzerine Netnografik Bir İnceleme, *Tüketici ve Tüketim Araştırmaları*, 8 (1): 73–92.

Ertekin, H.; Özaydın, Y. H.; Eren, İ. & Ertekin, H. Y. (2013). Fetişizm ve voyörizm birlikteliği: Bir olgu sunumu, *Türk Aile Hek.*, 17 (2): 70–72.

Fındıklı, B. E. (2016). Uğur Tanyeli'nin Mimarlık Tarihyazımında Barınma Kültürünün, Özel ve Kamusal Alanın Jeneolojisi, *Tarihin Peşinde - International Journal of History and Social Research*, 15, pp. 95–128.

Frandsen, K. (2012). Sports broadcasting journalism and the challenge of new media, *MedieKultur*, 53, pp. 5–21.

Friemel, N. T. (2014). The digital divide has grown old: Determinants of a digital divide among, *New Media & Society*, 1–19.

Geyran, Ç. P. & Uygur, N. (1996). Saldırgan Cinsel Davranış Motivasyonu Olarak Perversif Davranış, Adli Psikiyatrik Değerlendirme, *Düşünen Adam*, 9 (4): 47–53.

Green, P. S. (2017). To See and Be Seen: Reconstructing The Law of Voyeurism and Exhibitionism, Draft/June 2017.

Grieb, M. (2003). Transformations of The (Silver) Screen: Film After New Media, A Dissertation Presented to The Graduate School of The University of Florida in Partial Fulfillment of the Requirements for the Degree of Doctor of Philosophy, University of Florida.

Güz, N. & Yegen, C. (2017). *Ölü Canlar*'dan *Instagram*'a: Kimlik ve Statü Sahibi Olmanın Allomorfizmi Üzerine, *İletişim Kuram ve Araştırma* - Gazi University, 45, pp. 103–123.

Hammer, K. (2016). Motives for Engaging with the Kardashians' Reality Television Family, *Elon Journal of Undergraduate Research in Communications*, 7 (1): 88–96.

Hao Li, Z. H.; Squicciarini, A.; Rajtmajer, S.; Griffin, C.; Miller, D. & Caragea, C. (2016). Content-Driven Detection of Cyberbullying on the Instagram Social Network, Proceedings of the Twenty-Fifth International Joint Conference on Artificial Intelligence (IJCAI-16).

Holmes, M. R.; Tewksbury, R. & Holmes, T. S. (1998). Hidden JPGs: A Functional Alternative to Voyeurism, *The Journal of Popular Culture*, 32 (3): 17–29.

İncesu, C. (2004). Cinsel İşlevler ve Cinsel İşlev Bozuklukları, *Klinik Psikiyatri*, 3:3–13.

InstaTakip For Instagram. (2017). http://instatakip-for-instagram.indir.com/android, 05.11.2017.

Ju Su, P. (2012). Mediated voyeurism on social networking sites: The possible social needs and potential motivations of the voyeurs on Facebook, Thesis, Rochester Institute of Technology.

Kale, Ö. G. (2016). Marka İletişiminde Instagram Kullanımı, *The Turkish Online Journal of Design, Art and Communication - TOJDAC* April 2016, 6 (2): 119–127.

Lazer, D.; Baum, M., Grinberg, N.; Friedland, L.; Joseph, K.; Hobbs, W. & Mattsson, C. (2017). Combating Fake News: An Agenda for Research and Action May 2017, Conference held February 17–18, 2017 Organized by Matthew Baum (Harvard), David Lazer (Northeastern), and Nicco Mele (Harvard), Northeastern University, Harvard University.

Lim, J. Y. (2017). Decision to use either Snapchat or Instagram for Most Powerful Celebrities, *Research Journal of the Institute for Public Relations*, (February 2017), © 2017 Institute for Public Relations, 3 (2).

Marsh, J.; Brooks, G.; Hughes, J. Ritchie, L.; Roberts, S. & Wright, K. (2005). Digital beginnings: Young children's use of popular culture, media and new Technologies, Report of the 'Young Children's Use of Popular Culture, Media and New Technologies' Study, funded by BBC Worldwide and the Esmée Fairbairn Foundation Literacy Research Centre University of Sheffield.

Mills, E. J.; Romeijn-Stout, E.; Campbell, C. & Koester, A. (2015). Results from the Young Children, New Media, and Libraries Survey What Did We Learn?, Summer 2015 Children and Libraries, pp. 26–35.

Morris, J.; Rulyova, N. & Strukov, V. (2015). *New Media in New Europe-Asia*, University of Glasgow, Routledge.

O'Donohue, T. W. (2014). (Eds.) *Case Studies in Sexual Deviance: Toward Evidence Based Practice*, Routledge, Taylor & Francis Group.

Postmust, L. J. (2013). (Eds.). *Sexual Violence And Abuse: An Encyclopedia of Prevention, Impacts, and Recovery*, ABC-CLIO.

Rudinow, J. (1979). Representation, Voyeurism, and the Vacant Point of View, Philosophy and Literature, Fall 1979, 3 (2): 173–186.

Rye, J. B. & Meaney, J. G. (2005). Voyeurism - It Is Good as Long as We Do Not Get Caught, *International Journal of Sexual Health*, 2007, 19 (1).

Schäfer, S. M. & Taddicken, M. (2015). Mediatized Opinion Leaders: New Patterns of Opinion Leadership in New Media Environments?, *International Journal of Communication*, 9 (2015): 960–981.

Schneider, M. S. & Foot, A. K. (2004). The Web as an Object of Study, *Published in New Media and Society*, 6 (1): 114–122.

Sidi, H. & Midin, M. (2008). Voyeurism with Sexual Fantasy on Female Body Parts: A Subtype of Obsessive-Compulsive Disorder? – A Case Report, *The Malaysian Journal of Psychiatry*, 17 (1).

Show TV Main News Bulletin. (2017). "Sosyal Medya Hapishanesi - İnternet Bağımlıları Toplama Kampında", November 3, 2017.

Soldin, A. (2015). Seeing Feeling: Queer Voyeurism in À la recherche dutemps perdu, *French Studies*, 69 (3): 318–332.

Yesil, B. (2001). Reel Pleasures: Exploring the Historical Roots of Media Voyeurism and Exhibitionism, *The e-Journal of Culture and Communication*, (November 2001), 1(1).

Yurdigül, Y. & Elitaş, T. (2015). Sosyal Medyada Bir Mahremiyet Sorunu Olarak Voyörizm (Instagram Örneği Üzerinden), Proceedings of National Symposium on the Limits of Media and Privacy, Atatürk University Faculty of Communication Publication: Erzurum.

http://www.muhasebetr.com/asgari-ucret/. (2017). "Asgari Ücret 2017" (Turkey), 06.11.2017.

Güliz Uluç and Bilal Süslü

From Conventional to Digital: Vigilantism

I Introduction

'Disappointment in society for ensuring justice' sometimes stems from a reaction caused by the situation of those who have been punished unfairly in the judicial system, and sometimes where they cannot be punished for the offenses committed by the criminals. The emotions and thoughts in questions are transformed into vigilant acts, in which sanctions are imposed by the society to those who have been questioned and convicted by society.

Those who are found guilty in the context of vigilant acts now turn into an enemy to be removed from being a member of that community. From this point of view, the purpose of punishment of those people also gives legitimacy to the use of intentional punishment. According to the people who perform the vigilant actions, the society only will reach peace and security. They also justify their actions with the action that they think they have developed in response to the inaction of the authorities. In this regard, those who perform the vigilant acts, behave in a sense of duty while neutralizing those who contradict the established moral values. Nowadays, new media is a very suitable tool for this function. In digital media, those who confirm and spread each other's thoughts and convictions, can easily accomplish vigilant action by forming a common moral conscience. However, attempting to punish according to their own conscience and moral values, can result in the neglect of essential elements such as the intention and the necessity of evidence of punishment.

On the other hand, the ability of the internet to spread information that is inaccurate or right, unfortunately, works well in legitimizing the hateful and violent discourse of the digital vigilantes. This demonstrates the importance of verifying information in the digital media before sharing it quickly. This, therefore, brings to mind the question whether the vigilant actions provide the initial justification in the final point.

II Conventional Vigilantism

Vigilantism is the name given to the actions taken mostly by the insufficiency of the laws (and in this context justice) and the actions taken by those who put themselves in the position of justice practitioners, stems partly from the fact that citizens apply laws due to the feeling of insecurity for the competent organs of the state, and sometimes because of the feeling that they are responsible for establishing control.

There is a debate among the few authors who examine the vigilantism concept whether it is a real social movement or just a social response (Johnston, 1996: 221). The word of vigilante, means "watchman, guard", passed through English in 19th century originated from Spanish. Its origin based on the Latin word "vigilare", means "to keep awake" (Merriam-Webster, 2017).

Vigilantism: What types of vigilante activities are prevalent (individual, group, ad hoc, recurrent, patrol based, surveillance based)? Who engages in it (by age, sex, class, ethnicity, housing type, urban/rural residence)? What motivates participants' engagement (victimization, fear of victimization, revenge, frustration with the justice system)? How are vigilante groups organized? What is their longevity? How do they relate to local agencies such as the police? and so on. It is a concept that needs criminological analysis in the framework of the questions (Johnston, 1996: 233–234). Les Johnston (1996: 232), who tries to define vigilantism in his work "What is Vigilantism", is trying to establish vigilantism as a chronological concept and thus to establish the necessary bases for the empirical analysis of the future of the subject, vigilantism: "*is a social movement giving rise to premeditated acts of force. It arises as a reaction to the transgression of institutionalized norms by individuals or groups. Such acts are focused upon crime control and/or social control and aim to offer assurances of security both to participants and to other members of a given established order*". Therefore, an action should be intended to control action, crime or other social violations so that it can be characterized as vigilantism (Smallridge et al., 2016: 59).

According to Johnston (1996: 234), participation in vigilantism is much more prevalent in homogeneous communities in social and ethnic terms. Because gossip is a tool of social control in traditional societies. For example, in Medieval Venice citizens would give information about the bad actions of their neighbors (called "mouths of truth") by hanging unnamed notes on stone lions in public buildings (Marx, 2013: 57). The "Bald Knobber" movement, which emerged in Taney County, Missouri, USA, in the late 1800s, is an important example of vigilantism with its reputation for violent acts in order to reorganize the social order.

> "*The Bald Knobbers of Southwest Missouri were a vigilante organization that originated in Taney County, Missouri, in 1885, before spreading to adjacent Christian and Douglas counties in ensuing years. They began as a group dedicated to protecting life and property, aiding law enforcement officials in the apprehension of criminals, opposing corruption*

> *in local government, and punishing those who violated the social and religious mores of their community. In some places, the vigilantes gained much political influence, occupied key offices, and became effectively the ruling faction in local politics. They made many enemies, however, with whom they had several violent, sometimes fatal, confrontations."* (*Hernando, 2011:viii*).

On the other hand, Johnston states that every action in society should not be regarded as a vigilant, but that an action must have six characteristics in order to comply with the definition of vigilance. According to these six characteristics, the action (at least to a minimum) has been previously planned, designed and organized; realized by private agents that the private agents are not supported or authorized by the state; the use of force, or threats to the use of force; the motivation to trigger action, the perceived suicide, or the nature of response to social deviation; and the purpose of the action should provide guarantee of security (Smallridge et al., 2016: 65). It is useful to briefly mention these features, which Johnston describes in detail, in terms of the solution of the new forms that the vigilantism concept transforms today.[1]

(i) Vigilantism Involves Planning and Premeditation by those Engaging in it

Vigilantism requires planning (at least minimally) in the name of the provocative action of planning, premeditation, and organization which are the first features of vigilantism. For example, vigilante actors should be involved in a number of preparatory activities, such as observing a specific place or observing the person or people, who are thought to be victims. In this respect, self-defense is not regarded as vigilantism. Because the self-defense situation involves a momentary reaction, and therefore the immediate reaction of the victim to the assailant cannot be regarded as vigilantism (Johnston, 1996: 222). In other words, 'legitimate cases of self-defense is immediate self-preservation rather than retaliation against crime or social deviance'. Therefore, the action of a person who tries to defend himself against an aggressor without planning should not be considered as vigilant action (Smallridge et al., 2016: 58–66).

In his work, Johnston discusses the planning phase through a question of whether the case of a man who shoots the burglar with his gun that he has as a precaution at home could be treated as vigilant action or not. According to him, in fact, having a gun at home against any danger requires minimal preparation.

1 The six principles that Les Johnston listed as the characteristics of vigilantism are based on the author's study called "What Is Vigilantism".

However, Johnston states that shooting should not be seen as a part of planning and should be evaluated in the context of self-defense. According to Johnston, organization is the result of the planning. Some cases allow small groups to be mobilized, while others may require the organization of large groups or masses. In this regard, it is an important question whether vigilant action is a repetition or a one-time action (Johnston, 1996: 223–224). In other words, vigilantism (Smallridge et al., 2016: 62) sometimes requires the organization of relatively more organized civil groups on the basis of volunteerism and expertise, and sometimes short-lived groups acting in response to a certain injustice.

(ii) Vigilantism Participants Are Private Citizens Whose Engagement Is Voluntary

Vigilantism has to be undertaken by private agents. Even if these private agents are tried to be expanded to include the actions of public authorities such as state officials, state regimes or police, Johnston finds this argument flawed from various angles. Firstly according to him, police officers can take part in various actions in their own initiative at times when they are out of work (for example, the participation of police officers in Brazil and Colombian to "death squads"). Secondly, there may form other vigilant acts in the functions of police officer while doing illegitimate acts (as in the case of excessive use of force during arrest). Finally, when the police officer may follow a force policy inside that constitutes some form of violence or 'illegitimate coercion' (such an instance may arise when the police officer in charge of the task performs legitimate custody or harassment of a minority group of individuals), a third vigilant attendance can be occurred (Johnston, 1996: 224). Johnston's three reasons exclude public institutions from vigilant acts. In other words, the actions taken by police officers based on their institutional identities apart from working hours, should not be evaluated in the context of vigilantism. More explicitly, it is not possible for public institutions to participate in actions that can be described as vigilantism.

Vigilantism can be seen as an action of the purpose to restore the social order, which is supposed to have been deteriorated. In this regard, the action is not a crime on the basis of vigilantism unless it is in the form of a violation of criminal law; and therefore 'the vigilant act, should not be sanctioned by the government and agencies in the criminal justice system' (Smallridge et al., 2016:59). No legal duty will occur in this regard unless the actions taken during the vigilant act are inconsistent with the law in force against the public property or the third party, and there are no injurious behaviors.

(iii) Vigilantism is a Form of Autonomous Citizenship and, As Such, Constitutes a Social Movement

The third characteristic of vigilantism is that the vigilant action is a voluntary activity carried out by "active citizens" (special voluntary agents) without the authority or support of the state. Active citizenship helps understanding of documented vigilante activity (Johnston, 2016: 226). Vigilantism is an action that takes place under the umbrella of 'patriotism' that is a result of internal protection and requires active participation.

(iv) Vigilantism Uses or Threatens the Use of Force

Violence is a common feature of all vigilant acts, and this indicates that the use of force is a necessary component for any vigilant attitude (Johnston, 1996: 226). In other words, to put an action into the vigilantism class, it is necessary to use the force in order to have any vigilant manner. However, the violence component of vigilantism can be limited while evaluating in terms of digital vigilantism (Smallridge et al., 2016: 59). Because violence in digital vigilantism is mostly based on the psychological process. Notwithstanding, psychological violence can sometimes be more devastating than physical violence.

(v) Vigilantism Arises When an Established Order Is Under Threat from the Transgression, the Potential Transgression, or the Imputed Transgression of Institutionalized Norms

The aim of this feature of vigilantism is to determine vigilantism as a criminal concept rather than to derive it in another discourse. More clearly, a distinction is made between vigilantism focusing on one 'crime control' and vigilantism leading to 'social control' or, more specifically, protection of social, ethnic or sectarian order and values. However, it is not enough to define vigilantism only in the terminology of crime control and social control. Especially social control is a rather ambiguous concept, and if the definition of vigilantism is used only in this context, it encompasses a potentially limitless spectrum of socio-political activity. The solution to this problem – or at least a part – is the normative feature of vigilantism; vigilantism is a normative feature that justifies the fact that it is a criminal concept rather than just a political concept. Therefore, vigilantism may result in violations of established constitutional norms or violations of institutionalized norms (Johnston, 1996: 228–230). It is important that vigilant behavior emerges as a reaction to the violation of institutionalized norms; because it also identifies the boundaries of actions that can be considered as vigilant acts (Smallridge et al., 2016: 59). As a result of the establishment of social consensus on norms that become established, any change over time will become resistant to the demand, so a threat can occur that may result in a vigilant action.

(vi) Vigilantism Aims to Control Crime or Other Social Infractions by Offering Assurances of Security Both to Participants and to Others

The concept of security allows vigilantism to be perceived as a social movement. In this feature of vigilantism, security is marketed as a desirable social good, and the provision of this security may also require sacrifice (financial costs, limitations on privacy and so on). In this regard, vigilantism is not synonymous of social control. In the vast majority of vigilant forms there is a concern to reduce the threat to persons, property or values, and to reduce the fear associated with it. The purpose is to give assurance to the people that a system of public order will prevail; vigilantism is a popular strategy initiated by the public as a reaction to social deviation (Johnston, 1996: 230–231).

Johnston (1996: 232–233), who mentions two different definitions of vigilantism made by Richard Maxwell Brown (1975 'Strain of Violence' gtd. Johnston, 1996), who aims to conceptualize vigilantism at the end of his analysis that the first definition is 'organized extra legal movements' that take laws into their own hands and the other definition is 'relations where citizens come together to protect themselves under the conditions of turmoil'. In this respect, Brown's second definition refers to the security feature of vigilantism.

The six elements tried to be explained above are 'to be different from vigilantism to other acts that are often misleadingly given the vigilante classification' (Smallridge et al., 2016: 58). Therefore Smallridge et al. (2016: 64–66) state that the six elements that Johnston specifies needs to develop in order to be evaluated as vigilantism in the new period (reflecting the digital vigilantism in realization), which is shaped by digital developments.

III Digital Vigilantism

On the one hand, new media tools such as electronic magazines and databases are said to extend the framework of people and areas of which information can be accessed and are often said to speed up the communication (Kling, 2000: 221); on the other hand, there is a growing awareness of cyber places, social, economic and political boundaries that threaten traditional forms of governance and order (Wall, 1997: 208). Contrary to the notion of passive and one-sided approach to human technology, Mackay and Gillispie refer to the obvious superiority of man to technology (1992: 698–699): "*People are not merely malleable subjects who submit to the dictates of a technology. People may reject technologies, redefine their functional purpose, customize or even invest idiosyncratic symbolic meanings in them*".

Digital vigilantism with the active role of man in the face of technology is a coordinated retaliation process in which citizens are disturbed by the consequences of other citizen activities on mobile devices and social platforms. Such disturbing actions can range from a mild social rule violation to terrorist acts or rebellion. Certainly, events involving relatively high-profile crimes such as terrorist acts or rebellion must be separated from events that react to violations of more ordinary etiquette such as poor parking or the use of bad public transport (Trottier, 2016: 58). However, as a result, both cases can be evaluated in the context of vigilantism.

On the other hand, events that conform to the definition of digital vigilantism cannot be defined as 'digital vigilantism' everywhere. For example, the concept of "The human flesh search engine" refers to the digital vigilantism in China. According to Cheong and Gong (2010: 472) this concept refers to the mass movement of personal information or the truth behind certain events. In this move, where the Internet is used as a tool, information is collected in part through human effort. The "flesh" quality of the concept, as opposed to traditional algorithmic-based mechanics, indicates an important role of human intervention. Smallridge et al. (2016: 60) indicate that the 'human flesh search engine', which is used to punish those who cause perceived harm, also allows users to come around similar worries. According to them, such mass-based vigilante acts usually occur into two forms: (1) mass retributive actions against an identified wrongdoer, and (2) collective attempts to solve real-life crimes.

The nature of the vigilantists contains a view that the addresses of the targeted person, the working knowledge, etc., are published in a public platform. The feature of visibility, which is provided by digital vigilantism, is that it is unwanted (the target person usually does not want to be known), intense (content such as text, photos and videos can reach millions of users in a few days) and permanent (vigilantism campaign can turn into a cultural reference over time). Vigilant campaigns are also a combination of digital and physical domains as an element of privacy and violation of fundamental rights. Because digital vigilance is a product of user-contributed cultural practices and digital media platforms. At the same time, it is a practice of citizenship that both opposes and strengthens the state power and the police state. As a result, it is at the same time a 'surveillance' issue that makes a targeted individual visible in this way (Trottier, 2016: 55–57).

The surveillance word rooted from the verb 'surveiller' in French, 'refers to processes in which special note is taken of certain human behaviors that go beyond the idle curiosity' (Lyon, 2007: 13). Surveillance processes can be divided into the collection of personal data, the interpretation of these data, and the social consequences of this interpretation (Trottier, 2016: 65). Since digital vigilantism involves information transfer that users use to disseminate information and/or produce new texts on media platforms (Cheong and Gong, 2010: 474), this information transfer may result in collective intelligence. Collective intelligence refers to the ability of virtual communities to use knowledge through large-scale cooperation and negotiation (Jenkins, 2006: 281).

Although 'cyber space', which is produced by internet, has the potential to offset some of the boundaries between communities, it also opens doors to new types of crime and new forms of anti-social behavior (Wall, 1997: 208). At the same time, technological developments open the doors of migration of crime to the digital era, trespass becomes cyber-trespass, theft becomes cyber-theft, bullying becomes cyber-bullying; terrorism becomes cyber-terrorism" (Stratton et al., 2017:22–24). At first, the guilties attack machines with machines (such as activities or cyberattacks), then begin to attack people through machines. This development reveals criminologists a need of a discipline to examine and analyze criminal behavior in the cyber area. Through this discipline, however, crime and victimization occurring in the cyber environment can be examined (Jaishankar, 2011: 411). In this regard, the technological developments and the results related to crime and criminology of these developments can be shown in three periods: The 'pre-web' period from the beginning of the 1980s to the beginning of the 1990s, the 'global web' period from the beginning of the 1990s to the beginning of the 2000s, and the 'social web' period from the mid-2000s until today. Mobile web applications that collect synchronized data about us, our identities and everyday life and that are multiplied by smartphones and smart wear technology are important qualities of the social web age (Stratton et al., 2017: 19–21). We can add the qualities that Gary Marx (2013: 57) called as "digital record that never forgets" the ability to record and recall everything that digital technology requires. In addition, the digital revolution has the feature of spreading rapidly and facilitating sharing even beyond collecting and recording images (Tunç, 2017). In this regard, digital media also facilitates illegal purposes. Therefore, private organizations such as social platforms in digital vigilantism campaigns are seen as accomplice actors as long as they facilitate these aims. Digital vigilantism campaigns, however, not only pursue a goal at defining a targeted person or category of person, including expressing consensus on shared norms and values but also expressing a collective identity shaped by national, religious or ethnic solidarity forms (Trottier, 2016: 57–58). One of the most striking examples of this situation is the "YesYoureRacist" movement in response to Neo-Nazi demonstrations in Virginia in recent digital media:

> *"Reflections on both American and world politics are still ongoing after the racist demonstrations on Saturday, August 12 in Charlottesville, Virginia. As it is known. a neo-Nazi anti-racist demonstrator, killed a 32-year-old woman by his car and wounded 25 others ... After the events in Charlottesville, this group of anti-campaign began to hide their identities during the protests. "YesYoureRacist" Twitter address, which has long been idle and ruled by Logan Smith, has been swiftly circulated. At one point, at the address, which has 65,000 followers, increase 350,000 followers and the campaign began to make these people unable to get out of society. The photos of protest racists were seized, and the virtual communities were asked whether they recognized them or not. Even Hollywood actress Jennifer Lawrence, who is not known to keep her head down, participated in this campaign and tweeted the photo of the Neo-Nazis and said, "These are the hate faces. Look closely at them and let them know if you know them. You can not run away on the Internet! "Soon these efforts worked out. It turned out that the racist genocide was a student of Nevada University called Peter Cvjetanonic, was shouting angrily with a torch in his hands at the photo, which was sent to everyone. The racist teen became a global phenomenon in an incredibly short period of time. Peter's father said he could not come home without stopping his racist views. Virtual embarrassment and exposition did not end with it. One person lost his job in the restaurant, and the photos of others circulated in the virtual world hundreds of thousands of times. Each photo had the same question: "Do you know me? Are you my employer? I was on a Nazi walk in Charlotteville". The addresses of the racists carrying the torch, the names of the workplaces or the collections were used as t elements of opression in the social media. It seemed very difficult for these people to live in their neighborhoods, work and school surroundings and live an ordinary life as if nothing had happened" (Tunç, 2017).*

A more recent example of vigilantism is the demonstration of hundreds of protesters on the famous Hollywood Boulevard after accusations of harassment and sexual abuse in Hollywood. Those protesting that many accusations of harassment, including producers and famous actors, became a culture in Hollywood, have carried out vigilant campaigns with posters bearing '#MeToo' in their hands. Also in the campaign launched with #MeToo tag on Twitter following the outbreak of the scandal, many people tried to explain the harassment incidents they experienced (BBCTurkish, 2017a). As a result of these campaigns, many men in the past who were seen as "powerful and even untouchable" have had to acknowledge their claims, apologize and even leave work (BBCTurkish, 2018).

Digital Technologies with features such as portability, prevalence and ongoing communication that go beyond digital technology, cultural criminology and media-crime also enable the community to undertake new 'gatewatching' roles (e.g., Stratton et al., 2017: 25). For example, in the PyCon conference, Adria Richards, disturbed by the conversation of two male participants who made sexist jokes, showed his dissatisfaction by twinning with the expression "Not cool" by taking photograph of them. Since Richards had a significant number of followers on Twitter, this allowed his tweet to be quickly retweeted, and this caused one of the male participants lose his job. On the other hand, the fact that Richards posted them in the social media without giving them the opportunity to respond led to anger and later Richards lost his job (Bronwen, 2013).

The methods used to collect evidence about crime have also become very different from the digital era; because the rules which are developed to collect evidence in traditional methods are no longer valid for the digital period anymore. Especially cases such as physically participating criminal investigations and laws that balance privacy and public safety can lead to surprising results when common methods are applied using digital evidence investigations (Kerr, 2005: 280). Thus, when digital investigations are evaluated in relation with privacy and individual freedoms, new and important questions are raised about how the evidence is collected, protected and regulated (e.g., Stratton et al., 2017: 25). In other words, the control power of digital tools must be re-questioned in order to protect privacy and limits of individual freedom.

At first, the ability of police officers to identify suspects and broken law documents are intensive (Marx, 2013: 56), with digital developments, state institutions begin to interact with the public as never before. The use of social media by police and court causes disruptive effects on the traditional justice system, while encouraging access and participation from one side (eg, Stratton et al., 2017: 25). Because the digital evidence contains a feature that can be edited, manipulated, and can be distinguished from the truth. Moreover, the digital community also encourages 'informal' justice practices related to suicide. Some scientists express the concern about the 'unofficial' judicial process embracing the use of technology, stating that digital vigilantism with digital media can lead to harassment and violence against injustices that are allegedly guilty (Stratton et al., 2017: 25–26). For example, there are thousands of Facebook pages and independent websites, which are dedicated to accusing people of everything that might come to mind (Bronwen, 2013). In this respect, the 'happy slapping' phenomenon comes to the fore (as regards the evaluation of new media and communication technologies as a motivation of guilt); and the reason behind it seen as desire, reputation and gaining appreciation (e.g., Stratton et al., 2017: 23). Therefore, new media users, who are delighted to punish by means of criminal representations in the media, cause a potential lynching culture (Stratton et al., 2017: 24–25). This kind of vigilantism occur has legal and destructive consequences, such as the fact that after the murder of Sarah Payne at the age of 8 in the UK, some people were unfairly accused of pedophilia and attacked physically and verbally and sometimes have illegal and harmful consequences. Moreover, the campaign launched in the News of the World newspaper lists the names of pedophile convicts and accused them; they were protested by people due to living in the cities. This uprising (including violence) resulted in the stigmatization of innocent people as pedophiles; at least five innocent families left their homes and two suicides (Bull et al., 2006: 122). Therefore, it is understood that one of the important principles of modern law, which, when unfairly encountered, is the danger of infringement, which precedes the application of legal remedies. Because people can now easily identify and impersonate the target person or people with digital data. In other words, digital platforms can be transformed into a channel where executions are carried out without prosecution.

> *"As technical developments join mounting concern over terror and crime, the state is becoming increasingly covetous in its desire for information. Think of the new potential Google Glasses will bring to quietly record sound and images and then anonymously post them on the Internet or send them to the police"* (*Marx, 2013: 59*).

Despite these negative sides, new media offers its users the opportunity to participate in critical civilian discourses by structuring and sharing previously unreachable information on multiple digital platforms (Cheong and Gong, 2010: 473). For example, the social media networks, which have laid the starting process of the changing movement that has spread over the years have seriously contributed to the offline movement of protests that started online by strengthening ties between segments with different views and demands from current regimes. More recently, in the turmoil that has led to administrative changes in many Middle Eastern countries, called the Arab Spring, new communication technologies have been effective in involving millions of people in political actions and have led to an organized structure. Rioters from various sectors have fallen into various political and legal claims to governments. As a result, new communication technologies, which inspire the masses to organize and announce their voices, have weakened central control by breaking the information monopoly. For example, social media sites such as Facebook and Twitter, as well as individual blogs, helped political reform advocates to organize and overcome the regime's horror walls by overcoming the censorship indirectly that has been for many years of hardship in Egypt. However, new communication technologies are not only used by those who want to be democratized but also by dictator regimes. Therefore, it is worthwhile to note that new media technologies (it is a controversial issues that work more people or dictators) make monitoring of opposition movements easier (Telli, 2012: 85–86).

Digital vigilance campaigns can sometimes be seen as an expression of disappointment created by citizens' neglect and abuse of the police. Such campaigns support the enactment of laws and order (Trottier, 2016: 63). In this regard, digitally vigilantism is the name given to the movement of crowds, the official law enforcement officers, who have not investigated or cannot investigated the crimes and those who are in pursuit of justice through the use of the power they have by the technological equipment they possess and the innumerable online connections.

On the other hand, digital vigilance campaigns aims at other non-police institutional actors who are sometimes perceived as accomplices in the negligence of criminal justice. For example, in such situations like cyber bullying and so on. Social platforms and telecommunication companies may be subject to criticism and accusation against the grounds that they have not reacted to such crimes and that they have not taken any measures (e.g., Trottier, 2016: 63). Since 2016, Twitter has been the target of a day-long boycott campaign, due to shutting down theTwitter account of Rose McGowan, who has shared complaints about harassment and rape in Hollywood on her Twitter account. Many users who accused Twitter of sexism, including famous actors, participated in the boycott by supporting the #WomenBoycottTwitter tag (BBCTurkish, 2017b).

According to these data, re-evaluation of the vigilantism which has evolved into digital in the direction of the context of the six properties that Johnston identifies will make sense in order to better understand the digital appearance of vigilantism.

IV Traditional Vigilantism in the Context of Key Features – Digital Vigilantism Separation

In the study of Daniel Trottier called "Digital Vigilantism as Weaponisation of Visibility", he discriminates vigilantism as traditional-digital by interpreting the six principles that Johnston defined as the features of vigilantism. In other words, Trottier feels the need to adapt the six features defined by Johnston because vigilantism definition is insufficient to describe the new vigilant actions of the digital age. With this distinction, vigilant actions now lead the reaction to a deviant behavior spontaneously; it is a new era in which the vigilant actors cooperate with the state and public institutions, but the new regulations/limitations that are used as visual weapons are deemed necessary and in this regard the digital youth makes a premium.

Tab. 1: Key Features of Conventional and Digital Vigilantism

	Conventional Vigilantism	Digital Vigilantism
Planning	Premeditation	Facilitated spontaneity
Private agency	Distinguished from state and corporate actors	Possible connections with state and corporate actors
Autonomous citizenship	Self-protection	Asserting new boundaries
Use of force	Embodied	Visibility as weapon
Reaction to crime/ deviance	Threat of established order	Fusion of local and mediated norms
Personal and collective security	Policing localised territory	Mediated policing

This classification is based on Daniel Trottier's "Digital Vigilantism as Weaponisation of Visibility" according to the characteristics of vigilantism in Les Johnston's "What Is Vigilantism? British Journal of Criminology" (Trottier, 2016: 59).

i Planning

Minimal planning required being able to assess vigilant action in traditional vigilantism leaves spontaneity in digital vigilantism. Action is now a manifestation of the cultural codes that do not require a preliminary process, and the instant reaction that is created by established norms or stereotypes. Nevertheless, it seems that organization, which defines the limits of the effectiveness of the action, is an important process in digital vigilantism as well as in the traditional vigilantism.

ii Private Agency

In traditional form, vigilantism is undertaken by private agencies and vigilant acts are carried out by elites who also have corporate identity. When it comes to digital era, cooperation with institutional actors has become a condition for vigilantism to take action. The possibility of technological developments to gain a global dimension like a snowball has occasionally made it necessary for vigilant actions to be coordinated with institutions.

iii Autonomous Citizenship

While patriotism in traditional vigilantism is a phenomenon that resulted from the instinct of self-protection, digital vigilantism needs to redefine those boundaries. Because the initiation and participation of the vigilantism movement in the digital era can evolve into various dimensions. For example, participation in vigilant action in the new era; it may be the result of reaching the global dimension of reaction to a specific group, such as personal animosity, revenge, being liked by others, being appreciated by others, or the "YesYoureRacist" movement. With instant sharing, feedback, etc., social media networks, on the one hand, can lead to areas of intense involvement in actions aimed at the public order, and on the other, actions that can go as far as extrajudicial executions on the basis of revenge or disgrace. In this context, autonomous citizenship can emerge from the individual impulses that can later be transformed into collective action, and bring legal processes that require regulation, such as legal liability.

iv Use of Force

Vigilantism is framed as a kind of special form of violence that citizens try to prove their criminal law forms (e.g, Trottier, 2016: 62). When the state is considered to have a monopoly over coercion, citizens through vigilantism ignore this state monopoly in an attempt to legitimize their acts of violence. In the case of digital media, the legitimate attempt is made by sending text, photos and video content. Digital vigilantism "is distinguished from conventional vigilantism through its primary reliance on weaponized visibility for social change." In addition, the privacy and data protection rights of the target group are severely violated because personal information is transmitted to the public without the consent of the individual. Digital vigilantism, unlike traditional vigilantism, is expressed through digital media and is a medium that allows the infiltration of personal information of targeted criminals (Trottier, 2016: 62–68). The reflection of the legitimated power of violence to the level of violence occurs physically and psychologically.

v Reaction to Crime/Deviance

As Johnston mentions (1996: 229), vigilantism is a reaction to real or perceived deviance. Vigilant action in digital vigilance is often a response to cultural corruption through individual impulses such as revenge, threats, and collective impulses towards ethnic, sexual, political groups, while the behavior seen as a threat to the established order is the reason of vigilant action.

vi Personal and Collective Security

In traditional vigilantism, security is being promoted through regional/local police identity, while policing are carried out through mass media in digital vigilantism. Smartphones, wearable technological products and other mass media have made it necessary to redefine the concept of security. These tools have positive functions such as revealing the people against social order or arresting the criminals but these also include negative ones such as the violation of privacy, and therefore, individual and social security have been questioned in the context of new communication technologies. Thus, vigilant actions always contain the possibility of unintended consequences when carried out through the mass media. When messages, photos, hashtags, or comments shared without questioning their reality on the social media to be a member of vigilant campaign reach a level that harms the physical integrity of target person or people, it will be possible to observe that the term "security" becomes concrete rather than a virtual concept.

V Conclusion

In traditional vigilantism, an anti-social phenomenon can result in a vigilant action that creates an effect in a certain environment where it occurs from a heavy social disorder to terrorist acts and rebellion; however, digital vigilantism can interact with vast quantities without regard to the environmental limitations of the event.

As a result of shared images about anti-social behaviors detected by users, the new media, where millions of users can reach several texts, photos and videos that can be streamed around the world at any moment, have facilitated and spread the reactions and attacks against people who make mistakes. Considering the unlimited visa that technology applies to information, digital vigilantism is a phenomenon that occurs on a global scale, in large crowds of strangers rather than small communities of neighbors.

A significant part of digital vigilant actions are carried out in those who think that traditional legal instruments are technically inadequate in defending the victims of crimes committed through the Internet. When there is a problem, the usual way of applying for police for enforcing the law leaves the digital environment to try to enforce the law.

Digital vigilantism, on the other hand, has an important function in law enforcers that their power and authorities are questionable when they use these authorities. On the other hand, the law enforcers requested help from the people to identify the identities of those involved in some criminal cases. Therefore, law enforcers have the potential to become a partner in the effort to provide security to the public by directing the public interest to the necessary points – for example, the scanning of photographs of the scene of the incident. However, judging and accusing people, who do not have the right to defend themselves in social media, causes this function to be abused. As it occurs that the suspects are sometimes misidentified, it is also possible to develop a digital vigilant behavior without the right to answer or legal defense given to the accused person. While mobile devices and social platforms provide us with a great opportunity to expose the misuse of power and authority by public officials, it is also necessary to avoid a similar misbehavior. The fact that everyone can easily be accused of anything with unsubstantiated rumours can also have serious consequences for those who accuse, and those who are accused.

References

BBCTürkçe (2017a). "Yüzlerce kişi Hollywood'da cinsel tacize karşı yürüdü", http://www.bbc.com/turkce/haberler-dunya-41964290, E.T.: 15/01/2018.

BBCTürkçe (2017b). "Kadınlar Twitter'ı neden boykot ediyor?", http://www.bbc.com/turkce/haberler-dunya-41607541, E.T.: 15/01/2018.

BBCTürkçe (2018). "Türkiye'de kadınlar #SenDeAnlat (#MeToo) diyebiliyor mu?", http://www.bbc.com/turkce/haberler-turkiye-42666440, E.T.: 15/01/2018.

Bull, Ray; Bilby, Charlotte; Cooke, Claire and Grant, Tim (2006). Criminal Psychology: A Beginner's Guide. Oxford: Oneworld Publications.

Cheong Pauline Hope and Gong, Jie (2010). "Cyber vigilantism, transmedia collective intelligence, and civic participation", Chinese Journal of Communication, 3(4), 471–487.

Clune, Bronwen (2013). "Digital Vigilantism: Think Before Putting Pictures of 'Wrongdoing' Online", The Guardian, https://www.theguardian.com/commentisfree/2013/nov/29/digital-vigilantism-think-before-pictures-of-wrongdoing-online E.T.:23/08/2017.

Hernando, M. J. (2011). "The Bald Knobbers of Southwest Missouri, 1885-1889: A Study of Vigilante Justice in the Ozarks." Louisiana State University Doctoral Dissertations.

Jaishankar, K. (2011). Cyber Criminology: Exploring Internet Crimes and Criminal Behavior, CRC Press.

Jenkins, Henry (2006). Convergence Culture: Where Old And New Media Collide. New York and London: New York University Press.

Johnston, L. (1996). "What Is Vigilantism?" British Journal of Criminology, 36(2), 220–236.

Kerr, Orin S. (2005). "Digital Evidence and the New Criminal Procedure", Columbia Law Review, 105(1), 279–318.

Kling, Rob (2000). "Learning About Information Technologies and Social Change: The Contribution of Social Informatics", The Information Society, 16, 217–232.

Lyon, David (2007). Surveillance Studies: An Overview. Cambridge: Polity Press.

Mackay, Hughie and Gillespie, Gareth (1992). "Extending the Social Shaping of Technology Approach: Ideology and Appropriation", Social Studies of Science, 22(4), 685–716.

Marx, Gary T. (2013). "The Public as Partner? Technology Can Make Us Auxiliaries as Well as Vigilantes", Security & Privacy, 11(5), 56–61. http://web.mit.edu/gtmarx/www/marx-publicas.pdf

Merriam-Webster, (2017). "Vigilante/The Meaning and Origin of Vigilante" https://www.merriam-webster.com/dictionary/vigilante, E.T.: 03/12/2017.

Smallridge J., Wagner P. and Crowl J. N. (2016). "Understanding Cyber-Vigilantism: A Conceptual Framework", Journal of Theoretical & Philosophical Criminology, 8(1), 57–70.

Stratton G., Powell A. and Cameron R. (2017). Crime and Justice in Digital Society: Towards a 'Digital Criminology'? International Journal for Crime, Justice and Social Democracy 6(2): 17–33. DOI: 10.5204/ijcjsd.v6i2.355.

Telli, Azime (2012). "Mısır Devrimi'nde Sosyal Medyanın Rolü", Bilge Strateji, 4(7), 65–91.

Trottier, Daniel (2016). "Digital Vigilantism as Weaponisation of Visibility", Philos. Technol., Springer.

Tunç, Aslı (2017). "@EvetSenBirIrkçısın: Neo-Nazilerin Teşhiri", Platform24, http://platform24.org/yazarlar/2365/-evetsenbirirkcisin--neo-nazilerin-teshiri E.T.: 22/08/2017.

Wall, David (1997). "Policing the Virtual Community: The Internet, Cyberspace and Cyber-Crime", Francis P. et al. (eds.), Policing Futures, Hampshire, England: Palgrave Macmillan. 208–236.

Uğur Bakan and Ufuk Bakan

Face-ism in Online Profile Pictures: A Cross-Cultural Analysis of Adults' Visual Self-Presentation on Facebook

Introduction

In the last quarter of the twentieth century, the media industry has achieved significant economic strength by globally servicing via fiber optic technology and geostationary satellites. The concept of new media emerged with developing computer, internet and mobile technologies have created an environment in which the desired content can be selected, instant feedback can be obtained, and information can be stored easily and quickly. With the opportunities provided by the technology, the virtual information flow has reached an effective position at every stage of the communication process. This new media environment has created a virtual world in which people can develop their own tools, edit content, maintain existing links, as well as create new connections. This virtual world offers its users unique tools for self-presentation and effective management. Time and space dependence has been removed with the spread of web-based systems in the process of interpersonal communication. Increasing bandwidths and advances in Web 2.0 technology have to lead the development of social media environments designed for different purposes and structures. Another development that has made social media environments increasingly common is that almost everyone can afford smartphones. Initially starting with 1G first generation cellular technologies, 2G, 3G, 4G and 5G ongoing technological developments have created an infrastructure that allows users to share their content wherever they want. This new communication environment has also caused many changes in social and cultural structures. People now prefer virtual environment to physical environment for meeting and sharing their ideas. In a virtual environment, there is a virtual identity which has a lot of information about a person's character, qualities, and experience. This identity can reflect the person to a large extent and may also include information that does not correspond to the real identity. Virtual identities become a tool which creates a new collective role forming the desires and wishes wanted to have or never will have. People can show behaviors and thoughts that they cannot show in real life through virtual identity. Virtual identity, the importance of virtual identity in a techno-social structure and the creation of a new world with these identities will be seen as the main problems of the next century. At this point, the factors affecting the way in which people present themselves in the virtual environment and the reliability of personal information are analyzed in various ways in academic studies. In this study, identity presentation forms will be interpreted via profile photos of Facebook users, which reflect virtual identities most prominently. In this study, Facebook profile pictures of people in different cultures are examined using the face-ism index.

Self-Presentation on Online Social Media

With the widespread use of the Internet, a global network has created a heterogeneous social network that allows interaction between individuals and communities residing in different regions. In the process of communication in traditional media environments, the interaction between users is a one-way message transfer from a source on a linear scale, whereas the interaction process between users in a social media environment has reached a two-way dimension. This process, which is the nature of communication, is defined in the literature as "social networks", "communication networks" or "social media" as a structure that has many features. Also, it provides many concepts such as Tweet, Podcast, iCloud, Cookie, Hashtag and Wiki to come into our lives. Social media, which is the most important means of communication in terms of its scope and effects of our age, is shaped by relationships that individuals build with their virtual identities in accordance with their preferences. Social network sites, such as Facebook, Twitter, LinkedIn and Google Plus have significantly changed the creation and the distribution of information and social relationships with the development of Web 2.0 technology. On social networks, users can continue their social relationships with other stakeholders by joining a network and sharing any content with other members of the network (Mislove et al., 2007). Social networks provide opportunities for people to express their ideas by bringing people together, who have a common socio-cultural background, similar interests or common ideas in a virtual field. These environments offer an individual a flexible world where people can edit items such as video, text, photos whenever they want, how live performances, create stories, and use tags like hashtag to reach users by changing people's daily life. The fact that people on this virtual world have the option to view their photo, video and text-based items shared by other users and has feedback on their likes and comments as well as the ability to edit their own profile photos and status information allows users to display both real and ideal self.

Facebook, which is the first site to come to mind in terms of social media, allows users to connect together and groups, participate in groups and share sources with different contents at different levels of permissions in the networks created by its users (Gonzales & Vodicka, 2010). The profile page of the users on Facebook contains much more detailed information about users contrary to other platforms. These sections include: a profile page with personal preferences such as job, education, contact information, place of residence, date of birth; a "wall" where offline messages and images are located; message section where voice and video communication is available; sections that communicate with other groups and pages; and applications which include online software integrated with Facebook, especially for entertainment purposes by many developers.

It is tried to be explained that the Internet is considered to be the most important discovery of the 20th century; social networks, which is the next step in the virtual socialization process, and network theories developed to understand the structure of these networks in connection with the characteristics of the social structure, which creates the social networks. In general, social network analysis analyzes behaviors that arise from personal and social variables by examining the relationship between people on the network on a personal and social scale. As a result of the analyses, behaviors and tendencies are modeled mathematically and the personal and social changes in the network are followed. Social media studies have been a pioneer in studying the relationships between computers science, media and communication education, psychology, sociology, political science, and other disciplines (Miladi, 2016; Miller et al., 2016; Han & Keskin, 2016). As a result of the network analysis, a general judgment about a topic is made.

Social media has undoubtedly become an important tool in reflecting the identity and cultural values of people. One of the greatest innovations of social networks to our everyday life is that it allows people with common interests on the same platform to create their own groups and share each other without any space and time limitation. Using social media tools, people can present their identities, virtually. Self-presentation refers to how people present themselves in a variety of ways, tailoring the images they convey to others that vary according to situational demands (Bronstein, 2013, p. 163). The self-presentation consists of two sub-categories of assertive and defensive self-presentation, which show different behavioral patterns. Assertive self-presentation consists of behaviors used to draw attention to an identity such as ingratiation, intimidation and self-enhancement (Lee et al., 1999; Schutz, 1998). Defensive self-presentation, on the other hand, refers to behaviors used to defend or restore an identity such as self-handicapping, disclaiming the possibility of failure, and justification of negative behavior (Schutz, 1998). In his study "The Presentation of Self in Everyday Life" published in 1959, Goffman discusses daily social life as an interaction based on performance between the actor on stage and the audience watching the play. In his work, Goffman identifies daily social life as a theater stage in which there are interactions and communications between the actors and the audience. This theater stage is divided into two parts as downstage and upstage and people are playing different roles depending on the change of the scene. Goffman noted that one's self-presentation is a continual process of complex negotiation between these two positions, each of which may contain a multitude of presentation strategies that a person seeks to manage through a revolving process of interpreting audiences' goals and contexts (Smith, Sanderson 2015, p. 343). People employ varying self-presentation strategies both during face-to-face interactions and online (Schlenker & Pontari, 2000). He implies that people are playing many roles in everyday life, and depending on the role, they perform to make an impact in the mind of the audience. Goffman (2009) describes the concept of performance as the person's actions over audiences who follow and have an impact on him. The downstage is also seen as an area where the stage is hidden from any action that is not suitable for the player's own ideal image. Just as the existence of the other is needed for the formation of the identity, it is also needed for the presentation of the self. The upstage, which is the other part of the theater stage, is seen as an area where people move freely, do not need to act, display their true self and reveal their flaws.

Cross-Cultural Communication

Thousands of super cultures and subcultures are affecting each other with the widespread use of the internet and related technologies. Systematic studies on society show that all human communities are changing due to various factors. In each society, socio-cultural changes emerge in relation to different civilizations, cultures and religions within the historical development process (Porter et al., 1996). This change causes to disappear many values of social and cultural content, work, institutions and lifestyles or to differentiate as structure and function. The society, which exploits in cultural change, creates a society that exploits material beliefs and values as well as material cultures. Many studies have been conducted in this exchange process to determine how cultures and therefore communities are affected). Intercultural communication examines the interaction between individuals and groups from different cultures, the level of reflection of their cultural characteristics, and the effects of cultural specific behaviors and thought patterns on other cultures (Dodd, 1991). Most of the research on intercultural communication, systems, processes and schemes for understanding and interpreting various cultural practices and representations are created. Intercultural communication is the examination of the concept of communication, behavior, a particular idea or appearance in many cultures with the intent of comparison.

Hofstede, who is known for his pioneering work on intercultural communication, has identified four dimensions such as individualism and collectivism, masculinity and femininity, ambiguity avoidance, and social-organizational power distance in his comparative studies of different cultures in the 1980s (Hofstede, 1980). In general, Hofstede examines value preferences and critiques of survey-based approaches, especially in measuring their own context (Hofstede, 2001, p. 73). Hofstede's indices are based on responses of IBM's international staff to a survey of 116,000 people from 72 countries between 1967 and 1972 to determine how workplace values are influenced by culture. After exemplifying the variable field experimentally to measure cultural differences, Hofstede developed a unified measurement technique that uses a few indicators to create an index of variables. This measurement technique is based on the Power Distance Index (PDI), Individualism and Collectivism (IDV), Maternity and Fertility (MAS), Power Distance (PDI), Collectivism versus Individualism (IDV), Femininity versus masculinity), Uncertainty avoidance (UAI). The power distance, which forms the first dimension of cultural differences, distinctively shows the differences between superiors and subordinates in large cultures. Moreover, in cultures with wide power range, importance is given to title, degree, and position, and the differences among these positions are clearly displayed. On the contrary, in cultures where the power range is narrow, the title and the degree of the position are not important and the opinions of subordinates are also taken when the last word is given by the managers in the decision phase. Collectivism versus individualism (IDV), which constitutes the second dimension of Hofstede's work, expresses the extent to which the degree of interdependence among members of a society has been preserved. This top point of this dimension called individuality assesses the tendency of individuals' preference in comparison with other members of the culture. The position of a society in this dimension is determined by whether or not the image of the person is defined as "I" or "we". The data obtained from this study indicate that the cultures in which the interests and expectations of the society are held superior to the interests of the individuals are called as collectivist cultures and the vast majority of people belong to these cultures (Hofstede, 2006). In Collectivist cultures, individuals see themselves as a member of this great community or family, and the expectations of society and the individual seem to be parallel. The expectations and desires of the society in individual cultures do not overlap with the expectations of the individual, and therefore the bonds between individuals show a weak appearance since people in these societies do not see themselves as a member of the existing community. Femininity versus masculinity (MAS) in the third dimension examines behaviors and roles that are distributed by sex in terms of quality of life and level of achievement. In this dimension, individuality, ambition and aggressive competition prevail on the basis of masculinity, success and tangible assets. The masculine side represents financial possibilities such as success, heroism, and assertiveness which is necessary to achieve success. Feminism, on the other hand, attaches more importance to cooperation, humility, power balances and quality of life in a collective perspective. Femininity describes cultures, in which there are no gender-based role divisions, care about honesty and quality of life (Hofstede, 2006). According to the data obtained from a comprehensive study, Hofstede indicates that countries like Greece, Japan, Italy, Austria and Mexico are masculine cultures; Scandinavian countries like Sweden, Norway, Denmark and Finland are feminine cultures (Hofstede, 2004, p. 232). In the fourth dimension of the measurement technique called Uncertainty Avoidance (UAI), Hofstede measures to what extent a community feels threatened by certain or uncertain situations. The basic criterion here is the degree of preferring a society regulated by open rules than more ambiguous ones. This dimension shows to what extent an individual has benefited from written rules and other factors to remove ambiguities that are caused by fear, threats and anxiety. Cultures with a low degree of avoidance of ambiguity tend to accept the risk and prefer covert or flexible rules and guidelines. Members of a culture get points, perceiving uncertain or unknown situations as threats, taking precautions to avoid them, developing ideas and policies, and creating institutions. The similarities and studies used in cultural comparisons provide important information about how long and in what level different cultures interact.

Stereotypes

Because people are social beings, they live in groups instead of living alone and socialize by interacting with these groups. Classifying individuals into groups rather than alone is a matter of having both positive and negative meanings according to the group that they belong. Owing to the categorization, a number of schemes are formed in the mind, and with the help of these schemes, naming assignments to groups emerges as negative judgments, which are usually imputed in advance like prejudices or stereotypes, which can be both negative and positive "Public Opinion", published by Walter Lippmann in 1922, is regarded as the beginning of the studies related to stereotypes in the field of social sciences. Lippmann describes the stereotype as "pictures in our mind" (Lippmann, 1922, p. 3). Apart from Lippmann's definition, there are many other different definitions to describe the stereotype concept. Stereotypes are a set of concepts pertaining to a social category (Vinacke, 1957). Stereotypes are special schemes that regulate people's perceptions and attitudes towards other groups and distinguish them from others (Fiske & Taylor, 1991). According to the Oxford Advanced Learner's Dictionary, stereotype is a fixed idea or image that many people have of a particular type of person or thing, but which is often not true in reality" (Hornby, 1995, p. 1169). In the most general sense, common beliefs or thoughts accepted by a particular social group, department or ethnic community are considered as stereotypes. The concepts of stereotype and prejudice are often confused within the community or substitute each other. Stereotypes are common thoughts that they are not private and carried on by sharing within the same community. These values serve as a guide to generalize or define the groups as a stereotype by choosing among the numerous stimuli around individuals. Stereotypes used to describe individuals or communities can be built on a number of variables such as judgments, gender, minority classes, sexual preferences, professional communities, and ethnicity. But stereotypes take time to change in social life, despite rapid developments in the world of information and communication. Positive opinions about a community or a group can change over time and turn into a negative appearance in the eyes of the public.

Different types of media such as visual and print media are the platforms that people can meet different stereotypes. Advertisements, which charm people's interests and increase product sales, are among the platforms, in which stereotypes are most widely used. The advertisement is a paid announcement of information to the public about certain goods or services, to change opinions and attitudes towards them and to increase brand awareness and market share via such masses like television, newspaper, radio, billboard, magazine, cinema and internet. Advertisements also help companies to survive by setting new production policies and gaining an advantage over competitors in a competitive environment. It is seen in mass media in which the public is informed about the values, norms, beliefs, and clues about social stereotypes that dominate periodically. These clues are usually conveyed consciously and verbally, while prejudices have a wider range of expression. While stereotype has a more uniform quality, prejudice can encompass a large number of stereotypes (e.g., race, religion, sex stereotypes). Men and women often have one-sided and exaggerated images repeated frequently in daily life. Today, such images are found especially in social media tools such as pictures, videos and messages shared by the person. From this point of view, the presentation of the stereotypes such as origin, race, nation, class, sex can be determined in social networks.

Face-ism Theory

Archer and colleagues (1983) conducted five studies showing that women and men differ in terms of facial significance and discover Face-ism effect or facial prominence in American periodicals (Time, Newsweek, Ms., The San Francisco Chronicle, and The Santa Cruz Sentinel). In these studies, Archer et al.(1983) created a face-ism index to determine the relative position of a man or woman in a photo or a drawing. The notion of face-ism was examined, which is a cross-cultural form of gender stereotyping assuming that the focus of males is on his face and head, whereas females' bodies are more central (Reichart & Cooley, 2009). The method consists of two measurements in a photograph: the first is the numerator of measurement which includes from the head even hair to the end of the chin and the second part is the denominator of the measurement from the top of the head to the end of the body that appears in a photo (Archer, 1983, p. 726). Values in the face-ism index show the variation between the highest value of "1" if the only face is visible the other parts are not, and the lowest value "0" if the face is not visible. In the study of Archer and his colleagues, at least twelve numbers of each of the periodicals, 350 titles of each publishing, totally 1,750 photographs were selected and coded. In the second study, 3500 images were coded as sample sets, which provided suitable criteria from 13 publications in 11 different countries. It also shows that Face-ism plays a consistent role in describing men and women, interculturally. In the third study, Archer and colleagues examined the works of art include portraits and self-portrait that were painted from the 15th through 20th Century. Archer et al. (1983) attempted to explain sex differences in face-ism by suggesting that "the essence of a man is thought to reside in his face and head, whereas that of a woman is thought to reside more generally in her body" (Archer et al., 1983, p. 726).

Apart from above, various studies have been carried out on whether the politicians, governors, and senators working in different units within the political arena and the bureaucracy have a degree of importance at a different level than men in the visual and print press. In the first of these studies, Sparks and Fehlner (1986) analyzed the photos via content analysis method in the Time and Newsweek magazines published during the campaign between Geraldine Ferraro, the first female candidate; and the running mate, Walter Mondale. The result is that the female candidate is not shown more different than its competitor, or she has similar importance in terms of face. Another study on political figures is the study of Konrath and Schwarz (2007) on two different sets of samples. In the first study, the portraits of governors, senators, and representatives in the USA are used via the face-ism index identified by Archer. According to the results of the analysis of the coded photographs, male politicians in the USA have a more facial preference or face-index than female politicians. In the second study, Konrath and Schwarz (2007) conducted another study on the larger sample group in the parliaments of Canada, Norway, and Australia. In January 2005, photos were taken from different websites – 82 from Canada, 46 from Australia and 33 from Norway were coded. The analysis of the data shows that male politicians in Canada, Australia and Norway reflect themselves with more facial importance. In another study on Face-ism, Dodd and his colleagues (1989) try to investigate whether there is a face-ism index or not (ratio of face length to total body length) based on the assumption that body of females are front while the faces of males are more focused on the covers of American magazines. The study had two purposes: to determine whether the face-ism finding occurs on magazine covers and to determine the extent to which it is moderated by social role (Dodd et al., 1989). The cover pages of Time and Newsweek magazines published regularly between 1930 and 1980 were analyzed in the research. The covers examined for decennium under the scope of the study; however, the cover photograph of 1940 was excluded from the study due to the intense effects of the Second World War on the societies. When the results are examined, it can be seen that while there are fluctuations in the face-body ratio over the years, men (58) have a higher face-index (ratio of face length / total body length) than women (53). However, there was no significant increase or decrease between sexually determined periods.

In another study carried out by A. Copeland in 1989 on face-ism, it is indicated that women and men are shown differently from each other during the time zone when the most watched television shows, reality shows, entertainment, and matches are on air. Moreover, whether there is a differentiation between single and multi-camera programs not under the face-ism influence is another study area of this study. In order to prove the validity of the proposition given in the first hypothesis, the programs between June 14th and 27th were recorded and used in the selection of the sampling universe using a two-step random sampling technique. In the study, a total of 14 programs were selected (8 from ABC, 2 from CBS, 1 from Fox TV and finally 3 from NBC) (Copeland, 1989). Generally, in these programs scenes, which include over shoulder shoots and no person in the shooting plan, are coded. According to the data resulted from coding, there are 74 people totally participating in the study, and according to the distribution in terms of sex and age, the 48 of males and 26 of females are evaluated in terms of send and 61 of adults, 6 of young, 7 of children are evaluated in terms of age and constitute a set of samples. As in previous studies, this research has also achieved results that support face-ism. On the other hand, it is seen that women tend to care about their bodies and men care about their face. The values of close shots in which face is in the forefront in the prime times when the programs such as movies, reality shows, entertainment, and matches are on air and the TV channels have reached the highest number of people are higher than women (faces and body are seen) while in the wide shots the values of women are higher than men. In another study on broadcasting Iijima Hall and Crum recorded advertisements about sporting events at ABC, CBS, and NBC for four weeks between 1991–1992, and 59 different beer advertisements constituted the sample size of this work (Iijima Hall & Crum, 1994). Photographs of women that show a greater proportion of their bodies than their heads may serve to reinforce stereotypical images. They found a significant difference in the presentation of women's and men's bodies. For women, there were significantly more camera shots focusing on specific parts of the body (chests, buttocks, legs, crotches) than for men (Hall & Crum, 1994).

In the study conducted by Matthews, six popular journals in the United States (Newsweek, Time, Fortune, Money, People, and Sports Illustrated), created the sampling scale of the study. Each of the magazines in the mainstream media is aimed at a different mass of readers. For example, Newsweek and Time are news-focused, Fortune and Money are business-focused, and People and Sports Illustrated are entertainment-focused magazines. Four of the six magazines, totally 24 issues of each published in 2004 included in the study, and 779 photographs in the magazines were coded. The photographs in the research population are classified into two groups according to their professions in order to simplify and make the data meaningful. In the first group, the print media are classified as intellectual-focused photographs that represent intellectual-focused people in different professional groups such as politicians, businessmen, doctors and academicians. Photographs reflecting people who are famous for their physical features included construction workers, athletes, models and actors and are classified as physically focused photographs. Matthews assessed the relationship between professional status and gender between women and men in the photographs and found that gender differences between men and women have become more apparent when the occupational situation is taken into account (Matthews, 2007). When the photographs in the mainstream magazines are examined, it is seen that men who are shown in intellectually oriented careers (doctors, architects, politicians, etc.) have a higher face-ism index or more facial significance than females. On the other hand, it has been shown that men who have more physical-focused features (such as sportsmen, actors, and actors) are more likely to have a lower face-ism index or fewer facial features than females. In brief, when the characteristics related to the careers and certain qualities that reflect intellectuality come into prominence, men have more facial importance than females and when the physical characteristics are more specific or foreground, the result is that men have more physical orientation than females. When Matthews evaluates the results, he reaches the conclusion that the face-ism effect is still present in the magazines that are heading to different target groups, but it tends to show itself in a less specific way.

Method

Participants

This study examined the photographs and user-provided information at online social network sites (i.e., Facebook). A total sample of 413 profile pictures of female and male with faces of active Facebook users (ages 25–65+) from four separate nations (United States, Germany, Japan, and Turkey) were randomly selected by Facebook from September 12th to September 27th, 2017.

Procedure

Two independent raters coded each profile picture according to the coding scheme developed during the study. First, the data were coded by gender (217 males and 196 females), the age range of the sample (25-34, 35-44, 45-54, 55-64, and 65+), the number of Facebook friends (0-250, 251-500, 501-750, 751-1000, 1000+), part of the face (Left, Right, Center), mood (Positive, Negative, and Neutral), color type (black/white, color), eye contact (Yes, No) and the face-ism index. Then, the pictures were measured using the face-ism index (Cheek, 2016; Konrath & Schwarz, 2007; Smith & Cooley, 2012; Peng et al., 2008). Facebook profile pictures were selected based on the criteria suggested by Archer et al. (1983). The criteria used in the selection of the photos are as follows: (a) Profile pictures may only contain one person, other person or animal should be not involved in the photographs, (b) profile pictures that aim to capture a region of the body, a movement or a problem cannot be considered such as clothes or cosmetics, health products, using tools, in weight loss ads, etc., (c) profile pictures with a "co-subject" cannot be used such as a vehicle or structure, a person driving in a car, etc., (d) websites of celebrities or of a commercial nature were not included, and (e) photographs of disembodied heads are not counted. This is the reason why it has become the focus of face-ism researches.

Part of the Face

The position of the face is coded according to the camera viewpoint (L or R). If the face is equally spaced on both sides (L or R), this was coded as the center (C). At this point, it was not taken into consideration whether the head angle is up or down.

Eye Contact

If person's eye gaze was directed to someone looking at the photo, this was coded as the eye contact. Also, if the person has used sunglasses or something else preventing glance, it was coded as no eye contact.

Mood

Each picture was coded to the positive, negative, or neutral mood condition. The positive facial expressions category includes happy, smiling and surprised, etc. The negative facial expressions were assigned to high tension, low energy, anger, disgust, fear, and sadness, etc. The neutral facial expressions include all other expressions that were clearly positive or non-negative.

Face-ism Index

Face-ism was measured by the face-ism index measures the distance (in millimeters or any other unit). According to the method used by Archer and colleagues (1983), the face-ism index is the ratio of two measures: Measurement one is a linear measurement from the top of the head to the lowest point of the chin, and the second measurement is a linear measurement from the top of the head to the lowest visible part of the subject's body (see Fig. 1).

Research Questions/Hypothesis

The study aims to answer the following questions:

RQ1: What was the significant relationship between the gender and self-presentation tendency in the online profile pictures on Facebook?

RQ2: What role does nationality play in online self-presentation of the online profile pictures on Facebook?

Statistical Analysis

The data was compiled and analyzed by SPSS for frequency, percentages and ANOVA analysis. For each picture, the dependent variable is the face-ism index

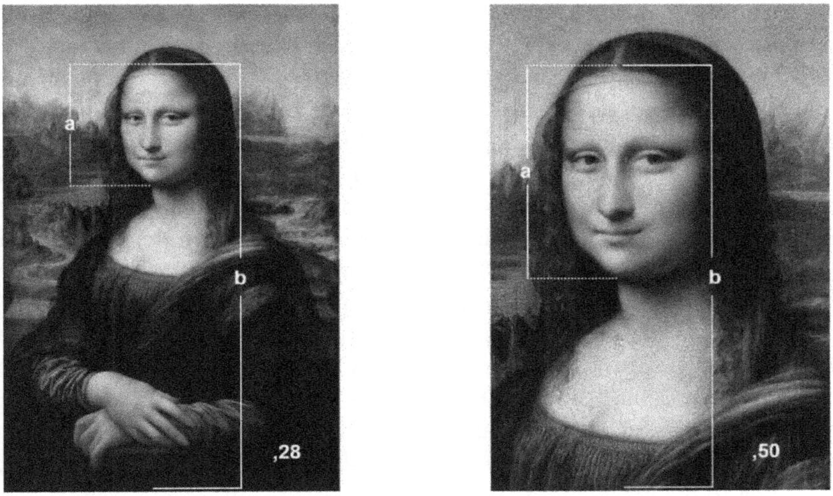

Fig. 1: Illustration of the measurement of the face-ism index by means of the picture of a female, with horizontal, vertical and diagonal lines indicated and additional lines for the face-ism calculation.

score. Age, gender, nationality and the number of Facebook friends are often treated as independent variables. Photoshop CC software was used to measure the dimensions of the photographs.

Results and Discussion

In this study, the self-presentation of the social media is analyzed according to various variables and firstly, a descriptive analysis is done. At the basis of the study, it was examined whether the relationship between gender and cultural differences is effective on self-presentation or not examined, according to the relational screening model with reference to the face-ism index. In the relational screening model, statistical analysis techniques are used to determine the relationships between dependent and independent variables. Therefore, while creating a sampling group, it is paid attention to ensure that the gender and the age distributions are balanced with countries. Intercultural communication has gained importance after the recognizing the features that distinguish subcultures from each other in many countries of the world and starting to investigate their effects on communication. Social networks have an important place in the social sciences because they have the opportunity to mobilize and use resources in any context by using the power of social relations established between individuals or groups. The network analysis method is used to examine the individual and group structure, the network structure is determined, the relations that are not easily observed among the communities are visualized and the existing connections are being modeled. When the distribution of the gender of participants is examined, it is seen that 217 (52.5 %) of them are men and 196 (47.5 %) of them are women. According to age groups, 100 people (24.2 %) are in the age range of 25–34 years, 89 people (21.5 %) are in the age range of 35–44 years, 122 people (15.3 %) are in the age range of 44–54 years, people (15.3 %) are in the age range of 63 and 39 people (9.4 %) are in age range of 65 and over. When the distribution according to countries is examined, 104 people (25.2 %) are from Turkey, 100 people (24.2 %) are from Germany, 107 people (25.9 %) are from the USA and 102 people (24.7 %) are from Japan in the study. The previous researchers were examined and the sample group is determined to have an age range of 25 and above. When the distributions of the profile pictures according to the main category and subcategories in Tab. 1 were examined, it is seen that 64.4 % of the participants used the center direction in their profile pictures, 19.9 % used the left side and 15.7 % used the right side. When the study of Musil and his colleagues (2017) are examined, it is determined that the direction of the face in the profile pictures is similarly in the center (n = 266,

Tab. 1: Online Profile Pictures on Facebook Coding Categories and Subcategories

Main category	Sub-category	Frequency	Percentage	Kappa[a]
Part of the face	L	82	19,9	0.89
	R	65	15,7	
	C	**266**	**64,4**	
Mood	Positive	**245**	**59,3**	0.70
	Negative	7	1,7	
	Neutral	161	39,0	
Color type	Color	**379**	**91,8**	0.24
	Black/white	34	8,2	
Eye contact	Yes	**331**	**80,1**	0,36
	No	82	19,9	
Face-ism index[b]		0.59	0.20	

Bold numbers represent modal values in each category.

[a] Fleiss' kappa

[b] Values in the columns are mean and standard deviation.

64.4 %). In the second dominant direction, it is differently seen that the direction is to the left side (n = 82, 19.9 %) in this study. In the profile pictures, some modes are defined as the reflection of the moods according to the person's facial expression.

In general, the mode is described as the moods in a "certain time". When the photographs are examined, positive (n = 245, 59.3 %) and neutral (n = 161, 39 %) modes are found to be at the forefront. In the distribution of sex, it is seen that women (65 %) are more positive than men (53.9 %). When the distribution of the modes according to the countries are examined, Turkey is the first in positive mode with 79 % and the USA is the second with 68.2 %. Japan is the lowest in the positive mood with 36.3 %. In negative mode, Japan is the first with 2.9 % and Germany is second with 2 %.

When the distribution between age and mode is examined, it is that age group of 25–34 (63 %) and 65 has (64.1 %) more positive view. Within the scope of the research, the relationship between the number of participants' friends in social media and the mood of participants are examined. According to the results, there is no connection between the increase in the number of friends and a positive or negative mood. While there are many psychological and sociological factors at the basis of people's moods, it is also assumed that they are influenced by the standards and rules called norms set by the individuals forming the society (Konrath, Au, and Ramsey, 2012). This situation, which emerges as compliance or obedience, reveals the result of behaviors as human beings should do not they want.

Facebook users also prefer black-and-white (n = 34, 8.2 %) images, although they usually use color pictures (n = 379, 91.8 %). While this is not a single cause, black and white photographs are considered to be nostalgic. According to the results of the study, it is seen that the users between the age of 45-54 years (n = 13, 10.7 %) and 65 and above (n = 4, 10.3 %) prefer black-and-white photographs more than other age groups. According to gender differences, there is no significant difference between the selected color in the photograph.

There are many studies on whether there is any variability in the degree of importance between the face and the body (face-to-body ratio) between women and men in the 21st century or not, if there is, in which level does this variability occurs. A meta-analysis of studies from 1974 to 2007 found women were more likely than men to have non-speaking roles, to be presented in a role dependent on others, to be presented in a domestic environment and role, and to give a non-scientific opinion on a product (Eisend, 2009). Psychological and sociological judgments about individuals and societies with different characteristics are obtained by index value with the result of these studies. In this study, face-ism

Tab. 2: T-test Results for the Face-ism Index by Gender

	N	Mean	SD	Source	SS	df	MS	t	p
Male	217	,6295	,20264	Between	,679	1	,679	16,679	,000
Female	196	,5483	,20074	Within	16,727	411	,041		
Total	413	,5910	,20554	Total	17,406	412			

index is taken into consideration according to the variables of gender, age and country. Moreover, the number of friends who determine the power of the network on Facebook and the point of view of the profile pictures are used as the other variables related to the face-ism index.

The difference between self-presentation and self-esteem is examined by using the face-ism index. According to the result of the independent group t-test, the difference between the arithmetic average of the groups is found to be statistically significant ($t = 16.667$; $p <.000$). According to the sex distribution, the face-ism index of males ($\bar{X} = 0.62$) is higher than females ($\bar{X} = 0.54$). According to these results, in the social media profiles, men want to get their faces to be seen even though women want their bodies to be seen. Those researchers in this domain have consistently found that generally, people have certain stereotypes of women, which are identified in three chief subgroups: homemaker, career woman, and sex object (Deaux et al., 1985; Six & Eckes, 1991). Collins (2011) explain that women are not only underrepresented in the media, but when women are represented they are often sexualized, shown in traditionally feminine rolls and "subordinated in various ways, as indicated by their facial expressions, body positions, and other factors".

In many studies, in which portraits are drawn in different periods, as well as visual presentations such as photographs and drawings in newspapers, magazines and TV programs, are examined; it is seen that men give more importance to the facial area (head-to-body ratio) than women (Dodd et al., 1989; Copeland, 1989; Matthews, 2007; Zuckerman, 1986). They have found that the mean face-to-body ratio for photos of men was 0.65 (more than half of the photo was dedicated to facial features), whereas the face-to-body ratio for photos of women was 0.45 (more than half of the photo was dedicated to the body). In this study as in previous studies, it is seen that the faces of the men and the physical characteristics of the women are in the forefront.

According to the results of one-way analysis of variance (ANOVA) the relationship between age and face-ism coefficients and the difference between the groups is found to be statistically significant ($F = 12,874$; $p <.000$). According to

Tab. 3: Analysis of Variance (ANOVA) between Face-ism Index and Ages

	N	Mean	SD	Source	SS	df	MS	F	p
25-34	100	,4886	,21162	Between	1,951	4	,488	12,874	,000
35-44	89	,5955	,18283	Within	15,455	408	,038		
45-54	122	,5971	,19778	Total	17,406	412			
55-64	63	,6579	,18031						
65+	39	,7164	,18717						
Total	413	,5910	,20554						

Turkey HSD results, there is a significant difference between those in the age range of 25-34 (X = 0.48) and those in the other age range. When the age distributions according to Face-ism index are examined in general, it is seen that there is no proportional link. In general, social media tools are used by younger generations who have become more adopted at the age of information technology. The mass media use individuals' sociological and psychological arguments to direct the information, preferences, attitudes, and behaviors of individuals. When the relationship between age and face-ism index is examined, it is determined that the highest average of age is 65 years and over ($\bar{X} = 0.72$). It is often seen that people in this age group prefer using portraits taken earlier to the photographs taken with mobile devices for their profiles.

Cultural symbols and practices also provide important clues to determine causes behind human behaviors. In cultures where emotions are less needed, repressed emotions are not sufficient in shaping behavior, because the frequency or intensity of individual emotions is low. Examining intercultural differences, Hofstede (1980) focuses on the relationship between strongest and weaknesses in the society via power distance (Power Distance). A classification is made between narrow power range and wide power range according to the degree of acceptance of equality in the power distribution of people who have lesser power in the institutions or establishments of a country. This dimension refers to how members of a society treat the inequalities of opportunity against the power centers and the role of organizations and institutions in the balance of this structure. The main issue here is how a society manages inequalities among people. The face-ism index is used to determine the cultural differences reflected in the social media. This index is also used for the systematic classification of the personality traits of people in different cultures. National identities consolidate the society by purifying the traditional and ethnic differences in different parts of society while at the same time strengthening the feelings of belonging

Tab. 4: Analysis of Variance (ANOVA) between Face-ism Index and Countries

	N	Mean	SD	Source	SS	df	MS	F	p
Turkish	104	,4972	,22463	Between	1,246	3	,415	10,508	,000
Germany	100	,6260	,20756	Within	16,160	409	,040		
USA	107	,6109	,20346	Total	17,406	412			
Japan	102	,6314	,15123						
Total	413	,5910	,20554						

to a society accepting individuals as unique have their origins in the past and members of an abstract family in which millions of people live together harmoniously. According to Tab. 4, people in the developed countries are more self-confident that they reflect these characteristics in profile pictures. According to Zuckerman and Kieffer (1994), previous studies suggested that higher facial prominence conveyed impressions of dominance and control while lower facial prominence conveyed warmth and compassion. Another factor that affects the face-ism index is the personality. The model also known as the Big Five is regarded as a measurement system developed from the language hypothesis that defines individual differences based on the first theoretical studies on the psychology of personality. According to the results of the studies, the five-factor personality model is defined as extraversion, agreeableness, conscientiousness, emotional stability and openness (Goldberg, 1990). The five-factor structure is accepted as a valid and important resource in determining personality traits of human beings. This model relies on the "feature-based approach" which handles behaviors at the individual dimension and benefits from "adjectives" that are used to describe people themselves and others. The manner in which the person presents himself / herself visually in social media is also considered to be indicative of personality (Sutherland et al., 2015; Walker & Vetter, 2016; Teijeiro-Mosquera et al., 2015). Faces with extremely high reliability appear positive mode; whereas the faces with an extremely high degree of domination appear physically strong.

When the distributions of face-ism index averages according to countries are analyzed, Germany is ($\overline{X} = 0.63$), USA ($\overline{X} = 0.61$), Japan is ($\overline{X} = 0.59$) and Turkey is ($\overline{X} = 0.50$). According to these results, one-way variance analysis (ANOVA) is performed to determine cultural differences ($F = 10{,}508$; $p < .000$). According to the Tukey HSD results, the face-ism index of the users in Turkey ($\overline{X} = 0.50$) is lower than the other countries. According to these results, the users in Germany, USA and Japan wanted their bodies more important.

Many network theories have been established in connection with the number of people in social networks. In general, the number of people in the network is considered as a value. Especially when the popularity of Facebook and other online social networks are considered, the populace in recent years with their billions of users, analysis interests of researchers with related social networks and issues have increased. There is no consensus on how to quantify the value of social networks while using several methods to quantitatively determine the transformation of social networks into a social capital. Fukuyama evaluates his study methods in two categories. The first is quantitative studies focusing on the quality of group membership in the society and on the number of groups; and the second, is a set of studies that determine trust and citizenship obligations via a set of developed scales (Fukuyama, 2001).

As shown in Tab. 5, the difference between the average of the groups is found to be statistically significant as a result of one (ANOVA) conducted to determine whether the average of the scores according to the frequency of social networking potential friends are significantly different from the face-ism index ($F = 4{,}057$; $p < {,}003$). According to the Tukey analysis to determine the source of the difference, the difference between the face-ism index averages of users with 0-250 friends ($\bar{X} = 0.64$) and 750-1000 friends ($\bar{X} = 0.51$) on Facebook is found significant. According to another result, face-ism index is found to be low compared to those with 751-1000 friends ($\bar{X} = 0.51$) and those with 1000 or more friends ($\bar{X} = 0.62$). Accordingly, users who have friends between 0-250, 1000 and above let their face more forefront than users who have friends between 751-1000.

On Facebook, users are always trying to be more effective when choosing profile pictures. Technically, those who have no specific background make only an aesthetic assessment. Benefit from the advantages of the angles to create

Tab. 5: Analysis of Variance (ANOVA) between Face-ism Index and the Number of Facebook Friends

	N	Mean	SD	Source	SS	df	MS	F	p
0-250	65	,6385	,21498	Between	,666	4	,166	4,057	,003
251-500	96	,5547	,19754	Within	16,740	408	,041		
501-750	64	,5827	,24529	Total	17,406	412			
751-1000	43	,5130	,21575						
1001+	145	,6206	,17339						
Total	413	,5910	,20554						

Tab. 6: Analysis of Variance (ANOVA) between Face-ism Index and the Part of the Face on Facebook

	N	Mean	SD	Source	SS	df	MS	F	p
Left	82	,5347	,17983	Model	,384	2	,192	4,626	,010
Right	65	,6320	,21134	Error	17,022	410	,042		
Center	266	,5983	,20871	Total	17,406	412			
Total	413	,5910	,20554						

impressive photos does not always give the desired results. Photographs taken at eye level are always ideal, but light, sharpness, the structure of the face, and head angle determine the effect of photography. In Tab. 6, the relationship between the face-ism index and the looking directions of profile pictures are examined. According to the analysis results, it is observed that the three-way (left, right, center) affects face-ism index ($F=4{,}626$; $p<{,}010$).

References

Archer, D., Iritani, B., Kimes, D. D., & Barrios, M. (1983). Face-ism: Five studies of sex differences in facial prominence. *Journal of Personality and Social Psychology*, *45*(4), 725–735. doi:10.1037/0022-3514.45.4.725

Bronstein, J. (2013). Personal blogs as online presences on the internet. *Aslib Proceedings*, *65*(2), 161–181. doi:10.1108/00012531311313989

Cheek, N. N. (2016). Face-ism and objectification in mainstream and LGBT magazines. *PLOS ONE*, *11*(4), e0153592. doi:10.1371/journal.pone.0153592

Collins, R. L. (2011). Content analysis of gender roles in media: Where are we now and where should we go? *Sex Roles*, *64*(3-4), 290–298. doi:10.1007/s11199-010-9929-5

Copeland, G. A. (1989). Face-ism and primetime television. *Journal of Broadcasting & Electronic Media*, *33*(2), 209–214. doi:10.1080/08838158909364075

Deaux, K., Winton, W., Crowley, M., & Lewis, L. L. (1985). Level of categorization and content of gender stereotypes. *Social Cognition*, *3*(2), 145–167. doi:10.1521/soco.1985.3.2.145

Dodd, D. K., Harcar, V., Foerch, B. J., & Anderson, H. T. (1989). Face-Ism and facial expressions of women in magazine photos. *The Psychological Record*, *39*(3), 325–331. doi:10.1007/bf03395884

Dodd, D. K. (1991). *Dynamics of intercultural communication*, A: Wm.C, Dubuque: Brown Publishers.

Eisend, M. (2009). A meta-analysis of gender roles in advertising. *Journal of the Academy of Marketing Science, 38*(4), 418–440. doi:10.1007/s11747-009-0181-x

Fiske, S., & Taylor, S. (1991). *Social cognition*. New York: McGraw-Hill.

Fukuyama, F. (2001). Social capital, civil society, and development, *Third World Quarterly, 22*(1), 2001, 7–20.

Goffman, E. (1959). *The presentation of self in everyday life*. New York: Anchor Books.

Goffman, E. (2009). Günlük Yaşamda Benliğin Sunumu (B. Cezar, Çev.). Metis Yayınları, İstanbul.

Goldberg L. R. (1990). An alternative "description of personality": the big-five factor structure. J. *Pers. Soc. Psychol. 59*, 1216. doi:10.1037/0022-3514.59.6.1216

Gonzales, L. & Vodicka, D. (2010). Top ten internet resources for educators, *Leadership, 39*(3), 8–37.

Han, T., & Keskin, F. (2016). Using a mobile application (WhatsApp) to reduce EFL speaking anxiety. *Gist Education and Learning Research Journal. 12*. 29–50.

Hofstede, G. (1980). *Culture's Consequences: International Differences in Work-Related Values*. Beverly Hills, CA: Sage.

Hofstede, G. (1993). Interkulturelle Zusammenarbeit: Kulturen - Organisationen - Management, Wiesbaden: Springer. doi:10.1007/978-3-322-90037-1

Hofstede, G. (2001). Culture's Consequences (2nd edition.), Thousand Oaks, CA: Sage Publications.

Hofstede, G. (2004). *Cultures and Organizations: Software of the Mind*2nd edition, , Boston: McGraw-Hill.

Hofstede, G. (2006). What did GLOBE really measure? Researchers' minds versus respondents' minds. *Journal of International Business Studies, 37*(6), 882–896. doi:10.1057/palgrave.jibs.8400233

Hornby, A. S. (1995). *Oxford Advanced Learner's Dictionary of Current English*. Oxford UK: Oxford University Press.

Iijima Hall, C. C., & Crum, M. J. (1994). Women and ? body-isms? in television beer commercials. *Sex Roles, 31*(5–6), 329–337. doi:10.1007/bf01544592

Konrath, S. H., & Schwarz, N. (2007). Do male politicians have big heads? Face-ism in online self-representations of politicians. *Media Psychology, 10*(3), 436–448. doi:10.1080/15213260701533219

Konrath, S., Au, J., & Ramsey, L. R. (2012). Cultural differences in face-ism: Male politicians have bigger heads in more gender-equal cultures. *Psychology of Women Quarterly*, *36*(4), 476–487. doi:10.1177/0361684312455317

Lee, S.-J., Quigley, B. M., Nesler, M. S., Corbett, A. B., & Tedeschi, J. T. (1999). Development of a self-presentation tactics scale. *Personality and Individual Differences*, *26*(4), 701–722. doi:10.1016/s0191-8869(98)00178-0

Lippmann, W. (1922). *Public Opinion*. New York: Harcourt, Brace and Company.

Matthews, J. L. (2007). Hidden sexism: Facial prominence and its connections to gender and occupational status in popular print media. *Sex Roles*, *57*(7–8), 515–525. doi:10.1007/s11199-007-9276-3

Miladi, N. (2016). Social media and social change. *Digest of Middle East Studies*, *25*(1), 36–51. doi:10.1111/dome.12082

Miller, D., Costa, E., Haynes, N., McDonald, T., Nicolescu, R., Sinanan, J., Spyer, J., Venkatraman, S., & Wang, X. (2016). *How the World Changed Social Media*. London: UCL Press.

Mislove, A., Marcon, M., Gummadi, K. P., Druschel, P. & Bhattacharjee, B. (2007). Measurement and analysis of online social networks, *Proceedings of the 7th ACM SIGCOMM Conference on Internet measurement*. San Diego, California, USA.

Musil, B., Preglej, A., Ropert, T., Klasinc, L., & Babič, N. Č. (2017). What is seen is who you are: Are cues in selfie pictures related to personality characteristics? *Frontiers in Psychology*, *8*: 1–13. doi:10.3389/fpsyg.2017.00082

Ndubisi, N. O., Gupta, O. K., & Ndubisi G. C. (2005). The Moguls' model of computing: Integrating the moderating impact of users' persona into the technology acceptance model, *Journal of Global Information Technology Management*, *8*(1): 27–47.

Peng, Z., Cher, M., Chan, C., Hsiao, H., & Li, X. (2008). *Gendered faces in MySpace: Self-inflicted face-ism in online profile photos*. Paper presented at The Association for Education in Journalism and Mass Communication, Chicago, IL.

Porter, R. E., & Samovar, L. A. (1996). *Cultural influences on emotional expression. Handbook of Communication and Emotion*, 451–472. doi:10.1016/b978-012057770-5/50019-9

Reichart, L. M., & Cooley, S. C. (2009). *Face-ism on Facebook: An analysis of self-inflicted face-ism in online profile pictures*. Paper presented at the annual meeting of the International Communication Association, Chicago, Illinois.

Schlenker, B. R., & Pontari, B. A. (2000). *The strategic control of information: Impression management and self-presentation in daily life*. In A.

Tesser, R. B. Felson, & J. M. Suls (Eds.), Psychological perspectives on self and identity, pp. 199–232.

Schutz, A. (1998). Assertive, offensive, protective, and defensive styles of self-presentation: A taxonomy. *The Journal of Psychology*, *132*: 611–628.

Six, B., & Eckes, T. (1991). A closer look at the complex structure of gender stereotypes. *Sex Roles*, *24*(1–2): 57–71. doi:10.1007/bf00288703

Smith, L. R., & Cooley, S. C. (2012). International faces: An analysis of self-inflicted face-ism in online profile pictures. *Journal of Intercultural Communication Research*, *41*(3): 279–296. doi:10.1080/17475759.2012.728771

Smith, L. R., & Sanderson, J. (2015). I'm going to Instagram it! an analysis of athlete self-presentation on Instagram. *Journal of Broadcasting & Electronic Media*. *59*(2): 342–358. doi:10.1080/08838151.2015.1029125.

Sparks, G. G., & Fehlner, C. L. (1986). Faces in the news: gender comparisons of magazine photographs. *Journal of Communication*, *36*(4): 70–79. doi:10.1111/j.1460-2466.1986.tb01451.x

Sutherland, C. A. M., Rowley, L. E., Amoaku, U. T., Daguzan, E., Kidd-Rossiter, K. A., Maceviciute, U., & Young, A. W. (2015). Personality judgments from everyday images of faces. *Frontiers in Psychology*, *6*: 1616. doi: 10.3389/fpsyg.2015.01616

Teijeiro-Mosquera, L., Biel, J.-I., Alba-Castro, J. L., & Gatica-Perez, D. (2015). What your face vlogs about: Expressions of emotion and big-five traits impressions in YouTube. *IEEE Transactions on Affective Computing*, *6*(2): 193–205. doi:10.1109/taffc.2014.2370044

Vinacke, W. E. (1957). Stereotypes as social concepts. *Journal of Social Psychology*, *46*(6): 229–243.

Walker, M., & Vetter, T. (2016). Changing the personality of a face: Perceived Big Two and Big Five personality factors modeled in real photographs. *Journal of Personality and Social Psychology*, *110*(4): 609–624. doi:10.1037/pspp0000064

Zuckerman, M. (1986). On the meaning and implications of facial prominence. *Journal of Nonverbal Behavior*, *10*(4): 215–229. doi:10.1007/bf00987481

Zuckerman, M., & Kieffer, S. C. (1994). Race differences in face-ism: Does facial prominence imply dominance? *Journal of Personality and Social Psychology*, *66*(1): 86–92.

Mehmet Fatih Yelmen

Social Media in Turkey in the Context of Futurism: Twitter Case

Introduction

Technological developments, one of the most fundamental dynamics of the twenty-first century, have played a decisive role in valuing time and life. These developments often take place in history with providing convenience about several subjects for people. Every invention that will make life easier has not happened yet. The above-mentioned phrase "to value time" is used to indicate the relation between man and time.

Some inventions, photography, television, etc., have actively taken place in life as a product of research, curiosity and effort that has lasted for decades or even centuries. For this reason, the primeval era is distinguished into the long periods of time that are completely different. The reason why primitive man is in such a slow developmental process can be found in the interaction between languages and mind because language is the carrier of thought, and therefore the thought created by mind can solely maintain its existence via the language and its opportunities. Also, it is unthinkable that the rational approach to perceive life is independent of this process. Developments that have occurred for quite a long time have rapidly increased as people begin to use their minds, time and economic power more intensely in technological areas. In fact, the human mind is in a race against time. It would be correct to look for the origin of this situation in the early twentieth century because the twentieth century is the period when mankind begins to have the benefits of the rational thought that has flourished in the enlightenment era. In particular, with the invention of the printing house, written and permanent works, in other words, the thoughts including science, literature and art have spread all over the world. As a result, not only individuals but also societies have become aware of technical innovations and all kinds of related developments. Developments have begun to become social and even global except individual one. An example of this situation: the artists of Futurism, which can be seen as a social movement including the period of First World War, represent quite archetypically how passionate modern people are related to technology due to their speed and technology obsessions. On the other hand, Jacques Attali (1999: 10), in regard to the examination of the future, emphasized that humanity gives much importance to technology, whereas traditions, social development and art are underestimated. However, the futurists refused the tradition and considered that the technology they admire was not separable from the social development and the modern world; on the contrary they thought that the development could occur to include all these things.

In this study, the argument that innovations in visual and discursive aspects are closely related to each other and the social aspect of technology are handled in the context of futurists; the effect of the use of Twitter in Turkey on the language will be examined.

Technology, Communication and Language

The word technology taken from French is defined as "The information that includes construction methods, equipment and tools related to a branch of industry" (TDK, 1992: 1443). While this definition is correct, the description, which is predominantly technique-based, is not adequate when social influences are considered because the sharing of knowledge at the social level has become meaningful through some practical processes. For example, transportation, in other words communication and transfer of information, was provided with animals till the invention of automobile. When technological developments are considered as a series of events, it is immediately recognized that transportation with the twenty-first century jet engines or the possibilities provided by internet technologies have a speed that is incomparable to the situation in the nineteenth or early twentieth century, and it can be clearly observed that this makes life easier. The futurists, who were the representatives of the literary artistic movement created by the young writers and a group of artists to revolt against the academia/academic tradition "in the most severe way" (Müller, 1972: 72), mainly perceived the technology in this sense. With the most pragmatic side, technology can be considered in terms of offering various possibilities in medicine. For example, on the one hand, there are significant technological developments in fighting diseases like cancer and finding solutions for these kinds of diseases, on the other hand airplanes, helicopters, etc., which are considered as the blessings of technology and provide speed and convenience, are utilized in a situation like organ transplantation that requires urgent intervention. In this context, it is seen that the interest of Futurism about technology, speed and dynamism is transferred into the subjects such as urban and industrial themes, airplanes, automobiles and trains rather than a new visual understanding at the formal level (Antmen, 2008: 67)."

When technology is imagined in the form of a ladder standing in a linear and vertical space in terms of process, it draws an attention that technology has a relation with previous steps, in other words, it is in an eclectic structure. Therefore, the face of technology is always futuristic and innovative as mentioned by Futurists. Even though the artistic and philosophical definition of the innovation in the Futurist manifest has not been written completely, it can be said that the approach of the Futurists about this issue is similar to Heidegger in philosophical sense.

According to Heidegger (Çüçen, 2000)

> "Technology in the true sense (techné) means to reveal the existing one, put forward, expose and demystify it. But technology has lost this meaning with modern philosophy, it becomes the one that control and determine the existing one instead of revealing it. Modern technology has covered the existing one as to surround it. (...) Science and technology are nothing more than the construction of a subject (human being) as a human product. Modern science does not allow the existing one or the Being to be revealed, but tells or determines how it should be. Generally, technology is defined as the application of science to the practical field. Science creates theoretical process and technology composes the practical process."

The above explanations of Heidegger exactly tally with those of the futurists trying to do in the literature. Based on the framing design in modern philosophy, they attacked the patterns of the existing language and made it necessary to overstep the boundaries of literary expression. The intervention in the language that forms the basis of oral and written communication will undoubtedly cause the social mentality to be affected. It is an important detail that the Italian futurists under the leadership of Tommaso Marinetti intervene in the literary language instead of the languages of science and philosophy, because literature is the organism of the society that has the longest permanence in communication in terms of the language. The lyricism, which Marinetti has idealized by glorifying against long and elaborated narratives, is based on the most effective expressions of enthusiasm and emotions with short sentences. Thus, there is a stronger and more sincere relation between the poet and the reader according to Futurist viewpoint. Marinetti explains his ideas to say that "The need to say short and wise words does not only correspond to the rate laws governing our life, but also to the old relations between the poet and the reader. These relations are similar to the friendship of two old friends who can explain what they think with a single word and a look (Batur, 2003: 77–78)." This metaphorical expression corresponds to a superior reception relation between the sender and the receiver when considering the communication processes, that is to say, the shortest, fastest and most efficient way of conveying the message is implemented between the receiver and the sender.

An idealized understanding of the language based on technology admiration is more organic in terms of the relation between society and communication, compared to technical processes. This relation can also be revealed from a historical perspective with more material examples. As it is known, "the communication process which becomes permanent with written expressions and writings carved into stones and wood, took an important step with the emergence of reproduction techniques in the writing. Books previously written in manuscripts began to publish rapidly (Önür, 2002: 76–77)." Technologically, the invention of the printing press, on the one hand, increases the number of people to whom a written work, in other words a thought, can reach, on the other hand, enables every element appealed to the mental structure of a society to be published as many as they can rapidly and easily go beyond the boundaries. The relation between the operations in the printing house and the technology can be understood that it is defined as "the work of writing, drawing paper, fabric, etc., with the help of a machine using the printed letters (Tdk, 1992: 144)", because the technology has a direct relation with mechanization.

In the years between 1909 and 1916 when Futurism effects were strongest, the printing press, in other words, the print media was the most effective communication tool all over the world. On the other hand, rapid developments in technology were about to provide new opportunities related to printing press and communication. Thus, "The print media was adequate in mass communication until the First World War, after this period, it could be seen that its effect gradually decreased with the development of electronic publishing. Electronic publishing became even better than the print media due to visual advantages (Önür, 2002: 176). "Electronic publishing was based on television and radio since the end of the twentieth century. Along with the millennium, changes began to take place in the definitions and practices of many concepts and values at the local and global level. The traditional media, especially defined as "communication media, communication tools" (www.tdk.gov.tr), acquired a brand new media. There were many different branches of this new media that started to emerge and develop with computer technologies, one of them is now called "social media".

Social Media as a Communication Tool

The present content of communication tools changed as it goes beyond traditional media. The new tools with these changing contents have expanded their fields, and accordingly, it became necessary to create different definitions than the definitions of communication tools belonging to the traditional media. As a result of this, the communication media was called "new media". Haluk Geray (2003: 20) defined the new media as "systems that mass audiences can be included as individual users, and users can interactively access content or applications at different times". One of the communication media in the new media is social media. As a term, social media represents "the tools, services and applications that enable users to interact by using network technologies." (Bostancı, 2010: 35) When it comes to network technologies, social media is just one example of this network in the broad areas of the Internet system because there are tools, services and practices that can be defined as network technologies and interaction but not exactly defined as social media. For example, "Social media tools with a basic understanding of sharing user-based content can emerge as sites that have different aspects like blogs, media sharing sites, social networks (İşlek, 2012: 21–22)." For this reason, the generality in the definition is appropriate but there are still many features about social media. Accordingly, social media sites have four common characteristics (Aktaran Bostancı, 2010: 37):

1. Users can create various kinds of media contents or contribute them.
2. Users can tag contents.
3. Users can evaluate the content either by active voting or by passive usage.
4. Users can create social media networks with other users by identifying common interests such as people and friends.

As mentioned above, the activeness of the users and its rating also provide a differentiation for social media as a communication tool. For example, in the media blogs, "a kind of blog where media workers or columnist comment the daily news in the web page of their institution or another blog site (İşlek, 2012: 28)", the web page of a newspaper allows its readers to comment if they are a member, and this indicates that readers/internet users participate in an active role with boundaries defined. These active readers/internet users can write comments on the news as well as read comments of the other users and communicate with them via the same environment when necessary. However, this type of interactive communication provided by the newspapers offers a quite limited opportunity compared to other examples of social media. When the internet user wants to write a comment in response to the news s/he reads, in other words, s/he wants to communicate more actively, s/he makes a comment using limited number of letters. However, this comment made by the reader is first read by the editor and published on the site if it is deemed appropriate (Photo 1). In this viewpoint, it can be come to a conclusion stated by Özkalp (2000: 122) that "the common point of all mass media is that it gives them a specific synthesis and a common mode of behavior by influencing the thoughts of people." On the other hand, the examination of reader thoughts by an editor indicates that

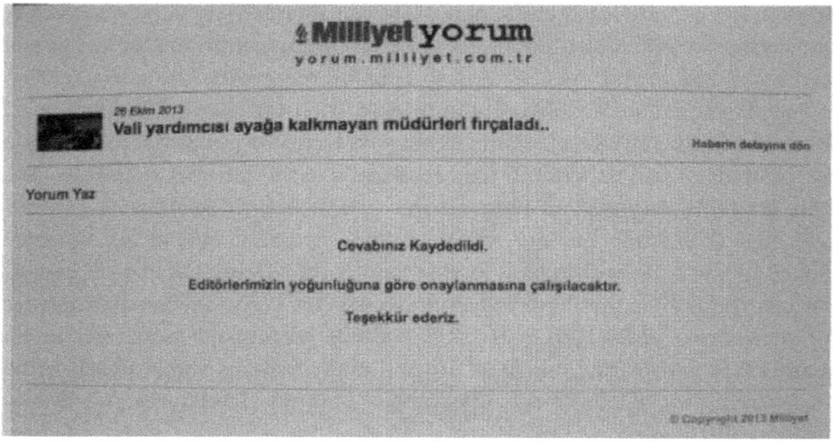

Photo 1: The page of the Milliyet newspaper that evaluates reader comments.

social perception, "the process of learning something about others and society in general" (Mutlu, 2012: 300), is kept under inspection. Thus, it is observed that some of the newspapers, constituting the press part of the social media, alternate between libertarian theory and social responsibility theory, because some newspapers may have the opportunity to express their own opinions, thanks to libertarian view that allows them to stand against the status quo, readers willing to comment are inspected, and this is not an appropriate approach for the social responsibility theory.

According to this theory; "Those who own mass media and those who manages this media should reflect the social goals and expectations by regarding the expectations of the society, and convey the political views with the producers and consumers in a balanced way" (Isik, 2012: 32). In other words, individuals should be able to express their opinions without any inspection. This kind of understanding is like the adaptation of Freud's psychoanalytic theory to the media in terms of the socialization of the individual. As it is known, "Freud dealt with the personality formation of the individual in the socialization as an emotional-motivational process (Tezcan, 1995: 48). According to this, the behaviors of the individuals are shaped by the coordination of three types of mechanisms: id, ego and superego. The editor of the newspaper in social media builds an authority over the individual by interfering in the comments of the individuals like a superego mechanism to censor as necessary or not publish. This includes the characteristics of the authoritarian theory within communication theories because all of the communication tools occur under the control and manipulation of the present authority according to this theory. Besides, it is reminded that the social media mentioned above are quite broad in terms of the definition and the functionality; there are social media tools like Twitter that authoritarian theory has less penetrated.

The Language of Social Media: Twitter Case

Twitter, one of the slightly libertarian areas of social media, functions as a natural flora for the language of an individual or her existence by constituting an environment of free expression. Where does the importance of the language come from, especially in the context of social media? If it is reminded at the beginning of the article that the futurists have a tendency to force their language patterns, the importance of the language can be better understood in terms of a culture of society. In this sense; "The language needs to be reformed, revolted, sometimes reversed like a glove, and sometimes renewed so that it can become a weapon of revolution or even a revolutionary fantasy (Waresquiel's administration, Gavi and Laudier supervised by Waresquiel, 2004: 190)." From the social point of view, the most effective way of renewing the language in its natural interaction is possible only if it is used actively. All kinds of print communication tools are used as a means of this situation, their effects can be observed slower and more difficult than social media. However, Twitter and similar communication tools in social media are faster and easier to observe.

Twitter is a social network and microblogging site. It is a communication tool within the new media that allows the users to write "tweet" or "chirping" called texts of 140 characters; it is stated in its official site that they admit as their mission to give everyone the power to create and share ideas and information instantly, without barriers (https://about.twitter.com/company Access Date: 13.07.2017). Users can use this power with texts of 140 characters in the context of the language and mind; therefore it means that they force to use the potentiality of the language at the same time because of a limited expression. This can only be understood by knowing the evolution of the language in the past and the present.

The relation between human communication and animal communication systems have been examined in some of the studies on the language evolution. Ernst Cassirer (1980: 112), "human speech (language) can be reduced to a basic instinct given to all living creatures. Fear, anger, pain or joys are not peculiar to human beings. We find them all around the animal world." He expressed his view with these sentences that human beings and animals have

Photo 2: The bird with the symbol of Twitter.

similarities in communication. One of the main reasons for this similarity is that humans and animals have phoneme-based expressions. Both species that can use sounds systematically, also use sounds as signs. Thus, Zeynel Kıran (1986: 19) defined the language, "A system of audible signs that allows people to express their feelings and thoughts and to communicate". With the audible signs, animals need to be taken as examples in terms of similarity. However, there have been many studies on the animals in the animal kingdom, which are the most similar animals in the context of voice and communication. Noam Chomsky (2011: 119) indicates that the characteristic W.H.Thorpe argues that non-human mammals seem to lack the ability to imitate human voices, and that it is not 'mammals', but birds (many of which have this ability) are expected to be the group that has the power to develop a language". When this determination is carefully examined, it is clear that the characteristics of the birds' voice and language are the basis of the structure of Twitter. One of them is the bird image that Twitter has used as a symbol (Photo 2). The bird in the logo is a male bird called Larry (http://www.ntv.com.tr/turkiye/twitter-kusunun-adi-larrymis,TFcN98OzbEmZdtrFI_UtdQ Access Date: 13.07.2017) and its color is blue, a universal masculine sign.

Besides, as mentioned above, the word "tweet" means "chirping" in English and, it seems that a similarity can be established between these sentences with 140 characters and a tweet that can be counted as an attempt to develop a language mentioned by Thorpe. As a detail, the expression referring to 140 characters is not the number of letters, for example, the statement "Ali came" does not consist of 7 letters except quotation marks, (Photo 3).

Apart from quotation marks, it is accepted as 9 characters by counting the space between two words and each punctuation mark. Each "tweet" on Twitter

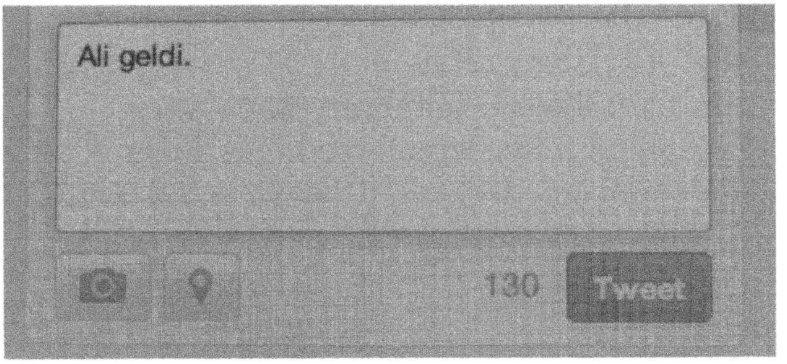

Photo 3: A 140-characters font on Twitter.

contains 140 characters with the space between the words and the punctuation marks. It is more significant that this technical characteristic is similar to the short and intermittent chirping of the birds.

From the point of view of possibilities, language is an unquestionable reality that allows expressing all kinds of feelings and thoughts in the most detailed way. However, it is necessary to have some discipline in using the language so that any situation, feeling or thought can be explained with 140 characters. The first one is the grammatical rules, because a comprehensible semantics of the language can only be created within the grammatical rules of that language. Besides, in a statement limited to 140 characters, anything can be expressed at once by choosing the shorter one of some synonyms. On the other hand, this kind of limitation necessarily affects the structure of mind because the mind will start to make the statements shorter. The philosopher Macit Gökberk (2008: 90), who wrote about Leibniz's thoughts on the German language and was also the head of the Turkish Language Institute, mentioned Leibniz's thought that the language was the mirror of the mind. In this context, it is seen in Leibniz's thought and in the example of Twitter that the limited expression will necessarily affect the mind and therefore, the language will be directly affected.

Punctuation marks, which have symbolic signs accepted as a language and have an international validity in grammar, have occasionally changed due to Twitter. One of the most noticeable one is the "..." sign written at the end of an unfinished statement. The signs of "+" or ">" can now be used in place of this sign that the statement is likely to continue or at least continue with another statement (Photo 4).

Photo 4: An example of the use of punctuation in Twitter.

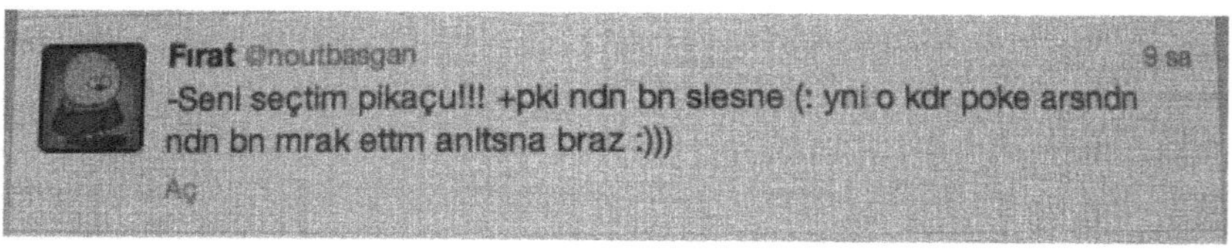

Photo 5: An example of the use of punctuation in Twitter.

Apart from this, a method of writing, which is used by young people especially in their texting with mobile phones by using only consonant letters, is occasionally used on Twitter (Photo 5).

Although it is considered that this kind of writing realizes the rule of gestalt psychology that the mind completes the missing parts, it seems that it contradicts the thoughts of finding language corrections and words appropriate to the spirit of the language which are emphasized in terms of preserving a language. In the context of Futurism, any form of expression, which overturns the grammar rules in Twitter, will be welcomed in this movement, as attacking the language patterns have become possible by breaking its relation to traditional expression. Therefore, a statement limited with 140 characters in

Twitter, which defines itself as a means of expression without borders, shakes the grammar and spelling rules of the language just as it is in Futurism. With the addition of new features, Twitter responds to innovative aspects and new desires of Futurists.

Conclusion

Humans need languages since their existence. It is possible to relate this situation to mental needs. The key of surviving and beat the nature is obtained to put the thoughts into practice in real life from pragmatic and practical perspectives. Such an effort is valuable in terms of individual and society because humans do not have a creation just for themselves; on the contrary, they are eager to be a member of a society as an example of unity and solidarity. For this reason, the language, which is the only way of expressing the thought that the mind creates, does not consist of the use of a single person. Language forms together with society, forms together with society and continues to live like a living being in line with the needs of that society. Lévi-Strauss emphasizes the importance of the relation between language and society in his words; whoever speaks about the people speaks about the language, and whoever speaks about the language speaks about the society (Aktaran: Yavuz, 2013: 17).

Societies that create several civilizations through language and carry them into the future with writing, a form of language, become the actors of an ongoing development process. This process has found the most active state in the technology due to the progress of the technique. All societies with technology have carried their conditions of their existences into another level compared to other societies. These societies, which have a privileged status of appearance, have overturned these processes uneasily; on the contrary, they attack the traditional patterns. The most concrete example of this is seen in the Futurism movement that emerged in Italy and Russia in the first half of the twentieth century. Futurist artists mentioned their admiration for technology, their passion of speed, and their aim of breaking the permanent patterns of language with the manifestos they published and the works they created.

While the technology, which broadens the boundaries of communication, directly related to the mental processes and the language, it is believed to be a contradiction at first sight to attack the language and being a technology enthusiast at the same time, this is not particularly the case for futurists; "Whether it is written, spoken, or silent, there is no doubt that the great movements in history, especially some of the revolutions, are based on breaking the words, violating the oaths, and tearing the contracts (Uygur, 1994: 49). In this respect, the Futurists

Photo 6: Twitter's slogan on "innovation".

did exactly what had to be done. A slogan on "innovation" used by Twitter also summarizes the passion of Futurists for the new.

In this slogan, Twitter is defined as "the best way to discover the new one" and therefore it is not just about communication. At the same time, it promises any kind of innovation that can be achieved with this kind of communication.

It is evident that the language and the mental process are the two factors that necessarily induce and grow each other. With the most rational use of the possibilities provided by both of them, technology has emerged due to technic. However, the technology tries to control the language and the mental process like Uranos, who is afraid of their own children and keeps them imprisoned and Kronos, keeps his children imprisoned in his stomach by eating them.

Kaynakça

Antmen, A. (2008). *20.yy Batı Sanatında Akımlar*. İstanbul: Sel.

Attali, J. (1999). *21.Yüzyıl Sözlüğü*. Kosta Sarıoğlu (Çev.). İstanbul: Güncel.

Batur, E. (2003). *Modernizmin Serüveni*. İstanbul: Yky.

Bostancı, M. (2010). *Sosyal Medyanın Gelişimi ve İletişim Fakültesi Öğrencilerinin Sosyal Medya Kullanım Alışkanlıkları*. (Yayımlanmamış yüksek lisans tezi). Kayseri: Erciyes Üniversitesi Sosyal Bilimler Enstitüsü Gazetecilik Anabilim Dalı.

Cassirer, E. (1980). *İnsan Üstüne Bir Deneme*. İstanbul: Remzi.

Chomsky, N. (2011). *Dil ve Zihin*. Ahmet Kocaman (Çev.). Ankara: Bilgesu.

Çüçen, A. K. (2000). *Heidegger'de Varlık ve Zaman*. Bursa: Asa.

Gavi, P., Laudier, B. (2004). *İsyankar Yüzyıl*. İsmail Yerguz (Çev.). İstanbul: Sel.

Geray, H. (2003). *İletişim ve Teknoloji Uluslararası Birikim Düzeninde Yeni Medya Politikaları*. Ankara: Ütopya.

Gökberk, M. (2008). *Değişen Dünya Değişen Dil*. İstanbul: Yky.

Işık, M. (2012). *Kitle İletişim Sistemleri*. Konya: Eğitim.

İşlek, M. S. (2012). *Sosyal Medyanın Tüketici Davranışlarına Etkileri: Türkiye'deki Sosyal Medya Kullanıcıları Üzerine Bir Araştırma*. (Yayınlanmamış yüksek lisans tezi). Karaman: Karamanoğlu Mehmetbey Üniversitesi Sosyal Bilimler Enstitüsü, İşletme Anabilim Dalı.

Kıran, Z. (1986). *Dilbilim Akımları*. Ankara: Erol Matbaası.

Milliyet.com.tr

Müller, J.E. (1972). *Modern Sanat*. Mehmet Toprak (Çev). İstanbul: Remzi.

Mutlu, E. (2012). *İletişim Sözlüğü*. Ankara: Sofos.

Önür, N. (2002). *İletişim ve Toplum*. Ankara: Alp.

Özkalp, E. (2000). *Sosyolojiye Giriş*. Eskişehir: T.C. Anadolu Üniversitesi Eğitim, Sağlık ve Bilimsel Araştırma Çalışmaları Vakfı.

Türk Dil Kurumu (1992). *Türkçe Sözlük 2 K-Z*. İstanbul: Milliyet.

Tezcan, M. (2008). *Kültürel Antropoloj*. Ankara: Maya.

Tezcan, M. (1995). *Sosyolojiye Giriş*. Ankara: Feryal.

Twitter.com/about

Uygur, N. (1994). *Dilin Gücü*. İstanbul: Kabalcı.

Yavuz, H. (2013). *Avrupa'nın Zihin Tarihi*. İstanbul: Timaş.

Rifat Becerikli

Cinema in Digimodern Era: Super Hero Films

Introduction

The changes in the social structure are reflected not only in institutions like economy, politics and science but also in culture and art. In the same way, cinema is influenced by the social developments like every art branch in the world. This interaction changes many aspects, from narrative features of the cinema to the construction process and distribution channels.

Alan Kirby (2006) states in the "The Death of Postmodernism And Beyond" that the postmodern era has ended and the world has entered a new turn. Kirby describes the period we live as "digimodern". Digimodern period has its own characteristics. Kirby especially points out the importance of developing computer technologies. Today's world is dominated by digitalization, which dominates every field. For this reason Kirby uses the digimodern as an abbreviation of the concept of digital modernism. Kirby mentions the influence of media such as cinema, television, video games in the digimodern period. He also mentions the changes observed in some of the characteristics of dominant media tools. In this regard, there are a number of developments in cinema art in the digimodern era. According to Alan Kirby, the digimodern description is "endless", in other words it is possible to make infinite number of additions and subtractions. Moreover in this period, digital visual effects are the predominant factor in the cinema films.

Superhero movies have been the most popular genre in the recent Hollywood movie industry. One of the main reasons of this case is developing technology and digitalization offers various advantages to reflect superhero stories on the white screen. Six Spider-Man series films between 2002 and 2017 and nine X-Men series films between 2000 and 2017 will be reviewed in the context of digimodern cinema/narrative features, which Alan Kirby describes and defines its basis.

Digitalization and Cinema

Although the cinema tool was invented in the late 19th century, it quickly became as important as the branches of art, such as music, painting and literature, which emerged years ago. Unlike other branches of art, cinema is more influenced by technological innovations and developments. Cinema is fundamentally and, above all, closely related to the technology (Bennett & Furstenau & Mackenzie, 2008: 6). Since its first emergence, cinema have constantly renewed itself with the inventions in camera, display systems, sound and light technologies of scientists, inventors and companies from different parts of the world. The cinema was produced without sound in its first years, then audio films, which were made later described as a revolution in this field. In the following years, the audiovisual films became ordinary.

As a result of technological developments, the tendency of digitalization is increasing today. Digitalization in the cinema starts at the beginning of the 1990s, but its development is in beginning of the 2000s. In 2002, Digital Cinema Initiatives (DCI) was established by Hollywood production companies (Disney, Fox, Paramount, etc.). DCI's primary goal is to create/document clear specifications that provide clear/regular technical rules, reliability and quality control for digital cinema (www.dcimovies.com). There is also an economic reason for the establishment of DCI. For example, with the digitalization of the cinema, the cost of film distribution is significantly reduced and this is attractive for the studios (Aveyard, 2009: 192). Digital cinema brings many different processes to analog film production and these are: digital capture or acquisition, computer synthesis, digital post-production, digital distribution, and digital sound production (Swartz, 2005: 3; Rodowick, 2007: 94). Digital capture allows an analog image to be digitally rendered or a digital image to be processed on a computer. Instead of 35 mm large film rolls, video images can be digitally processed using a computer. In digital post-production, all post-production arrangements (editing, visual effects, sound editing, etc.) can be made faster and better quality by the computer than analogue rotation. In digital sound production, computer technology is used for sound and music processing, which are the important elements of cinema. Computer synthesis offers the flexibility to combine or manipulate the base materials of the film (image, light, decor, etc.) digitally on a computer. In digital distribution, instead of big movie rolls, movies can be sent over the networks on the computer for miles away, and movie display can also be done via computer. So, this situation saves the analog films rollers from the costs of transportation, storage and return costs. Cinema is basically divided into pre-production, production and post-production. In the Digimodern period, the importance of post-production process has increased. Pre-production, generally referred as the preliminary period, is a period in which all film components (space, decor, costume, actor, transportation, etc.) are identified. In the production, camera shootings, which are designated before, are made. In the post-production, these shootings are edited, and make it ready to be presented to the audience by adding sound and music. It is not possible that any scenes, which are not take place during the production-shooting stage of the film. But in the Digimodern period, it is possible to rebuild the scenes that are forgotten or misplaced during the production stage with digital effects. Digital effects, on the other hand, are the elements, which are considered during the shooting of the film (blue/green screen). In this respect, the production stage and the post-production stage are intertwined.

Recently, especially digital technology has become widespread everywhere, audiences establish their own media and this deeply affects the American film industry (Sickels, 2011: 7). People use the conveniences of video hosting, video processing and video broadcasting. Platforms like Vimeo, which allow film producers to upload virtually unlimited videos at $ 60 per year, are good alternatives to Hollywood (Dixon, 2017b: 207). In this regard, people are able to bring their works to the public.

According to Lowenstein (2015), the latest cinema is a digital imagination machine. Cinema is defined as a fantasy machine on its basis and its birth. But today with the power of technical possibilities of digitalization, it exactly becomes a dream machine. In the short film by Lumière Brothers' called "L'Arrivée d'un train à La Ciotat" (1895), the audience escaped from the saloon in response to the coming of a train towards them. At the time of the first appearance of the cinema, people were surprised to see the elements that they had seen in their dreams before and could only revive in their minds. Nowadays, with visual effects/3D technologies, people see space crafts, dragons and giant creatures that come to them in their seats. Over time, cinema has become a normal form of enjoyment for people. Also visual effects are used in cinema in the analog period. However, it is possible to intervene much more than practicable using analog techniques with digitalization of the films (Wood, 2011: 289). The digimodern term abolishes the imaginary borderline, according to the Cambridge dictionary, the meaning of the imagination means: "the ability to form pictures in the mind" (dictionary.cambridge.org); According to the Oxford dictionary, "The ability to be creative or resourceful" (en.oxforddictionaries.com). In the Digimodern period, images are reflected on the screen that the audiences cannot imagine, cannot reach with imagination or dare to imagine. The heroes in the Avatar and Matrix films are between our ordinary reality and imagined universes (Zizek, 2010). The fact that Zizek refers as an ordinary reality in the example of Avatar and Matrix movies is that the corruptness, frustration of the world, where we live opposing to the created virtual universes. The world, where we live is no longer need to be seen, felt or experienced. Digimodern era refers a situation in which people reflect themselves to the screen, nets, or scene as a virtual reality based on the three-dimensional glasses and nicknames. The real world is not enough for the digital world; for this reason, the real world needs to be enhanced, accelerated, exaggerated, and amplified (Dixon, 2017: 5).

People are also getting used to the digital world. Audiences demand "flawless" from movies and cinema. In other words, the cinematic universe of the 21st Century is dominated by unreal fidelity (Dixon, 2017b: 207). This perfection is naturalized by the audience. In addition, the simulated areas of virtual reality surround the audience spatially (Spielmann, 2003: 65). When the audience watches the film, he/she has seen the place, where does not exist at any point in the world, the living creature or an object for the first time and get involved in the universe of the film. Visual effects raise the limit of their imagination by calling them spectacularly (Prince 2012: 226). In this regards, today's audience has an endless experience, because the perfect images that audiences see on the screen are produced by the computer and this is a privilege that no one has before. Computerization is effective in digimodern cinema; everything is built on the computer (www.dictionary.com). Even though video recording the, power mobility or technical features of the cameras are much more advanced than in the past, digital technology now produce the footage without recording. In this respect, the cinematic universe, where the camera is active/decisive is left behind. What is effective in cinema in the Digimodern era is that image processing via computer.

The development of the Internet and technology, at the same time its digitalization, makes it easy the facilitation and rapid diffusion of information on a global scale (Gaudreault & Marion, 2015: 45–46). Now the people across the world are aware of the contents of the films. This situation also allows the films to come out all over the world and they increase the interests of people to the cinema as well. In the past, when a film that came out in the US was distributed months after to the world but now, it is seen that it is now meeting with the audience almost at the same day. On the other hand, illegal film watching also increases in the digimodern period. However, the increasing use of the Internet and the speed of data transfer, as well as the establishing the websites like torrent, file host and so on, ease the piracy. The establishment of sites also facilitates piracy (Zand, 2015). Any data, which is inserted in the digital environment, is very rapidly distributed.

The decline of the prices in the technological field in the US film industry has resulted in a large increase in visual effects companies (Cram, 2012: 169–170). This led the increase in the use of visual effects. Companies now have the possibility to use a large number of visual effects in many films with the advantage of cheap. In 1999, about 200 cinematic visual effects were applied in the high-budget film "The Mummy", while in 2011, 225 visual effects were applied in the mid-budget production "Bridesmaids." Today, over 1,000 visual effects are applied to blockbuster (high- budget films) films (Cram, 2012: 170).

In other words, the use of visual effects in Hollywood cinema has grown and widespread. This is also a feature of the digimodern period. At the same time, the visual effects elements are now used for the basic elements of the cinema such as color and light corrections, space designs, decor and accessory placement as well as for fantastic spaces, space ships, various monsters or superheroes. People can be copied with digital effects; extensions or signs in the buildings, telephone poles, cords or other objects, which should not be seen by the audience can be removed in the film (Silverman, 2005: 24). For example, daytime shots can be arranged as if it was taken at night; accessories or decor elements, which must to be found on the stage, can be added. In this respects, visual effects have become a natural part of the cinema in the digimodern era. So it's impossible to see that no visual effect is used in any popular movie.

Blue/green screen and computer generated imagery (CGI) technologies are used in the construction of visual effects. Green screen is the process of adding/removing characters, objects and photos in the background of the film (Rouse, 2016). The reason for using the green color in the background is that the green is the farthest color to the human skin (www.lightcrafttech.com). CGI is that all the images in the film are created digitally by computer (Estrada, 2016).

Cinema in Digimodern Era

The use of digital effects has basically begun in the early 1990s. Terminator-2 (1991) is the first blockbuster (high-budget) production using extensive digital effects; the global box Office is over $ 500 million, and since then, especially with the production of Jurassic Park (1993), many films have begun to take advantage of digital technology (Prince, 2012: 24–25). In the Digimodern period, the use of CGI especially in movies has become widespread. Thus, all images on the film can be realized on a computer basis. And Kirby also accepts Steven Spielberg's Jurassic Park (1993) as the basis of CGI technology (Kirby, 2009: 173). The importance of the Jurassic Park is that it is the first film to create physical texture with characters in CGI technology (Semlyen, 2015). The main reason of the impression of realistic dinosaurs in the film is this.

In the Digimodern era, new technologies are used to make popular cinema interesting and traceable to young people (Kirby, 2009: 129). The fact that Internet and television facilities are basically a dominant element on the young target audience also leads the cinema to innovations. Especially young people are now waiting for fast fiction and action, amazing images and incredible visual effects as well as story and characters of the film. Audiences now make film preferences according to these criteria. As a result, the producers place a special emphasis on visual effects, and digital visual effects are increasingly being used in almost every film. When you look at the productions that attract the most audiences, it is seen that the productions with digital effects are at the top of the list (www.boxofficemojo.com).

Historical places, demolished buildings, destroyed buildings can be represented with visual effects via digital technology. CGI technology can revive the past; Gladiator (2000) is the best example of this (Kirby, 2009: 178). In addition, digital systems are used for the restoration of old films (Kirby, 2009: 182). For instance, the fans of Star Wars (1977) Film have found the 35 mm original copy of the film without any company and have been renewed it in square-frame digital format (www.thestarwarstrilogy.com, 2016).

The remakes of the films are developed using visual effects (Kirby, 2009: 182). Although the remakes are found in Hollywood cinema from its earliest years, the rise of these remakes in the digimodern period is closely related to the developing technology. The films that have previously met with the audience are being renewed with the effects of today's computerization. For example, The War of the Worlds (1953) has a remake, which was starred by Tom Cruise in 2005 and includes many digital visual effects. Similarly, the film *Willy Wonka & The Chocolate Factory* (1971), which has a lot of fantastic elements, has a lot of remakes in *Charlie And The Chocolate Factory* (2005), played by Johnny Deep, directed by Tim Burton.

One of the most important and basic forms of narrative in the Digimodern period is video games (Kirby, 2009: 171). Video games stand out because of being both online-interactive, using advanced technologies, and being a demonstration of digitalization, video games become prominent. On the other hand, films and video games are increasingly engaged (Kirby, 2009: 169). Many video games have been filmed, and many films have become video games. For instance, Max Payne, Hitman, Warcraft, Assassin's Creed, Resident Evil, and Lara Croft: Tomb Raider.

In Digimodern era, text and person are interacting (Kirby, 2009: 60). The audience can specify the logical mistakes in the film, the positive and negative elements in the story, and the player's preferences they desire via blogs, social networks, etc. People are interactively able to convey their thoughts about which film will be shot, which character will die, and which player will play that role, to producers, directors or writers through the internet and social media channels. For example, in 2006, James Bond fans launched a boycott campaign to protest the British actor Daniel Craig playing 007. Fans of the film set up a website called "www.craignotbond.com" to ask for new Bond Film for people to watch.

As a result of technological products become cheap and easily accessible in Digimodern era, it is seen that there are different technical, aesthetics and stories from Hollywood in cinema industry. In other words, non- mainstream movies are also improving. Kirby indicates the features of the following non-mainstream film specific to *The Blair Witch Project* (1999): different natural lighting, trembling hand camera shakes, blurred images, jump-cuts, rough images different from Hollywood technique (Kirby, 2009: 151).

Cinematic Characteristics of Digimodern Narrations

The Digimodern narrative term is intended to describe the characteristics of any text (film, TV program, video game) in today's digital world with the classification of Alan Kirby. The narrative features of the recent films differ from the previous years through the Digimodern era and cinema relation. Digimodern narrative does not finish, does not end (Kirby, 2009: 159). From this point of view, sequel films, remake films, various similar and overlapping story forms come to mind in this sense of cinema in this period. Digimodern narrative includes infinite debates, reordering, recombinations (Kirby, 2009: 159–160).

This narrative form is generally called "endlessness" narration. Digimodern narratives are constantly lack and in formation (Kirby, 2009: 52). In the Digimodern era, films start again but do not end; they repeat without an end. For instance; Batman Begins (2005) and Superman Returns (2006). The narrations of the films are developing by changing and articulating provided that the basic elements are remain the same. Digimodern narrative is a form of narrative that transfers a certain amount of material to the next scene (Kirby, 2009: 160). In this respect, the unity of narration occurs in the next episode, stage or film. The Digimodern text is unstable, and uncertain about the direction to go (Kirby, 2009: 52). The fact that the uncertainty and not known about the direction it evolves of the text allows many characters to be added or subtracted. For example, Rogue One: A Star Wars Story (2016) is basically mentioned about the previous of the Star Wars: Episode IV - A New Hope (1977), which is the first in the series.

The writer is unknown in the Digimodern narrative (unknown) (Kirby, 2009: 60). There are millions of content, ideas and comments on the Internet. In particular, popular movies, characters, and players are the main materials of this network communication. It is not clear who is influenced by whom in such an environment; the speed between the original idea of the original output source and the change process is far from ensuring the true ownership of the narrative.

In the Digimodern narrative, "mise en abyme", the feature of story within the story is dominant (Kirby, 2009: 173). The film creates other films or narratives in itself. For example, Inception (2010). In the film, it is seen that in the lower layers of the dreams, people travel between different stories in various place and different times (Byron, 2010).

In the Digimodern period, "spin-off" is also an important feature in narration. Spin-off is to create a new narrative from some of the character or story/situation in the movie that has been successful or not l (english.stackexchange.com, 2014). Recently spin-off has an important place in films in mainstream cinema (Hull, 2016). *Fantastic Beasts* and *Where to Find Them* (2016) are the Spin-offs of the Harry Porter (2016).

In Digimodern era, there are many climax, excitement/peak points in cinema films (especially due to the fact that the target audience is young). Flaxman (2012: 129), especially specific to Avatar (2012), says that the cinema uses an endless number of "money shots" and "climax" while presenting a digital show.

Super Hero Films in Digimodern Era

Superhero films have recorded a significant increase, especially after the 1990s. The unexpected number of the audiences of the X-Men's superheroes makes the emergence of productions such as Spider-Man and Daredevil easier (Burke, 2008: 100). Superheroes, which enters human life as a comic strips before, is started to be transferred into cinema in the digimodern period in the light of technological developments. The main reason of this is the economic success. It's a fact that Hollywood film industry will not spend hundreds of millions of dollars unless they make a profit. For this reason, after the enormous financial success of the first superhero movies (DVD, video game, toys, etc.) at the beginning of the 2000s, the producers has continued to invest in similar stories (Gray & Kaklamanidou, 2011: 3).

On the basis of Superman, superhero genre films can be said with ease: today, audiences want illusion and want to believe in that illusion (Evans, 2010: 590 In digital visual effects the elements of this illusion and credibility come at the forefront of texts, which can be created by superhero movies. Zizek (2010) also makes it clear that the main heroes in both the Matrix and Avatar films are willing to continue their lives in the digital world, not in the ordinary world. In the absence of a real body, cinematic superheroes exist digitally; they are a digital symbol of the fact that they are digital (Bukatman, 2011: 122). At the same time, this is a structure that reminds us of the necessity of being everything digital in the world. In *The Matrix*, Neo is literally a digital superhero (Rodowick, 2007: 5).

The superheroes, which are transferred from comics to the cinema, also have digital superheroes as well as their own characteristics.

In the Digimodern era, the narratives/stories about children are also seen obviously (Kirby, 2009: 126–128). Although superhero movies appeal to the entire audience, the main target audiences are children and teens. It is also suggested that the connection between film director and the child. The directors, who is entering the technological playground like the little boy, adore superheroes; for example, Sam Raimi Spider-Man (2002), Brett Ratner X-Men: The Last Stand (2006), Bryan Singer Superman Returns (2006) (Evans, 2010: 593). In this respect, it can be said that the directors who shoot the superhero films in the digimodern period have digitally do their heroes what they want, just like a child.

Analyses of X-Men and Spider-Man as the Examples of Digimodern Cinema:

Research Hypothesis: The films, which reflect the concept of cinema/narrative best in Digimodern era, are superhero films.
The Purpose of the Research: To determine how appropriate the superhero movies are for the cinematic/narrative understanding in the digimodern era.
Method of the Research: The narrative and character analyzes of Spider-Man and X-Men series, which are selected as a sample, will integrally be performed to determine the conformity of the digimodern cinema characteristics determined by Alan Kirby.
Sample of the research: The sample of the research is Spider-Man and X-Men series films, which are recorded and continued the first great success of the box office of the Digimodern superhero movies.

Narrative/Character Analysis of Spider-Man (2000)

Spider-Man likes his next door neighbor Mary Jayne Watson. Spider-Man is a short, weak and ribbed person in high school. Harry Osborn, the son of Oscorp's owner Norman Osborn, has to go to the state high school for being kicked out of private schools and becomes friends with Peter Parker. Peter is a photographer in the school journal. Peter and his schoolmates examine the spider species on the school science trip. After Peter got bitten by a spider, the power of the Spider-Man came to light.

Spider-Man lives with his aunt May and his wife. Norman Osborn tries the power serum, which is developed for the state by him and he turns into the evil Green Goblin. Spider-Man goes to the cage fighting to make money. They do not pay his money in the cage fighting, the place, where the cage fighting is done, is robbed; Peter does not catch the thieves because they do not pay their money, but the robber, who escaped kills his aunt's husband. After that Spider-Man promises to help everyone. Peter Parker finds a way to make money; he starts to work as a newspaper photographer to take the photograph of the Spider-Man. Harry and Marry Jane become lovers. Mary Jane attacks and Peter saves her. Mary Jane falls in love with the Spider-Man. The fight between Green Goblin and Spider-Man begins. Green Goblin kidnaps Mary Jean. Spider-Man kills Green Goblin by fighting. Harry sees his father killed by Spider-Man and vow vengeance against Spider-Man.

Narrative/Character Analysis of Spider-Man 2 (2002)

Spider-Man works at the pizza shop, and at the same time he studies at the college, Marry Jane is now a well-known actress/celebrity. Dr. Octavious is experimenting with fusion energy in the name of Oscorp. There is an explosion in the fusion experiment. Dr. Octavious turns into a creature with many arms, Octopus. Octopus, tries to kill aunt May; Spider-Man saves her. Harry hates Peter because his dead father likes Peter, Marry Jane likes Peter, and because Peter is smarter and intelligent than him. Peter Parker thinks that being a spider is very difficult for him and affects his life badly (his wife dies, takes bad marks from classes, separates from Mary Jane, fights Harry). The Spider-Man nets destroy and he loses his power then he stops being a spider. As soon as he becomes Peter Parker, he gets high marks from his lessons and he gets closer to Mary Jane. Harry looks for Spider-Man to take revenge. As Peter Parker, he saves the little boy, who is stranded in the building fire in but then there is another person who cannot save in the building, so he decides to be a spider again. Spider-Man attempts to jump from frenzy to regain his lost powers, but his powers do not come back. Octopus kidnaps Mary Jane, the Spider-Man forces come back to save Mary Jane. Harry and Octopus make a deal, and if Octopus brings the Spider-Man, Harry will be able to get the necessary thieryum material for the fusion.

Octopus catches the spider and brings him to Harry. Harry is about to kill him with a knife in his hand, and when he pulls out his mask, he sees that he is Peter Parker, he is shocked and drops the knife in his hand. Spider-Man stops the Octopus and saves Mary Jane. Spider-Man tells Mary Jane that he is Peter and they can never be together, but Mary Jane accepts Peter Parker as a Spider-Man. Harry wants to get revenge on the Spider-Man who finds things in his secret room left by his father as a green gin.

Narrative/Character Analysis of Spider-Man 3 (2002)

Mary Jane is a famous actress, and Peter Parker is still studying at the university. Peter Parker thinks about asking Mary Jane to marry with him. Harry attacks the Spider-Man with the technological equipment left from his father as New Green Goblin. Harry and Spider-Man fighting ends up by losing Harry's memory. Eddie Brock, who comes to Peter Parker as a rival photographer, is dating police chief's daughter (Gwen), and she is Peter Parker's schoolmate at the same time. Competition begins between two photographers to enter the payroll employment. Harry, who does not remember the past, gets closer to Peter. A criminal who flees from prison in a nuclear scientific experiment becomes a "sandman". It appears that the criminal who escaped from prison and become a sandman is the person who killed Peter's uncle. The Spider-Man starts looking for the sandman everywhere. The common form of coexistence, the black matter (Venom), captures the spider's body. The Spider-Man turns in black. Black matter makes Peter more nervous, aggressive, and getting worse. Harry's memory comes back and starts attacking Spider-Man with the remaining equipment of Green Evil. The alien organism captures Peter well. Harry and the black Spider-Man struggle, half of Harry's face become blurred. In Dr.Connors' research, it is determined that the alien organism's adherence increases aggression. The Spider-Man, who fights with Mary Jane and wants to be alone, goes up to the roof of a church and the noise and vibration of the bell in the church helps the alien organism to get out of Peter Parker.

The organism getting out of Peter enters the competing photographer Eddie Brock. Thanks to Venom, the photographer takes the power of the spider and makes plans to attack Spider-Man together with the sandman. The two kidnap Mary Jane and wait for the Spider-Man to appear. Peter Parker asks Harry's support against the sandman and black spider. Harry learns from the servant that the Spider-Man does not kill his father and the spider helps Spider-Man and fight against sandman and black spider together. When Venom kills the Spider-Man, Harry saves the Spider-Man by sacrificing him. Spider-Man destroys Venom.

Narrative/Character Analysis of The Amazing Spider-Man (2012)

The Spider-Man's parents leave him to his aunt when he is very young. The Spider-Man wants to know more information about his past and begins to research. When he goes to Oscorp technology to see his father's friend called Dr. Connors to ask his father's past, he mutates into a superhuman force and becomes a Spider-Man bitten by a spider during a secret experiment in the laboratory. One of the arms of Dr. Connors is cut from the elbow. Spider-Man gives the formula to Connors, which he finds in the secret part of his father's old suitcase. Gwen, who is lived by Spider-Man, is both his classmate and an Oscorp employee. May's wife dies because the Spider-Man cannot chase the thief. Gwen is the daughter of the police chief, who does not like Spider-Man. Dr. Connors wants to rebuild his arm; the genetic transition inherited from the reptilian organism fails and the physician turns into a powerful reptile (Lizard) and harms the surrounding. Dr. Connors wants to turn all people into reptiles with his formula. The Spider-Man prevents Lizard with the help of the loving Gwen, who also works in Oscorp.

Narrative/Character Analysis of The Amazing Spider-Man 2 (2014)

The Spider-Man's father experiments with spiders in the past. While his father is trying to be killed for the secret experiment results in the plane, the plane falls but he succeeds in backing up the secret experiment information to the computer. Peter Parker, who thinks he hurt his lover as a Spider-Man, leaves Gwen. Norman Osborn (owner of the Oscorp Company) heads his son, Harry Osborn to the company in his deathbed. Peter Parker and Harry Osborn meet since childhood. Max Dillon is a fan of Spider-Man, who has Spider-Man's photos, banners, scrolls, everywhere in his house and works at Oscorp. Max Dillon falls into the pool full of electrophorus electricus used in Oscorp experiments by receiving a high voltage, and turns into a fully electric "Electro" character.

Harry wants to make serum from the Spider-Man's blood to heal himself (he has a hereditary disease like his father and dies day by day). Spider-Man finds his father's secret lab at the subway station. Spider-Man learns the secret of his father, his father matches his own genetic code with spiders, for this reason Peter Parker, who has the same siblings with the spiders, is influenced. Harry uses the transformation serum but turns into bad Green Goblin. The Spider-Man, who fights with Electro and Green Goblin (Harry), defeats them, but the Spider-Man's lover Gwen is death during the fight.

Narrative/Character Analysis of Spider-Man: Homecoming (2017)

Spider-Man is a 15-year-old teenager and he does not know how to use his powers. He is not famous as a spider; he is not a famous person. He lives with aunt May; a middle-aged, single woman. He is under the control of Tony Stark (Iron Man); Tony Stark wants to know where he goes and what he does. The Stark industry and Mr. Stark produce the Spider-Man's costume. Vulture makes super-powerful tools by using the technological possibilities left by aliens (who come to invade the world in the Avengers film). In the film, the main enemy against Spider-Man is Vulture. Besides Vulture, there is also Shocker, Jackson Brice, from his own team. Vulture also supplies weapons to the world of crime by modifying and improving various space equipment. Spider-Man is in love with his classmate Liz. Liz is also the daughter of Vulture. The Spider-Man prevents Vulture from taking the equipment of the alien technology found in the Stark industry. In their last fight, Spider-Man defeats Vulture.

Narrative/Character Analysis of X-Men (2000)

There are mutant people with various powers that have evolved in the near future. Rogue escapes from the house (absorbs all powers of the mutant, which he touches), and finds Wolverine (regenerating his body itself, immortal, sharp blades can be out of his hands) luckily. Wolverine protects Rogue. Also Magneto (who can steer all iron as he wants) wants to capture Rogue; Storm (wind, rain, etc.) and Cyclops (laser beam out of your eyes) prevent this situation.

Rogue and Wolverine participate in the education center of Professor X for special children (head of X-Men with telepathic power able to manage every human/mutant). A senator who wants to destroy the mutants, is kidnapped by Mystique, and Magneto destroys the senator' gene and transforms him into mutant by giving him too much nuclear power. The government is against the mutants and wants to isolate the mutants by new laws. Magneto's aim is to kidnap Rogue and combine powers to make all people mutant like the senator. Professor X and his team prevent this situation by sending Magneto to prison.

In the film, there are totally 14 characters with super powers.

Narrative/Character Analysis of X-Men 2 (2003)

The government supports to Colonel Striker's mutant project. Magneto is held in a plastic prison. Colonel Striker, with the help of the drugs he has developed, obtains confidential information from Magneto and neutralizes his power. Colonel Striker also takes Professor X prisoner, who comes to visit Magneto in prison; then attacks the X-Men school to take control of the mutants.

Colonel Striker wants to use the professor's machine (Cerebro). The Colonel's son is also a mutant and can direct people with the power of thought like Professor X; with the help of his son, Professor X will connect to Cerebro and kill the mutants by identifying them. Mystique gets Magneto out of plastic prison. Magneto and Mystique help X-Men. Magneto and X-Men occupy the quarter of Colonel Striker, and this time Magneto directs Professor X to find and kill people. X-Men prevent this situation.

In the film, there are a totally 17 characters with super powers.

Narrative/Character Analysis of X-Men: The Last Stand (2006)

Professor X and Magneto go back Jean Gray's childhood 20 years ago (can do anything with power of thought) and educate him. The government develops a kind of medicine to suppress the mutant gene. They produce a vaccine from the mutant boy named Leech, which turns mutants into human beings. Power developmental sessions have caused personality disorder in Jean Gray's childhood; he develops good and bad personality. Bad personality captures Jean Gray. Jean Gray also kills Cyclops and Professor X. Jean Gray sides with Magneto. Magneto wants to set up his own brotherhood unity and take over the whole world, and he will also kill mutants, who are not from their brotherhood. Magneto wants to capture the mutant called Leech because of being a threat. Magneto and the X-Men teams start to fight. The bad side of Jean Gray is that he destroys everything around, and Wolverine stops him by killing him. X Men win the war.

In the film, there are totally 32 characters with super powers.

Narrative/Character Analysis of X-Men Origins: Wolverine (2009)

In 1845, Wolverine is a child and lives with his brother Sabretooth (very powerful and immortal). They fight together for many years (American Civil War, 1st World War and 2nd World War). During the Vietnam War, an American officer sets up a special team of gifted mutants. The team follows a stone. Wolverine prevents the colonel from persecuting the public to find out where the stone is. Wolverine leaves the group and starts to make woodcuts. In the following years, Colonel comes to Wolverine with a new task; his task is to catch his brother. Sabretooth kills the mutants on the colonel's side. Wolverine lives with a class teacher. Sabretooth kill the woman, who Wolverine loves. Wolverine swears for revenge. The Colonel experimented with Wolverine's whole body with the world's toughest material, adamantium. Wolverine steals away from him and the colonel tries to kill him. Then it comes out that Wolverine's boss and his class teacher are at the same side of the colonel. The colonel creates Deadpool, which is combined by different mutants' power. The colonel collects all the powers of the mutants he captures. Wolverine and Sabretooth fight together to kill Deadpool. Professor X leads Cyclops with brain power to save the mutants he collects to gain the powers of the colonel.

In the film, there are totally 11 characters with super power.

Narrative/Character Analysis of X-Men: First Class (2011)

Mystique and Professor X are friends since childhood; their youth passes away in the same house. Magneto also stays in Nazi prisons when he is a child. When he grows up, Magneto wants to kill the Nazi officer, who kills his mother, the Nazi colonel Sebastian Shaw is a mutant and wants to break out a world war with his team. He tries to put US nuclear rockets in Turkey, Soviet rockets in Cuba and starts the war between USA and Soviets. Magneto and Professor X, who saved him from death become friends. Magneto and Professor X fight together against their common enemy Sebastian Shaw. As a scientist, inventions of Beast for the CIA make Professor X's job easier. Professor X finds mutants with his abilities; he and Magneto establish a team with them. Magneto and Professor X prevent Sebastian Shaw. When the Soviet and the USA navies attack all the mutants, Magneto begins to abuse their power against the people. Agent MacTaggart shoots to stop Magneto but the bullet from Magneto hits the lead, and leads to Professor X's spinal cord. Magneto says "Do not hide, all the mutants will be free" and takes a few mutations to his side and goes.

In the film, there are totally 15 characters with super powers.

Narrative/Character Analysis of The Wolverine (2013)

In World War II, Wolverine is in the captivity in Japan. While atomic bombs are thrown into the city of Nagasaki in Japan, Wolverine saves a Japanese soldier from a nuclear disaster. Nowadays, Wolverine lives alone with animals in the wild. He goes to Japan to execute Yashida's (saved by him in the battle) last request. Wolverine has Viper (a power that poisons) has reduced healing ability and badly affects his strength. Yashida is about to die and wants Wolverine's healing power; he has armored himself with the material atamenium of Wolverine and lives in it. When the armor catches Wolverine, Wolverine starts pulling the healing power by putting several pins into his body. Yashida's grandson kills him by giving the injection and saves Wolverine.

In the film, there are a totally 3 characters with super powers.

Narrative/Character Analysis of X-Men: Days of Future Past (2014)

In the near future mutants are waiting for bad days. During mutant-human warfare, the watchers kill mutants. Wolverine tries to prevent the appearance of the watchers by going back to the past with the power of mutants. The watchers are based on the principle of the adaptation and development of the power of any mutant, from the genetics of Mystique and its transform into every form. It was the result of Mystique's capture by Dr. Task in 1973. With the help of Professor X and Magneto, Dr. Wolverine goes to the past. He tries to stop Task's watcher's project. But Professor X is sick and Magneto is in the prison. Wolverine, psychologically supports Professor X, and escapes Magneto from prison together with Professor X. Magneto tries to abolish Mystique, which is the basic structure of watchers. Professor X protects Mystique, who is old and best friend of him. Separating from Professor X, Magneto wants to give a lesson to the world by killing the president of the United States, who starts and finances the project. Mystique poses as president of USA and kills Magneto with plastic guns.

In the film, there are totally 22 characters with super powers.

Narrative/Character Analysis of X-Men: Apocalypse (2016)

The year 1983, the first mutant called Apocalypse, who is powerful and capable of all things, died as the king of Egypt, this appears in the excavation years later. Apocalypse wants to be the leader of the world, for this reason he collects mutants. Magneto is also in his side. Apocalypse speaks to the whole world with Professor X's brainpower. He wants to be by getting into the body of Professor X and use his power. The X-Men team escapes Wolverine from the prison. Under the leadership of Jean, all mutants unite against Apocalypse and destroy it.

In the film there are totally 21 characters with super powers.

Narrative/Character Analysis of Logan (2017)

Logan (Wolverine) is aged, his body is worn out because of the drink and he is a hired driver. Professor X and Caliban (knowing who is mutant and how strength mutants are) live in Mexico together, near the US border. Professor X is also aged and he is out of mental faculties. The government is close scientist Dr. Rice, who is a scientist close to the government, is experimenting with a group of children, to change their genes and assign mutant traits to them. Children escape from the place, where the experiments are done.

The little girl named Laura, has been given the same qualities of Wolverine, and the mutant machine. Also the replica of Wolverine called Wolverine X-24 is also created in the laboratory environment. Dr. Rice and his team chase Wolverine and the kids. They beat Rice's cyborgs and X-24 with the help of the little mutants' powers and Wolverine, Professor X and Logan (Wolverine) die.

In the film, there are totally 12 characters with super powers.

Narrative/Character Analysis of the Series of Spider-Man and X-Men

It is possible to examine the series of Spider Man films in three classifications. The first of these includes the first three series: Spider-Man (2002), Spider-Man-2 (2004), Spider-Man 3 (2007). In these films, Tobey Maguire plays starring as Spider man. The films in the second classifications are Amazing Spider-Man, which was published in 2012 and 2014 Andrew Garfield plays starring as Spider-Man. The third classification is the series in which Tom Holland plays as starring in 2017 and is said to be continued in 2019.

In the Spider-Man films, it is apparent that there is a love affair between Spider-Man and a girl. This love affair in the first trilogy is between Spider-Man and his next door neighbor Marry Jane, in serial films, the love is between his schoolmate Gwen (they had a short love to make Mary Jane jealous in the first trilogy), and line the first film of last trilogy the love with classmate Liz. These love affairs draw attention as the sorest point of the Spider-Man because the enemies of Spider-Man try to trap him by damaging the women he loves. Another point stands out in the love affairs of Spider-Man is that the fathers of the women he loves, dislike Spider-Man. For instance, the police chief in The Amazing Spider-Man (2012), and Venture the enemy he fights with in Spider-Man: homecoming (2017). In the first trilogy of the series, Dr. Connors is a physics teacher at Spider-Man's college, who does not have an arm until elbow, especially the person, who warns Spider Man by solving the contents of the organism coming from the space and identifies Venom. In biserial film *The Amazing Spider-Man*, Dr. Connors is the person, who studies on genetics interspecies and works with his father in Oscorp. When Spider-Man goes to see the doctor about his father, Connors takes a set of his fathers' studies and experiments it on his own via genetic codes getting from a reptile (to recreate the place with no arm). As a result of this, Dr. Connors turns into a reptile creature called The Lizard, and begins to struggle with the Spider-Man. In the film in 2017, Connors is never mentioned.

There are different approaches to Spider-Man's family life/relatives in the movies in the three classifications. Spider-Man's history, childhood, parents are hardly emphasized in first classifications, and his life with aunt May is told directly. In the second classifications there is cohesion in the narration of the Spider-Man, his parents and his childhood. In the first classifications, his father's profession is unclear and his is bitten luckily by a spider, while in the second-classifications his father is a respected scientist, who studies on mutations between species; it carries out a successful gene transfer interspecies but it can only be used by people from his family. In other words, it is not a chance to be a Spider-Man because of the characteristics of the Spider-Man's father and his family. In the film of 2017, Spider-Man's family is never mentioned. Another difference in this film is seen in aunt May. Aunt May is old and married in previous films, but in this film, she is reflected in middle ages, single, beautiful and charming.

In the first and second classification of Spider-Man and in *The Amazing Spider-Man*, Green Goblin is the most widely used evil hero. Green Goblin takes part in the entire first trilogy of 2002, 2004, 2007, Norman Osborn in 2002 (owns the Oscorp Company) becomes Green Goblin, and son Harry Osborn in 2004 and 2007 becomes New Goblin. In The Amazing Spider-Man 2 (2014) again son Harry Osborn becomes the evil hero Goblin. In the latest spider man film in 2017, Green Goblin is not mentioned.

In the publication of 2017, Spider-Man is still a child, which basically corresponds to the first description of the Spider-Man in the comic book. The same Spider-Man is on a short stage in Avengers: Captain America Civil War in 2015. Spider-Man needs the help of Tony Stark (Iron Man) as a reflection of this situation. Spider-Man's costume is manufactured in the Stark's industries, equipped with various powers; in this respect it reminds features of Iron Man. This also shows that the Spider-Man is affected by the digitalization. The costume on him is a totally digital costume, with many netting modes, perceiving sound waves, and doing target clarity. In the first two classifications, the costume is made by the Spider-Man himself, or it is not clear how it is made. Only in Spider-Man 3 (2007), Spider-Man is caught by organisms from space is done through the costume change. Spider-Man costume, which is caught by alien organisms, becomes increasingly black.

In accordance with the Digimodern narrative, the features of the people, especially Spider-Man, are constantly changing and renewing. While Spider-Man, who is known as the photographer in general, is the dominant photographic element in the first-classifications, and even in the Spider-Man 3 film, a rival photographer is included in the narrative, and the features of the alien organism are transferred to this photographer. In other classifications his feature a photographer is not mentioned. Similarly, working in newspapers related to photography and the character as a newspaper boss is effectively seen in the first classifications but this characterization is not available in subsequent films. In Digimodern narrative, the film carries many elements specific to children. Especially the second and third classification films address teenager groups. In the 2017 publication film, the Spider man character is at the age of 15 and reinforces its content pertain to children; In the film, a Spider-Man portrait is drawn, who makes lego and plays his toys.

Depending on the Digimodern narrative, Spider-Man films are both linked to each other and to other Marvel films. The series of the Spider-Man is seen in the *Avengers*, published in 2015. In 2018, Venom is expected to come out as the series of the alien creature-evil Spider-Man in the Spider-Man 3 (2007). The series of *Spider-Man Homecoming*, in which Tom Holland will play as starring again, is scheduled to come out in 2019. It is anticipated that the subsequent Spider-Man series will be linked to or series of the Avengers (Kennedy, 2017).

According to the digimodern narrative, Spider-Man is directly influenced by the fact that the audience is the dominant element or the views of the audience are important to the producer. Although The Amazing Spider-Man film in the second classification is planned as a trilogy but it is finalized in two films due to influence of the audiences. Denis Leary, who plays the role as the chief of police at The Amazing Spider-Man, also says that the third film of the series is also thought (Macy, 2015). The Amazing Spider-Man series (www.the-numbers.com) is the worst performer in terms of number of audience and profit. It has also been criticized in many social network platforms that Andrew Garfield is the starring of Spider-Man character.

In the second series films, which are called the incredible Spider-Man, not the Spider-Man, the change of the name is actually different from the first trilogy, indicates that the series start again. In the film in 2017, Spider-Man, who is previously seen in the film Captain America-Civil War (2015), stand by Iron man and his team. This shows that the narrative is influenced by other narratives and it is constantly renewed.

When X-Men films are examined, it is found that it includes an endless content based on adding's and subtractions of the digimodern narrative. In this regard, information about the past of many characters is revealed in coming X-Men series. For example, in the X-Men first class (2011) movie, the narratives and characters in previous films are explained. It turns out to be seen and performed by Sebastian Show how Professor X is founded the mutant school to protect the mutant children, and the reason of wearing magnetic helmet to avoid to be directed by Professor X with power of thought. It is seen that Professor X and Mystique, who are now enemies, are friends and grow up together in their youth. Since Mystique wants to live his own physiological elements freely, Magneto is influenced by the "no more hiding" motto of Magneto and leaves Professor X and sides with Magneto. The main reason of being black is that Magneto stays in the collecting camps in Nazi Germany, and his mother is killed in front of his eyes by a Nazi officer. Furthermore, while Magneto protects the world from the evils of Sebastian Shaw, the fact that people want to destroy

Magneto and the mutants cause Magneto to be an evil. Beast is a scientist, who works for CIA, invents a machine that allows Professor X to communicate with other mutants and people. Beast turns into a blue, feathered creature when he tries to transform himself into human being again from genetic features of Mystique. It is come to the fore in the film, in which Magneto avoids Professor X to be drowning and acts together as a friend then Magneto represents the bad side and Professor X represents the good side. In the X-Men Days of Future Past (2014), the entire timeline of future changes with the return of Wolverine to the past. When Wolverine turns back from past he sees that the superheroes who died before are still alive. In the X-Men Days of Future Past (2014) narrative, there is a break in the filmic universe of the X-Men series. The future after 1973 suffers a change. Wolverine, who changes the timeline by turning back to the past, causes the narrative to take on a new form and evolve. In this new future Rogue and Iceman are together and love each other. In future past, Iceman and Kitty Pryde are also in love. Beast is a teacher at Professor X's School. In the future past he is the minister of the Mutant. The biggest shock of Wolverine future he changed is that Jean Gray is alive. Jean Gray had already died in the future past. This makes it possible for the digimodern narrative to be able to be adding or subtraction like lego parts. Thus, multiple timelines occur in X-Men series (Burt, 2017).

In the X-Men Origins: Wolverine (2009), it is emphasized how the powers of Wolverine is developed in the laboratory, who meets with the audience in previous films. The hands of Wolverine made of bone are transformed into the toughest material in the world called adamantium in this film.

The story of Jean Gray, who died in X-Men: The Last Stand (2006), includes his childhood, personality division, and training with Professor X in the narrative of X-Men Apocalypse (2016). It turns out that Jean Gray's power of thought is even more influential than Professor X. In this regard, the version of the X-Men series that will be released in 2018 is X-Men Dark Phoenix (www.rottentomatoes.com). The name Phoenix is used for Jean Gray in the X-men comic series. Also, in the released trailer of the film, it is understood that the film will be based on Jean Gray. Likewise, in 2018, the new film of X-Men series called The New Mutants (www.movieinsider.com) will be released. New teenager mutants, who have not seen before in the X-Men films, will take place in the narrative. Another feature of Film is that it is a mixture of the superhero-horror film. The main feature of the Digimodern narrative is that it is constantly going on; in this regard the films which will similarly continue in the X-Men series in 2018 are in harmony with the basic elements of the digimodern narrative. Likewise, one or more of the New Mutants will enter the X-Men series and the possibilities of the narrative are getting increase. The words of Professor X at the end of the X-Men Apocalypse (2016) at the end of: "the past. After you passed it a new and uncertain world would begin. A world of endless possibilities and infinite outcomes. Countless choices define our fate. Each choice, each moment ... a ripple in the river of time. Enough rips and you change the tide. For the future is never truly set.", clarifies the features of the Digimodern narrative. It seems as if the digimodern in the X-Men series is evoking the narrative features. When X-Men films are examined, it is seen that the universe of X-Men narrative is constantly changing and developing between future and past.

When X-Men movies are examined, it is seen that X-Men Origins: Wolverine (2009), The Wolverine (2013) and Logan (2017). Although some of the heroes in the X-Men series are included in these films, the main narrative is about Wolverine. In the X-Men series, the character is a spin-off over Wolverine and the character is set aside for a separate narrative basis. At the end of Logan, Wolverine is dead. Although this is interpreted by some commentators that the Wolverine character is death, some commentator indicates that Hugh Jackman has stopped playing the character of Wolverine, but Wolverine character will continue with a new actor. In the movie, Wolverine's character dies, but it is said that the evil character of Wolverine called X-24, which is created in laboratory, is killed by shooting in head shows that the film will be able to continue with different face or it is talked that Wolverine has other cloned twins to (Chitwood, 2017). The Reavers Team in the film is the name of the semi-human, semi robot cyborg team, which is shown in the comics called Marvel (marvel.com). In this regard, it is adjusted structure of digimodern narrative, which incorporates many different elements.

The X-Men series reawaken the historical phenomena of digimodern era with digital effects. For example, the atomic bombing of Nagasaki in 1945 is shown to the audience via computer technology in The Wolverine (2013). Likewise, in the 1962 Cuban Missile Crisis (http://www.history.com) In X-Men First Class (2011), the moments of the encounter of Soviet and American navy is reproduced digitally (different from the reality).

The X-Men series are a basic example of the film with many superheroes. There are totally 147 heroes with super powers in the X-Men series and use these powers in the narrative of the film. In X-Men films, many visual effects are used based on CGI technology, specific to the digimodern narrative. Almost each character in X-Men's films has its own superhuman/supernatural power in the narrative. In this respect, it is necessary to use visual effects to describe the powers of each character and to transmit to the audience.

Conclusion

Digimodern cinema means that the digital elements are used for all kinds of needs of the cinema. It is also a structure that digital technologies are integrated with cinema. Digimodern cinema is a changing cinema that constantly renews itself. Because developing technology and production possibilities continuously occurs in it. The best films that reflect this developing technology are superhero movies. Superhero films not only follow technological developments in cinema but also constantly use technological developments in the narration.

In the Digimodern era, it appears that the films have been divided into two as single superhero movies and multi-superhero movies. In general, superhero movies such as Superman, Batman and Spider-Man etc. consist of a narration that main superhero protect the world from villains (people, creatures, aliens, and monsters). While single superhero movies dominate the history of cinema, multi-superhero movies become prominent in the Digimodern era. The reasons of this may be the expectation of thrill by the audience in the Digimodern era, the desire of the narration to affect the audience constantly, the appropriate cost of the technological innovations and the production of visual effects. There are also films about the struggle between good superheroes during the Digimodern period. For example, in Batman v Superman: Dawn of Justice (2016) and Captain America-Civil War (2016), several good superheroes have fought against each other because of misunderstanding, power relations, deception.

At the end of the study, it is seen that the basic and typical characteristics of the Digimodern narration defined by Alan Kirby are found in the Spider-Man and X-Men series. In these series, narration continuously evolves and changes. While every film in these series describes a situation in previous films, it also changes the other situation. From this viewpoint, the Spider-Man and X-Men series have an endlessness structure as Kirby mentions. In these films, the visual effects are used for character and scene design based on CGI, motion capture, and blue/green technologies. The continuous thrill of the Digimodern narration appears in these films. In the Spider-Man and X-Men series, there are many moments of continuous climax where the audience can focus carefully.

As a result, it is seen that the Digimodern era and superhero movies are rising at similar times and closely related. The study reveals that the Digimodern narrative characteristics described by Alan Kirby tally with superhero movies. It is also understood that digital and technological developments will increase in future years and accordingly, these developments will be evaluated within the art of cinema. In the same way, it is possible to say that many superhero movies will be produced and Digimodern narrative patterns will be mainly seen in these films.

References

Aveyard, Karina, (2009). 'Coming to a Cinema Near You?': Digitized Exhibition and Independent Cinemas in Australia, Studies in Australasian Cinema, 3:2, pp: 191–203.

Bennett, Bruce; Furstenau, Marc; and Mackenzie, Adrian (2008). Cinema and Technology Cultures, Theories, Practices, Palgrave Macmillan, New York.

http://www.boxofficemojo.com/alltime/?page=byrecord&p=.htm, All Time Box Office, Date of Access: 13.08.2017.

Bukatman, Scott (2011). Why I Hate Superhero Movies, Cinema Journal, 50:3 (Spring), pp: 118–122.

Burke, Liam, (2008). Superhero Movies, Pocket Essentials, Harpenden.

Burt, Kayti, (2017). http://www.denofgeek.com/us/movies/x-men/262623/explaining-the-x-men-movie-timelines, Explaining the X-Men Movie Timelines, Date of Access: 15.08.2017.

Byron, David, (2010). https://baroquepotion.com/2010/07/inception-a-meta-magical-matryoshka, Inception: A Meta-Magical Matryoshka, Date of Access: 10.09.2017

Chitwood, Adam, (2017). http://collider.com/logan-ending-explained/#james-mangold, 'Logan' Ending Explained: James Mangold Breaks Down the Wolverine's Conclusion, Date of Access: 09.09.2017.

Cram, Christopher, (2012). Digital Cinema: The Role of the Visual Effects Supervisor, Film History, 24:2, pp. 169–186

http://www.dcimovies.com, Digital Cinema Initiatives About DCI, Date of Access: 20.08.2017.

http://www.dictionary.com/browse/computerization,Computerize, Date of Access: 10.08.2017

https://dictionary.cambridge.org/dictionary/english/imagination, Imagination, Date of Access: 05.09.2017.

Dixon, Wheeler Winston, (2017). Synthetic Cinema: Mainstream Movies in the 21st Century, Quarterly Review of Film and Video, pp: 1–15.

Dixon, Wheeler Winston, (2017b). "Service Providers": Genre Cinema in the 21st Century, Quarterly Review of Film and Video, 34:3, pp: 201–21.

https://en.oxforddictionaries.com/definition/imagination, Imagination, Date of Access: 20.08.2017.

Estrada, Aaron, (2016). https://www.quora.com/What-is-the-difference-between-VFX-and-CGI, What is the difference between VFX and CGI?, Date of Access: 11.09.2017.

Evans, Nicola Jean (2010). Undoing the Magic? DVD Extras and the Pleasure Behind the Scenes, Continuum, 24:4, pp: 587–600.

https://english.stackexchange.com/questions/160878/whats-the-meaning-of-spin-off, What's the Meanining of Spin-off?, Date of Access: 29.08.2017.

Flaxman, Gregory (2012). Out of Field, Angelaki Journal of the Theoretical Humanities, 17:4, pp: 119–137.

Gaudreault, André and Marion, Philippe (2015). The End of Cinema?: A Medium Crisis in the Digital Age, Colombia University Press, New York.

Gray, Richard J and Kaklamanidou, Betty (2011). The 21st Century Superhero Essays on Gender, Genre and Globalization in Film, McFarland & Company, Inc., Publishers North Carolina, pp: 1–15.

http://www.history.com/this-day-in-history/cuban-missile-crisis, 1962-Cuban Missile Crisis, Date of Access: 29.08.2017

http://www.imdb.com/list/ls002144829/, Ranking All Superhero Movies Thus Far Seen, Date of Access: 21.09.2017.

Hull, Dustin (2016). http://www.tasteofcinema.com/2016/the-10-best-spin-off-movies-of-all-time/, The Ten Best Spin-off Movies of All Time, Date of Access: 13.08.2017.

Kennedy, Michael, (2017). https://screenrant.com/spider-man-homecoming-2-avengers-4-aftermath, Spider-Man Homecoming 2 Deals with the 'Aftermath' of Avengers 4, Date of Access: 23.08.2017.

Kirby, Allen, (2006). The Death of Postmodernism and Beyond, Philosophy Now, 58:34–37.

Kirby, Allen, (2009). Digimodernism How New Technologies Dismantle the Postmodern and Reconfigure Our Culture, The Continuum, New York.

http://www.lightcrafttech.com/support/doc/keying-interface/blue-vs-green-screen, The Green Advantage, Date of Access: 17.08.2017.

Lowenstein, Adam, (2015). Dreaming of Cinema: Spectatorship, Surrealism, and the Age of Digital Media, Columbia University Press, New York

Macy, Seth, (2015). http://www.ign.com/articles/2015/07/12/comic-con-2015-denis-leary-talks-what-could-have-been-in-spider-man-3, Comic-Con 2015: Denıs Leary Talks What Could Have Been In Amazıng Spıder-Man 3, Date of Access: 07.09.2017.

https://marvel.com/universe/Reavers, Reavers, Date of Access: 20.08.2017.

https://www.movieinsider.com/m13374/the-new-mutants, X-Men: The New Mutants, Date of Access: 03.09.2017.

Prince, Stephen, (2012). Digital Visual Effects ın Cinema-The Seduction of Reality, Rutgers Unıversıty Press, New Jersey

Rodowick, David, Norman (2007). The Virtual Life of Film, Harvard University Press, Massachusett.

https://www.rottentomatoes.com/m/x_men_dark_phoenix/, X-Men: Dark Phoenix- (2018), Date of Access: 27.08.2017.

Rouse, Margaret, (2016). http://whatis.techtarget.com/definition/green-screen-blue-screen, Green Screen (Blue Screen), Date of Access: 11.09.2017.

Sarkar, Samit, (2016). https://www.polygon.com/2016/10/13/13276092/fantastic-beasts-series-five-films-harry-potter, Harry Potter Spinoff Series Fantastic Beasts Will Now Have Five Movies, Date of Access: 21.08.2017.

Semlyen, Phil De, (2015). https://www.empireonline.com/movies/features/history-cgi, A History of CGI in the Movies, Date of Access: 23.08.2017.

Sickels, Robert, (2011). American Film ın the Digital Age, Praeger, California.

Silverman, Leon, (2005). The New Post Production Workflow: Today and Tomorrow, Swarz, Carles (Ed.), Understanding Digital Cinema: A Professional Handbook, pp: 15–57, Focal Press, Oxford.

Spielmann, *Yvonne*, (2003). *Elastic Cinema*: *Technological Imagery* in *Contemporary Science Fiction Films*, Convergence 9:3, pp: 56–73.

Swartz, Charles, (2005). İntroduction, Swartz, Charles, (Ed.), Understanding Digital Cinema: A Professional Handbook, pp:1–14, Focal Press, Oxford.

http://www.the-numbers.com/movies/franchise/Spider-Man#tab=summary, Box Office History for Spider-Man Movies, Date of Access: 27.08.2017.

http://www.thestarwarstrilogy.com/post/2016/01/15/team-negative-one-completes-35mm restoration-of-star-wars, Team Negative One Completes 35 mm Restoration of Star Wars Date of Access: 01.09.2017.

Wood, Aylish (2011). Digital afx: Digital Dressing and Affective Shifts in Sin City and 300, New Review of Film and Television Studies, 9:3, pp: 283–295.

Zand, Joel, (2015). https://www.similarweb.com/blog/the-most-popular-illegal-movie-downloads, The Most Popular Illegal Movie Downloads, Date of Access: 11.08.2017.

Zizek, Slavoj, (2010). https://www.newstatesman.com/film/2010/03/avatar-reality-love-couple-sex, Avatar: Return of the Natives, Date of Access: 13.08.2017.

Onur Keşaplı and Ahmet Dönmez

A Critical Approach to Digimodern Masterpiece: Production Relations and Public Representations in The Lord of the Rings Trilogy

Introduction

The Lord of the Rings trilogy, based on the book by John Ronald Reuel Tolkien and adapted to the cinema by Peter Jackson, is a series of movies in which peoples take place in an eternal war between the good and the evil that occurs in the fantasy world called the Middle-earth. Besides, these movies with computer generated imagery (CGI) indicate the latest examples of cinemaximum technology. The technology is important to convey the ideology and the plausibility of the film. When the peoples represented in the film are evaluated in terms of their physical properties, it shows the Westerners as good and the Easterners as evil. These preferences, in terms of production relations and representation of the people, show themselves in cinema adaptations. The approaches towards racism also stresses on the fact that the trilogy is more than a literary adaption and has ideological dimensions since it was released after 2001, the era when the Western coalition forces waged war on countries like Afghanistan and Iraq, under the leadership of the USA.

The Lord of the Rings, a fantasy literary trilogy written in 1954 by British linguist and author J.R.R. Tolkien, aroused great interest and curiosity in the twentieth century shaped by the political influences of Cold War, years following the First and Second World Wars. Translated into many languages, the book reached and still reaches readers worldwide. Tolkien, who enlarged the fantastic world created in his book Hobbit in 1937 with The Lord of the Rings trilogy, states that he writes "with a motivation stems from the lack of epic narration in British culture" (2010: 15). The series, which attribute to historical and current events, has received a global attention by its adaptation to the cinema by Peter Jackson. The films, which particularly have been successful at the Academy Awards, address to every age group in many countries with the help of special effects created by the latest developments in computer technology.

Alan Kirby, who introduced the concept of digital modernism and studied in the new media field, emphasizes that the series is one of the masterpieces of digital modernism (2009: 180). According to Kirby (2009), developments in new communication technologies are reshaping the culture and increasing plausibility. From this point of view, it is possible to say that the new technologies used in the communication are effective in the speed of spreading cultures and ideologies. On the other hand, developments in communication technologies serve to spread the dominant culture that holds the technology. According to Kirby, in the digimodern period, heroic stories based on ancient or medieval oral myths shine out in films such as The Lord of the Rings and Harry Potter (Kirby, 2009: 160). Considering these two films, especially the re-emergence of these fantastic and historical narrations in credible reality costs cheaper via digital technology. It is known that The Internet Movie Database (IMDb)'s top 100 fantasy films are produced after the year 2000 (www.imdb.com). The Lord of the Rings trilogy is also at the top of the list among others. In the movie, visual effects are not only used to describe the environment. In the trilogy, it is noteworthy to see the production process of the digital characters particularly. As an example, Gollum represents the perfect synthesis of live action and animation as an impressive CGI character (Prince, 2009: 132). Gollum is also the first motion capture (CG) character to interact directly with other actors (www.yahoo.com, 2014). This realistic presentation of Gollum also enables cinemagoers to identify with the characters easily.

In this study, Peter Jackson's three cinema movies will be discussed through modernity, peoples represented in the movie, and production relations by using the approach of Marxist film criticism. According to Kirby (2009), the movies are an example of digimodernist movies formalistically but in terms of content and ideology they are considered anti-modernist. Firstly, how Marxism interprets the terms to explain anti-modernist content and ideology and then Marxist film criticism, which was shaped in the art environment of the 1960s, will be discussed. In the study, the matter of how The Lord of the Rings Trilogy relates to The Industrial Revolution, the Enlightenment and modernity and what it offers instead via Marxist film criticism and ideological terms will be discussed.

From a Critical Perspective to Modernism, from the Death of Postmodernism to Digimodernism

Modernity emerged as a period that causes changes and transformations of many dynamics when the Enlightenment which came after the Renaissance and the Reform in a cultural and intellectual level get together with the Industrial Revolution, shaped by the scientific developments, especially in the western world. As a result of the spread of positive thoughts about modern in the nineteenth century, some other terms appeared such as "modernity", "modernism" and "modernist" and since then this term has been frequently used in art and other fields (Kovács, 2010: 11). Aytek Soner Alptan argumentatively defines the modernity which different ideological thoughts offer different solutions (2012: 21–22) as mentioned below:

> *"Considering the Marxist thought, we should say that modernism is a 'product' of capitalism. Basically, the development of capitalist production and property relations solved some of the traditional structures and evaporated them, as Marx said in the Manifesto of Communist Party, and absorbed some of them with a tendency somewhat out of the first anticipation of Marxism. As a result, a new social formation emerged. If we define social formation. If we describe the social formation as a whole which is rebuilt, reproduced and identify within the productive forces and existing productive relations, modernization is basically common term called as whole transformation that occurs in different aspects of social formation political, ideological, cultural, moral, etc.) developed by capitalist production relations".*

On the other hand, modernity, unlike a radical change, creates a new period that has not completely broken up with the traditional production relations, as the inequality still persists, but the form changes. It is the same period that creates Marxism.

> *"Marxism is a child of modernism. This, however, tells little about the relationship between Marxism and modernism as a total consequences of modernization. Marxism, neither can be read from modernist perspective, nor can it analyzed without modernism and previous movements. Marxism is one of the most comprehensive critics of modernism"* (Alphan, 2012: 25–26).

The class struggle, main argument of Marxism, is older than modernism; but Marxism became a theory with modernity. In this respect, Metin Çulhaoğlu's words are important to explain the relationship between Marxism and modernity; It is "a critic of modernism that is based on modernism, and its arguments and suggestions become meaningful on this basis" (qtd. in Alptan 2012: 26). The concept of postmodernism, which marks the postmodern era, has been mentioned most often since the 1950s. Postmodernism, a very difficult concept to define, unlike modernism "above all characterizes itself to challenge scientific dominance, especially a single theory of progress" (Mutlu, 2010: 237). In other words, contrary to modernism, which is based on rational and positivist thinking, postmodernism adopts irrationality. Great ideologies such as Marxist, liberal and capitalist ideology have lost their meaning by the emergence of postmodernism. Alan Kirby, however, claimed it for postmodernism in his article "The Death of Postmodernism and Beyond" in 2006. According to Kirby, the period of postmodernism has already ended and a new movement has begun in which computer-oriented digital technologies shape life and culture. Kirby describes this period as "Digimodernism". According to the Marxist point of view, it is possible to say that the dominant classes in the "Digimodernist" period have the technology that will facilitate the domination on lower classes. The dominant class, which holds the production means in the modern period, also reaches new communication technologies easily in the Digimodern period. The thoughts of a dominant class easily spread, thanks to these technologies. Thus, Hollywood cinema by using high technological equipment (such as CGI and IMAX) becomes prominent as an ideological culture and thought supporter and publisher.

On Marxist Film Criticism

In order to be described as a branch of art, cinema had to wait for a long time and the relationship between the criticisms through qualitative sense has also been delayed. Apart from the theoretical approach of the Soviet directors of the 1920s, cinema was generally considered under the title "entertainment". The criticisms influenced by Hollywood-based movements lacked ideological basis for a long time. However, this began to change in the 1960s, parallel with the anti-system movements at the point of action around the world, particularly in Europe. The film criticisms, which were just introductions of the films for a long time, systematically changed and gained a scientific dimension through interdisciplinary interaction via the studies of Andre Bazin.

> "*The concept of Marxist ideology, which had a positive development by changing the culture itself especially at the end of the 1960s, provided a theoretical framework for the rigorous re-evaluation and deconstruction of old Bazine aesthetics based on ontological realism in addition to other things. And discussions on film criticism, people can now comfortably talk about the fact that we have entered the post-Bazinian period of film aesthetics and today's discourse has a fragmented structure, but they do not have a complete idea what destroys old bases and how new bases are found*" (Macbean, 2006: 276).

At that time, traditional (commercial) cinema codes were criticized as a whole, especially thanks to the politicized avant-garde films of Jean Luc Godard, director of the French New Wave movement.

At this point of view, Marxist critic creates basic cinematographic dynamics via concepts of ideology and reality and follows a discourse method about production relations and class struggle. Godard describes the question of ideology in the cinema as "the bourgeois creates a world in its own image, but creates a copy of this world and perceives it as a reflection of reality" (qtd. in MacBean, 2006: 96). James Roy MacBean similarly argues that "ideological hocus pocus", the reflection of the reality used instead of "the reflection of bourgeois capitalism, does not only examine the methods to destroy the bourgeois capitalism as a problem but also constantly guarantees it in the re-emergence of "deceiving image of the reality".

According to Robin Wood, Hollywood, which is the center of conventional cinema that is claimed to be based on illusion and purification, leads to a definition on the audience that reforms should be enough to solve every problem in the existing system, rather than radical changes (qtd. in Yilmaz, 2008: 72). On the main movement identification approach, MacBean says:

> "*Cinema of bourgeois seems to ignore the existence of the audience, arranges what happened and told on the screen not to targeting the audiences, tells the lie that it is a reflection of reality, but this always exploits the feelings of the viewer / listener and benefits from the mechanisms of identification and reflection to persuade to participate in fantasies and dreams marketed by the capitalist society masterfully, insidiously and involuntarily*" (qtd. in Yilmaz, 2008: 72).

On the basis of reproduction of production relations via art, ruling ideology uses art as a tool and reinforces the sovereignty. In his work "German Ideology", Marx and Engels summarize this situation as follows:

> "*The ideas of the ruling class are in every age the ruling ideas, in other words the class which is the ruling material force of society, is at the same time its ruling intellectual force. The class which has the means of material production at its disposal, has control at the same time over the means of mental production, so that thereby, generally speaking, the ideas of those who lack the means of mental production are subject to it. The ruling ideas are nothing more than the ideal expression of the dominant material relationships, the dominant material relationships grasped as ideas; hence of the relationships which make the one class the ruling one, therefore, the ideas of its dominance*" (Marx ve Engels, 1992: 61).

According to Marxist criticism, the seventh art is serving the ruling class of today's society, the bourgeois, and presents illusions that reproduce the relations of production under its sovereignty. As Alan Kirby indicates, Hollywood cinema, which uses the latest developments of technology, increases the reality and credibility of this illusionary world (2009: 180). Through the visual effects, the audience is provided with a perfect display feast and the audience loses himself in this artificial world. Thus, the critic of the audience are invalid through the techniques used. According to Hess, Marxist criticism, like structuralist and critic of semiotics, totally rejects the autonomous structure of art and accepts it as an ideological link between individuals, groups, classes and societies. This makes the review of the social context of a film more important than that of the film (qtd. in MacBean, 2006: 278). MacBean, who finds this view partly correct but simplistic, argues that Marxism is ideologically brave unlike three mentioned criticisms, and contemporary Marxist criticism also tends to be cinematographic as a whole with the subject of the film, and Hess cannot see it (2006: 278).

Including different views, Marxist criticism, the ideology which is based on production relations and class struggle and the concept of reality accompanied by it, criticizes Hollywood-based cinema movement codes. Marxist criticism strives for the construction of a new understanding of cinema that will take an innovative approach with cinematographic items in film narration.

Narration in The Lord of the Rings Trilogy from the Critical Perspective

In The Lord of the Rings Trilogy, The One Ring, which is crafted and has the power of ruling over the Middle-earth, remains hidden for hundreds of years after Sauron's loss of the ring and his fall, and finally it comes to Frodo from the Hobbit. The ring must be destroyed to defeat Sauron, who gets back his power again and chases the ring for the ultimate attack. The One Ring could be destroyed only by the Fires of Mount Doom, in Mordor, the Land of Sauron, because that is the place it was made. The Hobbits, who resemble the English with their accents, names of their cities (like the Shire), tea and food habits, participated in an important duty, and that is quite appropriate for mainstream cinema which is based on the motivation of the achievement of a single character.

In the first film of the series, The Fellowship of the Ring, Frodo is accompanied by Men, Hobbits, wizards, Elves and the oldest people of Middle-earth, and Dwarfs form a fellowship to help him on this difficult and dangerous journey. In the second film, The Two Towers, the members of the Fellowship split into and give losses. Moreover, the battle of Fellowship against destroying the One Ring continues. In the third and last film, The Return of the King, both the evil are beaten by fighting against the Orcs, which represents the evil peoples and Sauron is also defeated forever by the destruction of the ring.

Firstly, the film must be analyzed formally in terms of cinematography before analyzing its interpretation of production relations, conservative structure, and racial and religious codes. Robin Wood says that the mainstream classical narrative of the US cinema acts with two basic premises and explains them below:

1. *The function of the classical narrative entirely is to validate and strengthen the male dominant order and its dependence on the woman.*
2. *The basic principles that determine the structure of classical narrative includes symmetry (especially symmetry of beginning and ending, 'final, initial response') and*

> *ending (approval of a number of values embodied in the system includes rewards and punishments of events discussed in narration such as moral issues, restoration of order, marriage and death respectively)* (qtd. in Yilmaz, 2008: 76).

The masculine order in the trilogy does not originate from the fact that almost all of the main characters are male. Furthermore, the way in which the female figures play in the narration helps in the occurrence of such system. For example, Arwen, the Elven, can forfeit her immortality for the sake of the wife of King Aragorn, who will eventually return to the throne and remains passive throughout the story. On the other hand, Eowyn from Rohan, who appears in the second film and gets emotionally close to Aragorn, acts as a strong character that holds guns and is able to fight. However, being such a strong character, she is not loved back by Aragorn, and in the end of the series, she is screened as a wife, who adapts to Gondorian Boromir's life. Galadriel, one of the most dominant female figures of the story, is an Elven queen. Apart from an asexual appearance, Galadriel's ambition of the One Ring led her to be appeared as black-hearted in the first film.

In the second principle, "Responding the ending to the beginning and providing the order restoration of all matter by solving them" takes place as a whole in The Lord of the Rings. The resurrection of the order and victories through deaths, and from the chaos, is rebuilt when Aragorn becomes king again and marries Arwen. Moreover, it provides unity to the story that Shire, the homeland of Hobbits, is shown in the same space and visual language at the end of the history that begins with a warm color.

In general, the adaption of the book based on heroism, morality and self-sacrifice to cinema turned especially into war movie, as a subtype in the last two episodes. In terms of Hollywood's war presentation, Ertan Yilmaz says:

> "*In Hollywood films, the war is not seen as a life experience that takes away many things from people, both physically, spiritually and morally, but presented as an experience that man finds himself, emerges morally, be a man and feels doing right things, so that the war, the American army and the fighting American soldier are blessed*" (Yılmaz, 1997: 30).

The battle scenes, which are the basic dynamics of The Two Towers and The Return of The Kings, is presented in the form of a violent visual scenes, creating emotions such as excitement, fear and enthusiasm with music, slow shots and preferred camera scales instead of excluding or alienating the audience from the war. The presentation of the enemy Orc armies seems to coincide with the Triumph of the Will, which is called the fascist cinema aesthetic filmed by Leni Reifenstahl about Hitler and Nazi troops. The camera scales and fictions of the balcony speech of Saruman, who is Sauron's greatest assistant and the evil wizard, is quite similar to Hitler and his troops. The Nazi antagonism is noteworthy in the book of Tolkien, who supported fascist Franko in the Spanish Civil War, but developed an opposite attitude towards Hitler and Nazis and more in the adaptation of director Jackson. This is supported by pointing the Uruk-Hai as a superior race by Saruman, destroying the trees that live in the land for thousands of years, massacring the Dwarfs and attacking the Orcs by the speech "the Men of the West", which is said stand chapter of the trilogy (Hoiem, 2005: 75). However, Rekin Teksoy says that at the time of shooting the film pullout, the United States acts in the opposite direction, in the direction of aggressive foreign policy focused on the Middle East.

> "*The trilogy, which is no more than a skillful use of image and sound effects, conveys battle between the good represented by blue-eyed whites and evil represented by the horrible-looking villains of dark-skinned like a fairy-tale and ends happily. In the trilogy based on domination of man, a world that excludes sexuality and glorifies war is praised. In a period when technological development brings the world to the brink of a conflict of civilizations, It is not surprising that the reflected perspective of this series is in a neo-Nazi direction. In the series people are discussed as "good and evil" refers to Walt Disney movies and also have the same opinion of administration of President Bush to classify the peoples of the world*" (Teksoy, 2005: 838).

Parallel to this, the sign "People of the West" verbally and visually stands against both the Nazis as the totalitarian regimes and the Soviet Union. The fact that Saruman controls over everything that grows in the land by using force, and captures individuals with a totalitarian regime against the libertarian West, leads us to think this (Markova and Hooker 2004: 165). Saruman's oppressive attitude as a visual counterpart refers the totalitarian regimes in the film.

Public Representations and Religious Signs in The Lord of the Rings Trilogy

The Lord of the Rings trilogy, both book and film, offers a remarkable structure considering the values, presentations, traditions and lifestyles of the peoples. In "The Clash of Civilizations" (1996: 183), which ideologically affects the cultural perception of the West, revealed by Samuel Huntington in the post–Cold War era, it finds itself particularly at the point of presentation of the people. In Middle-earth, where the Hobbits are a symbol of the English, the Elves are portrayed as the "oldest" and "perfect" people. In the first film, the Dwarfs, described by Saruman as greedy since they are fond of gold and because Balgrog, a monster, attempts to reach the surface while searching gold by digging deep down, and deterritorialized from his homeland, represents Jews in the fantasy

Middle-Earth (Suar, 2013: 16). People are enshrined with Rohan kingdoms, which is a reminiscence of Gondor and the Scandinavian mythology.

However, although they are human race and frequently mentioned in the book, the South Eastern "People of the Haradrim", who appears only in the battle scenes with the army of Sauron, turned into Sarasen who are described as Muslims by the West with their colors and dresses during the Crusades period (Sinex, 2010: 175). The Eastern peoples of the Middle-earth, Orcs, Goblins, Trolls, and Uruk-Hai, with their monstrous appearance and evil character correspond with the most primitive meaning of Eastern societies historically in the eyes of Western society. Moreover, it is told that these people represent the Huns as a whole (Lobdell, 2008: 264).

It is important to note how Sauron, the evil of the trilogy, is defined in Silmarillion in which Tolkien wrote as a resource and starting book for the universe of the Middle-Earth, while examining how the religious items of the series are indicated in the film. In the book, it is remarkable how Tolkien describes Sauron, who denied the existence of God by saying that "the One (God) is only a make-up of the jealous Valar in the West, and a trick based on his wishes" (Tolkien, 2010: 41). Despite the unbelief of the prime evil in the story, it is seen that the religious values are high in the characters that are described as "good" in the film. Mike Foster indicates that the author uses the Christian allegory (2007: 297). In this respect, Gandalf, the wizard, is the prominent character in the film. In the first episode, Gandalf, who sacrificed himself in order to save the Fellowship of the Ring in danger by fighting the Balrog, who returns to Earth because of the money ambition of the Dwarfs, in the Moria Kingdom, is the "sacrifice figure" of Catholic belief in the film. It is also thought that in this battle, Gandalf is often reflected on the screen by the use of Christian images, such as "the Crucifixion of Jesus" by the director Peter Jackson (Dawson, 2011: 212–213).

Similarly in The Two Towers, Gandalf, who returns like white and calm Jesus Christ after his death, rescues the Rohan King Theoden with a practice like exorcism, whose mind is captured by Saruman in the same episode (Bratman, 2011: 295). In the last film of the series, Gandalf, who talks about temporariness of this life and the beauty of the heaven to calm down Pippin from Hobbits that fears war and death, is shown as "the goody" with religious motivation against atheist Sauron.

Production Relations in The Lord of the Rings Trilogy

In terms of production relations, it is analyzed that the peoples of Middle-Earth are separated as the East and West. The Hobbits and Humans based on feudal associations and kingdoms, feudal lords and communities adopt the land-based production. Especially People, the Gondor and Rohan Kingdoms represent the feudalism in Europe with royal families, servants, citizens and knights. The Dwarfs are engaged in mining and producing gold besides land-based production. In this respect, it is a structure similar to the human kingdom as a form of government. The Elves, however, are not sampled in terms of production relations in the films, only shown engaging in forging in some of the scenes. Elves are seen as aristocrats who are as rigid as the People and the Dwarfs in a hierarchical structure. The Orcs and their variants, presented as Eastern peoples in the film, who are shown in a relatively equal position to everyone that does not have any hierarchy or an aristocratic class. One of the greatest distinctions is that Orcs and their variations do not have land-based production methods; they have pre-modern production relations in which they use different tools via machines (an advanced dam building in terms of the era described).

These new production relations are denigrated with the Orcs by Tolkien's pastoral frame. It is seen both in the book and film that the Hobbits' homeland Shire is presented as the source of joy and happiness (Act. Davis, 2008: 56), by the pastoral literary style that is defined by Tom Shippey as "nostalgic, conservative, and idealistic of the past" (qtd. in Davis, 2008: 56). Accordingly, whenever a possible destruction is expressed in the film, destroying of the Shire as a disaster scenario is shown.

Vaninskaya defines Tolkien as a neo-romantic writer, who brings a romantic critic of modern industrial society (qtd. in Dawson, 2011: 161). The Lord of the Rings is seen as an enemy of modernism since the industrial society and technology are shown by bad characters and races, and that the good are systematized with pre-industrial production relations. In this respect, Saruman is the most evil character in the film. Saruman has been involved in the first two films of the series and has found moves that could be described as the Industrial Revolution for Middle-Earth, while, Saruman, who is told in Tolkien's Silmarillion with his interests in forging while listing the negative qualities of him, shows things that could be described as the Industrial Revolution of the Middle-Earth by involving in the first two series of the film (2010: 613). Saruman, who makes various war equipment in places like factories, cuts the trees for the necessary raw materials and gets a dam built for the energy need, which are mentioned in the film together with fire and metal. In this production world called pre-modern era, not only the tools but also a personality, Uruk-Hai, have been introduced. This is a detailed information that is not in the book. Unlike the book, the Uruk-Hai, who are born vaginally, are "produced" in the film. Thus, the production mode based on industry and technology is shown to be able to create "a horrible" race even worse than Orcs. It was the nature itself that brought Saruman's end, who turns the tower and its surroundings in Isengard into a smoky factory. Ents presented as tree-like creatures, who defeats Saruman, the wizard, by declaring war, who destroys the natures by machines. Destroying the production halls and releasing the river by lifting the dam, Ents rescues Isengard and its surroundings. At this point, there is a major difference between the book and the film. In the film, Saruman, who is accused by the destruction of nature and trees at the level of "genocide", for this reason in the book he is criticized by all the peoples, especially with the Hobbits, in terms of the destruction of nature. The book tells in terms of the theme of human relation/conflict with the nature that all peoples of Middle-Earth harm the nature or will be. The greatest example of this is the collapse of the tree by the Hobbits. However, this part is not included in the film, and among all the people, only the Orcs are seen that destroy the nature and Saruman is the prime evil of this process (Davis, 2008: 60–68). This divergence between the book and the film can be interpreted as a critical mistake made to make the good-evil war of the film more visible.

Considering Tolkien's general approach of modernity and industrialization, equally with the evil forces and their actions (Dawson, 2011: 226), the interpretation of the director creates an inconsistency with author's world view.

Conclusion

The trilogy of the Lord of the Rings, which has the atmosphere of the First and Second World Wars and the Cold War, making references to these wars and attracts a great interest around the world, was adapted to the cinema by using new technologies in Hollywood Star Studios approximately fifty years after it was written, became a world-wide phenomenon with a trilogy which is based on the New World Order upon the terrorist attacks on September 11 and the invasion of Afghanistan and Iraq by the USA. According to Marxist film criticism, these films are accepted as an ideology, having a conservative structure in terms of production relations. The universe of The Lord of the Rings and the presentation of the peoples in Middle-Earth literally meet the approach of the Clash of Civilizations unveiled by Samuel Huntington, who unveils at the end of the Cold War and is widely accepted by the West. Both references from the original work and the films; the Trilogy, which clearly shows the Western and Eastern struggle, can be described as racist in which the "East" refers to a monstrous symbol of evil that is skeptical and tries to industrialize and the "West" symbolizes the "ethics", faith and goodness. Considering the attitude of Marxism towards modernity, unlike Marxism, the main ideological thought in the trilogy criticizes modernity which aestheticize the old orders which are shaken by the fundamentals of technology, industrialization and new production, except modernism. On the other hand, in terms of human–nature conflict, it seems that all peoples of the Middle-Earth are faulty in the book, while the film takes this criticism further by saying that only the peoples of East are at war with nature. At this point, another point that needs to be discussed in terms of Marxist criticism is the cinematographic language of the film. This trilogy, attempting to identify by using all the codes of the mainstream conventional cinema and the vast technological possibilities of the Digimodern period, creates a false reality presenting the form implicitly and has a conservative cinema language. Even though it is technically digimodernist, the aestheticization of the war scenes and the affirmation of the aristocracy and the feudal relations make the films of The Lord of the Rings one of the most conservative cinema narratives of the twenty-first century with its content and quality. It is also possible to say that with the technological innovations in the Digimodern era, Hollywood's cinema based on alluring people is getting stronger.

References

Alphan, A. S. (2012) Modernleşme ve Türkiye Solu: Bir Muhasebe Denemesi, Gelenek, Sayı: 117, 21–42.

Bratman D. ve DeTardo M. (2011) The Year's Work in Tolkien Studies 2008, Project Muse: Tolkien Studies, Sayı: 8, 243–295.

Davis G. D. (2008) Showing Saruman as Faber: Tolkien and Peter Jackson, Project Muse: Tolkien Studies, West Virginia University Press Sayı: 5, 55–71.

Dawson, D. A. (2011) The Ring Goes Ever On, Review Essay, Project Muse: Tolkien Studies, Sayı: 8, 143–241,

Foster, M. (2007) The Power of the Ring: The Spiritual Vision behind the Lord of the Rings "Book Reviews", Project Muse: Tolkien Studies, Sayı: 4, 293–297.

Hoiem, E. M. (2005) World Creation as Colonization: British Imperialism in 'Aldarion and Erendis, Project Muse: Tolkien Studies, Sayı: 2, 75–92.

Huntington, S. P. (1996) The Clash of Civilizations and Remaking of World Order, Simon and Schuster, New York.

Kirby, Alan, (2009) Digimodernism How New Technologies Dismantle the Postmodern and Reconfigure Our Culture, The Continuum International Publishing Group Inc.

Kovâcs, Andrâs Bâlint, (2010) Modernizmi Seyretmek, Avrupa Sanat Sineması 1950-1980, Ertan Yılmaz (çev), De Ki Yayınevi, Ankara.

Lobdell, J. (2008) Tolkien and Shakespeare: Essays on Shared Themes and Language, Book Reviews, Project Muse: Tolkien Studies, Sayı: 5, 257–265.

Macbean, J. R. (2006) Sinema ve Devrim, Ertan Yılmaz (çev), Kabalcı Yayınevi, İstanbul.

Markova, O. ve Hooker M. T. (2004) When Philology Becomes Ideology: The Russian Perspective of J. R. R. Tolkien, Project Muse: Tolkien Studies, Sayı: 1, 163–170.

Marks, K. ve Engels F. (1992) Alman İdeolojisi, Sevim Belli, Ahmet Kardam (çev), Sol Yayınları, Ankara.

Mutlu, Erol. (2010), İletişim Sözlüğü, Bilim ve Sanat Yayınları, Ankara.

Prince, Stephen. (2009), Digital Visual Effects in Cinema, Rutgers University Press, New Jersey.

Sinex, M. (2010) "Monsterized Saracens" Tolkien's Haradrim, and Other Medieval "Fantasy Products", Project Muse: Tolkien Studies, Sayı: 7, 175–196.

Süar, S. (2013) Orta Dünya'dan Ortadoğu'ya Uzanan Yol, Sol Gazetesi, 2 Ocak 2013.

Teksoy, R. (2005) Dünya Sinema Tarihi, Oğlak Yayınları, İstanbul.

Tolkien, J. R. R. (2010) Silmarillion, Berna Akkıyal (çev), İthaki Yayınları, İstanbul.

Yılmaz, E. (1997) Amerikan Sinemasında Savaş ve Vietnam Filmleri, Antrakt Sinema Kitapları, İstanbul.

Yılmaz, E. (2008) Sinema ve İdeoloji İlişkileri Üzerine, B. Bakır, Y. Ünal, S. Saliji (der), Sinema İdeoloji Politika, Orient Yayınları, Ankara.

Internet Sources

https://www.yahoo.com/entertainment/a-brief-history-of-motion-capture-in-the-movies-91372735717.html Erişim Tarihi-10.08.2017

http://www.imdb.com/list/ls009669258/ Erişim Tarihi-10.08.2017

Mikail Batu

The Reflections of Corporate Identity in Social Media

Introduction

If you are asked to introduce yourself, firstly you will say your name, the place you were born, your profession, age, education, etc. This is important for the first impression of a person. If it seems unreal, the "identity card" will be requested and the statements will be proved by a document approved by the competent authorities. From this point of view, the first way of recognizing a person is to listen the expressions about himself and second way is to acquire it by living together with the person or getting information by his surroundings. These two ways create a perceptive framework about any person.

This process is not different for corporations as well. It is possible to recognize a corporation not only by scrutinizing them closely, but also examining the corporation's activities or evaluating the image and reputations of its clients. Accordingly, the corporate identity, which has a great influence on the formation of the image and reputation of the corporation, has great importance in terms of introducing and understanding itself.

The appearance socially affects the formation of certain perceptions; a corporation is also recognized by corporate design and behavior, which are the elements of the corporate identity, and is evaluated by corporate philosophy/culture and corporate communication.

Together with new communication technologies, this analysis is conveyed by virtual platform not by words. Especially social media, which is favorite for sharing, was used by corporations to represent themselves with their corporate identities and places, where the customers can criticize them. Accordingly, corporations need to implement the strategic plan of how to use social media in terms of corporate identities. In this study, the details of carrying the corporate identity to the social media and studying the corporate identity in the social media are discussed.

Importance of the Corporate Identity for the Literature Review and Corporations

There is always a face of all corporations that provides it to distinguish from others and appears to be imprinted in the minds. This visible face is often referred to as corporate identity. However, the corporate identity is not limited to the visible face. Corporate identity is defined as a holistic concept of features that make the corporation unique (Balmer & Gray, 2003: 972). In addition, corporate identity is (Balmer & Stotvig, 1997: 170);

- Most often, corporate branding, the use of corporate culture, a symbolic presentation of the use of diversity of communication channels,
- A corporate brand or corporate culture, which is defined as the identification of a person or group. The composition of what the institution says and does can be defined in various ways as culture, communication, symbol, behavior, market, communication design, products and service strategy.

The term "corporate identity" is used for the whole message and associations expressing the corporate personality. In the same way, it relatively refers to all of the discrete elements like identity, logo, color, logo fonts, corporate behaviors, culture, philosophy, principles, values, missions, communication styles and communication relations with people or corporations (Theaker, 2006: 136). Corporate identity, as mentioned above, also includes the invisible aspects which are the source of these from the visible side of corporations. Thus, considering the corporate identity as a whole and evaluating it all in all will be helpful in understanding the personality of each corporation. Balmer's first literature researches reveal that the corporate identity has various uses as a helpful tool. These are listed below (Balmer, 2001: 263):

- Forming a strategy
- Culture-change
- Effective corporate communication
- Clearly stated corporate strategy
- Clearly stated/changed corporate culture
- Action as a field for corporate communication
- Providing visual corporate identity on a claimable way

When forming identity studies for a corporation, the agencies usually perform the tasks that are indicated in the frame below. However, these studies directly cover the visible signs and codes of corporations. In the basis of these signs and codes, the discrete studies based on these signs and codes such as behavior, culture, communication, corporate philosophy state the values that are asked by the managers and employees in the corporations, not by asking them from the agencies. Although the visible face of corporate identity can have similarities, cultural values often have differences.

The Aspects of Corporate Identity

- **Forming corporation and service name**
  - **Brand identity:** It includes finding a brand name, buying royalties, logo design, logotype design, using logo in different field, using brandings, advertising and similar applications.
- **Design of emblem, logo, logo type**: Some of items that form the identity of the corporation like typeface, scaling system and corporate colors should be correctly used.
- **Application of logos and emblems on different materials**: For example, reflecting the corporate identity via signboard and billboard, which are the indicators that have various usage areas and provide the corporations to be seen in the public permanently.
- **Design of the file, letterhead, envelope and business card**: The rules must be specified in detail and standardized in size and color to provide a strong and consistent identity.
- **Design of accounting documents**
- **Annual report and Corporate Introduction Book**
- **Design of valuable document, check, coupon**

Corporate Identity Studies in Agencies (www.matbaashow.com)

The current literature reveals many dimensions of the corporate identity. Schmidt (2000), for example, refers to the five main features of the corporate identity. These are listed below (Schmidt, 2000: 17):

1. Corporate culture
2. Corporate behavior
3. Products and services
4. Communication and design
5. Marketing conditions and strategies

The issues above are detailed below.

1 Corporate Culture

Corporate culture is a general term used to describe certain features or unique personality of a certain corporation, including certain elements such as values and beliefs, business ethics, rules of behaviors (Panda & Mandal, 2006: 446). Besides, corporate culture consists the sum of values, beliefs, assumptions, philosophy, mission, ethics, principles, guide, historical background, corporate nationality and subcultures of a corporation. In addition to this, corporate employees have a significant role in corporate culture (Cheng, Hines & Grime, 2008: 685).

The expression of corporate culture, shared among the members, is a relatively new phenomenon. Although, corporate culture for many sides decided to be basically a rational thought in terms of control and coordination of a group of people until the mid-1980s, it became an important subject for academics after the 1980s (Panda & Mandal, 2006: 447). Culture is generally a term that based on the relatively constant beliefs, values and manners of the corporation (Melewar & Akel, 2005: 50). In addition to unifying feature of corporate culture, other important features are given below (Robbins, 2004: 525):

- **Innovation and risk taking:** Measuring of encouraging employees on innovation and risk-taking.
- **Attention to the details:** The degree of expectation of employees to pay attention to the details and display sensitivity.
- **Output orientation:** The degree of the management to focus more on results or outputs than the processes and techniques that are used for success of the outputs.
- **Person orientation:** The degree of decisions of the management in which the outputs are effective on the employees.
- **Team orientation:** The degree of business practices that occur more with the team than individuality.
- **Being ambitious:** The degrees of people being competitive and ambitious rather than being easy-going.
- **Stability:** The degrees of emphasizing to defend the position unlike the development of the corporate practices.

While corporate culture expresses the sum of different values, it is based on communication strategy. The corporate culture based on this basis serves the individual development of employees and directly supports the development of the corporation. Thus, questioned and structured goals are laid instead of progressing within a single goal.

It is highly important to think of the corporation as a whole, evaluate the relations between the parts, but eventually design it as a whole. In this, the leader of the corporation has a big duty. The leader is a corporate architecture and constantly thinks about designing. Leaders, who are good in decision-making and problem-solving, are those who successfully design the learning processes.

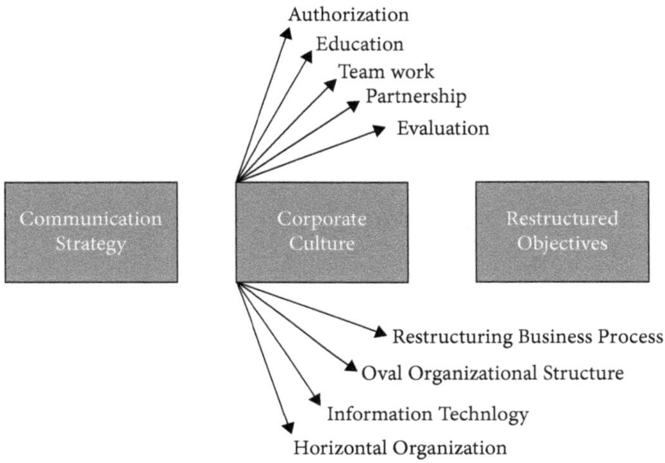

Fig. 1: Communication, Corporate Culture and Restructuring (Belmiro, et al, 2000: 292).

The leader, who assumes the teaching role in the learning corporations, should support the learning process of each in the corporation. They should help people to think systematically about problems and develop their concepts. The vision of the leaders should not prevent them to see the truths. Leaders should create and spread creative tendencies within the corporation (Keçecioğlu, 2003: 147).

The project of KaleWorks, the Common Corporate Culture and Sharing of Kaleseramik Dealer Sales Representatives (www.microsoft.com)

Kaleseramik has decided to carry out the project called KaleWorks for all sales representatives in its distributorship, which is first in the sector, in order to create a common corporate culture, to develop sense of belonging and to increase sales performances by watching them for all salesmen in its distributorship.

The first project in the sector is built on Microsoft Dynamics CRM to recognize the salesmen representatives closely, to get updated and accurate information about the products and to be informed on time from the campaigns organized by Kale Seramik and to increase strategic sales performances.

Kaleseramik, 1957 yılında kurulan ve Türkiye'nin ilk seramik karo üreticisi olan Çanakkale.

(continued on next page)

Kaleseramik was founded in 2000 by the integration of the companies, Çanakkale SeramikFactories A.Ş. founded in 1957 as Turkey's first ceramic tile producer and Kalebodur Seramik Sanayi A.Ş. founded in 1972 as Turkey's first local tile producer. Kaleseramik has 27.5 million square meters/year of wall tiles, 32 million square meters of floor tiles and 6.5 million square meters of granite ceramics, as well as 15,000 tons of insulators and 40,000 tons of frit production capacity. Kaleseramik produces 1,900 kinds of floor tiles in 50 different sizes and 2,200 kinds of wall tiles in 60 different sizes and presents approximately 200 kinds of new products to consumers every year.

Situation Analysis

Kale Seramik carries out its sales through a wide sales network with a large number of main and sub dealers. In particular, over 700 sales representatives working in the main dealers, who get in touch with the customers directly, have a very strategic importance. Despite this importance, they are not well recognized by Kaleseramik, are not contacted with directly, are asked about the product and position and could not be tested as often as desired. It was decided to create a sales club in order to get more information about these 700 sales representatives, follow and increase their sales performance, send messages given to the market in first hand, and share the problems they can live and find solutions. Kaleseramik began to work with the code name called **"Corporate Facebook"**, thinking that it would be easier and more effective to realize this project in a virtual platform for a healthy communication with sales representatives over wide range geographical area.

Solution

Kaleseramik started its studies under the management of Ayşen Uz, Branding and Communication Group Manager, and worked very hard to create an application that would be used actively by all sales representatives and contribute positively to the corporate culture and brand value without any disturbance by the dealer owners. The sales club content was determined in the direction of the focus group work consisting of the support of professional consultants and sales representatives. Profiles were defined by interviewing all sales representatives with the help of the call center. By looking at the answers of many different and detailed questions, the most accurate design was done.

As a result, a portal was decided to be used as front face of the sales club. The Microsoft Dynamics CRM substructure was approved to take profile of sales representative from portal, all kinds of sales activities, to design and supervise workflow, to collect reward points and to run the reward program. The developed portal application called KaleWorks, which can only be reached by sales representatives, provides technology services for each user to enter their own profile information, to enable representatives to communicate with each other, to get rapid information with instant messaging (chat) services, to record their performance and activities and sales.

In first sales, certain products are given surprise gifts via announced advertisements by daily portal. This provides sales representatives to focus quickly on specific targets. In addition, representatives who start the day by entering the portal to learn these surprise campaigns are immediately informed about updated information. Likewise, the salesmen of the month announced in this portal provides individual motivation as well as being useful for the sales representatives to recognize each other.

Benefits

Long-Term Association with Sales Representatives. It is a quite strategic priority for Kaleseramik that begin to recognize more closely the sales representatives who have direct strategic contacts. By monitoring the performance of their sales representatives, they provide the support needed to determine career goals and advance towards this goal. By recognizing the representatives, who have the experience on products and customers, they have the necessary information about the representatives to stay in the ecosystem of Kaleseramik. It enabled them to create individual development programs by documenting the sales representatives' resume samples recorded in Microsoft Dynamics CRM of successful representatives.

Personal Training

Individual targets and training plans for sales representatives, who shine out with their performance, began to be made. Parallel to these plans, personal thresholds were started to be determined for sales representatives. Even personal campaigns were organized to overcome these thresholds. Thus, it is provided that the representatives, who overcome with their personal thresholds, had increased motivation and bigger targets.

Objective Evaluation

The activities and sales performances of sales representatives in the sales club are measured as well as their sales relationship between customers. If there are any complaints or thanks from ultimate customers, these complaints will be added to the personal card in the CRM. In the short term, shadow customer application will be started and the representative performance will be observed on location and used as input in performance measurement.

2 Corporate Behavior

Corporate behavior is directly related to behavior of employees. Employees also adjust their behaviors according to the corporate culture. While the employees in corporations with authoritarian culture have developed more formal behavior, employees of democratic cultural corporations are more comfortable to express themselves. In this sense, corporate behavior can be decided as the reflection of corporate culture. On the other hand, since the corporate behavior can be defined as the perception, understanding and control of human behavior in corporations (Luthans, 1992: 7), it is possible for senior executives to be aware of some problems in the institution, especially by observing the body language of the employees of the corporation. Corporate behavior is also an indicator for customers to make judgments about the organization. Corporate behavior is also an indicator for customers to pass judgment about the corporation. A person who goes to a corporation for the first time is increasingly respectful for the corporation, pleased with the interest shown to him/her and also the one word of the gatekeeper of the corporation or the tone and style of the person who responds the telephone effects the image and prestige of the corporation within the scope of corporate behavior. In this respect, it can be said that the details are important of corporate behavior and can be seen as the sum of the details.

Comments of Two Customers for Corporate Behavior in Turkish Airlines

"After the overwhelming ground service at the Vienna Airport in Italy, we finally felt ourselves at home with good-humored hostesses THY plane even a 40-minute delay, to be at home wonderful, thanks to THY. (20 May 2007 - flight from Vienna to Istanbul)" (www.geldik.com).

(...)

"All of our esteemed friends in the THY family, from Mr. General Manager to the lowest level employees, both my individual domestic and overseas trips, I always prefer THY, our National Airlines due to security comfort, good-humor, and punctuality etc. (at least for me.) But; I fly to London in Business Class for the first time, departure May 31, 2007 and return to 03 June 2007, so I realized what COMFORT is it. The difference, which started at the airport, continued with various surprises and reached the peak of the Fast Track in Heathrow. Especially, the return to Istanbul is even better. I noticed that our hospitality was assimilated even by the English ground service staff working in THY, very nice" (www.geldik.com).

3 Products and Services

In the transfer of corporate identity, the product produced or sold is important as well as the process of the presentation of this product or presentation of the service as a discrete value plays an important role in transferring the corporate identity or corporate communication. While quality is the degree of a customer's satisfaction with a product or service, it plays an important role since the satisfaction is changed from person to person. Accordingly, the following example provides a content that should be considered in terms of product and service.

A Case Study on Products and Services of Vestel (www.halklailiskiler.com)

"We were delighted when the washing machine produced by Vestel, in the UK (there was no domestic production in the early days) (it was our first washing machine). After using it for many years, we had to remove it after the 1999 earthquake. In 2001 we replaced with Vestel, WMF69C FUZZY model washing machine. But we were not satisfied with like our first machine. After about 1.5 years, we called the service complaining about the terrible voices while squeezing. "The service crew squeezed the shock absorbers and left. The problem was not solved; the machine was still making noise - like the planes prepared to take off. We called the service again, and even once they took the machine to the service, the problem still remained. Our machine, which has a 3-year warranty, began to drain water close to the end of the warranty. The problem of jangling was still continuing. "We called the service again, and they told us the water hose came out. The hose is removed, cleaned and then reassembled. Neither sound was removed, nor was the water drain problem solved. Meanwhile, the warranty period of our machine is also full. We used it for 8-10 months more. We had no trust on Vestel Washing Machine. After for a while, we decided to buy another brand of machine. The gentleman, whom I met on the IETT bus (I apologize for not remembering the name) and learned that he was working at Zorlu Plaza General Directorate during the conversation, I told about my disappointment on Vestel's washing machine while trusting on Vestel's television, oven, fridge and gas heater. He asked me "why did you call Vestel consumer line?" Then I sent a complaint at www.vestel.com (thinking that they would not be interested in confessing). As far as I am concerned I was called by Avcılar Vestel Service. They came, checked the machine and brought it together service authority; They said no money would be asked. Mr. İbrahim, authority of the service, called me and asked me to go service building. According to the information when he called me, the money was stucked and broke the boiler. He asked me to come and see. After work, I went and see that the boiler was almost broken." He asked about 100 TL to mend it. I desperately accepted. But I came back home, my father said" I will pay 100TL and them it will be broken again, I prefer buying a new one". I asked for Mr. İbrahim to send it with a technical report. When I said "we will buy a new machine rather than pay it 100 TL, he said that they will call me after discussing with service manager.

A week later, Mr. Ibrahim called me and said that they will not request any money to mend the boiler. I asked him at least one-year warranty for this mending. They did not accept because said that the win will be renewed without receiving the repair money. I e requested at least a one-year warranty for this repair. They said that it was user error and could not be accepted. I insisted on not to accept it without warranty because they did not tell me anything when they brought my machine to the service. I felt stressed so immediately I called Mr. Gültekin, the service manager of the general management, and told him the current situation.

Mr. Gültekin kindly listened to me to the end. Having said that the only fault of their friends in the Avcılar service is that they did not open our machine before our

(continued on next page)

eyes; "we will defray all repair costs. But I cannot give you the one-year guarantee you want. "I said that I did not trust this machine, I do not believe in the quality of the first machines brought from England and unfortunately I would have to choose another brand in the washing machine when all the white goods in my house were Vestel. After Mr. Gültekin said that WMF69C FUZZY models were made in Korea, he called the Avcılar service authority with all courtesy; "I believe in the sincerity of Missis Ülkü. Vestel customer satisfaction comes first. Vestel regards its reliability, quality and satisfaction as its basic principles so send them WMK 1200 model new product, "he said. I heard all of them lively and I decided to do right I am preferring Vestel.

The WMK 1200 CL new Vestel Washing Machine took its place in our bathroom a week later. But there was a tiny problem; The warranty of the machine was not approved by Avcılar Service. They also wrote on the Warranty Document that the warranty period has expired since they sent this machine replacing the previous one.

I said that "I trusted the quality of Vestel, but I was very surprised that they could not guarantee a new-brand machine, and I ashamedly called Mr. Gültekin. Mr. Gültekin is very good at Vestel Service Manager, but I think he is a better public relations expert. He asked me to send warranty card and said that his signature is valid for all Vestel service dealers in Turkey and will guarantee that the machine will work for many years without any problems: "If you encounter any problem covered by the warranty, the problem will be solved by Vestel Thank you again Mr. Gültekin! Thank you, Vestel!"

4 Communication and Design

Corporate communication is the sum of communication activities of an institution developed by both internal and external stakeholders. In this sense, corporate communication is also directly related to the process of transferring corporate identity studies. Ten environmental forces contribute to the growth of corporate communications and corporate identity. These are expressed as below (Balmer & Gray, 2000: 256–261):

1. Accelerating product life cycle
2. Removing restrictive conditions (deregulation)
3. Corporatization programs
4. Increasing competition in non-profit making sectors and the public domain
5. Increasing competition in the service sector
6. Establishment of globalization and free trade area
7. Company marriages, acquisitions and elimination cases
8. Lack of high-skilled staff

Picture 1: The Old and New Logo of VakıfBank (www.yenimecra.org).

9. The expectations of the public for corporate social responsibility
10. The uncertainty of the boundaries of the corporation between internal and external entities

Corporate design is understood expressing a corporation itself visually. Corporations are distinguished from their counterparts by visual images and symbols, and try to stick in mind of the target groups (Okay, 1999: 65). For instance, although the old corporate logo, which is a reflection of Vakıfbank's corporate identity, is shown as a reflection of a stronger and more rigid corporate identity in terms of "fleshy" and "sharp" characters, "communication", " personal service" and "publicity" is emphasized with the new logos designed by more flexible and slim characters.

5 Marketing Conditions and Strategies

According to Philip Kotler, marketing is defined as "art and science" of customer acquisition, conservation and extension by selecting target markets, creating, presenting and contacting with superior customer values (Kotler, 2000: 8). In this respect, marketing refers to the post-consumption process of a product or service, which includes delivering to the customer and consumption and planning and watching post-consumption process. At this point, there are different situations for marketing. However, the priority is to take in consideration the analysis of customers' desires, needs and expectations.

According to the traditional marketing concept, the main purpose of the corporation is to sell more products goods, while modern concept relationship between the corporation and customers includes pre- and post-sales. The satisfaction of the customer is needed to have further communication. Today, most of the marketers agree that influencing a new customer is more expensive than influencing the current customer (Uzunoğlu, 2007: 12). Corporate working situations offer the following categories for marketing. This situation is prepared considering the stakeholders and also has been developed over time based on different conditions, especially considering the characteristics of typical groups in each market area (Payne, et al., 2005: 861).

Customer Markets: Customer Markets consist of buyers, intermediaries and end-users. Each intermediary or member of the supply chain may experience subdivision in an appropriate manner.

Referral Markets: There are two categories of referral sources: customer and non-customer referral resources. The customer category includes biased references and customer-based development. Non-client referrals are classified as general referrals, bilateral referrals, incentive-based referrals and staff-based referrals.

Procurement and Union Markets: Suppliers provide physical resources for business and can be classified as strategic, key, authorized and appointed suppliers. Collaborative suppliers often demonstrate more knowledge-based capacity than product-based capacity.

Effective Markets: These markets consist of many different components, such as financial and investor groups, unions, industry firms, regulatory firms, media, evaluation groups, political and government representatives.

Improvement Markets: Improvement markets include all potential employees. This market can be divided into functions, such as job role, seniority level and characteristics.

Domestic Markets: Domestic markets include situations that are used for potential employees, such as internal markets, job role, seniority level and characteristics.

Nowadays, a customer-centered strategy is being performed while making strategic decisions about marketing. This strategy, while demolishing the priority of "hot money", provides a long and profitable way to corporations. On the other hand, there are some internal and external threats. "Threats against the strategy are often caused by changes in technology or outside factors like behaviors of the competitors. However, the greater threats that threatening the strategy are caused by within the institution" (Keçecioğlu, et al., 1998: 29). Wrong decisions, lack of vision, wrong people to the right jobs and confusing the elements of risk can also interrupt the successful strategies. Thus, the issue should be held with a holistic view and the information of the strategy should be obtained as a result of realistic and well researches.

Corporate Identity and Social Media

Thanks to the 21st century technological advances, communication becomes an important tool for everybody; for this reason, the world becomes a "global village, stated in Alvin Toffler's works via information and communication technologies. In this global village, instead of traditional media, both people and corporations choose using new media because of the possibilities it has. Because new media has some advantages in terms of speed, ease of handling, simplicity of reaching the target masses and crystallizing these target masses, reducing capital expenditures and sharing flexibility. These advantages attract the attention of users on the one hand and provide attractive opportunities for the corporations on the other.

Thanks to the developing internet technologies, the efficiency on the web has been maximized, which is limited to user's reading and shopping before; it is possible to create and share content without any technical knowledge. Based on this interactivity and communication, the new web concept called "Web. 2.0", which enables quick and free sharing of information, was launched in 2004 by Tim O'Reilly, founder of O'Reilly Media" (Kahraman, 2010: 13). Web 2.0 allows to use web technologies differently by allowing the applications of participants.

"Comments, scores, content sharing tools, and services dedicated to sharing, just make this new media more powerful than ever before" (Kahraman, 2010: 14). Sharing of anyone from any part of the society like comment, photograph, video, etc., spreads worldwide. Despite cognitive complexity, the name of this new media, which seems to have limitless possibilities, is accepted as social media. The common characteristic of social media and traditional media is that they can reach small or large masses. For example, both a blog post and a television program can reach millions of people. The features that will help to identify the differences between social media and traditional media differ from the using methods. Some of these features are listed below (http://en.wikipedia.org):

1. **Access** – Both traditional and social media technology allow everyone to reach certain masses.
2. **Accessibility** – Production for traditional media is usually owned by private companies and governments; social media tools are generally available for everyone little or no cost.
3. **Availability** – Traditional media production often requires specialized skills and training. This is not necessary for most social media, so everyone can be in production.
4. **Innovation** – It can be said that the time difference (days, weeks, even months) in traditional media communications is long compared to social media, which is the immediate impact and reaction (participants decide the time interval of the reactions). However, traditional media is also adapting to social media practices, and this difference may soon disappear.

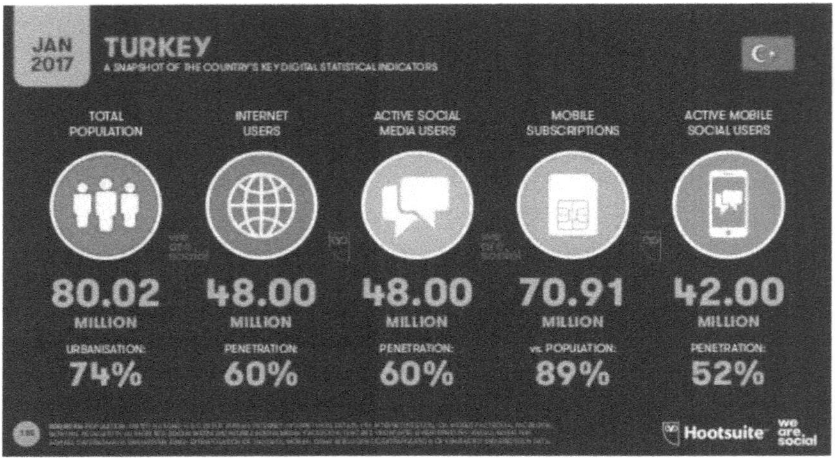

Fig. 2: Internet Usage in Turkey (https://wearesocial.com).

5. **Persistency** – Traditional media cannot be changed after it is produced (no changes can be made on the same article once a journal article is published and distributed), whereas social media can be changed immediately as a result of the comments or rearrangement.

Digital marketing agency, "We Are Social," published a report called "Digital in 2017 Global Overview" in cooperation with Hootsuite. A 106-page report prepared by the data from 238 countries includes up-to-date information, statistics and trends related to the development of the Internet and digital (www.digitalajanslar.com).

- More than half of the world now uses at least one smart phone.
- Almost two-thirds of the world's population owns at least one mobile phone.
- More than half of the web traffic worldwide is via mobile phone.
- Most of the mobile connections around the world are now "broadband".
- More than one-fifth of the world's population shopped over the last 30 days.

The numbers of Turkey are given below.

As you can see in the tables above, the social media usage in Turkey is parallel to the internet usage. Also, as of June 2017, the top three most popular social networks were YouTube, Facebook and Instagram. It is seen that YouTube has been used more than Facebook for the first time in Turkey.

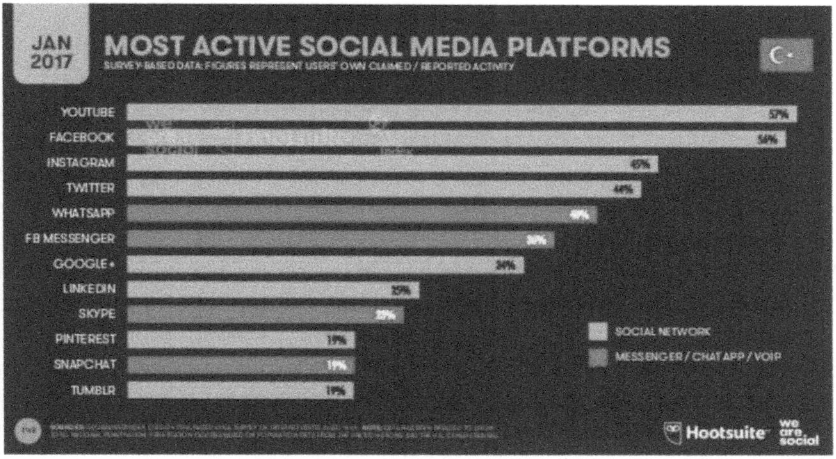

Fig. 3: The Most Active Social Media Platforms in Turkey (https://wearesocial.com).

Social Media Tools

Although the boundaries and content cannot be determined precisely, social media has several tools. These tools are in different content, designed for different applications and serve different purposes. In general, they are all in common based to share. These tools can be listed as social networks, blogs, wikis, social markup sites, content sharing sites and friend feed.

Social Networks: Social networks are applications that allow users to make personal websites to share personal information and messages. "While social networks sites use many technical features and applications, these sites are essentially based on the user profiles whose members are listed by friends using the same site" (Boyd & Ellison, 2007). In these profiles, users use accurate or inaccurate information about them and allow seeing their profiles by desired members. Facebook, MySpace and Orkut are examples of social networks.

Blogs: The concept of blog is based on the words "web log". The "web log" quickly merged into a single word as "weblog" and then turned into the word "blog". A blog is a web page that is constantly updated. Blogs, which usually come from short descriptions, are archived as a chronic order. Blogs are divided into five genres (Gilchrist, 2007: 123–129):

- Internal blogs (only targeting corporate employees),
- External blogs (focusing on the general public),

- Thematic blogs (linked to specific events or projects),
- Supported blogs,
- Personal blogs.

  Blogs especially consists of text, but many also have photo and multimedia possibilities. They can give a link to different web pages with their links. Blogs are the last forms of online communication and aim at a wide range of spread and form a basis to have a main idea quickly (Diane et al., http://home.comcast.net).

Wikis: Wikis is a system of web pages that can be produced, manipulated, edited and viewed in a simple way. In this term, wikis allows users to update or delete information on the site.

For some organizations, creating an item about an area where they are expert in virtual platform can give extremely important solutions. However, it is necessary to be careful not to advertise the corporation, the product or service in wikis. Blue (or purple, if clicked before) links indicate that there is a page with information on this topic. Red links indicate non-content-based pages (Scott, 2008: 108). The best example of wikis is Wikipedia, which is used as an internet encyclopedia and has millions of users.

Social markup sites: "Social markup allows users to save their bookmarks online to share with their friends by carrying the traditional markup one step further. In addition, the bookmarks of users are accessible from anywhere and on every computer" (Akar, 2010: 78). Mostly technological news, users can submit text, photos, videos and podcasts on the topics they want in these sites. By this way, they can share their videos, news, texts and photos based on their likes, laughs and have funs. People can make comments and share their ideas about the subject with other users in these shared entrances. With a simple interface, Digg is one of the social markup sites that take full advantage of the flexibility and usability of folksonomy, which is described by users as labeling. (http://www.vakaloji.com).

Content sharing sites: These are sites where certain types of content are organized, shared, and sharing of users continuously. While these sites concentrate on specific topics, they have regulations and practices in different formats. For example, Flicker has a content of photo sharing, while YouTube has just video content, and jamendo.com has only music sharing content.

FriendFeed: FriendFeed is a new-generation socialization tool that allows many apps to place comments, upload photo/video and the likes at the same time and follow the tracks noticed by their friends. It is possible to share anything on FriendFeed, whether from email, phone, blog or Facebook account.

Social media is important for the use of individuals as well as for the use of corporations. With the growing of social media day by day, corporations begin to wonder what they and their competitors are talking about brands, products and services, so they start to follow the comments made in this platform and determine strategies accordingly. Corporations discovering new communication channels like Facebook, YouTube, Twitter and WhatsApp have various advantages. Nowadays, activities of corporations in social media can be classified as to follow users directly by establishing corporate identity in social networks, communicate directly with the customer by increasing the number of followers, that is, ensuring to be followed or behave in accordance with viral marketing strategies. An organization may need to reconsider its communication strategies before starting corporate identity work in social media. Parallel to this, firstly, it is important to conduct an in-depth investigation. Then, what tools be used in the social media and what the long-term strategies are should be calculated. That is to say, the social media plan of the corporation needs to be prepared. This study is not a work done by anybody in the corporation but a strategic work that should be carried out by social media experts. Today, social media experts in Turkey are mentioned with a description parallel to the customer relationship manager.

Social Media Expert
DERİMOD Konfeksiyon Ayakkabı Deri San. and Tic. INC. Istanbul (Eur.) (www.kariyer.net)

General Qualities

- Graduated from university
- Has at least two years of agency experience
- Effective use of Turkish language to produce quality content for social media platforms
- Knows English enough to follow content from different platforms
- Knows fashion, follows trends, preferably previously experienced in the fashion & ready-made clothing sector
- Be creative and work hard to organize current content sharing in accordance with DERİMOD communication strategies in all digital media (Instagram, Facebook, Twitter, YouTube, Blog, etc.)
- Be experienced in reporting and analyzing performance and user data related to social media platforms
- Be careful and responsible for sending and responding to complaints and questions to the authorities and be able to undertake community management

(continued on next page)

Applicant Criteria

Experience: At least two years experienced applicants
Military Service: Done
Level of Education: University (Graduated)

Job Information

Corporation Sector: Textile, Leather, Footwear, Merchandising (Trade), Retail
Department: Communication, Public Relations, Service, Advertising, Design/ Graphics
The way of work: full time
Job level: Expert
Personnel number: 1
Country/City: İstanbul (Avr.)

It is important to be careful about corporations no matter how attractively involved in social media. Because involving in the virtual platform is sometimes not very difficult; but managing the virtual platform can be difficult in any case. In this respect, it may be necessary to consider the risks of this platform from the very beginning and establish teams that can direct and manage the communication plan both in normal communication process and crisis situation by taking into consideration these risks. The fact that these works constitute a team-oriented content stems from the continuity and the unpredictable spread of its limits.

An example of Expressing Corporate Identity via Social Media in Terms of the Corporate Communication: Ramon De Leon[1]

Corporations are getting interested in social media day by day, and social media works are increasing. Besides, in some corporations, the employees who have creative ideas provide the utmost benefit to the corporations.

One of them is Ramon De Leon, who opened the dealer of Domino's Pizza in Chicago. Latin American Ramon De Leon grows up in Chicago and continues his education until high school. After school, he works as a waiter at restaurants for years. Years later, he gets a chance to open a branch of Domino's Pizza in Chicago.

---

1 This part of the work was cited from Reyhan Çepik by a real Social Media Hero: Ramon De Leon, entitled "http://www.webrazzi.com/2010/11/05/sosyal-medya-ramon-de-leon/". The cited section was added to the study making necessary arrangements in content.

Firstly, in 1998, he established an online ordering system negotiating with an intermediary firm. In 2004, he started using AOL Instant Messenger as a micro blog before Twitter. In 2005, he established a Facebook group of Domino's Pizza branch operated by him with university students. Moreover, he was the pizza sponsor of many events in high schools and universities of surrounding. As a return, he asked students to share photos on the internet while eating pizza, waiting in the queue at the pizza stand and having fun. Ramon also participated in these events and served pizza with his own hands.

De Leon cannot share activity photos on Facebook himself. Because at that time Facebook only accepts members with "edu" extension. After a while, Ramon De Leon, who is demoralized by this situation and go to a nearby college and enrolls. He opens a Facebook profile with his e-mail address with "edu" extensions getting from the school and wears the badge seen on the next side to attract attention while working on the branch.

After opening his Facebook profile, he excitedly starts photo sharing.

(continued on next page)

Immediately after, De Leon sets up his own social media rules.

- Select the words and visuals that will best represent you and your business
- Show stability.
- Be transparent
- Be sincere and be sympathetic if you can.
- Be part of the communication
- Categorize your environment.
- Keep current customers and try hard to gain new customers.
- Never judge people and your competitors in social media, who do not buy products from yours.
- Follow your brand.

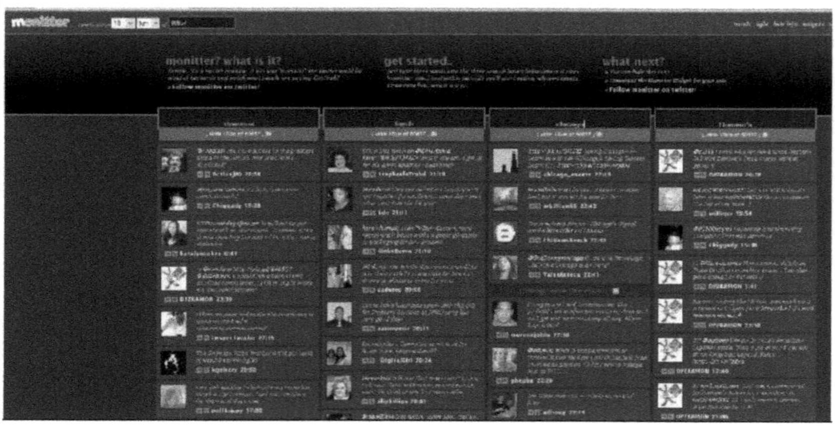

De Leon's follow-ups do not go for nothing and he comes across a negative tweet about his branch, which sends wrong pizza.

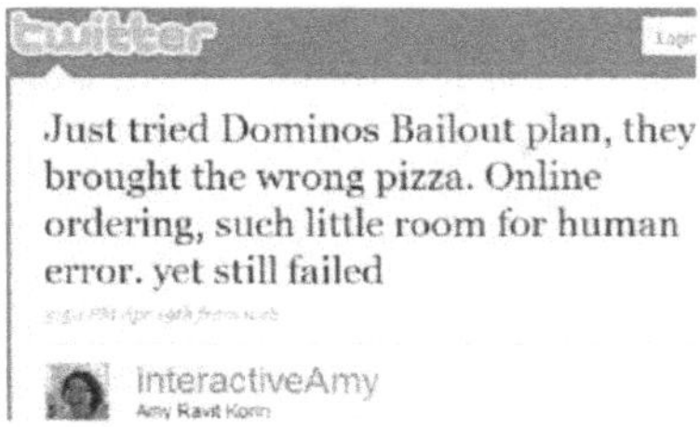

Then, Ramon De Leon immediately goes to the branch and finds the staff prepared the order at that hour. He prepares a video with an apology and streams it on the internet.

He shares this apology video in domestic and foreign press and marketing blogs. This video is watched 906,908 times. Ramon De Leon's social media studies are not limited to these. He used the most active Twitter followers' names on the brochures printed. With this study, Ramon gives the message to the customers: 'Follow us on Twitter', 'Join us', and 'Learn about our current campaigns'. Then he takes a photo of this brochure and shares it on TweetPhoto. In the other two brochure he shares the correspondences of his Twitter page. Thus, Ramon both shares the new menus and explains how they can make pizza ordering enjoyable using social media.

(continued on next page)

Another example; he shares the free slice pizza promotion for those who will order pizza from 16.00 to 20.00 on July 21 and receives feedback immediately.

This campaign is shared on Twitter by many customers immediately.

Ramon De Leon organizes activities not only on the internet, but also in his branch. In one of these activities, he hosts one of his customers every day and they cook their favorite pizza together. At the end of the activity, Ramon gives a gift token to the customers and shares the details of the activity on Twitter and Youtube.

Due to the traffic accident, a car drives into De Leon's pizza branch. The ordering area and waiting area become unserviceable. Ramon De Leon immediately prints a bill saying, "We take online orders."

(continued on next page)

Later, he becomes the sponsor of the activities sponsored by major brands like Google.

De Leon, who uses social media for years, eventually develops a great marketing strategy. He is awarded with the last bill. Here is that creative bill design.

Things to do on the bill while waiting for your pizza:

- Order addictive cheese bread.
- Take a photo of the poster standing next to you and share it on Facebook.
- Follow us on Twitter.
- Tell the world where you are and update your status.

Ramon's efforts are conclusive, and his Domino's branch is selected that offers the best customer service across America.

The story of Ramon De Leon is taught as an example project in universities and is invited to many marketing conferences as a speaker.

As it is known, being a dealer of a brand means operating it through the brand's rules and marketing policy. Therefore, the operator cannot add something himself. But De Leon manages to personalize the Domino's shop he operates in a way that is unlike known business rules, "I cannot earn money by selling one-dollar pizza, but I can make money by communicating with people", he says.

Conclusion

Corporation need to invest in their today's situations in order to have a permanent image in the eyes of the target groups and differentiate. This investment is based on the correct structuring and proper maintenance that form the keystones of corporate identity such as corporate communication, corporate behavior, corporate design, corporate philosophy and corporate culture elements.

The corporate identity has distinguished and important significance because the corporation is based on corporate personality. This importance, together with new communication technologies, is moving into virtual platform and shows itself once again in social media. In this sense, social media is far beyond just being a platform where people talk. This important media for corporations is an important area conveying and maintaining the identity values. Not every organization can conduct its own social media communications. Accordingly, professional agencies services have started to be established in Turkey. It takes great importance to get professional support for social media communication, both watch and direct the dialogues. Every corporation always has positive comments, and conveying a negative experience can be a bomb for other users. Therefore, the social media expert or his team should be aware of both the management of the social media and the business dynamics of the corporation.

Social media is not a place that corporations would call "let's create a page for ourselves". The page, which is opened without forming a necessary background and including the name of the corporation and employees, should be a planned and precious work because it represents the identity items of the corporation. There are many corporations which are successful or frustrated in social media. The common ground of successive corporations is to convey the work to the social media rightly by both texts and photos and watches it. It is also possible to turn into an opportunity by sharing the situations that may be caused by crisis in the corporation via social media, and if there is a problem, it should be conveyed to the social media in a transparent, open and questionable manner.

References

Akar, E. (2010), Sosyal Medya Pazarlaması Sosyal Webde Pazarlama Stratejileri, Ankara: Efil Yayınevi.

Balmer, J.M.T. & Stotvig, S. (1997), "Corporate Identity and Private Banking: A Review and Case Study", International Journal of Bank Marketing, 15(5).

Balmer, J.M.T. (2001), "Corporate Identity, Corporate Branding and Corporate Marketing Seeing through the Fog", European Journal of Marketing, 35(¾).

Balmer, J.M.T. & Gray, E.R. (2000), "Corporate Identity and Corporate Communications: Creating a Competitive Advantage", Industrial and Commercial Training, 32(7).

Balmer, J.M.T. & Gray, E.R. (2003), "Corporate Brands: What Are They? What of Them?", European Journal of Marketing, 37(7/8).

Belmiro T.R.; Gardiner, P.D.; Simmons, J.E.L.; Santos, F.C.A.; Rentes, A.F. (2000), "Corporate Communications within a BPR Context", Business Process Management Journal, 6(4).

Boyd, D.M., & Ellison, N.B. (2007). "Social Network Sites: Definition, History, and Scholarship". Journal of Computer-Mediated Communication, 13(1), http://jcmc.indiana.edu/vol13/issue1/boyd.ellison.html, (05/05/2017).

Cheng, R; Hines, T.; Grime, I. (2008), "Desired and Perceived Identities of Fashion Retailers", European Journal of Marketing, 42(5/6).

Gilchrist, A. (2007), "Can Web 2.0 be Used Effectively inside Organizations?", Bilgi Dünyası, 8(1).

Gerçek Bir Sosyal Medya Kahramanı: Ramon De Leon http://www.webrazzi.com/2010/11/05/sosyal-medya-ramon-de-leon/ (15/07/2017).

Güzel Bir Halkla İlişkiler Örneği / Bir Teşekkür (2005) http://www.halklailiskiler.com.tr/yazi.php?id=789, (10/07/2017).

İnternet ve Sosyal Medya Kullanıcı İstatistikleri (2017) http://www.dijitalajanslar.com/internet-ve-sosyal-medya-kullanici-istatistikleri-2017/ (15/08/2017).

Kahraman, M. (2010), Sosyal Medya 101, İstanbul: MediaCat Yayınları.

Kariyer.net (2017). "Sosyal Medya Uzmanı", http://www.kariyer.net/is-ilani/derimod-konfeksiyon-ayakkabi-deri-san-ve-tic-as-sosyal-medya-uzmani-1794762 (14/08/2017).

Keçecioğlu, T. (2003), Lider ve Liderlik, İstanbul: Okumuş Adam Yayınları.

Keçecioğlu, T.; Çelikel, E.; Çınarlı, S. (1998), Strateji Nedir? İzmir: Doğruluk Matbaacılık.

Kotler, P. (2000), Pazarlama Yönetimi, Çev. Nejat Muallimoglu, Millenium Baskı, İstanbul: Beta Yayınları.

Luthans, F. (1992), Organizational Behavior, Sixth Edition, New York: McGraw Hill Inc.

Melewar, T.C. & Akel, S. (2005), "The Role of Corporate Identity in the Higher Education Sector a Case Study", Corporate Communications: An International Journal, 10(1).

Okay, A. (1999), Kurum Kimliği, Ankara: MediaCat Yayınları.

Panda, K.C. & Mandal, M. (2006), "Corporate Culture in Libraries and Information Centers to Promote 'Knowledge-Based Business' in IT Era", Library Management, 27(6/7).

Payne, A.; Ballantyne, D.; Cristopher, M. (2005), "A Stakeholder Approach to Relationship Marketing Strategy: The Development and Use of the "Six Markets" Model", European Journal of Marketing, 39(7/8).

Sayımer, İ. (2010), "Sosyal Medya ve Blogların Halkla İlişkiler Sürecine Etkileri: Türkiye'de Halkla İlişkiler Uygulamacılarının Konuya Yaklaşımı", Editör: Sinem Yeygel Çakır, Teknolojinin Pazarlama İletişimine Etkileri, Ankara: Nobel Yayın Dağıtım.

Schiano, D.J.; Nardi, B.A.; Gumbrecht, M.; Swartz, L. "Blogging by the Rest of Us", http://home.comcast.net/~diane.schiano/CHI04.Blog.pdf, Erişim Tarihi: 05/01/2017.

Schmidt, K., (2000), The Quest for Identity: Corporate Identity, Strategies, Methods and Examples, London: Cassell.

Scott, D.M. (2008), Pazarlamanın ve *İletişimin* Yeni Kuralları, İstanbul: MediaCat Kitapları.

Stephen P. Robbins, (2004), Organizational Behaviour, 10. Baskı, India, Delhi: Pearson Education.

Sosyal Medya, http://tr.wikipedia.org/wiki/Sosyal_medya, Erişim Tarihi: 10/02/2017.

Theaker, A., (2006), Halkla İlişkilerin Elkitabı, Çev. Murat Yaz, İstanbul: MediaCat Yayınları.

Uzunoğlu, E. (2007), "Müşteri Odaklı Pazarlama Anlayışına Göre Değer Yaratma: Bir Model Olarak Değer İletim Sistemi", Eskişehir OsmanGazi Üniversitesi İİBF Dergisi, 2(1).

We Are Social, (2017), "Digital in 2017: Global Overview" https://wearesocial.com/special-reports/digital-in-2017-global-overview (08/08/2017).

We Are Social http://www.matbaashow.com/index.php/kurumsal-tasarm (09/12/2016).

We Are Social, http://www.microsoft.com/turkiye/casestudies/dynamics/kaleseramik. mspx (09/11/2016).

We Are Social, http://www.geldik.com/meslekler/22406-hostes-yer-hostesi-ucus-hostesi-hostes-yer-hostesi-ucus-hostesi-hakkinda.html (08/10/2015).

We Are Social, http://www.yenimecra.org/etiket/vak%C4%B1fbank-logo, (10/09/2016).

We Are Social, http://www.vakaloji.com/etiket/sosyal-isaretleme/ (12/02/2016).

Aslı Erciyeş Tosun

Visual Culture in Digimodern Era: An Analysis of Advertisement Photography

Introduction

The concept of digimodernism theorized by Alan Kirby since the mid-1990s represents the influence of computerization in all forms of art, culture and textuality. According to Kirby, digimodernism is the effect of computerization on cultural forms. It is the unique tendency to be gained from the new contexts and contents of cultural forms. It is a cultural change, a communicative revolution and a social organization. "Kirby argues that since the middle of the 1990s, digimodernism gradually left the postmodernism behind as the dominant cultural, technological, social and political expression of our time. Digimodernism can be expressed in seven words in the context of the effects of digitalization on the global forms of digitalization; progressive and motivational stimuli, imprecision and coincidence, disappearance of eyes, reformulation of textual roles, anonymous, multiple and social authorship, fluid-limited text, electronic-digitality". (Kirby, 2009: 2–53).

However, there are some difficulties in solving and reading the advertising photographs, which are the subject of this study in terms of digimodernism. As Kirby states, digimodernism is not primarily a visual culture (Kirby, 2009: 53). For this reason, digimodernism is inadequate in interpreting visual culture. The basic problem of this study is to find out how to read an image in a modern digital age and through which aesthetic items.

The basic claim of the study is that the postmodernism fills the deficiency of non-visual culture of digimodernism. Digimodernism carries its own aesthetic properties to postmodernism in terms of the visual cultural context, and today a photograph can only be read through the characteristics of postmodernism. Since Digimodernism is not basically a visual culture, it is useful to emphasize the similarities and differences between digimodernism and postmodernism.

Digimodernism and Visual Culture

Firstly, Kirby indicates that there is a coexistence of digimodernism and postmodernism, and it lives together (Kirby, 2009: 2). Therefore, an interaction between digimodernism and postmodernism is inevitable. Secondly, as Uslusoy expressed by referring Kirby digimodernism is not clearly distinguished from postmodernism, and digimodernism is seen as a technological development phase of postmodernism. (Uslusoy, 2016: 8). Thirdly, Alan Kirby treats digimodernism as a technologically inspiring vision. Digimodernism sees computers, mobile phones and interactive TVs as tools that move the changes in our culture and our everyday lives. These changes and alterations are based on similar superficialities of postmodernism and carry superficialities: for example, the author's approach to interactive media Roland Barthes conveys is the nonlinear, nonsequential nature of the postmodern narrative transfer from the writer to the reader. Does the Internet spell the diversity of Lyotardist little stories from beginning to end through their teleological-objective-non-spatial ways? Digital texts are always open-ended, as is the case with postmodern artifacts, as defined by Lyotard. Moreover, digimodernism from our cell phone cameras to reality TVs shares their concerns with the real category of postmodernism. Fourthly, contrary to modernism, digimodernism, like the postmodernism, is a division between high culture and low culture and is situated low culture over high culture. Digimodernism implements this positioning under postmodern anti-elitist values (Rudrum, Stavris, 2015: 271).

In the light of the four items mentioned above, digitalism does not mean the digitalization of modernism. If digimodernism is used in the sense of digitalization of modernism, then a phenomenon called postmodernism-post modernism is overtaken. For this reason, it is useful to re-emphasize the distinction between modernism and postmodernism which is emphasized by Ihab Hassan in the context of visual culture in order to prove that digimodernism is largely related to postmodernism than modernism.

When we look at the distinction between modernism and postmodernism in terms of visual culture, text and works of art such as advertising photography in the era of digimodernism, it is seen that digitalism is much more digitalized than the resurgence of digitalism with the modernization of digitalism. In addition to these, the characteristic features of degradation, hybridization, eclecticism (Collage, Bricolage), paste, which are seen in the postmodern life and in the work of art, are digitized. From this point of view, it is possible to make a distinction between postmodernism and digimodernism for visual culture and artworks such as photography, cinema, animation and video.

Tab. 1: Ihab Hassan, Modernism – The Postmodern, Simon Malpase Routledge Publisher, London and New York, s. 7–8. 2006.

Modernism	Postmodernism
Romanticism/Symbolism	Pataphysics/Dadaism
Form (conjunctive, closed)	Antiform (disjunctive, open)
Purpose	Play
Design	Chance
Hierarchy	Anarchy
Mastery/Logos	Exhaustion/Silence
Art Object/Finished Work	Process/Performance/Happening
Distance	Participation
Creation/Totalization	Decreation/Deconstruction
Synthesis	Antitez Antithesis
Presence	Absence
Centering	Dispersal
Genre/Boundary	Text/Intertext
Paradigm	Syntagm
Hypotaxis	Parataxis
Metaphor	Metonymy
Selection	Combination
Root/Depth	Rhizome/Surface
Interpretation/Reading	Against Interpretation/Misreading
Signified	Signifier
Lisible (Readerly)	Scriptable (Writerly)
Narrative/Grande Histoire	Anti-narrative/Petite Histoire
Master Code	Idiolect
Symptom	Desire
Genital/Phallic	Polymorphous/Androgynous
Paranoia	Schizophrenia
Origin/Cause	Difference-Difference/Trace
Metaphysics	Irony
Determinancy	Indeterminancy
Transcendence	Immanence

Tab. 2:

Postmodernism	**Digimodernism**
Antiform (disjunctive, open)	Digital Antiform (disjunctive, open)
Form (conjunctive, closed)	Antiform (disjunctive, open)
Play	Digital Play
Pataphysics/Dadaism	Digital Pataphysics/Dadaism
Chance	Digital Chance
Anarchy	Digital Anarchy
Participation	Digital Participation
Process/Performance/Happening	Digital Process/Performance/Happening
Decreation/Deconstruction	Digital Decreation/Deconstruction
Deterioration	Digital Deterioration
Pastiche	Digital Pastische
Collage	Digital Collage
Bricolage	Digital Bricolage
Hybridization	Digital Hybridization
Text/Intertext	Digital Text/Intertext
Rhizome/Surface	Digital Rhizome/Surface
Desire	Digital Desire
Signifier	Digital Signifier
Anti-narrative/Petite Histoire	Digital Anti-narrative/Petite Histoire
Idıolect	Digital Idıolect
Polymorphous/Androgynous	Digital Polymorphous/Androgynous
Schizophrenia	Digital Schizophrenia
Irony	Digital Irony

Analysis

Advertising Copy of Vogue Magazine

Picture 1: The Persephone Magazine Cover of Vogue Turkey Issue November 2012

Tab. 3: The Analysis of the Persephone Magazine Cover of Vogue Turkey Issue November 2012

The Persephone Magazine Cover of Vogue Turkey Issue November 2012		**Digital Array (Sentagmatic) Axis**	
Type of Digital Sign	**Digital Sign**	**Digital Signifier**	**Digital Signified**
Pictographic	Magazine	The cover of magazine	Showcase of magazine, a call to fantastic consumption area of the magazine
Pictographic	Tool	Chandelier	Criticism of Enlightenment, science, art, ethics, medicine, which are divided sagaciously, are together
Pictographic	Animal	Birds	Nature
Pictographic	Disharmonic Body	Mix of Human and Cadaver	Life and death, past and present chaos
Sign	Script	Black Vogue script on white platform	Vogue, brand of the magazine

Digimodern Reading through Postmodern Visual Culture

Digital Pataphysics and Digital Fragmentation: While the image of Persephone reflects a chaotic body fiction through a pataphysical universe in which the imagery and dream come true together, disharmony also opens to debate the concept of inconsistency.

Disharmony expresses the situation of opposing and incompatible phenomena such as death-life, good-bad, spiritual-natural, pleasure-reality and not being satisfied with human consciousness or subconsciousness side by side (Öcal; 2011: 254).

In the postmodern and digimodern stage, the disharmony goes one step further by the distinction of the separating lines between the dual contrasts that are formed by the categorization of the rational mind of the modernist mind, and it becomes intertwined with the incompatible and contradictory phenomena side by side. In photography, life and death are positioned at the same time and inwardly, so that the boundary between dualism and discriminative lines created by modernism through rational intelligence is ambiguous.

Death is positioned as an art object, which lives within the everyday life in the postmodern and digimodern era through photography of advertising. This resembles the relationship established with dead ancestors in primitive society. In primitive societies, the dead are not dead as we know. It is a death that does not break the relationship with his family. For the primitive man, who has never interfered with dead ancestors, the dead skull or any of his objects represents the dead and keeps him alive. Dead sometimes communicates via dreams. For a primitive society, there is no distinction between dream and reality, so dreams are also true. Just like dream and reality, death and life are not two separate phenomena for them. In modern society, the rational point of view of modernity based on enlightenment has reversed the relationship of primitive human with death, separates life from death. The most prominent feature of this situation is that the graveyards in modern cities, which represent the idea of death and death, are removed from the city centers, that is, the living spaces of the city.

In the modern era, the advertising photo takes the dead out of the graveyard in the cemetery, which is located outside the city, and in the postmodern and digimodern era, it is placed in the daily life of the city, in the center of the artistic and cultural world. Thus, through postmodern-digimodern zombies, life has participated in death, as well as tension to excitement, and life to death. In the primitive society, zombies that is the living deads were controlled by the magician, but in the postmodern and the digimodern era, they are controlled by the media and the transactional-information technologies, that is, by the transactional magicians. In fact, the western system, which is based on rational thinking, is now a zombie. Big narratives such as freedom, equality, brotherhood, development, justice and Marxism have also been zombied. The system plays zombie games via popular culture.

Your body is not in itself, nor is mine. In the world of postmodernism and digimodernism; cybernetics, calorie control, fat burning, diet, aerobics and drug use is under siege. The body has become a focal point for both academic and wider culture. The issue of the body continues to gain priority with the rise of new techniques of medical imaging of body and the resolution of sexualized identity norms. In the world, where the change becomes a norm, the change in the body seems inevitable at the same time. Everything is transformed if the body changes. The body is now at the point of resistance to the global imperatives of postmodernism and digimodernism, and the body must be first affected by these global imperatives. The claim of a clear classification between the singular body of modernity and the fragmented body of postmodern and digimodern becomes widespread. It is assumed that the enlightenment's body is disintegrated under the pressure of late capitalism. But the modern body of enlightenment is both fragmented and universalized (Mirzoeff; 2005: 3).

In the advertising photograph, the control of the government on both the dead and the living female body appears to be ineffective. In the modern West, power, which reproduces itself through elections, clearly distinguishes death from life.

In the postmodern and the digimodern world, the classic power does not know how to treat both hermaphroditic and bisexual subjects, both pro-freedom and pro-freedom, both male and female, at the same time. It is not possible to behave the body in the photograph neither like a dead nor a living creature. For this reason, the postmoderned-digimoderned sharp ideological borders, identities and policies were fragmented in power in the postmodern and digimodern era, and a hermaphrodite power conception becomes dominant.

Hence, the hierarchy between concepts such as death and life has been destroyed, and all concepts become equivalent. The disintegration of the body in the postmodern and digimodern world allows it to be reshaped in any way as if death and life are the same. The fragmentation caused by the uncertainty arising from the distinction between contradicting concepts also break the masculine view of the modernist thinker, who comes from the philosophy of the roots of enlightenment, as seen in the advertising photograph, which also dominates the female body. Having seen in the photo, both the dead and the living female body structure at the same time according to its structure emphasize the pluralistic, multi-faceted and fragmented identity of the postmodern and the digimodernist by nullifying the gender politics guided by the male dominant conception. This precisely corresponds to the notion Deleuze and Guattari's body without organ.

Body without organ does not express a body that cannot be organs. It is a body that is excluded from organizing, isolated from the identities of big narratives, fragmented, liberated, unjustly deserted and prone to be reformed in different ways (Best, Kellner; 2011: 117). Indeed, this is clearly seen when you look at the dead woman's arms in the advertising photo. Marxism, which is one of the greatest accounts of modernism, is transformed into a production-related limbic postmodern and digimodern world, which is relinquished to the arm, labor and production, but is now only reflected in itself.

Digital Schizophrenia: The photograph also refers to a schizophrenic post-modern and digimodern subject. Death emphasizes to past. However, photography and death, that is to say, past life, at breathing time by moving to the present time, both the past and present time, are now together. In a multi-layered time perception, the subject in the advertising photograph, which cannot construct its reality in a stable and steady manner over time and space relations, is now a digital-schizophrenic subject.

Digital Irony: Two-way lighting criticism is seen in advertising photography. Firstly, positive biology is criticized for its anatomy of enlightenment by using it against the enlightenment of its own origin, the womb. Thus, an enlightenment irony is realized by applying a visual autopsy to the modern subject of enlightenment. Secondly, there is a criticism of enlightenment via the chandelier. The

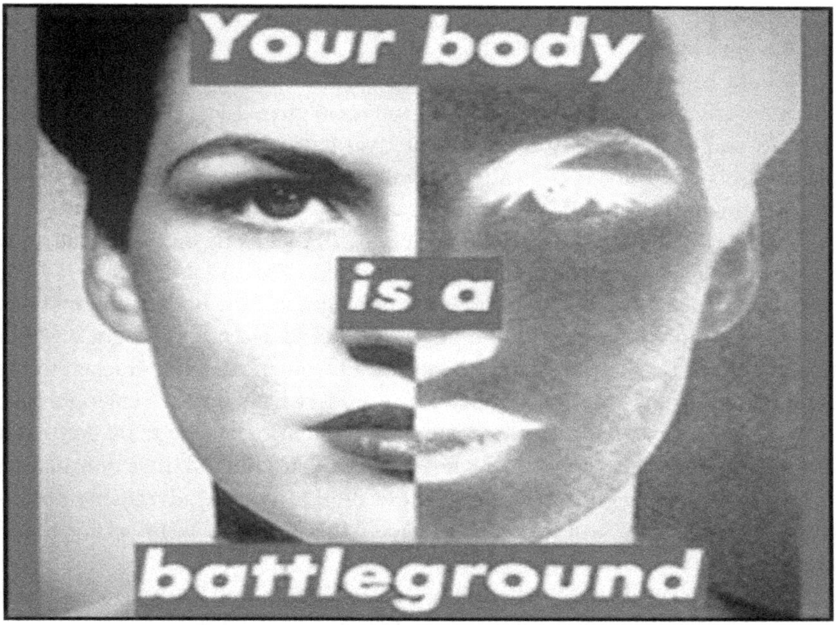

Picture 2: "Your Body is a Battleground", Barbara Kruger (www.dazeddigital.com. 12.01.2015)

chandelier, which represents the enlightenment, no longer shines, the utopian lights of modernism are off, and declares its own fragmentation by transforming the modern body of enlightenment. In this regard, the photograph makes a reference to a dystopic-dystopian world. The photo stands like missing report of the lost sight of the integral and unique subject of modern society and history.

Another point to consider in the advertising photo is the swallow birds. Unlike Alfred Hitchcock's birds, these birds do not attack live dead female bodies. The birds in the photo represent the pre-modern nature, where death and life are not separated from each other. In the photo, an ironic journey is made with a postmodern and digimodern body to nostalgia of nature in which the process of rational mind categorization has not yet begun.

Digital Intertextuality: The advertising photo makes a reference to "Your body is a battleground" by Barbara Kruger in 1989 in the context of digital intertextuality. Since the 1960s, feminist movements in art have increasingly felt themselves. Artists have historically produced artistic works from past to present, which deal with the social roles and positions of women and struggled against gender discrimination that has been bordering for centuries.

Barbara Kruger, Jenny Holzer, Sherrie Levine, Martha Rosler and Cindy Sherman are among the feminist postmodern-digitized art movements. Both feminist and contemporary art representatives have questioned and challenged the concepts of beauty, race, gender, sexuality, control, censorship, morality and ethics through the representations of human form[1].

Considering the advertising photograph in 2012 focusing on the female body and the way it separates the body parts, it seems like Barbara Kruger's work "The body is a battlefield" in 1989. In both the photographs, the body is fragmented, and the contradicting concepts of modernity, life and death are brought together side by side and disharmonically. The live dead woman's head is attached to the present time of life through the body; the subject has become schizophrenic. Both Kruger's work and the advertising photograph break the male dominant view over the modernized woman's body, allowing traditional male domination to be questioned. In consideration of these similarities, the advertising photograph leads Barbara Kruger's work to become pastische in terms of form and content.

Another piece of work, which advertising photograph makes reference is Damien Hirst's "The Virgin Mother" painted bronze sculpture in 2005.

The advertising photograph shows similarities between Hirst's statue of a virgin mother, except in the case of pregnancy. First of all, the sculpture has positioned the concepts of life and death simultaneously on the female body, just as it is in the advertisement photograph. Both works are formed through anatomy and take the body from the field of science and incorporate it into the field of art. In addition, the disharmonic structure is also remarkable in the Virgin Mother statue like in the advertising photo. Just as Hirst's sculpture, the advertisement photo breaks the view to the modernity of the masculine ruling regime, which has brought sexuality, beauty, health and the traditional gaze, and distributes the integrative subject paradigms. Both works are pornographic because of the removal of the skin on the body and the transparency of the body.

Another piece of Persephone advertisement photo is a cadaver sculpture called smoker of GuntherVon Hagens. The only difference is that the cadaver statue called the smoker is male. Hagens' sculpture made by human cadavers with a unique technique is called anatomy art examples (Dogan: 2001; 21).

---

1 http://artissues-bodyrepresentation.weebly.com/contemporary-views.html Access Date: March 2015.

Picture 3: The Virgin Mother. DamienHirst, 2005

GuntherVon Hagens has returned exitus letalis to life, which dies and relieved from medicine/science via art. When the cadaveric sculpture is taken into consideration, the remedy of immortality is ironically found in death. When we look at both works, death and life are side by side, almost intertwined. Life is articulated death, and death is to life. There is a Digital-Pathophysical universe in both works. Both advertising photograph and cadaver sculptures are covered

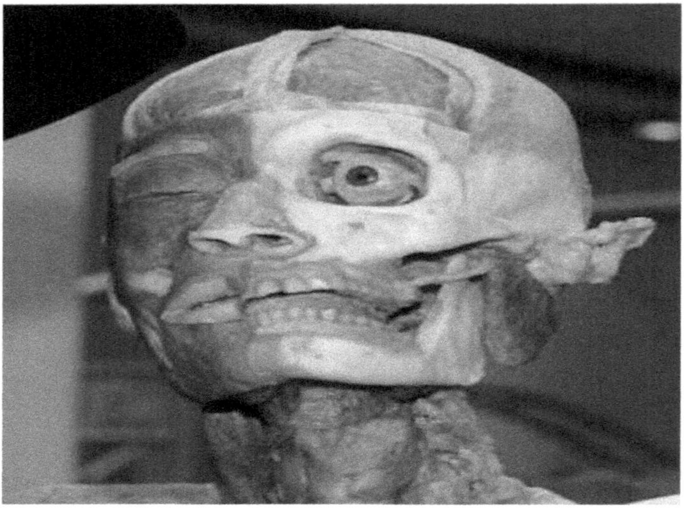

Picture 4: Smoker, Gunther Von Hagens, 2011

with chaos and anarchy. In the sense of time and reality, schizophrenia is at the extreme level in both of them. In both cadaver sculptures called "smoker" and "advertising photograph", there are unstable, decentralized, attracting attention to the subjects whose identities are largely ignored. In this respect, it is clearly seen that advertising photograph called Persephone makes cadaver sculpture called "smoker" to become digitally pastiche and imitates it out of gender.

Conclusion

As seen, it appears that the postmodernism of digitalism has evolved into a technological developmental stage in terms of the interaction, similarity and digitalization between digimodernism and postmodernism. From this point of view, it is not possible to consider digitalism as digital modernization. Digimodernism should not mean digitalization of modernization.

This work proves the claim, which has already been emphasized "postmodernism provides the non-visual digital culture that exists in the same side and the same time, digimodernism carries the aesthetics of postmodernism to its own account and nowadays a digital photograph can be read within the framework of the characteristics of postmodernism" after the analysis done. As Kirby's determination that digimodernism is not primarily a visual culture, the postmodernism of digitized aesthetics is used as the right choice for the advertising photographs being attempted to be solved. According to Alan Kirby, the cultural ration of the digitalism is honoring the truth as its dominant aesthetic; contrary to post-modern texts, digimodern texts mimic non-fiction forms to preserve the world; apparently true, the unconscious, the ironless, the unquestioned, the audience come without sign (Conway 2016; 182).

Advertising images and videos emerging at the intersection of photographic vision and digital vision have become important in terms of understanding the reality and consumption practices of the age. While photographs and digital images presented on the photographic image of the magazine put the individual in a more entertaining carnival world than the physical reality; digitally manipulated product influences consumer perceptions of purchasing practices and reality in terms of being more real, fun and enabling the individual's plural identity than his own physical reality. Digimodernism is a technological aspect of a society that has devoted itself to unbreakable consumerism, as Kirby sees (I Rud, Stavris: 2015; 272).

Reality has now become a consumer experience through images of advertisements via magazines. Digital manipulation and interventions, which are made in order to attract the attention of the consumer differentiate among the others in the magazine environment, where the product, brand and services are in competition, cause the reality to be defeated and leave the place to the images and images. Perceptions of the reality of the subject also cause fragmentation, deterioration, ambiguity and personal experiences. The reality based on digital images replaces the physical reality of the product and incorporates the essence into the world of different lifestyles, superficial identities and images. This draws the essence into a continuous purchase and hence the practice of identity, and this alienates itself from its vital realities and its surroundings by distracting from the reality.

The consumer, who creates meaning as a result of the relationship with reality, also recreates himself as a human being who desires. In the digimodern era, the main purpose of postmodern, digimodern advertising photographs, which have postmodern qualities and therefore can be described as postmodern-digimodern, is to construct human beings, who desire by manipulating the reality. Digimodern advertising signs and images are now confronted as a declaration of the loss of reality.

The more you consume, the more you are happier; manipulation of reality by advertisements and emptying of it is done in order to make more purchasing action. The whole symbol icon and reality images are manipulated in order to sell more, and the perceived vision needs and preferences of the consumer are manipulated and distorted. Advertising photographs, which are symbolic forms of reality, have lost the ability to rely on objective, reality and truth that has become almost operationally broken from the physical reference. While the forms of relationships we establish with our world also affect our perception of reality, the follower is trying to get into this relationship by trying to create meaning through digital manipulations created through photographs. Postmodern, digimodern advertising photography has no longer an innocent but dangerous role because of its creation. The digitalized postmodern-digimodern image actually approaches schizophrenically when presenting alternatives to consumer via photographs such as leaning over, establishing different relationships over time, and getting new identity and lifestyle models.

The objects that have the same function as the imperative paradigm of the capitalist system but are produced in a thousand kinds now confront us as anomic (irregular) antagonism with various appearances by doing metastasis and manipulating. Indicators now turn into appearances within the postmodern-digital advertisements. The objects are now transformed into image. The displays that contain postmodern images, which include digitalized images with digimodern advertisements therefore, take place in all of history, ideology, myths and narratives. These synthetic displays are dangerous because they determine our perception of reality and affect our consumption behavior. There are as many truths as the number of consciousness in the environment, and each person brings his or her past experience to a reality in the direction of their experiences and preferences, thanks to the advertising messages they meet.

References

Best, S., Kellner, D. (2011). Postmodern Teori. Mehmet Küçük (Çev.). İstanbul: Ayrıntı Yayınları.

Conway, J. (2016). After Politics/After Television: Veep, Digimodernism, and the Running Gag of Government Authors.

Doğan, H. (2001). Almanya'da Mannheım'da Prof. GuntherVonHagens'in Hayretler Uyandıran Anatomi San'atı Sergisi Hakkındaki Etik. Türkiye Klinikleri J Med Ethics, Sayı 9, No 1.

Erhat, A. (1996). Mitoloji Sözlüğü. İstanbul: Remzi Kitabevi.

Kirby, A. (2009). Digimodernism: How New Technologies Dismantle the Postmodern and Reconfigure Our Culture. London: The Continuum International Publishing Group.

Malpas, S. (2006). Modernizm – ThePostmodern. New York: Routledge Publisher.

Mirzoeff, N. (2005). Bodyscape Art, Modernity and the Ideal Figure. London: Routledge.

Öcal, O. (2011). Disharmonik Bir Varlık Olarak İnsan ve Sabahattin Ali'nin İçimizdeki Şeytan Romanı. TÜBAR XXX s. 253–268. http://www.tubar.com.tr/TUBAR%20DOSYA/cal_ouz%20253-268.pdf Erişim Tarihi: 22.10.2017.

Rudrum D., Stavris, N. (2015). Supplanting the Postmodern: An Anthology of Writings on the Arts and Culture of the Early 21st Century. Newyork: Bloomsbury.

Tosun, A. E. (2015). Postmodern Gerçekliğin Reklam Fotoğraflarında Yansıması. Ege Üniversitesi Sosyal Bilimler Enstitüsü, Yayımlanmamış Doktora Tezi.

Zöhre Akyol, Nevzat İnan and Selçuk Bazarci

Digital Images in Advertisements and "Technology" as Advertising Appeal

Introduction

Advertising had emerged long before the function of being a means of consumption as a result of the productive social movement, which emerged after the industrial revolution. Today, the tradition of commercial-oriented advertising (even though it did not represent a different state a hundred years ago) has a transition period when compared to features of the past both in terms of content and in an instrumental dimension. This transition has evolved in everyday life both in terms of social and individual dimensions and both in terms of consumption and popular culture. Especially in the 20th century, the function and diversity of advertising in the societies with liberal economic identity increased with the development of technology over the years. In this regard, although the validity of the traditional means of advertising still continues, it is observed that different presentations appear both in the strategies and discourses when the messages are generated and the symbols and images that are required to be shown with the computer and internet technologies.

Advertisements tell us various stories so that we can like and consume a product or an object. These stories are mostly exhibited through sections of social identity, which include statistical differences, represented by daily events and individuals. If the followers of these advertisements buy things or have feelings of identity or need as a result of this narration, this action will not only be a commerce but also it is seen as a change in the symbolic meaning of the image of the product or service at the same time. Given that visual elements used in the advertisements are delivered within a limited time and with a certain tool, it is expected that the images displayed in the advertisement should be a structure that supports the desired message as soon as possible. At this point, the visual language of the advertisement creates itself through the texts. In our age, it is inevitable that technology takes place in different advertising channels, either as an instrument or a content dimension. As a result of this situation, our research gives various definitions and interpretations of the concept of image in the section titled "Power of Viewing: A Conceptual View of Imaging". Also in the advertising literature, the concept of image view and image, which are considered as important, are evaluated. In the section entitled "Digital Imagery in Advertisements in The Context of Technology" of the study, the developing effects of visual images and the developing effects of consumption in society have been discussed through the advertising perspective. In the section entitled "Technology as an Advertisement Appeal", it is investigated how advertising, which uses digitalization not only as a technique but also as a visual element, builds "technology" appeal in their descriptions. In the section entitled "2017 Year in Turkey Lovemark selected Towards the Design of Social Media Sharing Analysis: Digital Image in Advertising Messages" of the study, the advertising shares, which are done via social media accounts of the brand applied to the selected samples via content analysis method, are studied.

The Power of Visualization: A Conceptual View of to the Image

There are many definitions and interpretations in the sources of different disciplines about the origin of the word "image". The word "image" comes from the Latin word "imago", which means imitation, emulation, copy, etc., although it has various specific sub-meanings from painting to poetry, from philosophy to cinema. Along with the expansion of the concept of meaning within the historical process; it gains other meanings in terms of concept, idea, and an impression that appear in the mind of the individual (adapted by Çağlar, 2013: 2). Image, imagination, dream, and image words are also used in our language as the synonyms of the image word. The word "image" also means new forms that are brought to mind by linking it with other things in order to express what is desired in a more vibrant, more effective, tangible, and expressible way. It can also be seen as the acquisition of images from the outside world, the appearance of the senses and impressions of the external world in the mind, as well as the acquisition of a pictorial value (Akbaytogan, 2006: 1).

According to Arnheim (2012: 34), a visual action is pointed as the basic environment of thinking by providing the richest information source for the outside world. In this regard, it is seen that there are different effects of the connotational narrations outstanding in the products offered by the advertisements in the consumer society. Each image has a form of vision. The images are created to visualize things that were not initially there. It has been understood that imagination is more permanent than what it has revived over time (Berger, 2010: 10).

Generally, images are the mental design of an object, entity, or phenomenon. This first picture of a sensation-derived entity becomes evident in different forms of consciousness that have hidden labyrinths, a magical derivative of copy, or design matter. The image is obtained either through perception in the mind, or through thinking, recalling, and imagining (Cengiz, 2009: 5). According to

Koçak (adapted by Akay, 2010: 8–9), each image expresses itself as an image or appearance. The image is that the person leaves the entity apart from the transparent appearance that brings the truth to itself. The image and its meaning is different in everyone's minds. For this reason, the more the mind is, the more images there are. The meaning of each image will be different. Perception of the image begins with a vision. Making sense of vision comes from human accumulation and culture; that is, the image belongs to the human, the center of the subject. The visual forms of human cannot be thought isolated from the fields like social environment, culture, politics, etc. From the moment when the sight has begun, people take and store a lot of visual input into their memory in their daily lives. Everything that people see and try to interpret needs the sum of their knowledge. On the other hand, there is a need for visual accumulation in order to reconstruct the acquired knowledge or image and reconstruct it. The effort of expressing the views of man is to create visual images (adapted by Çağlar, 2013: 13–14).

The aims of the images are to create an effect that corresponds to the human's visual hunger. Even if this effect is passive, a creation of this effect is the purpose of the image. For this purpose, they exist, change, document the truth, lie, or create an illusion (Akbaytogan, 2006: 25). The new visual media that is formed by the development of technology is expanding the field of vision of the people. This new visual media also offers people new digital imagery. So, people can get rid of the limitations of real life and enter into a complete illusion. They can experience new pleasures by living in this virtual environments (Aslan, 2003: 45). In this period when images are so dominant in the human mind, they draw more attention than the facts they represent (Parsa, 2004: 12). When we think the images are the innovations that digital culture has brought to daily life, imaginary production process seems to have the ability to convey a more message intensely. The visual effect created by images are being transferred via digital tools, thanks to display technologies and internet-based applications. This situation also plays an important a role in the acquisition of digital identity within the present social life.

Chatfield (2013: 11) says that all technologies have changed the communities, which use them. Telegraphs, telephones, radio, television, and other means of communication have provided massive changes in the field of communication. At this point, Burnett (2012: 30–32) suggests that no technology has emerged throughout history created by television in terms of the transmission and diversification and draws particular attention to the importance of television in the field of mass communication. Television can be regarded as one of the earliest image producers of the digital world that we live in today. The "watching culture" created by television, which is regarded as one of the ancestors of visual technologies, is now felt by the shows, which are created by the internet formed by the digital technologies. However, the watching culture created by the digital is similar to television as a form culture, but technically it separates from the television in terms of providing two-way communication. Any image shown on the screen, which is presented by television, computer, or mobile devices, actually shows people a created or representative world, not the real world (adapted by Parsa, 2004: 4). Cinema, serials, reality shows, and all kinds of content that can be varied with popular culture make use of the power of image production of mass media. At this point, advertisements are especially one of the frontrunners in this field. Consumption-oriented dynamics of the modern society often show us the reproducibility of the images, particularly in the field of advertising. Advertisements create a variety of meanings through the use of images apart from sales purposes and support the consumption of these meanings by people with their own products. These images can be created from all kinds of concepts of man and the world, and advertisements can create appeal through these concepts that they used. Moreover, today's examples show that the concept of "technology" is not only a media provider or communication channel for advertisements but also an image and an appeal that are shaped in advertising messages.

"Digital" Signs in Advertisements in the Context of Technology

Today, the image that technology evokes in people's minds is an image of modernity. Modernity, on the other hand, can be related to the future sustainability of the accumulation of knowledge acquired through all the social changes and developments that people have experienced in the historical process. In the context of technology-society, the events that take place in the historical process are evolving with a dynamic and developing structure. This dynamic structure makes it easy for modernity to become acceptable and usable for society by playing a role in the emergence and development of many technological elements. Especially give the commercial dimensions of new inventions and developed tools emerging with the developing technology become prominent with the massive effects of consumption [zr2] in our time. At this point, advertising is a field that utilizes both technology as an instrument and technology as a source of messages that is based on its meanings such as modernity and development.

Advertising is an important cultural event that shapes our lives. It is not possible to escape from images presented dominantly in modern urban environments and advertisements. In advertisements that can be regarded as real rituals of the modern age, ordinary objects and thoughts are almost sacred. It has been observed that even if the target says that they do not believe or ignore this spell-ritual mentioned, they behave in the form appropriate to that language. According to Marcuse (adapted by Batı, 2005: 175), "the target mass of the advertisement does not 'believe' the declaration of an operational concept, but the declaration itself takes action - in employment, sale or in purchase, in others, etc." The advertiser who is the source of the message knows all this. Thus, the advertising language becomes a form that uses imagery instead of concepts by reducing abstraction, development, linguistic form, and symbols as part of this ritual language (West, 2005: 175). For example, in advertisements, wealth is enriched with gold and environmentalism is enriched with green. For a building presented in gold color, in advertisements, the image of wealth has been tried to be reinforced both visually and verbally as "a life like gold". Similar applications are often confused as symbolic presentations in many advertisements.

If the images are regarded as a reproduction, then the advertisements also tell the target group that they must be like the images presented in advertisements. However, these images are not presented clearly but through visual indicators (Dumanli, 2011: 134). When the follower watches the advertisements, he chooses the product or service to behave as desired or to be as pretend in the advertisements by making sense of the imagery itself. The advertisement is the culture created by the consumer society. Thus society continues to have self-belief by multiplying through images (Berger, 2010: 139). Using an image gives an impression to the advertiser that he cannot tell. Notwithstanding, the image tracer is encouraging the advertiser to distinguish the intended meaning, that is, making the follower complicit in self-seduction (Rutherford, 1996).

The development of digital production and creation forms has led to more and more image production modes being carried to the cultural scene (Burnett, 2012: 82). The fact that the digital media focus itself on the basis of the participatory nature allows for a faster dissemination of the messages to be given. When we look at the perspective of advertisement, it is possible to see the positive effects of this situation. The rapid integration of digital opportunities into human life allows advertising activities to re-establish themselves within these transformations. The interaction expectation of consumers through communication systems like internet sites, social media, mobile applications, and tools leads businesses to be active in terms of digitalization.

"Technology" as Advertising Appeal

Today, it is possible to say that conventional efforts evolved into digital depending on the technological developments. Castells (2016: 9) emphasizes the impact of living changes on social structures. According to him, today's society is a network society. Looking at this construction, a certain power is provided to the society. Moreover, a community structure is thought to be designed to enhance the activity through a number of personal and organizational networks where communication with the internet and similar computer networks are carried out. In addition to Castells' views, Hebb (2015: 137) suggests that communications cannot be alone, but can be rephrased in a continuous communication process via a network. This state of affairs generally refers to communicative practices that constitute communication.

While Ron Burnett (2012: 177–186) interpreted Walter Benjamin's growth and development of passages in France at the beginning of the 20th century as virtual commerce centers devoted to simulation games, Burnett actually noted that Benjamin previously indicated the networked environments created in the 21st century culture. Because the shopping centers visited today are the same as the passages of the old periods. However, with the new shopping culture that has emerged with digital technologies, it is now seen that people do not meet physically but come together through virtual networks that connect with each other. Communication networks are partially concerned with autonomous relationships developed and maintained by machines which have sustained connections with requiring less human intervention. Of course, the machines do not talk each other in a real sense, they communicate digitally with each other, but it is assumed that this communication is mediated by people. Castells (2016: 13–14) states that individuals form and adopt a certain social domain idea via created social networks. According to him, social networks have a feature that is a direct influence on the transformation of the cultures.

Content creation of brands via digital media helps them to contact consumers, who they cannot reach through conventional channels, and to raise awareness about their products/services. In addition, features such as interactivity and synchronicity of digital media make it easier for consumers to communicate directly with brands. Moreover, consumer comments on brands' internet pages and social networks allow for quicker refinement of feedback on the results of the campaigns (Yılmaz and Erdem, 2016: 162). It is important for consumers to be able to access all kinds of information about the brand. Because emotional factors play an important role in the positive attitude of individuals towards brand.

It is possible to say the effectiveness of information communication technologies on ensuring continuous and organic communication of brands with consumers is important. Today's brand communication efforts behave in this direction with a consumer-centered understanding. New communication technologies which constantly evolve and develop provide a time flexibility in the formation of communication. Furthermore, brands communicate with existing and potential consumers through various digital advertisements, which are differentiated by digital media channels. Thus, while brands transmit rational and up-to-date information to the consumers, they diversify their strategic moves by taking advantage of the feedback received simultaneously from the other side.

Digital advertising applications offer a feature that makes it easy for marketers and advertisers to quick and efficient delivery of consumer-focused content through new media channels. As the changing consumer structure, which is shaped by a personal expectation from the brands, the advertising works evolve through new technology and new advertising platforms (www.marketo.com). It is possible for brands to establish long-term relationships with both current and potential customers and make their work more productive through digital advertisements (Dickey and Lewis, 2015: 5). It can be said that there are some opportunities for consumers to overcome easily the barriers against the advertisement with the diversification of digital media circles, especially depending on the consumer structure. Varnalı (2012: 63) emphasizes the effects of digital applications on both the advertising industry and the individual. According to him, digital advertisements provide real-time, detailed and more accurate data in their measurement systems. This leads advertisers to design their work in an accountable context. Namely, it is now possible for each user who participates in digital media to be able to report on the actions they take in the interactive sites, where they are involved with digital advertising. From this point of view, a digital advertisement can be quickly analyzed as to whether it has achieved the desired success or not. This seems to be an important step at the point where the creators of the advertisements carefully and meticulously coordinate each stage and the continuity of the business. Another factor is that individuals who are adapting to digital systems have succeeded in transferring the feelings they possess to the new world by creating a new virtual world outside of real life. At this point, it is important for a brand to comprehensively analyze the consumer profile in the virtual world so that a digital advertising effort can be positively concluded, and interact with the consumer at the right time by identifying the appropriate advertising medium in the digital environment.

The number of visual images in our life has also increased with the development of mass media. Throughout history, no society has been exposed to a mass of cluttered volumes of information and a heavy message falls in today's society. The connection of the images, which are fictionalized by messages of advertisements have a strange impression on us that we are used to but we are not careful. In advertisements, it is true that some kind of product of a company competes with others; however, it is more accurate that every commercial image accelerates the poverty. Advertising is not just a collection of messages competing against each other. Advertising is the process of generating appeal (Berger, 2010: 129–131).

Appeal refers to any content or vision, which are designed to attract attention, take attention or interest in an object or a person, and thereby have an impact on people. On the basis of attention and persuasion, many appeals of advertisements are preferred by advertisers. The appeal that varies depending on the behavior, feelings, and ideas of the individual with the influence of different variables (demographic characteristics, family, social group, etc.), and the culture which plays an important role in the consumer's attitude development towards a subject or a subject are affected by qualitative characteristics that changes depending on person (Elden and Bakır, 2010: 75–83). Clow and Baack (2016: 159–169) categorized the most successful advertising appeals into seven categories: fear, humor, sexuality, music, rationality, emotionality, and famine. Apart from this, it is also seen that appeals can be classified as adventure, beauty, collectivism, competition, convenience, politeness, economy, workability, entertainment, family, health, individuality, leisure time, magic, modernity, naturalness, ordination, care (compassion), patriotism, popularity, security, sexuality, social status, technology, tradition, uniqueness, richness, wisdom, work, and youth (Moon and Chan, 2005: 55–56). According to Oyelede and others (2009: 530), the main advertising appeals are hedonics, entertainment, luxury, wealth, prestige, sexuality, compassion, convenience, politeness, economy, education, efficacy, compassion, health, leisure, magic, modernity, quality, prestige, spirituality, superiority, technology, uniqueness, work, and youth.

Moon and Chan (2005: 55) define technology appeal in their study as having complex technical and engineering skills for the production of a particular product. However, on the basis of this study, the technology appeal was evaluated within the context of digital images (emojis, animated characters, robots, pictograms, and artificial intelligence) used in consumer presentation as an advertising attraction of the technology image.

V An Analysis of Social Media Sharings of Brands Selected as Lovemark in Turkey in 2017: Digital Images in Messages of Advertising

This research is based on Lovemark rewards, which has determined the most popular brands in Turkey since 2007 and has quite validity in terms of public. Lovemark is a brand appraisal criterion, which has been launched by MediaCat magazine since 2007 supported by different research companies, and it is a brand valuation criterion that suggests and rewards the brands, which Turkish consumers are "bonded by hearth". In this regard, social media sharings of Lovemark-selected brands in 2017 in Turkey were analyzed in this study. The visual images used in the Twitter accounts belonging to the brand identified within the scope of the research were interpreted with the content analysis method. The reason of choosing Twitter, which is one of the social media platforms in the research, is to provide a more regular content interface than others in the market and to provide significant advantages in linking with the consumer as an advertising tool for synchronicity.

Social media has long been used by brands as an effective tool at the point of digitization of narratives. Twitter has achieved to position itself as a digital space in the flow of daily life. Visual narratives that can be described as digital images according to the literature, which are found in the contents of the Twitter messages of the brands of Lovemark in 2017, are discussed in this study. These shares have been evaluated in terms of technology appeal, which is one of the themes that advertisement uses in the process of appeal.

Method

The Problem of the Research

What is the relationship between the general content used in twitter shots of selected brands in terms of technology appeal and image properties?

The Purpose of the Research

The purpose of this study is to examine the linkage of the brands with the technology by revealing the functioning, formal, and contextual features of the images which constitute the structure of the visual and verbal messages including the social media sharings.

The Pattern of the Research

In the research, it is tried to be understood the characteristics of the images in the sharings of the brands. In this regard, the phenomenological pattern that allows deeper examination of phenomena has been chosen within the scope of this research.

Data Analysis

The research data were obtained and evaluated by content analysis method. The main purpose of the content analysis is to form a framework that will reveal related concepts and relations to explain the obtained data (Yıldırım and Şimşek, 2013: 259). As a result of the analysis, the data, which have similarities, were arranged in a way that readers can understand by bringing concepts and themes together that include research.

One of the most valid reasons of using content analysis in the context of the research is that it is appropriate for the organic post acquisition of Twitter shares and it is a structural integrity appropriate to the research nature of the content analysis in defining the structural characteristics of the sharings. Content analysis also provides an accurate framework that guides the researcher a standardized and summarized format that reflects a meaningful composite of the data obtained.

Sample

The sample of this study includes selected brands as Lovemark in Turkey in 2017. However, only nine brands with technology focused (Arçelik, Bosh, Beko, Vodafone, Turkcell, İşcep, Garanti Bank, İş Bank, Yapı Kredi) are included in the research. Totally, 171 Twitter shares of these nine brands bearing the technology appeal between January 1, 2017 and September 30, 2017 were included in the study.

Creation of Categories and Collection of Data

The tables containing the appropriate code and theme list in line with the purpose determined within the scope of the research were created after the important studies accepted in the literature were evaluated. The categories shown in Tab. 1, which relate general content attributes in the sharings, have been adapted from Elden and Baker's study called "Advertising appeals: sexuality, humor, fear" (2010: 87–91). In order to identify the main factors that influence the formation of the message strategies in the sharing, the categories shown in Tab. 8 have been taken from Ashley and Tuten's (2015: 21) study called "Creative strategies in social media marketing: an exploratory study of branded social content and consumer engagement". Andesson's "Groupe μ and" the system of plastic form "for an evaluation" (2008) was used when creating Tabs. 5 and 6, which show image usage in sharing and the purpose of usage. Finally, categorization of creative strategies in sharing has been developed using Elden's study called "Advertising and advertising" (2013).

Results and Comment

Looking at Tab. 1, which shows the general content of the share, it is seen that about 41% of the contents of the brands' Twitter pages are attributed to the brand (product/service). In particular, it is possible to say that today's changing

Tab. 1: General Content of the Share

General Content of the Share	Frequency	Percentage
Success	4	2
Incentive	41	23,9
Challenge	1	0,6
Introducing the used item	3	1,7
Referencing the mark (product/ service)	71	41,1
Celebration	34	19,2
Other	20	11,5
Total	**171**	**100**

marketing communication efforts are formed in connection with the interaction strategies of the brand. The fact that it has a brand-focused strategic unity in terms of the fictional features of the shares designed around technology-based objects can be seen as a regular result at this point. In addition to this data, the preference of brands to promote a content concept that promotes 23.9% of the shares can be considered as a directive by the brand for the action of users, who have expectation of feedback.

As a social media platform, the unique feature that distinguishes Twitter from others is having a textual constrained messaging mechanism that works on concurrency. This allows Twitter to be a platform, which is used more plainly in the social media world and the message are given more directly. On the other hand, from brands' point of view, Twitter accounts routinely maintain their current share, and especially on special days during the year, content sharings also varies according to these special days. In Fig. 1, Garanti and Yapi Kredi banks share messages via Twitter accounts are seen. The robot character called Ugi in Garanti Bank's 2017 advertisement gives a holiday message. In the advertisement of Yapı Kredi Bank, humanoid robotics with artificial intelligence called Gary celebrates the Mother's Day with his friend in the commercial series. There is a candy emoji besides the message of Garanti "Everyone wish happy holidays!". There is a bunch of flower emoji besides the message of Yapı Kredi "Happy Mother's Day for our Mothers, who have never separated their compassionate look at us and have always added power to us!" These messages, supported by emoji, also carry a sample of another digital image presented in the message that is designed as a digital narrative.

The celebrations and contents for special day made by brands over a digital medium have enabled the spread of brand awareness among the masses in other

Fig. 1: (Left) Garanti Bank Advertising Character Ugi, "We Wish Everyone Good Feast", (Right) Yapı Kredi Advertising Character Gary, "Happy Anniversary Day". Source: (twitter.com)

Tab. 2: The Style of the Share

The Style of the Share	Frequency	Percentage
Humorous	37	21,6
Exaggeration	2	1,1
Satire	–	–
Metaphor (semantic)	1	0,6
Rational	87	50,8
Emotional	26	15
Other	18	10,5
Total	**171**	**100**

social medias such as Twitter. According to Tab. 1, it is seen that the celebration-based content feature is preferred in 19.2% of the shares. In order to achieve high interaction at this point, it can be stated that the brands have developed content for special days.

Social media circles are one of the digital media elements that brands use to reach consumers directly and effectively. It is important that the shares are formed according to both the consumer profile and the brand structural characteristics. In this regard, it is seen that in the messages of advertisement, which are arranged through technological images, the rhetoric of the consumers is emerged from the emotions towards the logical perceptions.

When we look at the table above, in which the style of shares is defined, it is seen that more than half (50.8%) of the shares prefer rational content expression (Tab. 2). In addition to this, it is observed that 21.6% of the messages use a humorous style whereas 15% of them use an emotional style.

Tab. 3: The Language of the Share

The Language Of The Share	Frequency	Percentage
Information Transfer	120	70,1
Emotion Transfer	41	24
Information-Emotion Transfer	10	5,9
Total	**171**	**100**

Tab. 4: Image Use in the Share

Image Use in The Share	Frequency	Percentage
Simple Image	144	84,2
Intense Image	27	15,8
Total	**171**	**100**

It is important that the visual and textual structure of the created share is constructed through a planned linguistic unity so that the brand has high user interaction and can create the desired perception in connection with other brand communication studies. In Tab. 3, which defines the language of the shares, shows that 70.1% of the messages are arranged with information-oriented narrative structure. While the emotional language features are preferred in 24% of the shares, only information and emotional expression features are seen in 5.9% of the shares.

It is possible to talk about the importance of some stages in the processes of consumer preference of brand. The relation between the consumer and the brand from past to present has been revealed in connection with the different elaboration processes. One of the important elements that basically starts the consumer's interest towards the brands is that drawing attention to the advertisement or advertised product the sustainability of this situation. The problem of confusion emerging from the current setbacks of time and information makes it necessary to direct consumer interest properly. At this point, it can be said that advertisement formats designed with a simple and clear expression are often preferred by content creators in accordance with digital media formations. Looking at the table above, it is seen that 84.2% of the shares use plain images and 15.8% of them use condensed images.

When we look at the message in Fig. 2, a daily snapshot, a street snapshot appears, but it is noteworthy that the cars and other vehicles on the road are designed in such a way as to be created by evoking to future not present. At this point, the viewer or the brand consumer, who watches the messages, interprets

Fig. 2: Bosch Turkey, "The City within the Next 10 Years, We Have Joined Forces with Daimler for Autonomous Vehicles You See on the Streets." Source: (twitter.com)

the time in this picture, which is shown through certain images by referring to the future. The visual is an example of image usage densely in Fig. 2. In intense imaginary message presentations, the details provide the visual sense of completeness and the imperceptions that will invoke different meanings or thoughts in the viewer's mind.

Changes in developing technology and information communication systems encourage brands to reshape their moves from digital platforms according to the consumer expectations and aspirations. At this point, connection with the brand's advertising and promotional efforts, companies not only try to manage brand-focused informations up-to-date in the message integrity but also they are making moves to attract consumer interest from one side to the other. Looking at Tab. 5 in the light of this information, it can be seen that the brands included in the sampling have made a reference to the brand in 69% of the shares they have made in the direction of image usage purpose. While the agenda is emphasized

Tab. 5: Purpose of Image Use in the Share

The Purpose of Image Use in the Share	Frequency	Percentage
Emphasis on The Agenda	37	21,7
Attribution to the Brand	118	69
Agenda-Brand Together	16	9,3
Total	**171**	**100**

Tab. 6: The character of the Image in the Share

The character of the Image in The Share	Frequency	Percentage
Animated Characters	48	28
Robot	105	61,5
Digital Templates (Emojis)	8	4,6
Pictogram	–	–
Artificial intelligence	–	–
Other	10	5,9
Total	**171**	**100**

in 21.7% of the posts, it is seen that both agenda-focused and brand-related contents are used together in 9.3% of them.

Developments in technology and the resulting differentiation in consumer expectations make it necessary for brands to enter into an evolutionary process in the name of survival. These change features direct the brands to use of technology as an attraction in the content of advertising and promotional activities. Looking at Tab. 6, which defines the characteristics of the images used by the brands in the research, it is seen that the robot character is preferred in 61.5% of the content creation. Robots are an image of development and modernity which often confronts us in social life or popular culture products. When we think of robot technologies independent from the point where they are today, robots are actually creatures based on human characteristics. At this point, the brand image looks sympathetic to the general acceptance of the robot image and is also relevant at the same time to the fact that robots are designed to communicate specifically in areas such as advertising and cinema.

Within the scope of this research, animated characters were used in the 28% of the shares of the brands included in the sampling. It is also observed that animated characters are actively used in general advertising campaigns. Animation emerges

Fig. 3: İşcep (İş Bankası) Sharing, the Bank's Advertising Face Cem Yilmaz as an Animated Character Comes Out Confrontation. Source: (twitter.com)

Tab. 7: Structural Characteristics of the Shared Post

Structural Characteristics of the Shared Post	Frequency	Percentage
Attention	72	42,1
News/Notification	76	44,4
Nostalgia	2	1,1
A Slice of Life	20	11,7
Witnessing	1	0,6
Comparison	–	–
Other	–	–
Total	**171**	**100**

as an effective technique at the point both the creation of its own language and the expression what is intended to be define. In Fig. 3, it is seen that there is an advertisement of the mobile application of the bank, which gives a message in which Cem Yılmaz, who is one of the most prominent comedians in Turkey "Those who want to grow their business investment purposes... That is in the pocket! İşCep Mobile Exchange explains in the show Cem Yılmaz new animations.

Tab. 8: Creative Strategies in the Share

Creative Strategies in the Share	Frequency	Percentage
Brand Name	5	0,3
General Strategy	62	36,2
Basic Selling Proposition	19	11,1
Strategy of Brand's Image	34	19,9
Positioning	22	12,9
Emotional Strategy	29	16,9
Total	**171**	**100**

Due to the structural characteristics of the information society, the sharings of the present communication process is rapidly consumed and the other elements take its place. At this point, brands need to shape the structural characteristics of the messages created by the content creators in the light of the correct variables in order to direct consumer interest to the brand and to maintain the update of this interest. In particular, in connection with consumer expectations, brands must have a dynamic understanding of reacting to current events and situations, not only promoting products and services as living organisms, but also with significant advantages in interacting with the target group. When we look at the table above, it is seen that 44.4% of the shares are characterized by news and announcements (Tab. 7). In addition to this, 42.1% of the messages have remarkable features. It is also observed that contents created in 11.7% of the shares have a feature f slice of life.

Defining the overall framework of the advertisement and setting an accurate creative strategy in order to regulate the content of the advertisement within a specific plan is important for the success of the advertisement. According to the table above, 36.2% of the shares of the brands prefer general advertising strategy. In 19.9% of the sharings, brands that set a strategy for brand image are observed to have an emotional advertising strategy in 16.9% of the messages.

It can be said that the contents created on the digital platforms are quickly consumed by the users. In this regard, it is important to organize the messages that are used in the context of the brand's communication strategies in a dimension that overlaps with the brand identity and to diversify in relation to the structural characteristics of the target group, which is important achieving the desired success of sharing. For this reason, messages that reflect content integrity on a specific strategic basis bring with it certain conveniences in controlling the prospective moves of the brand.

Tab. 9: Message Strategies in the Share

Message Strategies in the Share	**Frequency**	**Percentage**
Integrated Content	9	5,2
Interaction	16	9,4
Functional Appeal	59	34,5
Emotional Attractiveness	29	17
Experimental Appeal	11	6,4
Unique Selling Proposition (USP)	4	2,3
Comparison	1	0,6
Resonance	–	–
User's Image	–	–
Social Impact	13	7,6
Exclusivity	19	11,1
Brand Ambassador	10	5,9
Other	–	–
Total	**171**	**100**

Looking at Tab. 9, in which the message strategies are defined, it is seen that message features that provide functional appeal are preferred in 34.5% of the shares. In addition to this, it is determined that 17% of the messages are directed to emotional appeals. Consumers often seem to prefer the interaction strategy (9.4%) in order to include the generated share and provide an active participation practice. The content of the message to be delivered becomes more important for brands at this point.

Conclusion

Advertising creates itself out of the competitions, which emerges as a result of the production and consumption relations of the brands, and the differences that are evident with the fictions of this competition on media tools. Advertising in today's society; radio, press, and especially television, is different from the past in terms of synchronicity and interactivity as a result of internet-based technologies and related innovations. It is seen that developing technologies have been influencing the advertising field every day to keep up with the times and to be open to innovations. From large global brands to local businesses, advertising on a wide scale is seen as a need. Technology is used both as a tool and as a

Fig. 4: René Magritte, "The Treachery of Images", 1929. Source: (www.renemagritte.org)

rhetoric for advertising because it is predicted that the communication process established with the consumer will have a quality to cover the majority of the current discourse. At this point, it is observed that technology comes into prominence as a feature in brand, product, or advertising discourse. Although technology plays an instrumental role in the people's daily life practices, it is possible at this point to put an imaginative meaning within the language of advertising, and the data taken from the result of the study support this view.

It can be said that the messages of the brands are presented in a realistic structure because the technological elements and images used in the advertising messages examined in the research are shaped mostly through rational practices. In this regard, it is possible to say that technologically, as an attractive element, the brands are designed in message discourses to make the contents created in this context more rational to the consumers.

At the beginning of the study, we talked about the fact that the images exist itself in different disciplines through different examples. At this point, it would be appropriate to refer to the famous work by the Belgian painter René Magritte, which gave us the opportunity to discuss the semantic breadth and perception of the image concept (Fig. 4), entitled "Betrayal of The Images" (La trahison des images). In Magritte's painting, there is a plain and clearly articulated pipe, and under the picture of the pipe, there is a handwritten "This is not a pipe" (Ceci n'est pas une pipe).

Michel Foucault, one of the most important philosophers of the 20th century, debated the meaning of this picture corresponding with Magritte. Later, Foucault published a book about the letters and passages between Magritte and him. The general interpretation of this painting in the book is that; what is seen in the painting is not a pipe both visually and imaginary but it is a reproduced image of a canvas or a picture (Foucault, 2001). When examined through this example, it is observed that the images have a very meaningful and flexible structure that shows itself in every area of life. It is not possible to make a definite judgment about the images, whether in the art of painting or in advertising, as the interpretation of the image differs from the perception and reception of each person. However, it is possible to say that in advertising messages, which point to a particular phenomenon, such as "technology," and are designed to evoke this theme, the images will lead to a general appreciation or a positive perception.

References

Akay, P. (2010), "Çağdaş Sanatta Politik İmge ve Sembol Uygulamaları", Gazi Üniversitesi Güzel Sanatlar Enstitüsü, Yüksek Lisans Tezi, Ankara.

Akbaytogan, K. (2006), "Günümüzde İmge ve Farkındalık", Yeditepe Üniversitesi, Sosyal Bilimler Enstitüsü, Yüksek Lisans Tezi, İstanbul.

Andersson, F. (2008), "Groupe μ and "The System of Plastic Form" — For an Evaluation". https://lup.lub.lu.se/luur/download?func=downloadFile&recordOId=

Arnheim, R. (2012), "Görsel Düşünme", (Rahmi Öğdül, çev.) MetisYayıncılık, İstanbul.

Ashley, C. ve Tuten, T. (2015), "Creative Strategies in Social Media Marketing: An Exploratory Study of Branded Social Content and Consumer Engagement", Psychology and Marketing, Cilt: 32, Sayı: 1, ss. 15–27.

Aslan, T. (2003), "Görsel Olanı Okumak: Eleştirel Görsel Okur-Yazarlık", Ankara Üniversitesi İletişim Araştırmaları Dergisi, Vol. 1 Issue: 1, pp. 39–64.

Batı, U. (2005), "Bir Anlam Yaratma Süreci ve İdeolojik Yapı Olarak Reklamların Göstergebilim Bir Bakış Açısıyla Çözümlenmesi", Cumhuriyet Üniversitesi Sosyal Bilimler Dergisi Cilt: 29 Sayı: 2, 175–190.

Berger, J. (2010), "Görme Biçimleri", (Yurdanur Salman, çev.) Metis Yayıncılık, İstanbul.

Burnett, R. (2012), "İmgeler Nasıl Düşünür?", (Güçsal Pusar, çev.) Metis Yayıncılık, İstanbul.

Castells, M. (2016), "İletişimin Gücü", (Ebru Kılıç, çev.) Anahtar Kitaplar Yayınevi, İstanbul.

Çağlar, C.F. (2013), "Güncel Sanatta İmge Dönüşümleri", Gazi Üniversitesi Güzel Sanatlar Enstitüsü, Yüksek Lisans Tezi, Ankara.

Cengiz, M. (2009), " İmge Nedir?", Şiirden Yayınları, İstanbul.

Chatfield, T. (2013), "Dijital Çağa Nasıl Uyum Sağlarız", (Levent Konca, çev.) Sel Yayıncılık, İstanbul.

Clow, K. E. ve Baack, D. (2016), "Bütünleşik Reklam, Tutundurma ve Pazarlama İletişimi" (Gülay Öztürk, çev.) Nobel Kitapevi, Ankara.

Dickey, I. ve Lewis, W. (2015), An Overview of Digital Media and Advertising, Eastin ve diğ. (Ed.), Handbook of Research on Digital Media and Advertising: User Generated Content Consumption (ss. 1–31), Information Science Reference, New York.

Elden, M. (2013), "Reklam ve Reklamcılık", Say Yayınları, İstanbul.

Elden, M. ve Bakır, U. (2010), "Reklam Çekicilikleri: Cinsellik, Mizah, Korku", İletişim Yayınları, İstanbul.

Foucault, M. (2001), Bu Bir Pipo Değildir, (Selahattin Hilav. çev.) Yapı Kredi Yayınları, İstanbul.

Hebb, A. (2015), "Medyatikleşen Kültürler", (Çiğdem Bozdağ-Elif P. Devrani. çev.) Dipnot Yayınları, Ankara.

Lamb, C., Hair, J. veMcDaniel, C. (2009). "Marketing", South-Western Cengage Learning, Ohio.

Moon, Y. ve Chan, K. (2005). "Advertising Appeals and Cultural Values in Television Commercials: A Comparison of Hong Kong and Korea", International Marketing Review, Vol. 22 Issue: 1, pp. 48–66.

Oyedele, A., Minor, M. ve Ghanem, S. (2009). "Signals of Global Advertising Appeals in Emerging Markets", International Marketing Review, Vol. 26 Issue: 4/5, pp. 521–541.

Parsa, A.F. (2004), "İmgenin Gücü ve Görsel Kültürün Yükselişi", Fotoğrafya Dergisi, Sayı: 19.

Rutherford, P. (1996), "Yeni İkonalar: Televizyonda Reklam Sanatı" (M. Gerçeker. cev.) Yapı Kredi Yayınları, İstanbul.

Varnalı, K. (2013), "Dijital Tutulma: Pazarlama İletişimi ve İnsan", MediaCat Yayınları, İstanbul.

Yıldırım, A. ve Şimşek, H. (2013), "Sosyal Bilimlerde Nitel Araştırma Yöntemleri", Seçkin Yayınları, Ankara.

Yılmaz, R. ve Erdem, N. (2016), "150 Soruda Geleneksel ve Dijital Reklamcılık", Umuttepe Yayınları, Kocaeli.

Yılmaz, R. ve Erdem, N. https://www.marketo.com/definitive-guides/digital-advertising/, ErişimTarihi: 17.10.2017

Yılmaz, R. ve Erdem, N. https://twitter.com/boschturkey/, Erişim Tarihi: 17.10.2017

Yılmaz, R. ve Erdem, N. https://twitter.com/garanti/, Erişim Tarihi: 19.10.2017

Yılmaz, R. ve Erdem, N. https://twitter.com/YapiKredi/, Erişim Tarihi: 19.10.2017

Yılmaz, R. ve Erdem, N. https://twitter.com/iscepisbankasi/, Erişim Tarihi: 19.10.2017

Yılmaz, R. ve Erdem, N. https://www.renemagritte.org/the-treachery-of-images.jsp Erişim Tarihi: 29.12.2017